Numerical Methods

for Computer Science,
Engineering,
and Mathematics

Numerical Methods

for Computer Science,
Engineering,
and Mathematics

JOHN H. MATHEWS

California State University, Fullerton

PRENTICE-HALL, INC.

Englewood Cliffs, New Jersey 07632

Library of Congress Cataloging-in-Publication Data

Mathews, John H.
    Numerical methods for computer science, engineering,
and mathematics.

    Bibliography: p.
    Includes index.
    1. Numerical analysis.   2. Electronic data processing
—Mathematics.   3. Engineering mathematics.   I. Title.
QA297.M39   1987   519.4   86-16999
ISBN 0-13-626656-8

Editorial/production: Nicholas C. Romanelli
Manufacturing buyer: John Hall
Cover design: 20/20 Services, Inc.

Printed in the United States of America

10   9   8   7   6   5   4   3   2

ISBN 0-13-626656-8

Prentice-Hall International, *London*
Prentice-Hall of Australia Pty. Limited, *Sydney*
Editora Prentice-Hall do Brasil, Ltda., *Rio de Janeiro*
Prentice-Hall Canada, Inc., *Toronto*
Prentice-Hall Hispanoamericana, S.A., *Mexico*
Prentice-Hall of India Private Limited, *New Delhi*
Prentice-Hall of Japan, Inc., *Tokyo*
Prentice-Hall of Southeast Asia Pte. Ltd., *Singapore*

To my wife, Frances, and my
three sons, Robert, Daniel, and James

Contents

Preface

This book provides an elementary presentation of numerical methods or elementary numerical analysis and is suitable for an undergraduate course for students in mathematics, science, engineering, and computer science. The text can be used for a junior level course or for an advanced sophomore level course. It contains enough material so that topics can be selected for a one semester course or one lasting two quarters. The instructor has considerable flexibility in choosing topics which are appropriate for his/her course. The mathematics prerequisite is one year of calculus that includes limits and techniques of differentiation and integration of functions of one variable. An introduction to the topics in linear algebra and ordinary differential equations would be helpful but is not necessary. Adequate introductory material is presented so that these topics are not prerequisites. It is assumed that the student has taken a computer programming course such as FORTRAN, PASCAL, or BASIC.

Students of all backgrounds enjoy numerical methods. This is a practical book for the student who likes to see results, and it has something for everyone. The standard topics are covered: iteration, root finding, linear systems of equations, interpolation and approximation, least-squares curve fitting, differentiation, minimization of a function, integration, nonlinear systems of equations, and the solution of ordinary differential equations. Several sections are optional and can be used for independent study projects.

Emphasis is placed on understanding how the methods work. For each topic, the underlying theory is included. Detailed examples guide the student through the calculations necessary to understand the algorithm(s). If repetitive iterations are required, then a table of further calculations is included. Limitations imposed by the mathematical assumptions and computational difficulties inherent in computers are usually pointed out. The error analysis is described in a manner suitable to the student's level and interest. Carefully written proofs are included wherever necessary. Proofs use elementary results from calculus and serve to expand the student's understanding of numerical analysis.

For applications to computer programming, the presentation using pseudo-algorithms was chosen. The algorithms are stated so that the student familiar with structured programming will be able to write nice looking programs. Algorithms are generally written in the most straightforward manner and include sufficient details so that they are easily translated into FORTRAN 77, PASCAL, or BASIC. This gives the student a chance to develop skills in software engineering. Moreover, printed listings of the computer programs in FORTRAN 77, PASCAL, and BASIC are available from the publisher. The programs are also available on a floppy disk.

Each section contains a variety of problems. Some exercises emphasize routine calculations and can be done with a calculator or simple program. The instructor should anticipate that these problems could require considerable effort. Some problems require simple proofs which are based on skills that the student learned in calculus. Some problems allude to a library research projects involving topics not developed in the text. The exercises were evolved over the last five years and have been extensively class-tested. Selected answers to the problems are to be found at the end of the text.

The use of software packages and libraries are encouraged. Sometimes the phrase "use a computer" occurs in the problems. This must be interpreted in view of a school's particular learning environment. Software libraries such as IMSL could be used. The floppy disk available as a supplement to this text could be used. These materials can serve as tools for the student to run various "numerical experiments."

ACKNOWLEDGEMENTS

Many people deserve thanks for helping with the development of this book. I thank the students at California State University, Fullerton, and my colleagues, Edward Sabotka and Soo Tang Tan, who used manuscript copies in their courses. They located errors and made helpful comments. I also thank James Friel, Chairman of the Mathematics Department at CSUF, for his encouragement. I extend my appreciation to my colleagues, William Gearhart, Stephen Goode, Mathew Koshy, and Harris Shultz, who read portions of the manuscript and made valuable suggestions for improvement.

I express my gratitude to the reviewers who made recommendations for improvement of the manuscript: Walter M. Patterson, III, Lander College; George B. Miller, Central Connecticut State University; Peter J. Gingo, The

University of Akron; Michael A. Freedman, The University of Alaska, Fairbanks; and Kenneth P. Bube, University of California, Los Angeles.

Finally, I wish to express my appreciation to the staff at Prentice-Hall, especially David Ostrow, mathematics editor, and Nicholas Romanelli, production editor, for their assistance and encouragement.

The author would appreciate receiving correspondences regarding the book and its supporting software. Suggestions for improvement are always welcome.

J.H.M.

Numerical Methods

for Computer Science,
Engineering,
and Mathematics

1

Mathematical Preliminaries

1.1 Introduction

We human beings do arithmetic using the decimal number system. Most computers do arithmetic using the binary number system. It may seem otherwise, because we communicate with the computer (input/output) in base 10 numbers. This transparency does not mean that the computer uses base 10. In fact, it converts our inputs to base 2 (or perhaps base 16), then performs base 2 arithmetic, and finally, translates the answer into base 10 before it prints it out to us. Some experimentation is required to verify this. One computer with nine decimal digits of accuracy gave the answer

(1)
$$\sum_{k=1}^{100,000} 0.1 = 9{,}999.99447.$$

Here the intent was to add the number $\frac{1}{10}$ repeatedly 100,000 times. As everyone knows, the mathematical answer is exactly 10,000. One goal is to understand the reason for the computer's apparently flawed calculation. We shall see that something is lost when the computer translates the decimal fraction $\frac{1}{10}$ into a binary fraction.

1.2 Binary Numbers

For ordinary purposes we use base 10 numbers. For illustration, the number 1,563 is expressible as

$$1{,}563 = 1 \times 10^3 + 5 \times 10^2 + 6 \times 10^1 + 3 \times 10^0.$$

In general, let N denote a positive integer; then the digits $a_0, a_1, a_2, \ldots, a_K$ exist so that N has the base 10 expansion

$$N = a_K \times 10^K + a_{K-1} \times 10^{K-1} + \ldots + a_2 \times 10^2 + a_1 \times 10^1 + a_0 \times 10^0,$$

where the digits a_k are chosen from $\{0, 1, \ldots, 8, 9\}$. Thus N is expressed in decimal notation as

(2) $$N = a_K a_{K-1} \ldots a_2 a_1 a_0 \text{ ten} \qquad \text{(decimal)}$$

If it is understood that 10 is the base, we write (2) as

$$N = a_K a_{K-1} \ldots a_2 a_1 a_0.$$

For example, we understand that $1{,}563 = 1{,}563_{\text{ten}}$.

Using powers of 2, the number 1,563 can be written

(3) $$1{,}563 = 1 \times 2^{10} + 1 \times 2^9 + 0 \times 2^8 + 0 \times 2^7 + 0 \times 2^6 + 0 \times 2^5$$
$$+ 1 \times 2^4 + 1 \times 2^3 + 0 \times 2^2 + 1 \times 2^1 + 1 \times 2^0.$$

This can be verified by performing the calculation

$$1{,}563 = 1{,}024 + 512 + 16 + 8 + 2 + 1.$$

In general, let N denote a positive integer; the digits $b_0, b_1, b_2, \ldots, b_J$ exist so that N has the base 2 expansion

(4) $$N = b_J \times 2^J + b_{J-1} \times 2^{J-1} + \ldots + b_2 \times 2^2 + b_1 \times 2^1 + b_0 \times 2^0,$$

where each digit b_j is either 0 or 1. Thus N is expressed in binary notation as

(5) $$N = b_J b_{J-1} \ldots b_2 b_1 b_0 \text{ two} \qquad \text{(binary)}.$$

Using the notation (5) and the result in (3) yields

$$1{,}563 = 11000011011_{\text{two}}.$$

Remarks. We will always use the word "two" as a subscript at the end of a binary number. This will enable us to distinguish binary numbers from the ordinary base 10 usage. Thus 111 means one hundred eleven, whereas 111_{two} stands for seven.

It is usually the case that the binary representation for N will require more digits than the decimal representation. This is due to the fact that powers of 2 grow much more slowly than do powers of 10.

An efficient algorithm for finding the base 2 representation of the integer N can be derived from equation (4). Dividing both sides of (4) by 2, we obtain

$$\frac{N}{2} = b_J \times 2^{J-1} + b_{J-1} \times 2^{J-2} + \ldots + b_2 \times 2^1 + b_1 \times 2^0 + \frac{b_0}{2}.$$

Hence the remainder, upon dividing N by 2, is the digit b_0. Next we find b_1. If we write $N/2 = Q_0 + b_0/2$, then

(6)
$$Q_0 = b_J \times 2^{J-1} + b_{J-1} \times 2^{J-2} + \ldots + b_2 \times 2^1 + b_1 \times 2^0.$$

Now divide both sides of (6) by 2 to get

$$\frac{Q_0}{2} = b_J \times 2^{J-2} + b_{J-1} \times 2^{J-3} + \ldots + b_2 \times 2^0 + \frac{b_1}{2}.$$

Hence the remainder, upon dividing Q_0 by 2, is the digit b_1. This process is continued and generates sequences $\{Q_k\}$ and $\{b_k\}$ of quotients and remainders, respectively. The process is terminated when we find the first integer J such that $Q_J = 0$. The sequences obey the following formulas:

(7)
$$N = 2Q_0 + b_0$$
$$Q_0 = 2Q_1 + b_1$$
$$Q_1 = 2Q_2 + b_2$$
$$\vdots$$
$$Q_{J-2} = 2Q_{J-1} + b_{J-1}$$
$$Q_{J-1} = 2Q_J + b_J \qquad (Q_J = 0).$$

Algorithm 1.1 [*Binary Expansion for Integers*] Let N be a positive integer. Form the quotients $\{Q_k\}$ and remainders $\{b_k\}$ given in (7) until the smallest integer J is found for which $Q_J = 0$. Then

$$N = b_J b_{J-1} \ldots b_2 b_1 b_0 \text{ two}.$$

```
INPUT   N                            {Input the integer N}
Q(0) := INT( N/2 )                   {Compute the quotient}
B(0) := N − 2*Q(0)                   {Compute the remainder}
J := 0                               {Initialize the counter}


WHILE   Q(J) ≠ 0   DO
    J := J + 1                       {Increment the counter}
    Q(J) := INT( Q(J−1) / 2 )        {Compute the quotient}
    B(J) := Q(J−1) − 2*Q(J)          {Compute the remainder}


PRINT "The integer is " N            {Output}
PRINT "The binary representation is
FOR   K = J   DOWNTO  0   DO
    PRINT   Q(K)
```

Example 1.1 Show how to obtain $1{,}563 = 11000011011_{\text{two}}$.

Solution. Start with $N = 1{,}563$ and construct the quotients and remainders according to the equations in (7):

$$
\begin{aligned}
1{,}563 &= 2 \times 781 + 1, & b_0 &= 1 \\
781 &= 2 \times 390 + 1, & b_1 &= 1 \\
390 &= 2 \times 195 + 0, & b_2 &= 0 \\
195 &= 2 \times 97 + 1, & b_3 &= 1 \\
97 &= 2 \times 48 + 1, & b_4 &= 1 \\
48 &= 2 \times 24 + 0, & b_5 &= 0 \\
24 &= 2 \times 12 + 0, & b_6 &= 0 \\
12 &= 2 \times 6 + 0, & b_7 &= 0 \\
6 &= 2 \times 3 + 0, & b_8 &= 0 \\
3 &= 2 \times 1 + 1, & b_9 &= 1 \\
1 &= 2 \times 0 + 1, & b_{10} &= 1.
\end{aligned}
$$

Thus the binary representation for 1,563 is

$$1{,}563 = b_{10}b_9b_8 \ldots b_2b_1b_0 \text{ two} = 11000011011_{\text{two}}.$$

Sequences and Series

When fractions are expressed in decimal form, it is often the case that infinitely many digits are required. A familiar example is

(8)
$$\tfrac{1}{3} = 0.33333333\overline{3} \ldots .$$

Here the symbol $\overline{3}$ means that the digit 3 is repeated forever to form a decimal. It is understood that 10 is the base in (8). Moreover, it is the mathematical intent that (8) is the shorthand notation for the infinite series

(9)
$$S = 3 \times 10^{-1} + 3 \times 10^{-2} + 3 \times 10^{-3} + \ldots + 3 \times 10^{-n} + \ldots = \tfrac{1}{3}.$$

If only a finite number of digits are displayed, only an approximation to $\tfrac{1}{3}$ is obtained. Suppose that we write

$$\frac{1}{3} \approx 0.333333 = \frac{333{,}333}{1{,}000{,}000}.$$

Then the error in this approximation is $1/3{,}000{,}000$ and the reader can verify that

$$\frac{1}{3} = 0.333333 + \frac{1}{3{,}000{,}000}.$$

We should try to understand the nature of the expansion in (9). A naive approach is to multiply the right side by 10 and then subtract.

$$10S = 3 + 3 \times 10^{-1} + 3 \times 10^{-2} + 3 \times 10^{-3} + \ldots + 3 \times 10^{-n} + \ldots$$
$$- \quad S = \quad -3 \times 10^{-1} - 3 \times 10^{-2} - 3 \times 10^{-3} - \ldots - 3 \times 10^{-n} - \ldots$$
$$\overline{9S = 3 + 0 \times 10^{-1} + 0 \times 10^{-2} + 0 \times 10^{-3} + \ldots + 0 \times 10^{-n} + \ldots}$$

Therefore, $S = \frac{3}{9} = \frac{1}{3}$. The theorems necessary to justify the calculation above can be found in most calculus books. We shall review a few of the concepts and the reader may want to refer to a standard reference on calculus to fill in all the details.

A *sequence* $\{a_n\}$ is a function whose domain is the set of all positive integers. The function values $a_1, a_2, \ldots, a_n, \ldots$ are called *terms* of the sequence and a_n is the *general term*. For example, $a_n = n^2/2^n$ is the general term for the sequence

$$\frac{1}{2}, \frac{4}{4}, \frac{9}{8}, \frac{16}{16}, \frac{25}{32}, \frac{36}{64}, \frac{49}{128}, \frac{64}{256}, \frac{81}{512}, \ldots.$$

The sequence $\{a_n\}$ *converges* and has the limit L if for every $\epsilon > 0$ there exists an integer N so that

(10) $$n \geq N \quad \text{implies that} \quad |a_n - L| < \epsilon.$$

If $\{a_n\}$ converges to L, we denote this by

$$\lim_{n \to \infty} a_n = L.$$

It is not necessary to review all results from calculus. However, numerical methods are a computer-oriented treatment of some of the topics in calculus. Hence it is helpful to be able to recall ideas from calculus. We leave it for the reader to check to see that L'Hôpital's rule can be used to show that

$$\lim_{n \to \infty} a_n = \lim_{n \to \infty} \frac{n^2}{2^n} = \lim_{n \to \infty} \frac{2n}{\ln (2) 2^n} = \lim_{n \to \infty} \frac{2}{[\ln 2]^2 2^n} = 0.$$

For our purposes we want to see how calculus is implemented on a computer and we might try calculating some terms. When we arrive at $a_{100} \approx 7.888609 \times 10^{-27}$ we could suspect that the terms are going to zero, and this can be proved by using theorems about limits.

Let $\{a_n\}$ be a sequence. Then a new sequence can be formed by successively adding up the terms of the given sequence. Consider the sequence of partial sums $S_1, S_2, \ldots, S_n, \ldots$ defined as follows:

$$S_1 = a_1$$
$$S_2 = a_1 + a_2$$

(11) $$S_3 = a_1 + a_2 + a_3$$
$$\vdots$$
$$S_n = a_1 + a_2 + a_3 + \ldots + a_n.$$

The sum of all the terms a_j is called an *infinite series*:

(12) $$a_1 + a_2 + a_3 + \ldots + a_j + \ldots \quad \text{or} \quad \sum_{j=1}^{\infty} a_j.$$

The infinite series (12) *converges* provided that the sequence of partial sums (11) has a limit S, that is,

(13) $$S = \lim_{n \to \infty} S_n.$$

In this case we say that the series has S as its sum and we write

(14) $$S = \sum_{j=1}^{\infty} a_j.$$

If the sequence $\{S_n\}$ does not converge, then we say that the series in (12) *diverges*.

   Remark. It is not always convenient to start the index of summation with 1. For ease of notation we usually start the index with an integer that is appropriate to the problem at hand.

Definition 1.1   [Geometric series] The infinite series

(15) $$\sum_{k=0}^{\infty} cr^k = c + cr + cr^2 + \ldots + cr^n + \ldots,$$

where $c \neq 0$ and $r \neq 0$, is called a *geometric series* with ratio r.

Theorem 1.1   [Geometric series] The geometric series in (15) has the following properties:

(16) $|r| < 1$ implies that the series converges, and $\sum_{k=0}^{\infty} cr^k = \dfrac{c}{1-r}.$

(17) $|r| \geq 1$ implies that the series diverges.

   Proof. The summation formula for a finite geometric series is

(18) $$S_n = c + cr + \ldots + cr^n = c\frac{1 - r^{n+1}}{1 - r} \quad \text{for } r \neq 1.$$

To establish (16) we observe that

(19) $$|r| < 1 \quad \text{implies that} \quad \lim_{n \to \infty} r^{n+1} = 0.$$

Taking the limit as $n \longrightarrow \infty$, we use (18) and (19) to get

$$\lim_{n \to \infty} S_n = \frac{c}{1-r}\left(1 - \lim_{n \to \infty} r^{n+1}\right) = \frac{c}{1-r}(1 - 0).$$

By (13) the limit above establishes (16).

   When $|r| \geq 1$, the sequence $\{r^{n+1}\}$ does not converge. Hence the sequence $\{S_n\}$ in (18) does not tend to a limit. Therefore (17) is established.

Example 1.2 Show that

(20)
$$1 + \tfrac{1}{4} + \tfrac{1}{16} + \tfrac{1}{64} + \tfrac{1}{256} + \ldots = \tfrac{4}{3}.$$

Solution. We must first observe that $c = 1$ and $r = \tfrac{1}{4}$. Then using (16) we conclude that

$$1 + \frac{1}{4} + \left(\frac{1}{4}\right)^2 + \left(\frac{1}{4}\right)^3 + \ldots + \left(\frac{1}{4}\right)^n + \ldots = \frac{1}{1 - \tfrac{1}{4}} = \frac{4}{3}.$$

Binary Fractions

Binary fractions can be expressed as sums involving negative powers of 2. If R is a real number that lies in the range $0 < R < 1$, then there exist digits $d_1, d_2, \ldots, d_n, \ldots$ so that

(21)
$$R = d_1 \times 2^{-1} + d_2 \times 2^{-2} + d_3 \times 2^{-3} + \ldots + d_n \times 2^{-n} + \ldots.$$

We usually express the quantity on the right side of (21) in the binary fraction notation

(22)
$$R = 0.d_1 d_2 d_3 \ldots d_n \ldots_{\text{two}}.$$

Since 2 is the base, the digits d_j are chosen from the set $\{0, 1\}$.

There are many real numbers whose binary representation requires infinitely many digits. The fraction $\tfrac{7}{10}$ can be expressed as 0.7_{ten}, yet its base 2 representation requires infinitely many digits:

(23)
$$\tfrac{7}{10} = 0.1011001100110011\overline{0011}\ldots_{\text{two}}.$$

The binary fraction in (23) is a repeating fraction where the group of four digits 0011 is repeated forever. To see what is happening, we could try to measure 0.7 inch on a ruler that is graduated in $\tfrac{1}{2}, \tfrac{1}{4}, \tfrac{1}{8}$, and $\tfrac{1}{16}$ inch. The result will look like this:

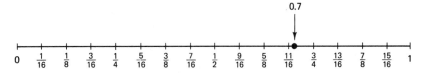

Thus 0.7 is located between $\tfrac{11}{16}$ and $\tfrac{3}{4}$ and satisfies the relation $\tfrac{11}{16} < \tfrac{7}{10} < \tfrac{3}{4}$. Using (21) and (22), it is easy to verify the expansions

$$\tfrac{11}{16} = \tfrac{1}{2} + \tfrac{0}{4} + \tfrac{1}{8} + \tfrac{1}{16} = 0.1011_{\text{two}} \quad \text{and} \quad \tfrac{3}{4} = \tfrac{1}{2} + \tfrac{1}{4} + \tfrac{0}{8} + \tfrac{0}{16} = 0.1100_{\text{two}}.$$

Hence we have the binary relation

$$0.1011_{\text{two}} < \tfrac{7}{10} < 0.1100_{\text{two}}.$$

A ruler graduated in $\tfrac{1}{256}$ of an inch would reveal that

$$\tfrac{179}{256} < \tfrac{7}{10} < \tfrac{45}{64}.$$

The reader can verify that this leads to the binary relation

$$0.10110011_{two} < \tfrac{7}{10} < 0.10110100_{two}.$$

When the limit process is applied we get

$$\tfrac{7}{10} = 0.1011001100\overline{110}\ldots {}_{two}.$$

We now develop an algorithm for finding base 2 representations. If both sides of (21) are multiplied by 2, the result is

(24) $$2R = d_1 + (d_2 \times 2^{-1} + d_3 \times 2^{-2} + \ldots + d_n \times 2^{-n+1} + \ldots).$$

The quantity in parentheses on the right side of (24) is a positive number and is less than 1. Therefore, we can take the integer part of both sides of (24) and obtain

$$d_1 = \text{int}\,(2R),$$

where int $(2R)$ is the integer part of the real number $2R$. To continue the process, take the fractional part of (24) and write

(25) $$F_1 = \text{frac}\,(2R) = d_2 \times 2^{-1} + d_3 \times 2^{-2} + \ldots + d_n \times 2^{-n+1} + \ldots,$$

where frac $(2R)$ is the fractional part of the real number $2R$. Multiplication of both sides of (25) by 2 results in

(26) $$2F_1 = d_2 + (d_3 \times 2^{-1} + \ldots + d_n \times 2^{-n+2} + \ldots).$$

Now take the integer part of (26) and obtain

$$d_2 = \text{int}\,(2F_1).$$

The process is continued, possibly ad infinitum, and two sequences $\{d_k\}$ and $\{F_k\}$ are recursively generated:

(27)

$$
\begin{aligned}
d_1 &= \text{int}\,(2R), & F_1 &= \text{frac}\,(2R) \\
d_2 &= \text{int}\,(2F_1), & F_2 &= \text{frac}\,(2F_1) \\
d_3 &= \text{int}\,(2F_2), & F_3 &= \text{frac}\,(2F_2) \\
&\;\;\vdots & &\;\;\vdots \\
d_n &= \text{int}\,(2F_{n-1}), & F_n &= \text{frac}\,(2F_{n-1}). \\
&\;\;\vdots & &\;\;\vdots
\end{aligned}
$$

The binary representation of R is given by the convergent series

$$R = \sum_{j=1}^{\infty} d_j \times 2^{-j}.$$

Algorithm 1.2 [*Binary Fractions*] Let R be a positive real number in the interval $0 < R < 1$. The first M terms of the sequence $\{d_k\}$ in (27) is recursively generated, and R has the representation:

$$R = 0.d_1 d_2 d_3 \ldots d_M \ldots {}_{two}.$$

```
      Max := 32                              {Bits in the mantissa}
      M := 1                                 {Initialize}
      INPUT R                                {Input the number R}
      X := R


FOR   K = 1  TO  Max  DO
   │  Y := 2*X
   │  D(K) := INT( Y )                       {Integer part}
   └  X := Y − D(K)                          {Fractional part}


FOR   K = Max  DOWNTO  1  DO                 {Eliminate trailing zeros}
   │  IF  D(K) ≠ 0 THEN
   └     └  M := K , K := 1


PRINT "The number is" R                      {Output}
PRINT "The binary representation is"
FOR   K = 1  TO  M  DO
   └     PRINT  D(K)
```

Example 1.3 Show how to obtain the representation

$$\tfrac{7}{10} = 0.101100110011001\overline{10}\ldots_{\text{two}}.$$

Solution. Start with $R = 0.7$ and use the formulas in (27) to get

$$2R = 1.4, \qquad d_1 = 1 = \text{int }(1.4), \qquad F_1 = 0.4 = \text{frac }(1.4)$$
$$2F_1 = 0.8, \qquad d_2 = 0 = \text{int }(0.8), \qquad F_2 = 0.8 = \text{frac }(0.8)$$
$$2F_2 = 1.6, \qquad d_3 = 1 = \text{int }(1.6), \qquad F_3 = 0.6 = \text{frac }(1.6)$$
$$2F_3 = 1.2, \qquad d_4 = 1 = \text{int }(1.2), \qquad F_4 = 0.2 = \text{frac }(1.2)$$
$$2F_4 = 0.4, \qquad d_5 = 0 = \text{int }(0.4), \qquad F_5 = 0.4 = \text{frac }(0.4)$$
$$2F_5 = 0.8, \qquad d_6 = 0 = \text{int }(0.8), \qquad F_6 = 0.8 = \text{frac }(0.8)$$
$$2F_6 = 1.6, \qquad d_7 = 1 = \text{int }(1.6), \qquad F_7 = 0.6 = \text{frac }(1.6)$$
$$2F_7 = 1.2, \qquad d_8 = 1 = \text{int }(1.2), \qquad F_8 = 0.2 = \text{frac }(1.2)$$
$$2F_8 = 0.4, \qquad d_9 = 0 = \text{int }(0.4), \qquad F_9 = 0.4 = \text{frac }(0.4)$$
$$\vdots \qquad\qquad \vdots \qquad\qquad \vdots$$

A repeating pattern is emerging: $d_k = d_{k+4}$ and $F_k = F_{k+4}$ for $k = 2, 3, \ldots$. Thus the binary representation for $\tfrac{7}{10}$ is

$$\tfrac{7}{10} = 0.d_1d_2d_3d_4d_5d_6d_7d_8\ldots_{\text{two}} = 0.1011\overline{0011}0\ldots_{\text{two}}.$$

Geometric series can be used to help find some binary fractions.

Example 1.4 Use geometric series to show that
$$\tfrac{1}{3} = 0.01010101010101\overline{01}\ldots_{\text{two}}.$$

Solution. If the result of Example 1.2 is used and 1 is subtracted from both sides of (20), then we have

(28)
$$\tfrac{1}{3} = \tfrac{1}{4} + \left(\tfrac{1}{4}\right)^2 + \left(\tfrac{1}{4}\right)^3 + \ldots + \left(\tfrac{1}{4}\right)^j + \ldots.$$

Equation (28) can be rewritten in the form of (21):
$$\tfrac{1}{3} = 0 \times 2^{-1} + 1 \times 2^{-2} + 0 \times 2^{-3} + 1 \times 2^{-4} + \ldots + 0 \times 2^{-2j+1} + 1 \times 2^{-2j} + \ldots$$

or

$$\tfrac{1}{3} = 0.01010101010101\overline{01}\ldots_{\text{two}}.$$

Binary Shifting

If we are to find the rational number that is equivalent to an infinite repeating binary expansion, then a shift in the digits can be helpful. For example, let S be given by

(29)
$$S = 0.0001100110011001100110011\overline{0011}\ldots_{\text{two}}.$$

Multiplication of (29) by $32 = 2^5$ will shift the binary point five places to the right and $32S$ has the form

(30)
$$32S = 11.00110011001100110011001\overline{10011}\ldots_{\text{two}}.$$

Similarly, when S is multiplied by 2, the result is

(31)
$$2S = 0.00110011001100110011001\overline{10011}\ldots_{\text{two}}.$$

Since fractional parts in (30) and (31) agree, they will cancel out when we form the difference $32S - 2S$, that is,

(32)
$$32S - 2S = 11.00000000000000000000000000000\ldots_{\text{two}}.$$

Using $11_{\text{two}} = 3$ in (32), we obtain $30S = 3$. Therefore, $S = \tfrac{1}{10}$.

Scientific Notation

A standard way to present a real number, called *scientific notation*, is obtained by shifting the decimal point and supplying an appropriate power of 10. For example,

$$0.0000747 = 7.47 \times 10^{-5},$$

$$31.4159265 = 3.14159265 \times 10,$$

$$9,700,000,000 = 9.7 \times 10^9.$$

In chemistry, an important constant is Avogadro's number, which is 6.02252×10^{23}. It is the number of atoms in the gram atomic weight of an element. In computer science, $1K = 1.024 \times 10^3$.

Machine Numbers

Computers use a normalized floating-point binary representation for real numbers. This means that the mathematical quantity x is not actually stored in the computer. Instead, the computer stores a binary approximation to x:

(33)
$$x \approx \pm q \times 2^n.$$

The number q is the mantissa and it is a finite binary expression satisfying $\frac{1}{2} \leq q < 1$. The integer n is called the *exponent*.

In a computer, only a small subset of the real number system is used. Typically, this subset contains only a portion of the binary numbers suggested by (33). The number of binary digits is restricted in both the numbers q and n. For example, consider the subset of positive numbers

(34)
$$0.d_1 d_2 d_3 d_4 \text{ two} \times 2^n,$$

where $d_1 = 1$ and d_2, d_3, d_4 are either 0 or 1 and n is chosen from the set $\{-3, -2, -1, 0, 1, 2, 3, 4\}$. There are eight choices for the mantissa and eight choices for the exponent in (34), and this produces a set of 64 numbers:

(35)
$$\{0.1000_{\text{two}} \times 2^{-3}, \ 0.1001_{\text{two}} \times 2^{-3}, \ \ldots, \ 0.1110_{\text{two}} \times 2^4, \ 0.1111_{\text{two}} \times 2^4\}.$$

The decimal form for these 64 numbers is given in Table 1.1. It is important to learn that when the mantissa and exponent in (33) are restricted, the computer has a limited number of values it chooses from to store as an approximation to the real number x.

TABLE 1.1 Decimal equivalents for set of binary numbers with a 4-binary-bit mantissa and exponent of $n = -3, -2, \ldots, 3, 4$

Mantissa	Exponent:							
	$n = -3$	$n = -2$	$n = -1$	$n = 0$	$n = 1$	$n = 2$	$n = 3$	$n = 4$
0.1000_{two}	0.0625	0.125	0.25	0.5	1	2	4	8
0.1001_{two}	0.0703125	0.140625	0.28125	0.5625	1.125	2.25	4.5	9
0.1010_{two}	0.078125	0.15625	0.3125	0.625	1.25	2.5	5	10
0.1011_{two}	0.0859375	0.171875	0.34375	0.6875	1.375	2.75	5.5	11
0.1100_{two}	0.09375	0.1875	0.375	0.75	1.5	3	6	12
0.1101_{two}	0.1015625	0.203125	0.40625	0.8125	1.625	3.25	6.5	13
0.1110_{two}	0.109375	0.21875	0.4375	0.875	1.75	3.5	7	14
0.1111_{two}	0.1171875	0.234375	0.46875	0.9375	1.875	3.75	7.5	15

Let us see what would happen if a computer had only a 4-bit mantissa and was required to perform the computation $\left(\frac{1}{10} + \frac{1}{5}\right) + \frac{1}{6}$. We assume that the computer rounds all real numbers to the closest binary number in Table 1.1.

At each step the reader can look in the table to see the best approximation is being used.

(36)
$$\begin{aligned}
\tfrac{1}{10} &\approx 0.1101_{\text{two}} \times 2^{-3} = 0.01101_{\text{two}} \times 2^{-2} \\
\tfrac{1}{5} &\approx 0.1101_{\text{two}} \times 2^{-2} = \underline{0.1101_{\text{two}} \times 2^{-2}} \\
\tfrac{3}{10} && \qquad 1.00111_{\text{two}} \times 2^{-2}.
\end{aligned}$$

The computer must now decide how to store the number $1.00111_{\text{two}} \times 2^{-2}$. We assume that it is rounded to $0.1010_{\text{two}} \times 2^{-1}$. The next step is

(37)
$$\begin{aligned}
\tfrac{3}{10} &\approx 0.1010_{\text{two}} \times 2^{-1} = 0.1010_{\text{two}} \times 2^{-1} \\
\tfrac{1}{6} &\approx 0.1011_{\text{two}} \times 2^{-2} = \underline{0.01011_{\text{two}} \times 2^{-1}} \\
\tfrac{7}{15} && \qquad 0.11111_{\text{two}} \times 2^{-1}.
\end{aligned}$$

The computer must decide how to store the number $0.11111_{\text{two}} \times 2^{-1}$. Since rounding is assumed to take place, it stores $0.1000_{\text{two}} \times 2^{0}$. Therefore, the computer's solution to the addition problem is

(38)
$$\tfrac{7}{15} \approx 0.1000_{\text{two}} \times 2^{0}.$$

The error in the computer's calculation is

(39)
$$\tfrac{7}{15} - 0.1000_{\text{two}} = 0.46666667 - 0.50000000 \approx -0.03333333.$$

Expressed as a percentage of $\tfrac{7}{15}$, this amounts to 7.14%.

Computer Accuracy

To store numbers accurately, computers must have floating-point binary numbers with at least 24 binary bits used for the mantissa; this translates to about seven decimal places. If a 32-binary-bit mantissa is used, numbers with nine decimal places can be stored. Let us return to the difficulty encountered in (1) when a computer added $\tfrac{1}{10}$ repeatedly.

Suppose that the mantissa q in (33) contains 32 binary bits. The condition $\tfrac{1}{2} \leqq q$ implies that the first digit is $d_1 = 1$. Hence q has the form

(40)
$$q = 0.1d_2 d_3 d_4 \ldots d_{31} d_{32 \text{ two}}.$$

When fractions are represented in binary form, it is often the case that infinitely many digits are required. An example is

(41)
$$\tfrac{1}{10} = 0.0001100110011001100110011\overline{0011}\ldots_{\text{two}}.$$

When the 32-bit mantissa is used, truncation occurs and the computer uses the internal approximation

(42)
$$\tfrac{1}{10} \approx 2^{-3}(\underbrace{0.11001100110011001100110011001100_{\text{two}}}).$$

$$\text{32 significant binary digits}$$

The error in the approximation in (42) is

(43)
$$2^{-35}(0.1100110011\overline{0011}\ldots_{\text{two}}) \approx 2.328306437 \times 10^{-11}.$$

Because of (43), the computer must be in error when it sums the 100,000 addends of $\frac{1}{10}$ given in (1). The error must be greater than $100,000 \times 2.328306437 \times 10^{-11} = 2.328306437 \times 10^{-6}$. Indeed, there is a much larger error. Occasionally, the partial sum could be rounded up or down. Also, as the sum grows, the latter addends of $\frac{1}{10}$ are small compared to the current size of the sum, and their contribution is truncated more severely. The compounding effect of these errors actually produced the error $10,000 - 9,999.99447 = 5.53 \times 10^{-3}$.

Computer Floating-Point Numbers

Computers have both an integer mode and a floating-point mode for representing numbers. The *integer mode* is used for performing calculations that are known to be integer value and have limited usage for numerical analysis. *Floating-point numbers* are used for scientific and engineering applications. It must be understood that any computer implementation of equation (33) places restrictions on the number of digits used in the mantissa q, and that the range of possible exponents n must be limited.

Computers that use 32 bits to represent single-precision real numbers use 8 bits for the exponent and 24 bits for the mantissa. They can represent real numbers whose magnitude is in the range

$$2.938736E - 39 \quad \text{to} \quad 1.701412E + 38$$

(i.e., 2^{-128} to 2^{127}) with six decimal digits of numerical precision (e.g., $2^{-23} = 1.2 \times 10^{-7}$).

Computers that use 48 bits to represent single-precision real numbers, might use 8 bits for the exponent and 40 bits for the mantissa. They can represent real numbers in the range

$$2.9387358771E - 39 \quad \text{to} \quad 1.7014118346E + 38$$

(i.e., 2^{-128} to 2^{127}) with 11 decimal digits of precision (e.g., $2^{-39} = 1.8 \times 10^{-12}$).

If the computer has 64-bit double-precision real numbers, it might use 11 bits for the exponent and 53 bits for the mantissa. They can represent real numbers in the range

$$5.562684646268003 \times 10^{-309} \quad \text{to} \quad 8.988465674311580 \times 10^{307}$$

(i.e., 2^{-1024} to 2^{1023}) with about 16 decimal digits of precision (e.g., $2^{-52} = 2.2 \times 10^{-16}$).

Exercises for Binary Numbers

1. Use a computer to accumulate the following sums.

   (a) $10,000 - \sum_{k=1}^{100,000} 0.1$

   (b) $10,000 - \sum_{k=1}^{80,000} 0.125$

Here the intent is to have the computer do repeated subtractions. Do *not* use the multiplication shortcut.

2. Use equations (4) and (5) and convert the following binary numbers to decimal form (base 10).
   (a) 10101_{two}
   (b) 111000_{two}
   (c) 1101101_{two}
   (d) 10110111_{two}

3. Use equations (21) and (22) and convert the following binary fractions to decimal form (base 10).
   (a) 0.11011_{two}
   (b) 0.10101_{two}
   (c) 0.1010101_{two}
   (d) 0.11011011_{two}

4. Convert the following binary numbers to decimal form (base 10).
   (a) 1.0110101_{two}
   (b) 11.0010010001_{two}

5. The fractions in Exercise 4 are approximately $\sqrt{2}$ and π. Find the error in these approximations, that is, find
   (a) $\sqrt{2} - 1.0110101_{two}$
       Use $\sqrt{2} = 1.41421356237309\ldots$
   (b) $\pi - 11.0010010001_{two}$
       Use $\pi = 3.14159265358979\ldots$

6. Use Algorithm 1.1 and convert the following to binary numbers.
   (a) 23   (b) 75   (c) 360   (d) 1766

7. Use either Algorithm 1.2 or equations (21) and (22) and convert the following to a binary fraction of the form $0.d_1 d_2 \ldots d_{n\,two}$.
   (a) $\frac{7}{16}$ (b) $\frac{13}{16}$ (c) $\frac{23}{32}$ (d) $\frac{75}{128}$

8. Use Algorithm 1.2 and obtain the expansions.
   (a) $\frac{1}{10} = 0.0001100110\overline{011}\ldots_{two}$
   (b) $\frac{1}{3} = 0.01010101010\overline{01}\ldots_{two}$
   (c) $\frac{1}{7} = 0.001001001\overline{001}\ldots_{two}$
   Hint. For parts (b) and (c) use rational arithmetic, for example,
   $$\text{int}\left(\tfrac{4}{3}\right) = 1 \quad \text{and} \quad \text{frac}\left(\tfrac{4}{3}\right) = \tfrac{1}{3}.$$

9. For the following seven-digit binary approximations, find the error in the approximation $R - 0.d_1 d_2 d_3 d_4 d_5 d_6 d_7\,_{two}$.
   (a) $\frac{1}{10} \approx 0.0001100_{two}$
   (b) $\frac{1}{3} \approx 0.0101010_{two}$
   (c) $\frac{1}{7} \approx 0.0010010_{two}$
   Hint. First convert the binary expression to decimal.

10. For the expansion $\frac{1}{7} = 0.001001001\overline{001}\ldots_{two}$, show that this is equivalent to $\frac{1}{7} = \frac{1}{8} + \frac{1}{64} + \frac{1}{512} + \ldots$. Then use Theorem 1.1 to establish the expansion.

11. For the expansion $\frac{1}{5} = 0.001100110\overline{011}\ldots_{two}$, show that this is equivalent to $\frac{1}{5} = \frac{3}{16} + \frac{3}{256} + \frac{3}{4,096} + \ldots$. Then use Theorem 1.1 to establish the expansion.

12. Prove that any number 2^{-N}, where N is a positive integer, can be represented as a decimal number that has N digits, that is,
    $$2^{-N} = 0.c_1 c_2 c_3 \ldots c_N$$
    Hint. $\frac{1}{2} = 0.5, \frac{1}{4} = 0.25$.

13. Write a report on hexademical numbers.

14. Use Table 1.1 and find what happens when a computer with a 4-bit mantissa performs the following calculations.
    (a) $\left(\frac{1}{3} + \frac{1}{5}\right) + \frac{1}{6}$
    (b) $\left(\frac{1}{10} + \frac{1}{3}\right) + \frac{1}{5}$
    (c) $\left(\frac{3}{17} + \frac{1}{9}\right) + \frac{1}{7}$
    (d) $\left(\frac{7}{10} + \frac{1}{9}\right) + \frac{1}{7}$

15. (a) Prove that when 2 is replaced by 3 in all formulas in (7), the result is a method for finding the base 3 expansion of a positive integer.
    Express the following integers in base 3.
    (b) 10   (c) 23   (d) 49   (e) 123

16. (a) Prove that when 2 is replaced by 3 in all formulas in (27), the result is a method for finding the base 3 expansion of a positive number R that lies in the interval $0 < R < 1$.
    Express the following numbers in base 3.
    (b) $\frac{1}{3}$ (c) $\frac{1}{2}$ (d) $\frac{1}{10}$

17. (a) Prove that when 2 is replaced by 5 in all formulas in (7) the result is a method for finding the base 5 expansion of a positive integer.
    Express the following integers in base 5.
    (b) 10   (c) 32   (d) 49   (e) 144

18. (a) Prove that when 2 is replaced by 5 in all formulas in (27) the result is a method for finding the base 5 expansion of a positive number R that lies in the interval $0 < R < 1$.
    Express the following numbers in base 5.
    (b) $\frac{1}{3}$ (c) $\frac{1}{2}$ (d) $\frac{1}{10}$

19. Investigate floating-point real numbers on your computer.
    (a) How many digits in base 10 are used for the mantissa of a real floating-point number?
    (b) What is the range for the base 10 exponent of a real floating-point number?
    (c) Use functions in your computer to find approximations to the real numbers $2^{1/2}, 2^{1/3}, e, \pi$, and ln (3). Formulas for BASIC, FORTRAN, and Pascal are given in the table.

Real Number	BASIC Formula	FORTRAN Formula	Pascal Formula
$2^{1/2} \approx 1.4142135623730950488$	SQR(2)	SQRT(2)	SQRT(2)
$2^{1/3} \approx 1.2599210498948731648$	2^(1/3)	2**(1/3)	EXP(LN(2)/3)
$e \approx 2.7182818284590452353$	EXP(1)	EXP(1)	EXP(1)
$\pi \approx 3.1415926535897932384$	4*ATN(1)	4*ATAN(1)	4*ARCTAN(1)
ln (3) $\approx 1.0986122886681096913$	LOG(3)	ALOG(3)	LN(3)

20. Investigate how your computer stores real numbers.
    (a) How many bytes (and bits) are used for the mantissa of a real floating-point number?
    (b) What is the range for the base 2 exponent of a real floating-point number?
    (c) Write a program that will find the smallest positive integer N for which $1 + 2^{-N-1} = 1$ is a true expression for your computer.

21. In computer terminology, what are "exponent overflow" and "exponent underflow"?

22. When a computer translates a given real number x that is input to a stored binary machine number, what are the meanings of the phrases "computer rounding" and "computer chopping"?

23. In computer terminology, what is a "guard digit"?

Hexadecimal Numbers. When the base of the number system is 16, the digits are 0, 1, 2, 3, 4, 5, 6, 7, 8, 9, A, B, C, D, E, F. Here $A = 10_{ten}$, $B = 11_{ten}$, $C = 12_{ten}$, $D = 13_{ten}$, $E = 14_{ten}$, and $F = 15_{ten}$.

24. Convert the following hexadecimal numbers to decimal (base 10) form.
(a) 213_{16} (b) $7C9_{16}$ (c) $1ABE_{16}$ (d) $F09C_{16}$
(e) 0.2_{16} (f) 0.99_{16} (g) $0.A4B_{16}$ (h) $0.F0B_{16}$

25. Adapt the formulas in (7) so that the base 16 expansion of a positive integer N can be found. Express the following integers in base 16.
(a) 512 (b) 2,001 (c) 51,264 (d) 91,919

26. Adapt the formulas in (27) so that the base 16 expansion of real number R that lies in the interval $0 < R < 1$ can be found. Express the following real numbers in base 16.
(a) $\frac{1}{3}$ (b) $\frac{1}{2}$ (c) $\frac{1}{10}$ (d) $\frac{1}{15}$

27. Write a report on how to convert hexadecimal numbers to binary numbers and vice versa.

28. Write a report on the 1's complement and 2's complement of a binary number.

29. Write a report on interval arithmetic.

30. Write a report on L'Hôpital's rule.

1.3 Taylor Series and Calculation of Functions

Limit processes are the basis of calculus. For example, the derivative

$$f'(x) = \lim_{h \to 0} \frac{f(x + h) - f(x)}{h}$$

is the limit of the difference quotient where both numerator and denominator go to zero. *Taylor series* illustrate another type of limit process. In this case an infinite number of terms are added together by taking the limit of certain partial sums. An important application is their use to represent the elementary functions $\sin(x)$, $\cos(x)$, $\exp(x)$, $\ln(x)$. Table 1.2 gives several of the common Taylor series expansions. The partial sums can be accumulated until an approximation to the function is obtained which has the accuracy that is specified. Series solutions are used in the areas of engineering and physics.

We want to learn how a finite sum can be used to obtain a good approximation to an infinite sum. For illustration we shall use the exponential series in Table 1.2 to compute the number $e = \exp(1)$, which is the base of the natural logarithm and exponential functions. Here we choose $x = 1$ and use the series

$$\exp(1) = 1 + \frac{1}{1!} + \frac{1^2}{2!} + \frac{1^3}{3!} + \frac{1^4}{4!} + \cdots + \frac{1^k}{k!} + \cdots.$$

The definition for the sum of an infinite series in Section 1.1 requires that the partial sums S_n tend to a limit. The values for several of these sums are given in Table 1.3.

TABLE 1.2 Taylor series expansions for some common functions

$$\sin(x) = x - \frac{x^3}{3!} + \frac{x^5}{5!} - \frac{x^7}{7!} + \cdots \qquad \text{for all } x$$

$$\cos(x) = 1 - \frac{x^2}{2!} + \frac{x^4}{4!} - \frac{x^6}{6!} + \cdots \qquad \text{for all } x$$

$$\exp(x) = 1 + x + \frac{x^2}{2!} + \frac{x^3}{3!} + \frac{x^4}{4!} + \cdots \qquad \text{for all } x$$

$$\ln(1+x) = x - \frac{x^2}{2} + \frac{x^3}{3} - \frac{x^4}{4} + \cdots \qquad -1 < x \leq 1$$

$$\arctan(x) = x - \frac{x^3}{3} + \frac{x^5}{5} - \frac{x^7}{7} + \cdots \qquad -1 < x \leq 1$$

$$(1+x)^p = 1 + px + \frac{p(p-1)}{2!}x^2 + \frac{p(p-1)(p-2)}{3!}x^3 + \cdots \quad \text{for } |x| < 1$$

TABLE 1.3 Partial sums S_n used to determine e

n	$S_n = 1 + \dfrac{1}{1!} + \dfrac{1}{2!} + \cdots + \dfrac{1}{n!}$
0	1.0
1	2.0
2	2.5
3	2.666666666666...
4	2.708333333333...
5	2.716666666666...
6	2.718055555555...
7	2.718253968254...
8	2.718278769841...
9	2.718281525573...
10	2.718281801146...
11	2.718281826199...
12	2.718281828286...
13	2.718281828447...
14	2.718281828458...
15	2.718281828459...

A natural way to think about the power series representation of a function is to view the expansion as the limiting case of polynomials of increasing degree. If enough terms are added, then an accurate approximation will be obtained. This needs to be made precise. What degree should be chosen for the polynomial, and how do we calculate the coefficients for the powers of x in the polynomial? Theorem 1.2 answers these questions.

Theorem 1.2 [Taylor polynomial] If $f(x)$ has a continuous derivative of order $1, 2, \ldots,$ $N + 1$ on the interval $[a, b]$ containing the variable x and the fixed value x_0, then $f(x)$ has the representation

(1)
$$f(x) = \sum_{k=0}^{N} \frac{1}{k!} f^{(k)}(x_0)(x - x_0)^k + E_N(x),$$

where the remainder term $E_N(x)$ has the form

(2)
$$E_N(x) = \frac{f^{(N+1)}(c)(x - x_0)^{N+1}}{(N + 1)!},$$

and c is a value that lies somewhere between x and x_0.

Proof. The proof is left as an exercise (see Exercises 20 and 21).

Remark. The finite series in (1) is the Nth-degree Taylor polynomial

(3)
$$P_N(x) = \sum_{k=0}^{N} \frac{1}{k!} f^{(k)}(x_0)(x - x_0)^k = a_0 + a_1(x - x_0) + \ldots + a_N(x - x_0)^N,$$

and the coefficients $\{a_k\}$ are computed using the definition

(4)
$$a_k = \frac{f^{(k)}(x_0)}{k!}.$$

Relation (3) indicates that the coefficients a_k of the Taylor polynomial will be calculated using equation (4). Although the error term (2) involves a similar expression, notice that $f^{(N+1)}(c)$ is to be evaluated and that the number c may differ from the number x_0. For this reason we do *not* try to evaluate $E_N(x)$; it is used to determine a bound for the accuracy of the approximation.

Example 1.5 Show why 15 terms are all that are needed to obtain the 13-digit approximation $e \approx 2.718281828459$ in Table 1.2.

Solution. Expand $f(x) = \exp(x)$ in a Taylor polynomial of degree 15 using the fixed value $x_0 = 0$ and involving the powers $(x - 0)^k = x^k$. The derivatives required are $f'(x) = \exp(x), f''(x) = \exp(x), \ldots, f^{(16)}(x) = \exp(x)$. The first 15 derivatives are used to calculate the coefficients $a_k = \exp(0)/k! = 1/k!$ and are used to write

(5)
$$P_{15}(x) = 1 + x + \frac{x^2}{2!} + \frac{x^3}{3!} + \ldots + \frac{x^{15}}{15!}.$$

Setting $x = 1$ in (5) gives the partial sum $S_{15} = P_{15}(1)$. The remainder term is needed to show the accuracy of the approximation

(6)
$$E_{15}(x) = \frac{f^{(16)}(c)x^{16}}{16!}.$$

Since we chose $x_0 = 0$ and $x = 1$, the value c lies between them (i.e., $0 < c < 1$), which implies that $e^c < e^1$. Notice that the partial sums in Table 1.3 are bounded above by 3; this permits us to write the crude inequality $e^1 < 3$. Combining these two inequalities yields $e^c < 3$, which is used in the following calculation

$$|E_{15}(1)| = \frac{|f^{(16)}(c)|}{16!} \leq \frac{e^c}{16!} < \frac{3}{16!} = 1.433843 \times 10^{-13}.$$

Therefore, all the digits in the approximation $e \approx 2.718281828459\ldots$ are correct, because the actual error (whatever it is) must be less than 2 in the thirteenth decimal place.

Instead of giving a rigorous proof of Theorem 1.2, we shall discuss some of the features of the approximation; the reader can look in any standard reference on calculus for more details. For illustration, we again use the function $f(x) = \exp(x)$ and the value $x_0 = 0$. From elementary calculus, we know that the slope of the curve $y = f(x)$ at the point $(x, f(x))$ is $f'(x) = \exp(x)$. Hence the slope at the point $(0, 1)$ is $f'(0) = 1$. Therefore, the tangent line to the curve at the point $(0, 1)$ is $y = 1 + x$. This is the same formula that would be obtained if we use $N = 1$ in Theorem 1.2, that is, $P_1(x) = f(0) + f'(0)x/1! = 1 + x$. Therefore, $P_1(x)$ is the equation of the line tangent to the curve. The graphs are shown in Figure 1.1.

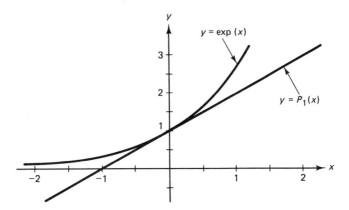

Figure 1.1. The graphs $y = \exp(x)$ and $y = P_1(x) = 1 + x$.

Observe that the approximation $\exp(x) \approx 1 + x$ is good near the center $x_0 = 0$ and that the distance between curves grows as x moves away from 0. Notice that the slopes of the curves agree at $(0, 1)$. In calculus we learned that the second derivative indicates whether a curve is concave up or concave down. The study of curvature* shows that if two curves $y = f(x)$ and $y = g(x)$ have the property that $f(x_0) = g(x_0)$, $f'(x_0) = g'(x_0)$, and $f''(x_0) = g''(x_0)$,

---

*The curvature K of a graph $y = f(x)$ at (x, y) is defined by

$$K = |f''(x_0)|(1 + [f'(x_0)]^2)^{-3/2}.$$

they have the same curvature at x_0. This property would be desirable for a polynomial function that approximates $f(x)$. Corollary 1.1 shows that the Taylor polynomial has this property for $N \geq 2$.

Corollary 1.1 If $P_N(x)$ is the Taylor polynomial of degree N defined in equation (3), then

$$P_N(x_0) = f(x_0)$$
$$P_N'(x_0) = f'(x_0)$$
(7)
$$P_N''(x_0) = f''(x_0)$$
$$\vdots$$
$$P_N^{(N)}(x_0) = f^{(N)}(x_0).$$

Proof. Set $x = x_0$ in equation (3) and the result is $P_N(x_0) = f(x_0)$. Differentiate both sides of (3) and get

(8) $$P_N'(x) = a_1 + 2a_2(x - x_0) + 3a_3(x - x_0)^2 + \ldots + Na_N(x - x_0)^{N-1}.$$

By definition (4), $a_1 = f'(x_0)$. Use this in (8) with $x = x_0$ to obtain $P_N'(x_0) = f'(x_0)$. Successive differentiations of (8) with the appropriate use of definition (4) will establish the other identities in (7). The details are left as an exercise.

Applying Corollary 1.1, we see that $y = P_2(x)$ has the properties $f(x_0) = P_2(x_0), f'(x_0) = P_2'(x_0)$, and $f''(x_0) = P_2''(x_0)$; hence the graphs have the same curvature at x_0. For example, consider $f(x) = \exp(x)$ and $P_2(x) = 1 + x + x^2/2$. The graphs are shown in Figure 1.2 and it is seen that they are curved up in the same fashion at $(0, 1)$.

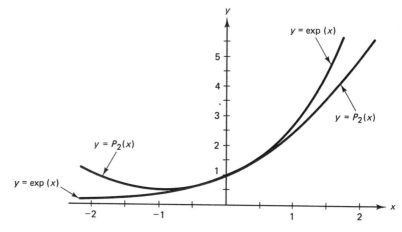

Figure 1.2. The graphs $y = \exp(x)$ and $y = P_2(x) = 1 + x + x^2/2$.

In the theory of approximation, one seeks to find an accurate polynomial approximation to the analytic function* $f(x)$ over $[a, b]$. This is one technique used in developing computer software. The accuracy of a Taylor polynomial is increased when we choose N large. The accuracy of any given polynomial will generally decrease as the value of x moves away from the center x_0. Hence we must choose N large enough and restrict the maximum value of $|x - x_0|$ so that the error does not exceed a specified bound. If we choose the interval width to be $2R$ and x_0 in the center (i.e., $|x - x_0| < R$), the absolute value of the error satisfies the relation

(9)
$$|\text{error}| = |E_N(x)| \leqq \frac{MR^{N+1}}{(N+1)!}.$$

where $M \leq \{\max |f^{(N+1)}(z)|\}$ and z ranges throughout the interval $x_0 - R \leqq z \leqq x_0 + R$. If N is fixed, and the derivatives are uniformly bounded, the error bound in (9) is proportional to $\dfrac{R^{N+1}}{(N+1)!}$ and decreases if R goes to zero. For R fixed, the error bound is inversely proportional to $(N+1)!$ and it goes to zero as N gets large. Table 1.4 shows how the choices of these two parameters affect the accuracy of the approximation $\exp(x) \approx P_N(x)$ over the interval $|x| \leqq R$. The error is smallest when N is largest and R smallest. Graphs for P_2, P_3, and P_4 are given in Figure 1.3.

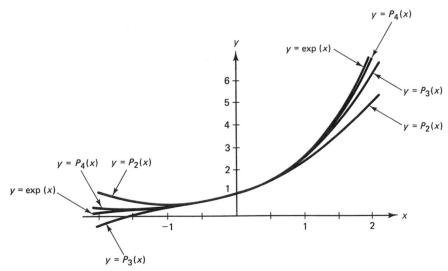

Figure 1.3. The graphs of $\exp(x)$, $P_2(x)$, $P_3(x)$, and $P_4(x)$.

---

*The function $f(x)$ is analytic at x_0 if it has continuous derivatives of all orders and can be expressed as a Taylor series in an interval about x_0.

TABLE 1.4   Values for the error bound $|\text{error}| < e^R R^{N+1}/[(N+1)!]$
using the approximation $\exp(x) \approx P_N(x)$ for $|x| \le R$

	$R = 2.0$, $\|x\| \le 2.0$	$R = 1.5$, $\|x\| \le 1.5$	$R = 1.0$, $\|x\| \le 1.0$	$R = 0.5$, $\|x\| \le 0.5$
$\exp(x) \approx P_5(x)$	0.65680499	0.07090172	0.00377539	0.00003578
$\exp(x) \approx P_6(x)$	0.18765857	0.01519323	0.00053934	0.00000256
$\exp(x) \approx P_7(x)$	0.04691464	0.00284873	0.00006742	0.00000016
$\exp(x) \approx P_8(x)$	0.01042548	0.00047479	0.00000749	0.00000001

Example 1.6 Establish the error bounds for the approximation $\exp(x) \approx P_8(x)$ on the intervals $|x| \le 1.0$ and $|x| \le 0.5$.

Solution. If $|x| \le 1.0$, then $|f^{(9)}(c)| = |e^c| \le e^{1.0}$ implies that

$$|\text{error}| = |E_8(x)| \le \frac{e^{1.0}(1.0)^9}{9!} \approx 0.00000749.$$

For the interval $|x| \le 0.5$, $|f^{(9)}(c)| = |e^c| \le e^{0.5}$ implies that

$$|\text{error}| = |E_8(x)| \le \frac{e^{0.5}(0.5)^9}{9!} \approx 0.00000001.$$

Example 1.7 If $f(x) = \exp(x)$, then show that $N = 9$ is the smallest integer, so that $|\text{error}| = |E_N(x)| \le 0.0000005$ for x in $[-1, 1]$. Hence $P_9(x)$ can be used to compute approximate values of $\exp(x)$ that will be accurate in the sixth decimal place.

Solution. We need to find the smallest integer N so that

$$|\text{error}| = |E_N(x)| = \frac{e^c(1)^{N+1}}{(N+1)!} \le 0.0000005.$$

In Example 1.6 we saw that $N = 8$ is too small, so we try $N = 9$ and discover that $|E_N(x)| \le e^1(1)^{9+1}/[(9+1)!] \le 0.000000749$. This value is slightly larger than desired; hence we would be likely to choose $N = 10$. But we used $e^c \le e^1$ as a crude estimate in finding the error bound. Hence 0.000000749 is a little larger than the actual error. Figure 1.4 shows a graph of $E_9(x) = \exp(x) - P_9(x)$. Notice that the maximum vertical range is about 3×10^{-7} and occurs at the right endpoint $(1, E_9(1))$. Indeed, the maximum error on the interval is $E_9(1) \approx 3.024 \times 10^{-7}$. Therefore, $N = 9$ is justified.

When series are used to approximate the elementary functions it is worthwhile to take advantage of the special properties of the functions. Suppose that we wanted to compute $\exp(86.3)$ using series. Then we would have to be patient and use $N = 150$ to obtain

(10)
$$\exp(86.3) \approx P_{150}(86.3) \approx 3.01726735 \times 10^{37}.$$

This sum was accumulated using a computer that has floating-point numbers with nine digits of accuracy.

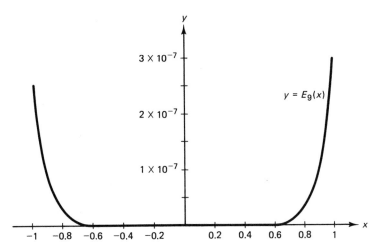

Figure 1.4. The graph of the error $y = E_9(x) = \exp(x) - P_9(x)$.

We could save a considerable amount of effort if we use the identity $e^{86.3} = e^{86}e^{0.3}$ and the fact that we have already found e^1 accurate to 13 decimal places. We need only to compute $\exp(0.3) \approx P_9(0.3) \approx 1.34985881$ and use the formula

$$
\begin{aligned}
e^{86}e^{0.3} &\approx (2.718281828459)^{86}(1.34985881) \\
&= 2.23524660 \times 10^{37}(1.34985881) = 3.01726732 \times 10^{37}.
\end{aligned}
$$

(11)

The products in (11) were computed in double-precision arithmetic. If only nine significant digits are carried, a larger error results:

$$
\begin{aligned}
2.71828183^{86}(1.34985881) &= 2.23524670 \times 10^{37}(1.34985881) \\
&= 3.01726744 \times 10^{37}.
\end{aligned}
$$

This time the answer has seven significant digits of accuracy because the error in the approximation to e is propagated in the multiplications.

Theorem 1.3 [Alternating series] Suppose that $c_k > 0$ and that $c_k \geq c_{k+1} > 0$ for $k = J$, $J + 1, \ldots$ and that $\lim_{k \to \infty} c_k = 0$. Then

(12) $$ S = \sum_{k=J}^{\infty} (-1)^k c_k \quad \text{is a convergent infinite series.} $$

If the partial sum S_N is used to approximate S, then the truncation error $E_N = S - S_N$ is no larger than the magnitude of the next term:

(13) $$ |E_N| = |S - S_N| \leq c_{N+1} \quad \text{(see the figure below).} $$

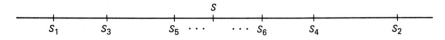

Another way to state the result in (12) is to say that the partial sums converge to S, that is,

(14)
$$S = \lim_{N \to \infty} S_N = \lim_{N \to \infty} \sum_{k=J}^{N} (-1)^k c_k = \sum_{k=J}^{\infty} (-1)^k c_k.$$

As a note of caution, we investigate what happens when an alternating series is used to compute $\exp(-11)$. Theorem 1.3 implies that the truncation error for $P_{50}(-11)$ is no larger than $11^{51}/51! = 8.324955 \times 10^{-14}$. A computer with nine decimal digits of accuracy was used to accumulate this sum and it gave the answer $P_{50}(-11) = 1.60363964 \times 10^{-5}$. The mathematical error analysis might lead us to believe that all nine digits were correct. However, the computer's built-in exponential function gives the answer $\exp(-11) = 1.67017008 \times 10^{-5}$. Therefore, $P_{50}(-11)$ has roughly two significant digits of accuracy! So far, this is the worst discrepancy that we have encountered.

The reason for the flawed calculation is that each term has its own round-off error, and the error in the total will be as big as the largest error in one of its addends. Theorem 1.3 discusses the accuracy of the mathematical approximation

$$P_{50}(-11) = 1 - 11 + \frac{11^2}{2} - \frac{11^3}{6} + \ldots - \frac{11^{49}}{49!} + \frac{11^{50}}{50!}$$

$$= \frac{34{,}694{,}835{,}499{,}645{,}991{,}522{,}891{,}035{,}426{,}510{,}983{,}460{,}551{,}595{,}052{,}804{,}990{,}261}{2{,}077{,}323{,}488{,}949{,}756{,}030{,}572{,}543{,}416{,}847{,}535{,}608{,}522{,}480{,}812{,}032{,}000{,}000{,}000{,}000}.$$

The computer actually calculated a different sum:

$$P_{50}(-11) \approx 1 - 11 + 60.5 - 221.833333 + \ldots + 0.00000000000385975186$$

$$\approx 0.00000160363964.$$

The maximum error occurs with the tenth and eleventh terms in the series when the computer uses $11^{10}/10! \approx 7147.65889$ and the round-off error is 5.7782×10^{-6}. Table 1.5 gives some of the terms, the computer's approximation, and the round-off error for the term.

If an accurate approximation for $\exp(-11)$ is needed, it can be found by using $x = 11$ and computing $P_{50}(11) = 59{,}874.1417$. The error for each term in the series is less than 0.0000057782, and the total error is less than $51 \times 5.7782 \times 10^{-6} = 0.00029469$. This is a small fraction of the total $59{,}874.1417$, so that eight significant digits are guaranteed. Now the reciprocal can be taken to obtain

$$\exp(-11) \approx \frac{1}{59{,}874.1417} = 1.67017008 \times 10^{-5}.$$

TABLE 1.5 Errors for some approximation terms in the $\exp(-11) \approx P_{50}(-11)$

k	$(-11)^k/k!$	Computer Approximation	Computer's Round-Off Error
0	1.	1.	0.0000000000000000
1	−11.	−11.	0.0000000000000000
2	60.5	60.5	0.0000000000000000
3	−221.8333333333333333. . .	−221.833333	−0.0000003333333333
4	610.0416666666666666. . .	610.041667	−0.0000003333333333
5	−1,342.0916666666666666. . .	−1,342.09167	0.0000033333333333
10	7,147.6588957782186948. . .	7,147.65889	0.0000057782186948
15	−3,194.4100699966086664. . .	−3,194.41007	0.0000000033913335
20	276.5216160254471009. . .	276.521616	0.0000000254471009
25	−6.9850810562949810. . .	−6.98508106	0.0000000037050189
30	0.0657840306839339. . .	0.0657840307	−0.0000000000160660
35	−0.0002719639449612. . .	−0.000271963945	0.0000000000000387
40	0.0000005547053290. . .	0.000000554705329	0.0000000000000000
50	0.0000000000003859. . .	0.0000000000003859	0.0000000000000000

Example 1.8 Show that $\ln(2) = 1 - \frac{1}{2} + \frac{1}{3} - \frac{1}{4} + \frac{1}{5} - \frac{1}{6} + \ldots$ is a convergent series. How many terms are needed so that the truncation error is less than 2×10^{-4}?

Solution. In general, $c_k = 1/k$ and it is easy to verify that

$$c_k = \frac{1}{k} \geq \frac{1}{k+1} = c_{k+1} \quad \text{and} \quad \lim_{k \to \infty} c_k = \lim_{k \to \infty} \frac{1}{k} = 0.$$

By Theorem 1.3 we conclude that the series is convergent. To make $|E_N| < 2 \times 10^{-4}$ we seek the smallest integer N so that $c_{N+1} < 2 \times 10^{-4}$ and find $N = 5,000$. Calculation reveals that

$$S_{4,999} \approx 0.693247197 \quad \text{with} \quad E_{4,999} \approx -0.000100016$$

$$S_{5,000} \approx 0.693047197 \quad \text{with} \quad E_{5,000} \approx 0.000099984$$

$$S_{5,001} \approx 0.693247157 \quad \text{with} \quad E_{5,001} \approx -0.000099976.$$

Notice that the series is slowly converging and that the accuracy of the partial sums $S_{5,000}$ and $S_{5,001}$ is not much better than $S_{4,999}$. In the exercises there will be a series that will converge faster.

Methods for Evaluating a Polynomial

There are several mathematically equivalent ways to evaluate a polynomial. Consider, for example, the function

(15)
$$f(x) = (x - 1)^8.$$

The binomial formula can be used to expand $f(x)$ in powers of x, and the result is

(16)
$$g(x) = x^8 - 8x^7 + 28x^6 - 56x^5 + 70x^4 - 56x^3 + 28x^2 - 8x + 1.$$

A third way to evaluate $f(x)$ is *Horner's method*, which is also called *nested multiplication*. It is explained in detail in Section 2.4. When applied to formula (16), nested multiplication permits us to write

(17) $$h(x) = (((((((x - 8)x + 28)x - 56)x + 70)x - 56)x + 28)x - 8)x + 1.$$

To evaluate $f(x)$ requires one subtraction and an eighth-power calculation. Notice that the eighth power can be computed by seven multiplications. The function $g(x)$ can be evaluated using powers of x in the calculations, but this could be time consuming. If the powers of x were stored, then $g(x)$ could be computed using seven multiplications and eight additions or subtractions. The function $h(x)$ is also seen to require seven multiplications and eight additions or subtractions.

Algorithm 1.3 uses the idea of storing the terms $(x - x_0)^k/k!$. If this is done in the sequential order

$$\frac{(x - x_0)}{1!},$$

$$\frac{(x - x_0)^2}{2!} = \frac{(x - x_0)(x - x_0)}{2},$$

$$\vdots$$

$$\frac{(x - x_0)^k}{k!} = \frac{[(x - x_0)^{k-1}/(k - 1)!](x - x_0)}{k},$$

a larger value of k can be used before the computer encounters an exponent overflow or underflow.

Algorithm 1.3 [*Computer Evaluation of a Taylor Series*] Evaluate

$$P(x) = \sum_{k=0}^{\infty} f^{(k)}(x_0)\frac{(x - x_0)^k}{k!} = \sum_{k=0}^{\infty} D(k)\frac{(x - x_0)^k}{k!},$$

where the kth derivative of $f(x)$ at x_0 is $D(k) = f^{(k)}(x_0)$.

Only a finite number of terms can be accumulated. Hence the summation process is terminated when consecutive partial sums differ by less than the preassigned value Tol.

READ Tol	{Termination criterion, e.g., 10^{-7}}
READ N	{Maximum degree, e.g., 100}
READ X0	{Point of expansion}
READ D(0) , . . . , D(N)	{Get the derivatives of f(x)}
Close := 1	{Closeness of consecutive partial sums}

```
            K := 0                                   {Initialize the counter}
            Sum := D(0)                              {Initialize the variable}
            Prod := 1                                {Variable that holds (x − x₀)ᵏ/k!}
            INPUT  X                                 {The independent variable}
            IF  X = X0  THEN  CLOSE := 0
    WHILE   Close ≥ Tol and K < N  DO
            K := K + 1
            Prod := Prod*(X−X0)/K                    {Make the factor (x − x₀)ᵏ/k!}
            WHILE  D(K) = 0 and K < N  DO            {If D(K) = 0, then continue
                   K := K + 1
                   Prod := Prod*(X−X0)/K              to build up (x − x₀)ᵏ/k!}
            Term := D(K)*Prod
            IF  Term ≠ 0  THEN  Close := |Term|
            Sum := Sum + Term


    IF      Close < Tol and K ≤ N  THEN             {Output}
            PRINT  'The sum of the Taylor polynomial is'  Sum
                   'Consecutive partial sums are closer than'  Close
       ELSE
            PRINT  'The current partial sum is'  Sum
                   'Convergence has NOT been achieved.'
    ENDIF
```

We end this section with two theorems about series. They can be found in several standard references and should be consulted before any complicated application is made using series.

Theorem 1.4 [Taylor series] Suppose that f is analytic and has continuous derivatives of all orders $1, 2, \ldots, N, \ldots$ on an interval (a, b) containing x_0. If the Taylor polynomials tend to a limit

(18)
$$S(x) = \lim_{N \to \infty} S_N(x) = \lim_{N \to \infty} \sum_{k=0}^{N} \frac{1}{k!} f^{(k)}(x_0)(x - x_0)^k,$$

then $f(x)$ has the Taylor series expansion

(19)
$$f(x) = \sum_{k=0}^{\infty} \frac{1}{k!} f^{(k)}(x_0)(x - x_0)^k.$$

Proof. This follows directly from the definition of convergence of series in Section 1.2. The limit condition is often stated by saying that the error term must go to zero as N goes to infinity. Therefore, a necessary and sufficient condition for (19) to hold is that

(20) $$\lim_{N \to \infty} E_N(x) = \lim_{N \to \infty} \frac{f^{(N+1)}(c)(x - x_0)^{N+1}}{(N+1)!} = 0.$$

where c depends on N and x.

Theorem 1.5 [Radius of convergence] For every power series

(21) $$\sum_{k=0}^{\infty} a_k(x - x_0)^k$$

there exists a number R with $0 \le R \le +\infty$, called the *radius of convergence* of the series. It has the properties:

(22)   If $|x - x_0| < R$, then the series (21) converges.

(23)   If $|x - x_0| > R$, then the series (21) diverges.

The value R can often be found by using the *ratio test*:

(24) $$R = \lim_{n \to \infty} \frac{|a_n|}{|a_{n+1}|} \quad \text{when this limit exists.}$$

Exercises for Taylor Series and Calculation of Functions

1. Let $f(x) = \sin(x)$ and apply Theorem 1.2.
   (a) Use $x_0 = 0$ and find $P_5(x)$, $P_7(x)$, and $P_9(x)$.
   (b) Use the polynomials in part (a) and evaluate $P_5(0.5)$, $P_5(1.0)$, $P_7(0.5)$, $P_7(1.0)$, $P_9(0.5)$, and $P_9(1.0)$ and compare with $\sin(0.5)$ and $\sin(1.0)$. [Evaluate $\sin(x)$ with x in radians.]
   (c) Show that if $|x| \le 1$, then the approximation

   $$\sin(x) \approx x - \frac{x^3}{3!} + \frac{x^5}{5!} - \frac{x^7}{7!} + \frac{x^9}{9!}$$

   has the error bound $|E_9| < 1/10! = 0.000000275573.\ldots$
   (d) Use $x_0 = \pi/4$ and find $P_5(x)$, which involves powers of $(x - \pi/4)$.
   (e) Use the polynomial in part (d) and evaluate $P_5(0.5)$ and $P_5(1.0)$ and compare with $\sin(0.5)$ and $\sin(1.0)$.

2. Let $f(x) = \cos(x)$ and apply Theorem 1.2.
   (a) Use $x_0 = 0$ and find $P_4(x)$, $P_6(x)$, and $P_8(x)$.
   (b) Use the polynomials in part (a) and evaluate $P_4(0.5)$, $P_4(1.0)$, $P_6(0.5)$, $P_6(1.0)$, $P_8(0.5)$, and $P_8(1.0)$ and compare with $\cos(0.5)$ and $\cos(1.0)$. [Evaluate $\cos(x)$ with x in radians.]
   (c) Show that if $|x| \le 1$, then the approximation

   $$\cos(x) \approx 1 - \frac{x^2}{2!} + \frac{x^4}{4!} - \frac{x^6}{6!} + \frac{x^8}{8!}$$

   has the error bound $|E_8| < 1/9! = 0.00000275573.\ldots$

(d) Use $x_0 = \pi/4$ and find $P_4(x)$, which involves powers of $(x - \pi/4)$.

(e) Use the polynomial in part (d) and evaluate $P_4(0.5)$ and $P_4(1.0)$ and compare with $\cos(0.5)$ and $\cos(1.0)$.

3. Does $f(x) = x^{1/2}$ have a Taylor series expansion about $x_0 = 0$? Justify your answer.

4. (a) Find the Taylor polynomial of degree $N = 5$ for $f(x) = 1/(1 + x)$ expanded about $x_0 = 0$.

(b) Find the error term for the polynomial in part (a).

5. Find the Taylor polynomial of degree $N = 3$ for $f(x) = \exp(-x^2/2)$ expanded about $x_0 = 0$.

6. Use $f(x) = (2 + x)^{1/2}$ and apply Theorem 1.2.

(a) Find the Taylor polynomial $P_3(x)$ expanded about the value $x_0 = 2$ that involves powers of $(x - 2)$.

(b) Use $P_3(x)$ to find an approximation to $3^{1/2}$.

(c) Find the maximum value of $|f^{(4)}(c)|$ on the interval $1 \leqq c \leqq 3$ and find a bound for $|E_3(x)|$.

7. (a) Use the geometric series

$$\frac{1}{1 + t^2} = 1 - t^2 + t^4 - t^6 + t^8 - \ldots \qquad \text{for } |t| < 1,$$

and integrate both sides term by term to obtain

$$\arctan(x) = x - \frac{x^3}{3} + \frac{x^5}{5} - \frac{x^7}{7} + \ldots \qquad \text{for } |x| < 1.$$

(b) Use $\pi/6 = \arctan(3^{-1/2})$ and the series in part (a) to show that

$$\pi = 3^{1/2} \times 2\left(1 - \frac{3^{-1}}{3} + \frac{3^{-2}}{5} - \frac{3^{-3}}{7} + \frac{3^{-4}}{9} - \ldots\right).$$

(c) Use the series in part (b) to compute π accurate to eight digits. *Fact.* $\pi \approx 3.14159265358979323384. \ldots$

8. Use $f(x) = \ln(1 + x)$ and $x_0 = 0$, and apply Theorem 1.2.

(a) Show that $f^{(k)}(x) = (-1)^{k-1}[(k - 1)!]/(1 + x)^k$.

(b) Show that the Taylor polynomial of degree N is

$$P_N(x) = x - \frac{x^2}{2} + \frac{x^3}{3} - \frac{x^4}{4} + \ldots + \frac{(-1)^{N-1}x^N}{N}.$$

(c) Show that the error term for $P_N(x)$ is

$$E_N(x) = \frac{(-1)^N x^{N+1}}{(N + 1)(1 + c)^{N+1}}.$$

(d) Evaluate $P_3(0.5)$, $P_6(0.5)$, and $P_9(0.5)$. Compare with $\ln(1.5)$.

(e) Show that if $0.0 \leq x \leq 0.5$, the approximation

$$\ln(x) \approx x - \frac{x^2}{2} + \frac{x^3}{3} - \frac{x^4}{4} + \ldots + \frac{x^7}{7} - \frac{x^8}{8} + \frac{x^9}{9}$$

has the error bound $|E_9| \leq (0.5)^{10}/10 = 0.00009765. \ldots$

9. (a) Let N go to infinity in Exercise 8(b) and obtain

$$\ln(1 + x) = x - \frac{x^2}{2} + \frac{x^3}{3} - \ldots + \frac{(-1)^{n-1}x^n}{n} + \ldots.$$

(b) Change the variable in part (a) and obtain the expansion

$$\ln(1 - x) = -x - \frac{x^2}{2} - \frac{x^3}{3} - \ldots - \frac{x^n}{n} - \ldots.$$

(c) Subtract the series in part (b) from the series in part (a) to obtain

$$\ln\left(\frac{1 + x}{1 - x}\right) = 2\left(x + \frac{x^3}{3} + \ldots + \frac{x^{2n-1}}{2n - 1} + \ldots\right).$$

(d) Set $x = \frac{1}{3}$ in part (c) and show that

$$\ln(2) = 2\left(3^{-1} + \frac{3^{-3}}{3} + \frac{3^{-5}}{5} + \ldots + \frac{3^{-2n+1}}{2n - 1} + \ldots\right).$$

(e) Use the ratio test and show that the series in parts (a) and (b) converge when $|x| < 1$.

10. *Binomial Series.* Let $f(x) = (1 + x)^p$ and $x_0 = 0$.
    (a) Show that $f^{(k)}(x) = p(p - 1)\ldots(p - k + 1)(1 + x)^{p-k}$.
    (b) Show that the Taylor polynomial of degree N is

$$P_N(x) = 1 + px + \frac{p(p - 1)x^2}{2!} + \ldots + \frac{p(p - 1)\ldots(p - N + 1)x^N}{N!}.$$

(c) Show that $E_N(x) = p(p - 1)\ldots(p - N)x^{N+1}/[(1 + C)^{N+1-p}(N + 1)!]$.
(d) Set $p = \frac{1}{2}$ and compute $P_2(0.5)$, $P_4(0.5)$, and $P_6(0.5)$. Compare with $(1.5)^{1/2}$.
(e) Show that if $0.0 \le x \le 0.5$, then the approximation

$$(1 + x)^{1/2} \approx 1 + \frac{x}{2} - \frac{x^2}{8} + \frac{x^3}{16} - \frac{5x^4}{128} + \frac{7x^5}{256}$$

has the error bound $|E_5| \le (0.5)^6(21/1024) = 0.0003204\ldots$.
(f) Show that if $p = N$ is a positive integer,

$$P_N(x) = 1 + Nx + \frac{N(N - 1)x^2}{2!} + \ldots + Nx^{N-1} + x^N.$$

Notice that this is the familiar binomial expansion.

11. Let $f(x) = x^3 - 2x^2 + 2x$. Find the Taylor polynomial $P_3(x)$ expanded about $x_0 = 1$ that involves powers of $(x - 1)$.

12. Show that

$$f(x) = 1 + 3(x - 1) + 3(x - 1)^2 + (x - 1)^3$$
$$g(x) = -1 + 3(x + 1) - 3(x + 1)^2 + (x + 1)^3$$

are two representations for the polynomial $p(x) = x^3$.

13. Assume that the Taylor polynomial for $f(x)$ of degree $N = 3$ expanded about the value $x_0 = 2$ is given by

$$P_3(x) = 2 + 3(x - 2) - 5(x - 2)^2 + 4(x - 2)^3.$$

    (a) Find $P_3(2.1)$ and $P_3(1.9)$.   (b) Find $[P_3(2.1) - P_3(1.9)]/0.2$.

    (c) Show that $f'(2) = 3$.   *Remark.* An approximation to $f'(2)$ is given in part (b).

14. Finish the proof of Corollary 1.1 by writing down the expression for $P_N^{(k)}(x)$ and showing that

$$P_N^{(k)}(x_0) = f^{(k)}(x_0) \qquad \text{for } k = 2, 3, \dots, N.$$

15. Write a report on the proof of Theorem 1.3 (you can look it up in a calculus book). Be sure to include a discussion about the even partial sums $\{S_{2k}\}$ and the odd partial sums $\{S_{2k+1}\}$. In your own words, describe why the limit exists.

16. Modify Algorithm 1.3 so that it will find the sum of a power series.

17. In your own words, describe the similarities and differences between Taylor series and power series.

    The Taylor polynomial of degree $N = 2$ for a function $f(x, y)$ of two variables x and y expanded about the point (a, b) is

$$P_2(x, y) = f(a, b) + f_x(a, b)(x - a) + f_y(a, b)(y - b)$$

$$+ \frac{f_{xx}(a, b)(x - a)^2}{2} + f_{xy}(a, b)(x - a)(y - b) + \frac{f_{yy}(a, b)(y - b)^2}{2}.$$

18. (a) Show that the Taylor polynomial of degree $N = 2$ for $f(x, y) = y/x$ expanded about $(1, 1)$ is

$$P_2(x, y) = 1 - (x - 1) + (y - 1) + (x - 1)^2 - (x - 1)(y - 1).$$

    (b) Find $P_2(1.05, 1.1)$ and compare with $f(1.05, 1.1)$.

19. (a) Show that the Taylor polynomial of degree $N = 2$ for $f(x, y) = (1 + x - y)^{1/2}$ expanded about $(0, 0)$ is

$$P_2(x, y) = 1 + \frac{x}{2} - \frac{y}{2} - \frac{x^2}{8} + \frac{xy}{4} - \frac{y^2}{8}.$$

    (b) Find $P_2(0.04, 0.08)$ and compare with $f(0.04, 0.08)$.

    Exercises 20 and 21 form a proof of Taylor's theorem.

20. Let $g(t)$ and its derivatives $g^{(k)}(t)$, for $k = 1, 2, \dots, N + 1$, be continuous on the interval (a, b), which contains x_0. Suppose that there exist two distinct points x and x_0 such that $g(x) = 0$, $g(x_0) = 0$, $g'(x_0) = 0$, $g''(x_0) = 0, \dots, g^{(N)}(x_0) = 0$. Prove that there exists a value c that lies between x_0 and x such that $g^{(N+1)}(c) = 0$.

    Remark. $g(t)$ is a function of t and the values x and x_0 are to be treated as constants with respect to the variable t.

    Hint. Use Rolle's Theorem* on the interval with endpoints x_0 and x to find the number c_1 such that $g'(c_1) = 0$. Then use Rolle's theorem applied to the function $g'(t)$

Rolle's Theorem. Let f be continuous on $[a, b]$ and differentiable on (a, b). If $f(a) = f(b)$, then there exists a number c in (a, b) such that $f'(c) = 0$.

on the interval with endpoints x_0, c_1 to find the number c_2 such that $g''(c_2) = 0$. Inductively repeat the process until the number c_{N+1} is found such that $g^{(N+1)}(c_{N+1}) = 0$.

21. Use the result of Exercise 20 and the special function

$$g(t) = f(t) - P_N(t) - E_N(x)\frac{(t - x_0)^{N+1}}{(x - x_0)^{N+1}},$$

where $P_N(x)$ is the Taylor polynomial of degree N to prove that the error term $E_N(x) = f(x) - P_N(x)$ has the form

$$E_N(x) = f^{(N+1)}(c)\frac{(x - x_0)^{N+1}}{(N + 1)!}.$$

Hint. Find $g^{(N+1)}(t)$ and evaluate it at $t = c$.

1.4 Fundamental Iterative Algorithm

A fundamental principle in computer science is *iteration*. As the name suggests, it means that a process is repeated until an answer is achieved. Iterative techniques are used to find roots of equations, solutions of linear and nonlinear systems of equations, and solutions of differential equations. In this section we study the process of iteration using repeated substitution.

A rule or function $g(x)$ for computing successive terms is needed together with a starting value p_0. Then a sequence of values $\{p_k\}$ is obtained using the iterative rule $p_{k+1} = g(p_k)$. The sequence has the pattern

$$
\begin{aligned}
&p_0 \quad \text{(starting value)}\\
&p_1 = g(p_0)\\
&p_2 = g(p_1)\\
&\quad \vdots\\
&p_k = g(p_{k-1})\\
&p_{k+1} = g(p_k)\\
&\quad \vdots
\end{aligned}
$$

(1)

What can we learn from an unending sequence of numbers? If the numbers tend to a limit, then we feel that something has been achieved. But what if the numbers diverge to infinity? The next example addresses this situation.

Example 1.9 Discuss the divergent iteration

$$p_0 = 1 \quad \text{and} \quad p_{k+1} = 1.001p_k \quad \text{for } k = 0, 1, \ldots.$$

Solution. The first 100 terms look as follows:

$$p_1 = 1.001p_0 = 1.001 \times 1.000000 = 1.001000,$$

$$p_2 = 1.001p_1 = 1.001 \times 1.001000 = 1.002001,$$

$$p_3 = 1.001p_2 = 1.001 \times 1.002001 = 1.003003,$$

$$\vdots \qquad \vdots \qquad \vdots$$

$$p_{100} = 1.001p_{99} = 1.001 \times 1.104012 = 1.105116.$$

The process can be continued indefinitely and it is easily shown that $\lim_{n \to \infty} p_n = +\infty$. In Chapter 9 we will see that the sequence $\{p_k\}$ is a numerical solution to the differential equation $y' = 0.001y$. The solution is known to be $y(x) = \exp(0.001x)$. Indeed, if we compare the 100th term in the sequence with $y(100)$, we see that $p_{100} = 1.105116 \approx 1.105171 = \exp(0.1) = y(100)$.

In this section we are concerned with the type of functions $g(x)$ that produce convergent sequences $\{p_k\}$.

Finding Fixed Points

Definition 1.2A [Fixed point] A *fixed point* of a function $g(x)$ is a value P such that $P = g(P)$.

Definition 1.2B [Fixed-point iteration] The iteration $p_{n+1} = g(p_n)$ for $n = 0, 1, \dots$ is called *fixed-point iteration*.

Theorem 1.6 Let $g(x)$ be continuous and $\{p_n\}$ be a sequence generated by fixed-point iteration. If $\{p_n\}$ converges to P, then P is a fixed point of $g(x)$.

Proof. Suppose that

(2)
$$\lim_{n \to \infty} p_n = P.$$

Then we can use the relation $p_{n+1} = g(p_n)$ and (2) to write

(3)
$$P = \lim_{n \to \infty} p_{n+1} = \lim_{n \to \infty} g(p_n) = g(\lim_{n \to \infty} p_n) = g(P).$$

Therefore, $P = g(P)$, that is, P is a fixed point of $g(x)$.

Example 1.10 Discuss the convergent iteration

$$p_0 = 0.5 \quad \text{and} \quad p_{k+1} = \exp(-p_k) \qquad \text{for } k = 0, 1, \dots.$$

Solution. The first 10 terms are obtained by the calculations

$$p_1 = \exp(-0.500000) = 0.606531, \qquad p_2 = \exp(-0.606531) = 0.545239$$

$$p_3 = \exp(-0.545239) = 0.579703, \qquad p_4 = \exp(-0.579703) = 0.560065$$

$$\vdots \qquad\qquad\qquad \vdots$$

$$p_9 = \exp(-0.566409) = 0.567560, \qquad p_{10} = \exp(-0.567560) = 0.566907.$$

The sequence is converging and further calculations reveal that

$$\lim_{n \to \infty} p_n = 0.567143. \ldots$$

Thus we have found the fixed point of the function $\exp(-x)$, that is,

$$0.567143\ldots = \exp(-0.567143\ldots).$$

We now state a theorem that can be used to determine whether the iteration (1) will produce a convergent or a divergent sequence.

Theorem 1.7 [Fixed-point iteration] Let $g(x)$ and $g'(x)$ be continuous on an interval (a, b) that contains the fixed point P and suppose that the starting value p_0 is chosen sufficiently close to P.

(4)   If $|g'(P)| < 1$, then the iteration $p_{k+1} = g(p_k)$ converges to P. {local convergence}

(5)   If $|g'(P)| > 1$, then the iteration $p_{k+1} = g(p_k)$ does not converge to P. {local divergence}

Remark. Theorem 1.7 does not tell how close p_0 must be to P for statement (4) to hold. It is assumed that $p_0 \neq P$ for statement (5), and the "lucky guess" $g(p_0) = P$ is ruled out. The theorem describes the situation when we are very close to P. The proof, given at the end of the section, will supply more details.

Graphical Interpretation of Fixed-Point Iteration

Since we seek a fixed point P to $g(x)$, it is necessary that the graph of the curve $y = g(x)$ and the line $y = x$ intersect at (P, P). Two simple types of convergent iteration, monotone and oscillating, are illustrated in Figures 1.5(a) and (b), respectively.

To visualize the process, start at p_0 on the x-axis and move vertically to the point $(p_0, p_1) = (p_0, g(p_0))$ on the curve $y = g(x)$. Then move horizontally from (p_0, p_1) to the point (p_1, p_1) on the line $y = x$. Finally, move vertically downward to p_1 on the x-axis. The recursion $p_{n+1} = g(p_n)$ is used to construct the point (p_n, p_{n+1}) on the graph, then a horizontal motion locates (p_{n+1}, p_{n+1}) on the line $y = x$, and vertical movement ends up at p_{n+1} on the x-axis. The situation is shown in Figure 1.5.

If $|g'(P)| > 1$, then the iteration (1) produces a sequence that diverges away from P. The two simple types of divergent iteration, monotone and oscillating, are illustrated in Figures 1.6(a) and (b), respectively.

Example 1.11 Investigate the nature of the iteration (1) when the function $g(x) = 1 + x - x^2/4$ is used.

Solution. The fixed points can be found by solving the equation $P = 1 + P - P^2/4$ and two solutions $P = -2$ and $P = 2$ are found. The derivative of the function is $g'(x) = 1 - x/2$, and there are two cases to consider.

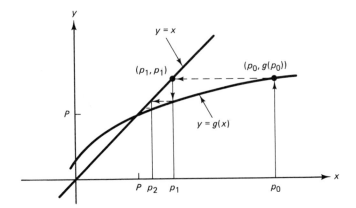

Figure 1.5(a). Monotone convergence when $0 < g'(P) < 1$.

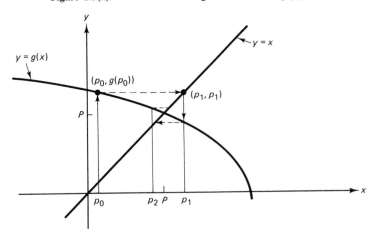

Figure 1.5(b). Oscillating convergence when $-1 < g'(P) < 0$.

Case (i) $P = -2$
Start with $p_0 = -2.05$
then get $p_1 = -2.100625$
$p_2 = -2.20378135$
$p_3 = -2.41794441$
$\vdots$

$\lim_{n \to \infty} p_n = -\infty$

Since $g'(-2) = 1 + 2/2 = 2$, by Theorem 1.7, the sequence will not converge to $P = -2$.

Case (ii) $P = 2$
Start with $p_0 = 1.6$
then get $p_1 = 1.96$
$p_2 = 1.9996$
$p_3 = 1.99999996$
$\vdots$

$\lim_{n \to \infty} p_n = 2$

Since $g'(2) = 1 - 2/2 = 0$, Theorem 1.7 says that the sequence converges to $P = 2$.

Theorem 1.7 does not state what will happen when $g'(P) = 1$. The next example has been specially constructed so that the sequence $\{p_k\}$ converges whenever $p_0 > P$ and it diverges if we choose $p_0 < P$.

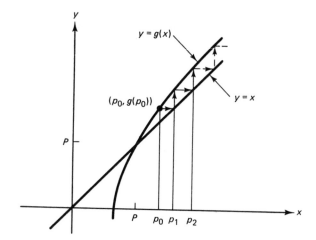

Figure 1.6(a). Monotone divergence when $1 < g'(P)$.

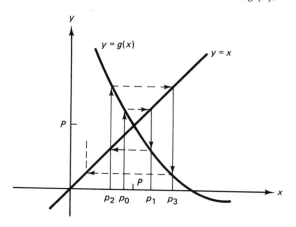

Figure 1.6(b). Divergent oscillation when $g'(P) < -1$.

Example 1.12 Investigate the nature of the iteration (1) when

$$g(x) = 2(x - 1)^{1/2} \qquad \text{for } x \geq 1.$$

Solution. Only one fixed point $P = 2$ exists. The derivative is $g'(x) = 1/(x - 1)^{1/2}$ and $g'(2) = 1$, so that Theorem 1.7 does not apply. There are two cases to consider, when the starting value lies to the left or right of $P = 2$.

Case (i) Start with $p_0 = 1.5$,
then get $p_1 = 1.41421356$
$p_2 = 1.28718851$
$p_3 = 1.07179943$
$p_4 = 0.53590832$

$p_5 = 2(-0.46409168)^{1/2}$

Case (ii) Start with $p_0 = 2.5$,
then get $p_1 = 2.44948974$
$p_2 = 2.40789513$
$p_3 = 2.37309514$
$p_4 = 2.34358284$
$\vdots$

$\lim_{n \to \infty} p_n = 2$

Since p_4 lies outside the domain of $g(x)$, the term p_5 cannot be computed.	This sequence is converging slowly to the value $P = 2$; indeed, $p_{1,000} = 2.00398714$.

Absolute and Relative Error

In Example 1.12 case (ii), the sequence converges slowly and after 1,000 iterations the three consecutive terms are

$$p_{1,000} = 2.00398714 \qquad p_{1,001} = 2.00398317 \qquad p_{1,002} = 2.00397921.$$

This should not be too disturbing; after all, we could compute a few more thousand terms and find a better approximation! But what about a criterion for stopping the iteration? Notice that if we use the difference between consecutive terms,

$$|p_{1,001} - p_{1,000}| = |2.00398317 - 2.00398714| = 0.00000397.$$

Yet the absolute error in the approximation $p_{1,000}$ is known to be

$$|P - p_{1,000}| = |2.00000000 - 2.00398714| = 0.00398714.$$

This is about 1,000 times larger than $|p_{1,001} - p_{1,000}|$ and it shows that closeness of consecutive terms does not guarantee that accuracy has been achieved. But it is usually the only criterion available and is often used to terminate an iterative procedure.

Definition 1.3 Let p_0 be an approximation to P. The difference

(6) $$E_0 = P - p_0 = (\text{exact value}) - (\text{approximate value})$$

is called the *error of the approximation*.

Hence if E_0 is added to p_0, the result is P:

(7) $$P = p_0 + E_0.$$

Fixed-point iteration involves a sequence of approximations $\{p_k\}$ where corresponding error terms $E_k = P - p_k$.

Definition 1.4 Suppose that p_0 is an approximation to P and $P \neq 0$. The error E_0 divided by P is called the *relative error*:

(8) $$R_0 = \frac{P - p_0}{P}.$$

The relative error gives the portion of p_0 that is in error, as a fraction of P, and can be adapted to determine the number of significant digits in the approximation. Consider the following three cases:

Case (i)	Case (ii)	Case (iii)
$P = 3.141592$	$P = 1000000$	$P = 0.000012$
$p_0 = 3.14$	$p_0 = 999{,}991$	$p_0 = 0.000009$
$E_0 = 0.001592$	$E_0 = 9$	$E_0 = 0.000003$
$R_0 = 0.000507$	$R_0 = 0.000009$	$R_0 = 0.25$

In case (i) there is not too much difference in E_0 and R_0 and either could be used to determine the accuracy of p_0. In case (ii) the value P is of the magnitude 10^6 and the error E_0 is large and the relative error R_0 is small. We would most likely call p_0 a good approximation. In case (iii), P is of magnitude 10^{-6} and the error is the smallest of all three cases, but the relative error is the largest. In terms of percentage, p_0 is 25% in error and is a bad approximation. Observe that as $|P|$ moves away from 1, either larger or smaller, the relative error is a better indicator of the accuracy of the approximation. The relative error criterion is the one that should be used.

Algorithm 1.4 [*Computer Implementation of Fixed-Point Iteration*] Find a fixed point of $g(x)$. Start with p_0 and compute $p_{k+1} = g(p_k)$. Terminate the iteration process when the relative error estimate, $2|p_{k+1} - p_k|/(|p_{k+1}| + \text{Small})$, in consecutive iterates p_k and p_{k+1} differs by less than the preassigned value Tol.

```
Tol   := 0.000001                              {Termination criterion}
Max   := 200                          {Maximum number of iterations}
Small := 0.000001                          {Initialize the variable}
K     := 0                                  {Initialize the counter}
RelErr := 1                                {Initialize the variable}
INPUT  Pterm                             {The initial approximation}
Pnew  := g(Pterm)                              {The first iteration}


WHILE  RelErr ≥ Tol   and   K ≤ Max      DO
  │    K := K+1                                {Increment the counter}
  │    Pold := Pterm                       {Previous iterate p_{k-1}}
  │    Pterm := Pnew                           {Current iterate p_k}
  │    Pnew := g(Pterm)                   {Compute new iterate p_{k+1}}
  │    Dg := Pnew − Pterm                      {Difference in g(x)}
  │    Delta := |Dg|                             {Absolute error}
  └─── RelErr := 2*Delta/[|Pnew|+Small]          {Relative error}
       Dx := Pterm − Pold                       {Difference in x}
       Slope := Dg/Dx                              {g'(p_k)}


PRINT 'The computed fixed point of g(x) is'  Pnew            {Output}
PRINT 'Consecutive iterates are within'  Delta
IF  |Slope| > 1 THEN
     PRINT "The sequence appears to be diverging."
ELSE
     PRINT "The sequence appears to be converging."
ENDIF
```

Proof of Theorem 1.7, statement (4). The condition $|g'(P)| < 1$ and the continuity of $g'(x)$ can be used to find a small interval (a, b) about P and a constant K with $0 < K < 1$ so that

(9)
$$|g'(c)| \leq K \qquad \text{for } a \leq c \leq b.$$

Select p_0 in (a, b); then p_1 must also lie in (a, b) because

(10)
$$|P - p_1| = |g(P) - g(p_0)| \leq |g'(c)| \, |P - p_0| < |P - p_0|.$$

The relationships between P, p_0, and p_1 are illustrated in Figure 1.7. The details involved in relation (10) are important. The error in the initial approximation p_0 is expressed as

(11)
$$E_0 = P - p_0.$$

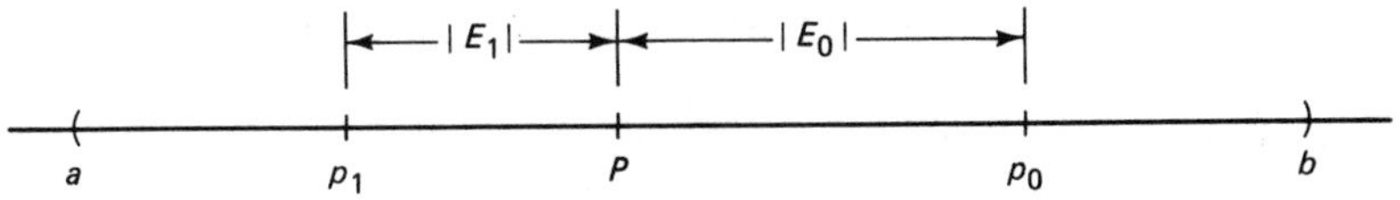

Figure 1.7. The relationships between P, p_0, p_1 for fixed-point iteration when $|g'(x)| < 1$ and $|E_1| < |E_0|$.

The fixed-point property $P = g(P)$ and one step of iteration $p_1 = g(p_0)$ are used to express the next error E_1 as follows:

(12)
$$E_1 = P - p_1 = g(P) - g(p_0).$$

The *Mean Value Theorem** can be used to find a value c_0 so that

(13)
$$g(P) - g(p_0) = g'(c_0)(P - p_0).$$

Using (11)–(13) we obtain the relationship

(14)
$$E_1 = g(P) - g(p_0) = g'(c_0)(P - p_0) = g'(c_0)E_0.$$

Taking absolute values in (14) and using inequality (9), we get

(15)
$$|E_1| = |g'(c_0)| \, |E_0| \leq K|E_0|.$$

Notice carefully that relation (15) involves the absolute values of the error terms E_0 and E_1. Figure 1.7 illustrates the relationships between P, p_0, p_1, E_0, and E_1.

Similarly, one obtains a relationship between E_1 and E_2:

(16)
$$E_2 = g(P) - g(p_1) = g'(c_1)(P - p_1) = g'(c_1)E_1.$$

Properties of inequalities and (15) and (16) imply that

(17)
$$|E_2| = |g'(c_1)| \, |E_1| \leq K|E_1| \leq K^2|E_0|.$$

---

*Mean Value Theorem. If $g(x)$ and $g'(x)$ are continuous on an interval containing P and Q, then there exists a value c that lies betweeen P and Q so that

$$\frac{g(P) - g(Q)}{P - Q} = g'(c).$$

We often use the equivalent expression

$$g(P) - g(Q) = g'(c)(P - Q).$$

Mathematical induction establishes the general pattern

(18)
$$|E_N| = |g'(c_{N-1})||E_{N-1}| \le K|E_{N-1}| \le K^N|E_0|.$$

Since $0 < K < 1$, K^N goes to zero as N goes to infinity. Hence

(19)
$$0 \le \lim_{N \to \infty} |E_N| \le \lim_{N \to \infty} K^N|E_0| = 0.$$

The limit of $|E_N|$ is squeezed between zero on the left and zero on the right, so we can conclude that $\lim_{N \to \infty} E_N = 0$. Therefore, statement (4) of Theorem 1.7 is proven.

Exercises for the Fundamental Iterative Algorithm

When it is asked to implement the iteration (1), either a calculator or computer should be used.

1. Investigate the nature of the iteration (1) when

$$g(x) = -4 + 4x - \frac{x^2}{2}.$$

(a) Show that $P = 2$ and $P = 4$ are fixed points.
(b) Use starting value $p_0 = 1.9$ and compute p_1, p_2, and p_3.
(c) Use starting value $p_0 = 3.8$ and compute p_1, p_2, and p_3.
(d) Find the errors E_k and relative errors R_k for the values p_k in parts (b) and (c).
(e) What conclusions can be drawn from Theorm 1.7?

2. Graph the curve $y = g(x) = (6 + x)^{1/2}$ and the line $y = x$, and plot the fixed point $(3, 3)$. Start with $p_0 = 7$ and construct $p_1 \approx 3.61$ and $p_2 \approx 3.10$ as indicated in Figure 1.5(a). Will fixed-point iteration converge?

3. Graph the curve $y = g(x) = 1 + 2/x$ and the line $y = x$, plot the fixed point $(2, 2)$. Start with $p_0 = 4$ and construct $p_1 \approx 1.5$ and $p_2 \approx 2.33$ as indicated in Figure 1.5(b). Will fixed-point iteration converge?

4. Let $g(x) = 0.4 + x - 0.1x^2$ and consider the iteration (1).
(a) Start with $p_0 = 1.9$ and find $p_1, p_2, \ldots, p_5$. Will this sequence converge to the fixed point $P = 2$? Why?
(b) Start with $p_0 = -1.9$ and find $p_1, p_2, \ldots, p_5$. Will this sequence converge to the fixed point $P = -2$? Why?
(c) Find the errors E_k and relative errors R_k for the values p_k in parts (a) and (b).

5. Let $g(x) = x^2 + x - 4$. Can fixed-point iteration (1) be used to find the solution to the equation $x = g(x)$? Why?

6. Suppose that $g(x)$ and $g'(x)$ are defined and continuous on (a, b) and that p_0, p_1, p_2 lie in this interval and $p_1 = g(p_0)$, $p_2 = g(p_1)$. Also, assume that there exists a constant K so that $|g'(x)| < K$. Show that $|p_2 - p_1| < K|p_1 - p_0|$. *Hint.* Use the Mean Value Theorem.

7. Graph the curve $y = g(x) = x^2/3$ and the line $y = x$, and plot the fixed point $(3, 3)$. Start with $p_0 = 3.5$ and construct $p_1 \approx 4.08$ and $p_2 \approx 5.56$ as indicated in Figure 1.6(a). Will fixed-point iteration converge?

8. Graph the curve $y = g(x) = 2 + 2x - x^2$ and the line $y = x$, and plot the fixed point $(2, 2)$. Start with $p_0 = 2.5$ and construct $p_1 = 0.75$ and $p_2 \approx 2.94$ as indicated in Figure 1.6(b). Will fixed-point iteration converge?

9. Suppose that $g(x)$ and $g'(x)$ are continuous on (a, b) and that $|g'(x)| > 1$ on this interval. If the fixed point P and the initial approximations p_0 and p_1 lie in the interval, then show that $p_1 = g(p_0)$ results in $|E_1| = |P - p_1| > |P - p_0| = |E_0|$. Hence statement (5) of Theorem 1.7 is established. {local divergence}

10. Let $g(x) = x - 0.0001x^2$ and $p_0 = 1$ and consider iteration.
    (a) Show that $p_0 > p_1 > \ldots > p_n > p_{n+1} \ldots$.
    (b) Show that $p_n > 0$ for all n.
    (c) Since the sequence $\{p_n\}$ is decreasing and bounded below, it has a limit. What is the limit?
    (d) Use a calculator or computer and find $p_1, p_2, \ldots, p_5$.

11. Let $g(x) = 1.5 + 0.5x$ and $p_0 = 4$ and consider iteration.
    (a) Show that the fixed point is $P = 3$.
    (b) Show that $|P - p_n| = |P - p_{n-1}|/2$ for $n = 1, 2, \ldots$.
    (c) Show that $|P - p_n| = |P - p_0|/2^n$ for $n = 1, 2, \ldots$.

12. Suppose that $g(x) = x/2$ and $p_k = 2^{-k}$.
    (a) Find the quantity $|p_{k+1} - p_k|/|p_{k+1}|$.
    (b) Discuss what will happen if the relative error stopping criterion above is used in Algorithm 1.4.

13. Investigate the nature of the iteration (1) when $g(x) = 3(x - 2.25)^{1/2}$.
    (a) Show that $P = 4.5$ is the only fixed point.
    (b) Use starting value $p_0 = 4.4$ and compute $p_1, p_2, p_3,$ and p_4. What do you conjecture about the limit?
    (c) Use starting value $p_0 = 4.6$ and compute $p_1, p_2, p_3,$ and p_4. What do you conjecture about the limit?

14. For the fixed-point iteration (1), discuss why it is an advantage to have $g'(P) \approx 0$.

15. Verify that $P = 3$ is a solution to the equation $P = g(P)$ and find the other real solutions for
    (a) $g(x) = (x + 9/x)/2$
    (b) $g(x) = 18x/(x^2 + 9)$
    (c) $g(x) = x^3 - 24$
    (d) $g(x) = 2x^3/(3x^2 - 9)$
    (e) $g(x) = 81/(x^2 + 18)$

16. Find $g'(x)$ and $g'(3)$ for the functions $g(x)$ in Exercise 15.

17. Start with $p_0 = 3.1$ and use the iteration (1) to find p_1 and p_2 for the functions $g(x)$ in Exercise 15. Will the iteration converge to the fixed point $P = 3$? Why?

18. Write a report on the contraction mapping theorem.

2

The Solution of
Nonlinear Equations f(x) = 0

2.1 Bracketing Methods for Locating a Root

Consider the familiar topic of interest. Suppose that you save money by making regular monthly deposits P and the annual interest rate is I; then the total amount A after N deposits is

(1)
$$A = P + P\left(1 + \frac{I}{12}\right) + P\left(1 + \frac{I}{12}\right)^2 + \ldots + P\left(1 + \frac{I}{12}\right)^{N-1}.$$

The first term on the right side of equation (1) is the last payment. Then the next-to-last payment, which has earned one period of interest, contributes $P(1 + I/12)$. The second-from-last payment has earned two periods of interest and contributes $P(1 + I/12)^2$, and so on. Finally, the first payment, which has earned interest for $N - 1$ periods, contributes $P(1 + I/12)^{N-1}$ toward the total. Recall that the formula for the sum of the first N terms of a geometric series is

(2)
$$1 + r + r^2 + r^3 + \ldots + r^{N-1} = \frac{1 - r^N}{1 - r}.$$

We can write (1) in the form

$$A = P\left[1 + \left(1 + \frac{I}{12}\right) + \left(1 + \frac{I}{12}\right)^2 + \ldots + \left(1 + \frac{I}{12}\right)^{N-1}\right],$$

and use the substitution $r = (1 + I/12)$ in (2) to obtain

$$A = P \frac{1 - (1 + I/12)^N}{1 - (1 + I/12)}.$$

This can be simplified to obtain the annuity-due equation,

(3)
$$A = \frac{P}{I/12}\left[\left(1 + \frac{I}{12}\right)^N - 1\right].$$

The following example uses the annuity-due equation and requires a sequence of repeated calculations to find an answer.

Example 2.1 You save \$250 per month for 20 years and desire that the total value of all payments and interest is \$250,000 at the end of the 20 years. What interest rate I is needed to achieve your goal?

Solution. If we hold $N = 240$ fixed, then A is a function of I alone, that is, $A = A(I)$. We will start with two guesses, $I_0 = 0.12$ and $I_1 = 0.13$, and perform a sequence of calculations to narrow down the final answer. Starting with $I_0 = 0.12$ yields

$$A(0.12) = \frac{250}{0.12/12}\left[\left(1 + \frac{0.12}{12}\right)^{240} - 1\right] = 247{,}314.$$

Since this value is a little short of the goal, we next try $I_1 = 0.13$:

$$A(0.13) = \frac{250}{0.13/12}\left[\left(1 + \frac{0.13}{12}\right)^{240} - 1\right] = 283{,}311.$$

This is a little high, so we try the value in the middle, $I_2 = 0.125$:

$$A(0.125) = \frac{250}{0.125/12}\left[\left(1 + \frac{0.125}{12}\right)^{240} - 1\right] = 264{,}623.$$

This is again high and we conclude that the desired rate lies in the interval $[0.12, 0.125]$. The next guess is the midpoint $I_3 = 0.1225$:

$$A(0.1225) = \frac{250}{0.1225/12}\left[\left(1 + \frac{0.1225}{12}\right)^{240} - 1\right] = 255{,}803.$$

This is high and the interval is now narrowed down to $[0.12, 0.1225]$. Our last calculation uses the midpoint approximation, $I_4 = 0.12125$:

$$A(0.12125) = \frac{250}{0.12125/12}\left[\left(1 + \frac{0.12125}{12}\right)^{240} - 1\right] = 251{,}518.$$

Further iterations can be done to obtain as many significant digits as required. The purpose of this example was to find the value of I that produced a specified level L of the function value, that is to find a solution to $A(I) = L$. It is standard practice to place the constant L on the left and solve the equation $A(I) - L = 0$.

Definition 2.1   [Root of an equation, zero of a function] Let $f(x)$ be a continuous function. Any number r for which $f(r) = 0$ is called a *root* of the equation $f(x) = 0$. Also, we say that r is a *zero* of $f(x)$.

For example, the equation $2x^2 + 5x - 3 = 0$ has two real roots $r_1 = 0.5$ and $r_2 = -3$, whereas the corresponding function $f(x) = 2x^2 + 5x - 3 = (2x - 1)(x + 3)$ has two real zeros, $r_1 = 0.5$ and $r_2 = -3$.

In this section we develop two methods for finding a zero of a continuous function. We must start with an initial interval $[a, b]$, where $f(a)$ and $f(b)$ have opposite signs. Since the graph $y = f(x)$ of a continuous function is unbroken, it will cross the x-axis somewhere in the interval (see Figure 2.1). The first method systematically moves the endpoints of the interval closer and closer together until we obtain an interval of arbitrarily small width that brackets the zero.

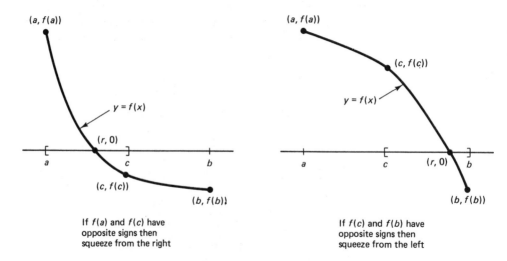

If $f(a)$ and $f(c)$ have opposite signs then squeeze from the right

If $f(c)$ and $f(b)$ have opposite signs then squeeze from the left

Figure 2.1. The decision process for the bisection method.

The Bisection Method of Bolzano

If $f(x)$ is continuous on the interval $[a, b]$ and $f(a)$ and $f(b)$ have opposite signs, then the *Intermediate Value Theorem** of calculus shows that there exists at least one root r of the equation $f(x) = 0$ in the interval. We seek an interval half as wide that contains r. If we choose the midpoint $c = (a + b)/2$, then three possibilities arise:

**Intermediate Value Theorem.* Suppose that $f(x)$ is continuous on the closed interval $[a, b]$. Let L be any number between $f(a)$ and $f(b)$, so that either $f(a) < L < f(b)$ or $f(b) < L < f(a)$. Then there exists at least one number c in $[a, b]$ so that $f(c) = L$.

(4) $f(a)$ and $f(c)$ have opposite signs and a root lies in $[a, c]$,

or

$f(c)$ and $f(b)$ have opposite signs and a root lies in $[c, b]$,

or

(5) $f(c) = 0$ and you found a root.

If either of the cases in (4) occur, then we have found an interval half as wide as the original interval that contains the root, that is, we are squeezing down on the root (see Figure 2.1). To continue, relabel the new smaller interval $[a, b]$ and repeat the process until the interval width is as small as desired.

Convergence of the Bisection Method

The function $f(x)$ must be continuous on the starting interval $[a_0, b_0]$, and the values $f(a_0)$ and $f(b_0)$ must have opposite signs. The decision process that is mentioned above is used to construct a sequence of intervals $\{[a_n, b_n]\}$ each of which brackets the root. In the construction process the interval width is reduced by a factor of $\frac{1}{2}$ at each step. Indeed, we will show that

(6) $$\lim_{n\to\infty} |b_n - a_n| = \lim_{n\to\infty} |b_0 - a_0|/2^n = 0.$$

Since the root r lies somewhere in the interval $[a_n, b_n]$, if we select the midpoint $c_n = (a_n + b_n)/2$ as the nth approximation, the error $e_n = r - c_n$ will be at most one-half the width of this interval (i.e., $|e_n| \leq |b_n - a_n|/2$), as shown in the figure.

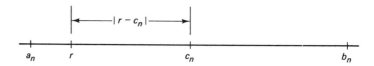

This leads to the relation

(7) $$|e_n| \leq \frac{|b_n - a_n|}{2} = \frac{|b_0 - a_0|}{2^{n+1}}.$$

To prove (7), first observe that $|b_1 - a_1| = |b_0 - a_0|/2$; then $|b_2 - a_2| = |b_1 - a_1|/2 = |b_0 - a_0|/2^2$. Finite mathematical induction is used to conclude that $|b_n - a_n| = |b_0 - a_0|/2^n$. When this is combined with the inequality $|e_n| \leq |b_n - a_n|/2$, we obtain (7). Therefore, the sequence $\{c_n\}$ converges to a root r, that is,

$$\lim_{n\to\infty} c_n = \lim_{n\to\infty} (r - e_n) = r - \lim_{n\to\infty} e_n = r.$$

Example 2.2 The function $h(x) = x \sin(x)$ occurs in the study of undamped forced oscillation. Find the value of x that lies in the interval $[0, 2]$, where the function takes on the value $h(x) = 1$.

Solution. We use the bisection method to find a zero of the function $f(x) = x \sin(x) - 1$. Starting with $a_0 = 0$ and $b_0 = 2$, we compute

$$f(0) = -1.000000 \quad \text{and} \quad f(2) = 0.818595,$$

so that a root of $f(x) = 0$ lies in the interval $[0, 2]$. At the midpoint $c_0 = 1$, we find that $f(1) = -0.158529$. Hence the function changes sign on $[c_0, b_0] = [1, 2]$.

To continue, we squeeze from the left and set $a_1 = c_0$ and $b_1 = b_0$. The midpoint is $c_1 = 1.5$ and $f(c_1) = 0.496242$. Now, $f(1) = -0.158529$ and $f(1.5) = 0.496242$ imply that the root lies in the interval $[b_1, c_1] = [1.0, 1.5]$. The next decision is to squeeze from the right and set $a_2 = a_1$ and $b_2 = c_1$. In this manner we obtain a sequence $\{c_k\}$ that converges to $r \approx 1.114157141$. A sample of the calculations is given in Table 2.1.

Table 2.1 Bisection method solution of $x \sin(x) - 1 = 0$

k	Left Endpoint, a_k	Midpoint, c_k	Right Endpoint, b_k	Function Value, $f(c_k)$
0	0.	1.	2.	-0.158529
1	1.0	1.5	2.0	0.496242
2	1.00	1.25	1.50	0.186231
3	1.000	1.125	1.250	0.015051
4	1.0000	1.0625	1.1250	-0.071827
5	1.06250	1.09375	1.12500	-0.028362
6	1.093750	1.109375	1.125000	-0.006643
7	1.1093750	1.1171875	1.1250000	0.004208
8	1.10937500	1.11328125	1.11718750	-0.001216
.	.	.	.	.
.	.	.	.	.
.	.	.	.	.

A virtue of the bisection method is that formula (7) provides an estimate for the accuracy of the computed solution. In Example 2.2 the width of the starting interval was $b_0 - a_0 = 2$. Suppose that Table 2.1 were continued to the thirty-first iterate; then by (7) the error bound would be $|e_{31}| \leq (2 - 0)/2^{32} \approx 4.656612 \times 10^{-10}$. Hence c_{31} would be an approximation to r with nine decimal places of accuracy.

The number N of repeated bisections needed to guarantee that the Nth midpoint c_N is an approximation to a root and has an error less than the preassigned value Delta is

$$(8) \qquad N = \text{int}\left(\frac{\ln(B - A) - \ln(\text{Delta})}{\ln(2)}\right).$$

The proof of this formula is left as an exercise (see Exercise 19).

Algorithm 2.1 [*Bisection Method*] Find a root of $f(x) = 0$ in the interval $[a, b]$ given that $f(a)$ and $f(b)$ have opposite signs.

```
Delta := 10⁻⁶                              {Tolerance for width of interval}
Satisfied := "False"                       {Condition for loop termination}

INPUT  A, B                                {Input endpoints of interval}
YA := F(A),  YB := F(B)                     {Compute function values}
Max := 1+INT([ln(B−A)−ln(Delta)]/ln(2))    {Calculation of the
                                            maximum number of iterations}

IF  SIGN(YA) = SIGN(YB)  THEN              {Check to see if
      PRINT 'The values f(a) and f(b)'      the bisection
      PRINT 'do not differ in sign.'             method
      TERMINATE THE ALGORITHM                   applies}

DO  FOR  K = 1  TO  Max  UNTIL  Satisfied = "True"
      C := (A+B)/2                          {Midpoint of interval}
      YC := F(C)                            {Function value at midpoint}

        IF YC = 0 THEN
            A := C  and  B := C             {Exact root is found}
        ELSEIF SIGN(YB)=SIGN(YC)  THEN
            B := C                          {Squeeze from the right}
            YB := YC
        ELSE
            A := C                          {Squeeze from the left}
            YA := YC
        ENDIF
      IF  B − A < Delta  THEN              {Check for early
        Satisfied := "True"                      convergence}
END

PRINT 'The computed root of f(x) = 0 is' C  {Output}
PRINT 'The accuracy is + −' B−A
PRINT 'The value of the function f(C) is' YC
```

Regula Falsi Method

Another popular algorithm is the *method of false position* or the *regula falsi method*. It was developed because the bisection method converges at a fairly slow speed. As before, we assume that $f(a)$ and $f(b)$ have opposite signs. The bisection method used the midpoint of the interval $[a, b]$ as the next iterate. A better approximation is obtained if we find the point $(c, 0)$ where the straight line L joining the points $(a, f(a))$ and $(b, f(b))$ crosses the x-axis (see Figure 2.2). To find the value c, we write down two versions of the slope m of the line L:

(9)
$$m = \frac{f(b) - f(a)}{b - a},$$

where the points $(a, f(a))$, $(b, f(b))$ are used, and

(10)
$$m = \frac{0 - f(b)}{c - b},$$

where the points $(c, 0)$ and $(b, f(b))$ are used.

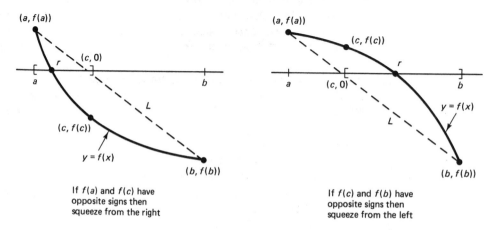

If $f(a)$ and $f(c)$ have
opposite signs then
squeeze from the right

If $f(c)$ and $f(b)$ have
opposite signs then
squeeze from the left

Figure 2.2. The decision process for the false position method.

Equating the slopes in (9) and (10) we have

$$\frac{f(b) - f(a)}{b - a} = \frac{0 - f(b)}{c - b},$$

which is easily solved for c to get

(11)
$$c = b - \frac{f(b)(b - a)}{f(b) - f(a)}.$$

The three possibilities are the same as before:

(12) $f(a)$ and $f(c)$ have opposite signs and a root lies in $[a, c]$,
        or
        $f(c)$ and $f(b)$ have opposite signs and a root lies in $[c, b]$,
        or

(13) $f(c) = 0$ and you found a root.

Convergence of the False Position Method

The decision process implied by (12) together with (11) is used to construct a sequence of intervals $\{[a_n, b_n]\}$ each of which brackets the root. At each step the approximation to the root r is

(14)
$$c_n = b_n - \frac{f(b_n)(b_n - a_n)}{f(b_n) - f(a_n)},$$

and it can be proven that the sequence $\{c_n\}$ will converge to r. But beware, although the interval width $b_n - a_n$ is getting smaller, it is possible that it may not go to zero. If the graph $y = f(x)$ is concave near $(r, 0)$, one of the endpoints become fixed and the other one marches into the solution (see Figure 2.3).

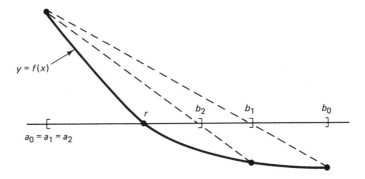

Figure 2.3. The stationary endpoint for the false position method.

Now we rework the solution to $x \sin(x) - 1 = 0$ using the method of false position and observe that it converges faster than the bisection method. Also, notice that $b_n - a_n$ does not go to zero.

Example 2.3 Use the false position method to find the root of $x \sin(x) - 1 = 0$ that is located in the interval $[0, 2]$.

Solution. Starting with $a_0 = 0$ and $b_0 = 2$, we have $f(0) = -1.00000000$ and $f(2) = 0.81859485$, so that a root lies in the interval $[0, 2]$. Using formula (14), we get

$$c_0 = 2 - \frac{0.81859485(2 - 0)}{0.81859485 - (-1)} = 1.09975017 \quad \text{and} \quad f(c_0) = -0.02001921.$$

The function changes sign on the interval $[c_0, b_0] = [1.09975017, 2]$, so we squeeze from the left and set $a_1 = c_0$ and $b_1 = b_0$. Formula (14) produces the next approximation:

$$c_1 = 2 - \frac{0.81859485(2 - 1.09975017)}{0.81859485 - (-0.02001921)} = 1.12124074 \quad \text{and} \quad f(c_1) = 0.00983461.$$

Now $f(x)$ changes sign on $[a_1, c_1] = [1.09975017, 1.12124074]$, and the next decision is to squeeze from the right and set $a_2 = a_1$, $b_2 = c_1$. A summary of the calculations is given in Table 2.2.

Table 2.2   False position method solution of $x \sin(x) - 1 = 0$

k	Left Endpoint, a_k	Point of Intersection, c_k	Right Endpoint, b_k	Function Value, $f(c_k)$
0	0.00000000	1.09975017	2.00000000	−0.02001921
1	1.09975017	1.12124074	2.00000000	0.00983461
2	1.09975017	1.11416120	1.12124074	0.00000563
3	1.09975017	1.11415714	1.11416120	0.00000000

The termination criterion used in the bisection method is not useful for the false position method and may result in an infinite loop. The closeness of consecutive iterates and the size of $|f(c_n)|$ are both used in the termination criterion for Algorithm 2.2. In Section 2.2 we discuss the reasons for this choice.

Algorithm 2.2 [*False Position Method*] Find a root of $f(x) = 0$ in the interval $[a, b]$ given that $f(a)$ and $f(b)$ have opposite signs.

```
Delta := 10⁻⁶                                        {Closeness for consecutive iterates}
Epsilon := 10⁻⁶                                          {Tolerance for the size of f(C)}
Max := 199                                            {Maximum number of iterations}
Satisfied := "False"                                    {Condition for loop termination}


INPUT  A, B                                          {Input the endpoints of the interval}
YA := F(A),  YB := F(B)                                  {Compute the function values}

DO   FOR K = 1 TO Max UNTIL Satisfied = "True"
       DX := YB*[B − A]/[YB − YA]                                      {Change in iterate}
       C := B − DX                                                          {New iterate}
       YC := F(C)                                                       {Function value}
           IF YC = 0   THEN
                 Satisfied = "True"                                {Exact root is found}
             ELSEIF  SIGN(YB)=SIGN(YC)    THEN
                 B := C,  YB := YC                              {Squeeze from the right}
             ELSE
                 A := C,  YA := YC                               {Squeeze from the left}
           ENDIF
       IF  |DX|<Delta   AND   |YC|<Epsilon   THEN                         {Check for
           Satisfied = "True"                                          convergence}
END
```

PRINT 'The computed root of f(x) = 0 is' C   {Output}
PRINT 'Consecutive iterates differ by' DX
PRINT 'The value of the function f(C) is' YC

Exercises for Bracketing Methods

In Exercises 1–3, find an approximation for the interest rate I that will yield the total annuity value A if 240 monthly payments P are made. Use the two starting values for I and compute the next three approximations using the bisection method.

1. Monthly payment $P = \$275$, annuity value $A = \$250{,}000$.
   Use the starting values $I = 0.11$ and 0.12.

2. Monthly payment $P = \$325$, annuity value $A = \$400{,}000$.
   Use the starting values $I = 0.13$ and 0.14.

3. Monthly payment $P = \$300$, annuity values $A = \$500{,}000$.
   Use the starting values $I = 0.15$ and 0.16.

4. Find an interval $a \leq x \leq b$ so that $f(a)$ and $f(b)$ have opposite signs for the following functions.
   (a) $f(x) = \exp(x) - 2 - x$
   (b) $f(x) = \cos(x) + 1 - x$ (x is in radians)
   (c) $f(x) = \ln(x) - 5 + x$
   (d) $f(x) = x^2 - 10x + 23$

In Exercises 5–8, start with the interval $[a_0, b_0]$ and use the bisection method to find an interval of width 0.05 that contains a solution of the given equation.

5. (a) $0 = \exp(x) - 2 - x,\quad [a_0, b_0] = [1.0, 1.8]$
   (b) $0 = \exp(x) - 2 - x,\quad [a_0, b_0] = [-2.4, -1.6]$

6. $0 = \cos(x) + 1 - x,\quad [a_0, b_0] = [0.8, 1.6]$ (x is in radians)

7. $0 = \ln(x) - 5 + x,\quad [a_0, b_0] = [3.2, 4.0]$

8. (a) $0 = x^2 - 10x + 23,\quad [a_0, b_0] = [3.2, 4.0]$
   (b) $0 = x^2 - 10x + 23,\quad [a_0, b_0] = [6.0, 6.8]$

9. Use a computer program and find the roots of $f(x) = 0$ accurate to 5×10^{-6} using the bisection method. Test your program using the functions in Exercise 4.

In Exercises 10–13, start with $[a_0, b_0]$ and use the false position method to compute $c_0, c_1, c_2,$ and c_3. Be sure to compute the function carefully.

10. $\exp(x) - 2 - x = 0,\quad [a_0, b_0] = [-2.4, -1.6]$

11. $\cos(x) + 1 - x = 0,\quad [a_0, b_0] = [0.8, 1.6]$ (x is in radians)

12. $\ln(x) - 5 + x = 0,\quad [a_0, b_0] = [3.2, 4.0]$

13. $x^2 - 10x + 23 = 0,\quad [a_0, b_0] = [6.0, 6.8]$

14. Use a computer program and find the roots of $f(x) = 0$ accurate to 5×10^{-6} using the false position method. Test your program using the functions in Exercise 4.

15. Denote the intervals that arise in the bisection method by $[a_0, b_0], [a_1, b_1], \ldots,$ $[a_n, b_n], \ldots$.

(a) Show that $a_0 \leq a_1 \leq \ldots \leq a_n \leq a_{n+1}$ and that $b_{n+1} \leq b_n \leq \ldots \leq b_1 \leq b_0$.

(b) Show that $b_n - a_n = (b_0 - a_0)/2^n$.

(c) Let the midpoint of each interval be $c_n = (a_n + b_n)/2$. Show that

$$\lim_{n \to \infty} a_n = \lim_{n \to \infty} c_n = \lim_{n \to \infty} b_n.$$

Hint. Review convergence of monotone sequences in your calculus book.

16. What will happen if the bisection method is used with the function $f(x) = 1/(x - 2)$ and
   (a) the interval is $[3, 7]$?   (b) the interval is $[1, 7]$?

17. What will happen if the bisection method is used with the function $f(x) = \tan(x)$ (x is in radians) and
   (a) the interval is $[3, 4]$?   (b) the interval is $[1, 3]$?

18. Suppose that the bisection method is used to find a zero of $f(x)$ in the interval $[2, 7]$. How many times must this interval be bisected to guarantee that the approximation c_N has an accuracy of 5×10^{-9}?

19. Establish formula (8) for determining the number of iterations required in the bisection method. *Hint.* Use $|B - A|/2^{n+1} <$ Delta and take logarithms.

20. Show that formula (14) for the false position method is algebraically equivalent to

$$c_n = \frac{a_n f(b_n) - b_n f(a_n)}{f(b_n) - f(a_n)}.$$

21. What are the differences between the bisection method and the regula falsi method?

22. Write a report on the modified regula falsi method (sometimes referred to as the "modified false position" method).

2.2 Initial Approximations and Convergence Criteria

The bracketing methods depend on finding an interval $[a, b]$ so that $f(a)$ and $f(b)$ have opposite signs. Once the interval has been found, no matter how large, the iteration will proceed until a root is found. Hence these methods are called *globally convergent*. However, if $f(x) = 0$ has several roots in $[a, b]$, then a different starting interval must be used to find each root. It is not easy to locate these smaller intervals on which $f(x)$ changes sign.

In Section 2.3 we develop the Newton–Raphson method and the secant method for solving $f(x) = 0$. Both of these methods require that a close approximation to the root be given to guarantee convergence. Hence these methods are called *locally convergent*. They usually converge more rapidly than do global ones. Some hybrid algorithms start with a global method and switch to a local method when the iteration gets close to a root.

If the computation of roots is one part of a larger project, then a leisurely pace is suggested and the first thing to do is graph the function. If this is done by hand, then it may be easier to rearrange the equation $f(x) = 0$ into an equivalent form $g(x) = h(x)$, where the graphs of $y = g(x)$ and $y = h(x)$ are easy to sketch.

For example, the curve $y = f(x) = x^2 - \exp(x)$ is difficult to sketch [see Figure 2.4(a)], but the functions $g(x) = x^2$ and $h(x) = \exp(x)$ are easy to graph and the abscissa of their point of intersection is the desired root p [see Figure 2.4(b)].

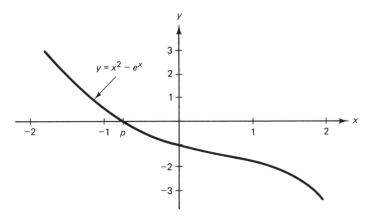

Figure 2.4(a). Graph of $f(x) = x^2 - \exp(x)$ and the root $p \approx -0.7$.

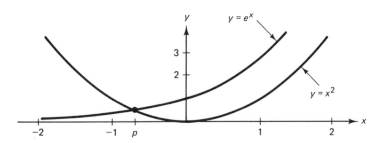

Figure 2.4(b). Graphs of $g(x) = x^2$ and $h(x) = \exp(x)$.

Suppose that a computer is used to graph $y = f(x)$ on $[a, b]$. The interval must be partitioned $a = x_0 < x_1 < \ldots < x_N = b$ and the function values $y_k = f(x_k)$ computed. Then the line segments between consecutive points (x_{k-1}, y_{k-1}) and (x_k, y_k) are plotted for $k = 1, 2, \ldots, N$. This process could take a significant amount of time, but there is something to gain. We can view the graph $y = f(x)$ and make decisions based on what it looks like. But more important, if the coordinates of the points (x_k, y_k) are stored, then they can be analyzed and the approximate location of roots determined.

We must proceed carefully. There must be enough points so that we do not miss a root in a portion of the curve where the function is changing rapidly. If $f(x)$ is continuous and two adjacent points (x_{k-1}, y_{k-1}) and (x_k, y_k) lie on opposite sides of the x-axis, then the *Intermediate Value Theorem* implies that at least one root lies in the interval $[x_{k-1}, x_k]$ and $(x_{k-1} + x_k)/2$ is an approximate root.

Near two closely spaced roots or near a double root, the line segment between (x_{k-1}, y_{k-1}) and (x_k, y_k) may fail to cross the axis. If $|f(x_k)|$ is smaller than a preassigned value ϵ, then x_k is a tentative approximate root. But the graph may be close to zero over a wide range of values near a double root, and we may get too many root approximations. Hence we add the requirement that the slope changes sign near (x_k, y_k), that is,

$$m_{k-1} = \frac{y_k - y_{k-1}}{x_k - x_{k-1}} \quad \text{and} \quad m_k = \frac{y_{k+1} - y_k}{x_{k+1} - x_k}$$

must have opposite signs. Since $x_k - x_{k-1} > 0$ and $x_{k+1} - x_k > 0$, it is not necessary to use the difference quotients and it will suffice to check to see if the differences $y_k - y_{k-1}$ and $y_{k+1} - y_k$ change sign. In this case x_k is the approximate root. Unfortunately, we cannot guarantee that this starting value will produce a convergent sequence. If the graph $y = f(x)$ has a local minimum (or maximum) that is extremely close to zero, then it is possible that x_k is reported as an approximate root when $f(x_k) \approx 0$, although x_k may not be close to a root.

Example 2.4 Find the approximate location of the roots of $x^3 - x^2 - x + 1 = 0$ on the interval $[-1.2, 1.2]$.

Solution. For illustration, choose $N = 8$ and look at Table 2.3.

Table 2.3 Finding approximate locations for roots

x_k	Function Values		Differences in y		Significant Changes in $f(x)$ or $f'(x)$
	y_{k-1}	y_k	$y_k - y_{k-1}$	$y_{k+1} - y_k$	
−1.2	−3.125	−0.968	2.157	1.329	
−0.9	−0.968	0.361	1.329	0.663	f changes sign in $[x_{k-1}, x_k]$
−0.6	0.361	1.024	0.663	0.159	
−0.3	1.024	1.183	0.159	−0.183	f' changes sign near x_k
0.0	1.183	1.000	−0.183	−0.363	
0.3	1.000	0.637	−0.363	−0.381	
0.6	0.637	0.256	−0.381	−0.237	
0.9	0.256	0.019	−0.237	0.069	f' changes sign near x_k
1.2	0.019	0.088	0.069	0.537	

The three abscissas for consideration are −1.05, −0.3, and 0.9. Since $f(x)$ changes sign on the interval $[-1.2, -0.9]$, the value −1.05 is an approximate root; indeed, $f(-1.05) \approx -0.210$. Although the slope changes sign near −0.3, we find that $f(-0.3) = 1.183$; hence −0.3 is *not* near a root. Finally, the slope changes sign near 0.9 and $f(0.9) = 0.019$, so that 0.9 is an approximate root (see Figure 2.5).

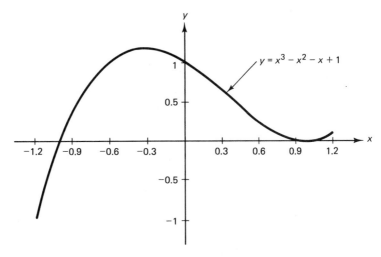

Figure 2.5. The graph of $y = x^3 - x^2 - x + 1$.

Algorithm 2.3 [*Approximate Location of Roots*] Given that $f(x)$ is continuous on $[a, b]$ and that $y_k = f(x_k)$ has been computed at the equally spaced values $x_k = a + hk$ and $h = (b - a)/N$ for $k = 0, \ldots, N$, the approximate location of the roots is found using the criterion $y_{k-1}y_k < 0$ [i.e., $f(x_{k-1})$ and $f(x_k)$ have opposite signs] or $|y_k| < $ Epsilon and $(y_k - y_{k-1})(y_{k+1} - y_k) < 0$ i.e., $|f(x_k)|$ is small and the slope of the curve $y = f(x)$ changes sign near $(x_k, f(x_k))$.

```
Epsilon := 10⁻²                          {Tolerance for |f(xₖ)|}
INPUT  N                                 {Number of subintervals}
INPUT  A, B                              {Endpoints of interval}
H := (B−A)/N                             {Subinterval width}

For  K = 0  TO  N  DO
      X(K) := A + H*K                    {Compute the abscissas}
      Y(K) := F(X(K))                    {Compute the ordinates}
M := 0                                   {Counter for the number of roots}
Y(N+1) := Y(N)                           {This permits one loop}

FOR  K = 1  TO  N  DO
      IF  Y(K−1)*Y(K) ≤ 0  THEN          {Check for a change in
            M := M+1                         sign or finds a root}
            R(M) := [X(K−1) + X(K)]/2     {Approximate root found}
      S := [Y(K)−Y(K−1)]*[Y(K+1)−Y(K)]
      IF  |Y(K)| < Epsilon AND S < 0 THEN {Small function value
            M := M+1                         and slope changes sign}
            R(M) := X(K)                   {Tentative root found}
```

PRINT 'The approximate location of the roots are'
FOR   K = 1   TO   M   DO
└──── PRINT   R(K)

Checking for Convergence

A graph can be used to see the approximate location of a root, but an algorithm must be used to compute a value p_n that is an acceptable computer solution. Iteration is often used to produce a sequence $\{p_k\}$ that converges to a root p, and a termination criterion or strategy must be designed ahead of time so that the computer will stop when an accurate approximation is reached. Since the goal is to solve $f(x) = 0$, the final value p_n should have the property that $f(p_n) \approx 0$.

The user can supply a tolerance value ϵ for the size of $|f(p_n)|$ and then an iterative process produces points $P_k = (p_k, f(p_k))$ until the last point P_n lies in the horizontal band bounded by the lines $y = +\epsilon$ and $y = -\epsilon$, as shown in Figure 2.6(a). This criterion is useful if the user is trying to solve $h(x) = L$ by applying a root-finding algorithm to the function $f(x) = h(x) - L$.

Another termination criterion involves the abscissas, and we can try to determine if the sequence $\{p_k\}$ is converging. If we draw the vertical lines $x = p + \delta$ and $x = p - \delta$ on each side of $x = p$, then we could decide to stop the iteration when the point P_n lies between these vertical lines, as shown in Figure 2.6(b).

The latter criterion is often desired, but it is difficult to implement, because it involves the unknown solution p. We adapt this idea, and terminate further calculations when the consecutive iterates p_n and p_{n-1} are sufficiently close or if they agree within M significant digits.

Sometimes the user of an algorithm will be satisfied if $p_n \approx p_{n-1}$ and other

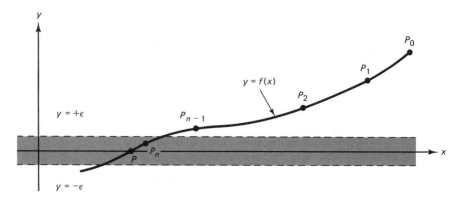

Figure 2.6(a). The horizontal convergence band for locating a solution to $f(x) = 0$.

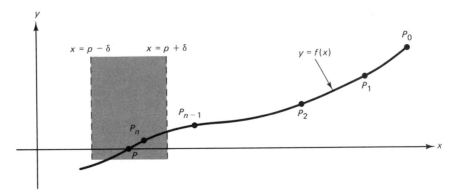

Figure 2.6(b). The vertical convergence band for locating a solution to $f(x) = 0$.

times when $f(p_n) \approx 0$. Correct logical thinking is required to understand the consequences. If we require that $|p_n - p| < \delta$ and $|f(p_n)| < \epsilon$, then the point P_n will be located in the rectangular region about the solution $(p, 0)$, shown in Figure 2.7(a). If we stipulate that $|p_n - p| < \delta$ or $|f(p)| < \epsilon$, then the point P_n could be located anywhere in the region formed by the union of the horizontal and vertical strips, as shown in Figure 2.7(b).

The size of the tolerances δ and ϵ are crucial. If the tolerances are chosen too small, iteration may continue forever. They should be chosen about 100 times larger than 10^{-M}, where M is the number of decimal digits in the computer's floating-point numbers. The closeness of the abscissas is checked with one of the criteria

$$|p_n - p_{n-1}| < \delta \quad \text{(estimate for the absolute error)}$$

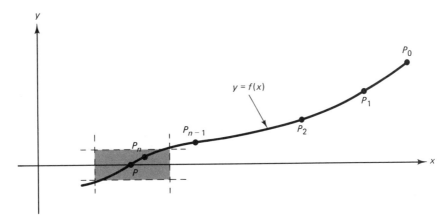

Figure 2.7(a). The rectangular region defined by $|x - p| < \delta$ and $|y| < \epsilon$.

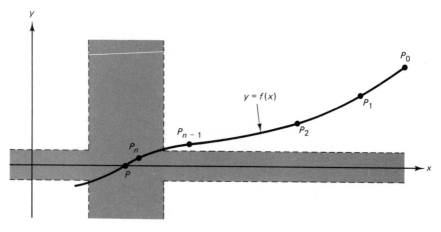

Figure 2.7(b). The unbounded region defined by $|x - p| < \delta$ or $|y| < \epsilon$.

or

$$\frac{2|p_n - p_{n-1}|}{|p_n| + |p_{n-1}|} < \delta \qquad \text{(estimate for the relative error)}.$$

The closeness of the ordinate is usually checked by

$$|f(p_n)| < \epsilon.$$

Troublesome Functions

A computer solution to $f(x) = 0$ will almost always be in error due to round-off and/or instability in the calculations. If the graph $y = f(x)$ is steep near the root $(p, 0)$, then the root-finding problem is well conditioned (i.e., a solution with several significant digits is easy to obtain). If the graph $y = f(x)$ is shallow near $(p, 0)$, then the root-finding problem is ill-conditioned (i.e., the computed root may have only a few significant digits). This occurs quite often when $f(x)$ has a multiple root at p. This is discussed further in the next section.

Exercises for Initial Approximations and Convergence Criteria

In Exercises 1–6, determine functions $y = g(x)$ and $y = h(x)$ so that $g(x) = h(x)$ is equivalent to $0 = f(x)$; sketch the two curves $y = g(x)$ and $y = h(x)$ and determine the approximate location of the roots of $0 = f(x)$ graphically. If a computer plotter is available, then plot $y = f(x)$ directly.

1. $f(x) = x^2 - \exp(x)$ for $-2 \le x \le 2$.
2. $f(x) = x - \cos(x)$ for $-2 \le x \le 2$.
3. $f(x) = \sin(x) - 2\cos(2x)$ for $-2 \le x \le 2$.
4. $f(x) = \cos(x) + (1 + x^2)^{-1}$ for $-2 \le x \le 2$.

5. $f(x) = (x - 2)^2 - \ln(x)$ for $0.5 \leq x \leq 4.5$.

6. $f(x) = 2x - \tan(x)$ for $-1.4 \leq x \leq 1.4$.

7. A computer program that plots the graph of $y = f(x)$ over the interval $[a, b]$ using the points $(x_0, y_0), (x_1, y_1), \ldots, (x_N, y_N)$ usually scales the vertical height of the graph and a procedure must be written to determine the minimum and maximum values.

(a) Write an algorithm that will find the values

$$Y_{\max} = \max_k \{y_k\} \quad \text{and} \quad Y_{\min} = \min_k \{y_k\}.$$

(b) Write an algorithm that will find the approximate location of the extreme values of $f(x)$ on the interval $[a, b]$.

2.3 Newton–Raphson and Secant Methods

Slope Methods for Finding Roots

If $f(x), f'(x)$, and $f''(x)$ are continuous near a root p, then this extra information regarding the nature of $f(x)$ can be used to develop algorithms that will produce sequences $\{p_k\}$ that converge faster to p than either the bisection or false position methods. The Newton–Raphson (or simply Newton's) method is one of the most useful and best known algorithms that rely on the continuity of $f'(x)$ and $f''(x)$. We shall introduce it graphically and then give a more rigorous treatment based on the Taylor polynomial that was introduced in Section 1.3.

Assume that the initial approximation p_0 is near the root p. Then the graph $y = f(x)$ intersects the x-axis at the point $(p, 0)$ and the point $(p_0, f(p_0))$ lies on the curve near the point $(p, 0)$ (see Figure 2.8). Define p_1 to be the point of intersection of the line tangent to the curve at the point $(p_0, f(p_0))$ and the x-axis. Then Figure 2.8 shows that p_1 will be closer to p than p_0 in this case. An equation relating p_1 and p_0 can be found if we write down two versions for the slope of the tangent line L;

(1)
$$m = \frac{0 - f(p_0)}{p_1 - p_0},$$

which is the slope of the line through $(p_1, 0)$ and $(p_0, f(p_0))$, and

(2)
$$m = f'(p_0),$$

which is the slope at the point $(p_0, f(p_0))$. Equating the values of the slope m in equations (1) and (2) and solving for p_1 results in

(3)
$$p_1 = p_0 - \frac{f(p_0)}{f'(p_0)}.$$

The process above can be repeated to obtain a sequence $\{p_k\}$ that converges to p. We now make these ideas more precise.

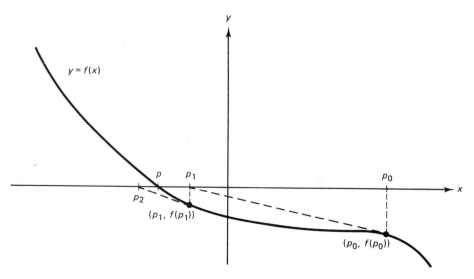

Figure 2.8. Construction of p_1, p_2 for the Newton-Raphson method.

Theorem 2.1   [Newton-Raphson method] If $f(x), f'(x)$ and $f''(x)$ are continuous near the root p, then we define $g(x)$ as follows:

(4)
$$g(x) = x - \frac{f(x)}{f'(x)} \qquad \text{(Newton-Raphson iteration function).}$$

If $f'(p) \neq 0$, and p_0 is chosen close enough to p, then the sequence $\{p_k\}$ defined by the recursive iteration formula

(5)
$$p_k = g(p_{k-1}) = p_{k-1} - \frac{f(p_{k-1})}{f'(p_{k-1})} \qquad \text{for } k = 1, 2, \ldots,$$

will converge to p.

Discussion. The geometric construction of p_1 given in Figure 2.8 does not help in understanding why p_0 needs to be close to p or why the continuity of $f''(x)$ is essential. Consider the Taylor polynomial representation of degree $N = 1$ with remainder term

(6)
$$f(x) = f(p_0) + f'(p_0)(x - p_0) + f''(c) \frac{(x - p_0)^2}{2},$$

where c lies somewhere between p_0 and x. Substituting $x = p$ in equation (6) and using the fact that $f(p) = 0$ results in

(7)
$$0 = f(p_0) + f'(p_0)(p - p_0) + f''(c) \frac{(p - p_0)^2}{2}.$$

If p_0 is close enough to p, then the last term on the right side of equation (7) will be small compared to the sum of the first two terms. Hence it can be neglected and we use the approximation

(8)
$$0 \approx f(p_0) + f'(p_0)(p - p_0).$$

Solving for p in equation (8), we get $p \approx p_0 - f(p_0)/f'(p_0)$. This value is used to define the next approximation, p_1:

(9)
$$p_1 = p_0 - \frac{f(p_0)}{f'(p_0)}.$$

When p_{k-1} is used in place of p_0 in (9), the general rule (5) is obtained. We postpone the proof of convergence until later.

Remark. Newton's method finds a zero of $f(x)$, but the iteration uses the Newton–Raphson function $g(x)$ given in equation (4). This distinction must be understood.

Finding the square root of a number is a common calculation, and an algorithm for its computation is hardwired into all modern computers. The next result illustrates this important special case (c.f. Exercise 29).

Corollary 2.1 [Newton's square-root algorithm] Let $A > 0$ be a real number and let $p_0 > 0$ be an initial approximation to $A^{1/2}$. Define the sequence $\{p_k\}$ using the recursive rule

(10)
$$p_k = \frac{p_{k-1} + A/p_{k-1}}{2} \qquad \text{for } k = 1, 2, \ldots.$$

Then the sequence $\{p_n\}$ converges to $A^{1/2}$, that is,

(11)
$$\lim_{k \to \infty} p_k = A^{1/2}.$$

Outline of the Proof. Start with the function $f(x) = x^2 - A$, and notice that the roots of the equation $x^2 - A = 0$ are $\pm A^{1/2}$. Now use $f(x)$ and the derivative $f'(x)$ in formula (4) and write down the Newton–Raphson iteration function

(12)
$$g(x) = x - \frac{f(x)}{f'(x)} = x - \frac{x^2 - A}{2x}.$$

This formula can be simplified to obtain

(13)
$$g(x) = \frac{x + A/x}{2}.$$

When $g(x)$ in (13) is used to define the recursive iteration in (5), the result is formula (10). It can be proven that the sequence that is generated by (10) will converge for any starting value $p_0 > 0$. The details are left for the exercises.

An important point of Corollary 2.1 is the fact that the iteration function $g(x)$ involved only the arithmetic operations $+$, $-$, $\times$, and $/$. If $g(x)$ had involved the calculation of a square root, then we would be caught in the circular reasoning that being able to calculate the square root would permit you to recursively define a sequence that will converge to $A^{1/2}$. For this reason, $f(x) = x^2 - A$ was chosen, because it involved only the arithmetic operations.

Example 2.5 Use Newton's square-root algorithm to find $5^{1/2}$.

Solution. Starting with $p_0 = 2$ and using formula (13), we compute

$$p_1 = \frac{2 + 5/2}{2} = 2.25$$

$$p_2 = \frac{2.25 + 5/2.25}{2} = 2.236111111$$

$$p_3 = \frac{2.236111111 + 5/2.236111111}{2} = 2.236067978$$

$$p_4 = \frac{2.236067978 + 5/2.236067978}{2} = 2.236067978.$$

Further iterations produce $p_k \approx 2.236067978$ for $k > 4$, so we see that convergence accurate to nine significant digits has been achieved.

Algorithm 2.4 [*Newton–Raphson Iteration*] Find a root of $f(x) = 0$ given one initial approximation p_0 and using the iteration

$$p_k = p_{k-1} - \frac{f(p_{k-1})}{f'(p_{k-1})} \qquad \text{for } k = 1, 2, \ldots, n.$$

```
Delta := 10⁻⁶, Epsilon := 10⁻⁶, Small := 10⁻⁶        {Tolerances}
Max := 99                                  {Maximum number of iterations}
Cond := 0                                  {Condition for loop termination}
INPUT  P0                                  {P0 must be close to the root}
Y0 := F(P0)                                {Compute the function value}


DO  FOR K := 1 TO Max UNTIL Cond ≠ 0
      Df := F'(P0)                         {Compute the derivative}
        IF  Df = 0  THEN                   {Check division
              Cond := 1                        by zero}
              Dp := 0
          ELSE
              Dp := Y0/Df
        ENDIF
      P1 := P0 − Dp                        {New iterate}
      Y1 := F(P1)                          {New function value}
      RelErr := 2∗|Dp|/[|P1|+Small]        {Relative error}
      IF  RelErr<Delta AND |Y1|<Epsilon  THEN   {Check for
        IF Cond ≠ 1 THEN Cond := 2              convergence}
      P0 := P1, Y0 := Y1                    {Update values}
```

```
        PRINT 'The current k-th iterate is' P1                    {Output}
        PRINT 'Consecutive iterates differ by' Dp
        PRINT 'The value of f(x) is' Y1
IF      Cond = 0  THEN
        PRINT 'The maximum number of iterations was exceeded.'
IF      Cond = 1  THEN
        PRINT 'Division by zero was encountered.'
IF      Cond = 2  THEN
        PRINT 'The root was found with the desired tolerances.'
```

Now let us turn to a familiar problem from elementary physics and see why the location of a root is an important task. Suppose that a projectile is fired from the origin with an angle of elevation b_0 and initial velocity v_0. In elementary courses air resistance is neglected and we learn that the height $y = y(t)$ and the distance traveled $x = x(t)$ measured in feet obey the rules

(14)
$$y = v_y t - 16t^2 \quad \text{and} \quad x = v_x t,$$

where the horizontal and vertical components of the initial velocity are $v_x = v_0 \cos(b_0)$ and $v_y = v_0 \sin(b_0)$, respectively. The mathematical model expressed by the rules (14) are easy to work with but tend to give too high an altitude and too long a range for the projectile's path. If we make the additional assumption that the air resistance is proportional to the velocity, the equations of motion become

(15)
$$y = f(t) = (Cv_y + 32C^2)\left[1 - \exp\left(-\frac{t}{C}\right)\right] - 32Ct$$

and

(16)
$$x = r(t) = Cv_x\left[1 - \exp\left(-\frac{t}{C}\right)\right],$$

where $C = m/k$ and k is the coefficient of air resistance and m is the mass of the projectile. A larger value of C will result in a higher maximum altitude and a longer range for the projectile. The graph of a flight path of a projectile when air resistance is considered is shown in Figure 2.9. This improved model is more realistic but requires the use of a root-finding algorithm for solving $f(t) = 0$ to determine the elapsed time until the projectile hits the ground. The elementary model in (14) does not require a sophisticated procedure to find the elapsed time.

Example 2.6 A projectile is fired with an angle of elevation $b_0 = 45°$ and $v_y = v_x = 160$ ft/sec and $C = 10$. Find the elapsed time until impact and find the range.

Solution. Using formulas (15) and (16), the equations of motion are $y = f(t) = 4,800[1 - \exp(-t/10)] - 320t$ and $x = r(t) = 1,600[1 - \exp(-t/10)]$. Since $f(8) = 83.220972$ and $f(9) = -31.534367$, we will use the initial guess $p_0 = 8$.

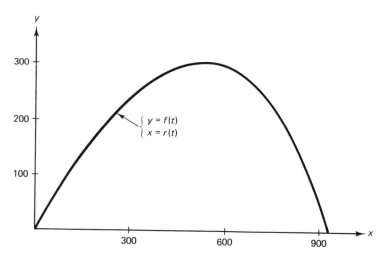

Figure 2.9. Path of a projectile with air resistance considered.

The derivative is $f'(t) = 480 \exp\left(-t/10\right) - 320$, and its value $f'(p_0) = f'(8) = -104.3220972$ is used in formula (5) to get

$$p_1 = 8 - \frac{83.22097200}{-104.3220972} = 8.797731010.$$

A summary of the calculations is given in Table 2.4.

Table 2-4 Finding the time when the height $f(t)$ is zero

k	Time, p_k	$p_{k+1} - p_k$	Height, $f(p_k)$
0	8.00000000	0.79773101	83.22097200
1	8.79773101	−0.05530160	−6.68369700
2	8.74242941	−0.00025475	−0.03050700
3	8.74217467	−0.00000001	−0.00000100
4	8.74217466	0.00000000	0.00000000

The value p_4 has eight decimal places of accuracy and the time until impact is $t \approx 8.74217466$ sec. The range can now be computed using $r(t)$ and we get

$$r(8.74217466) = 1{,}600[1 - \exp\left(-0.874217466\right)] = 932.4986302 \text{ ft.}$$

The Division-by-Zero Error

One obvious pitfall of the Newton–Raphson method is the possibility of division by zero in formula (5), which would occur if $f'(p_{k-1}) = 0$. Algorithm 2.4 has a procedure to check for this situation, but what use is the final value p_{k-1} in this case? It is quite possible that $f(p_{k-1})$ is sufficiently close to zero and that p_{k-1} is an acceptable approximation to the root. We now investigate this situation and will uncover an interesting fact, namely, how fast the iteration converges.

Definition 2.2 [Order of a root] Suppose that $f(x)$ and its derivatives $f'(x), \ldots, f^{(M)}(x)$ are defined and continuous near p. We say that $f(x) = 0$ has a root of order M at p [or $f(x)$ has a zero of order M at p] if and only if

(17) $$f(p) = 0, \quad f'(p) = 0, \quad \ldots, \quad f^{(M-1)}(p) = 0, \quad f^{(M)}(p) \neq 0.$$

A root of order $M = 1$ is often called a *simple root*, and if $M > 1$ it is called a *multiple root*. A root of order $M = 2$ is sometimes called a *double root*, and so on. The next result will help us understand these concepts.

Lemma 2.1 If $f(x) = 0$ has a root of order M at p, then there exists a continuous function $h(x)$, and $f(x)$ can be expressed as the product

(18) $$f(x) = (x - p)^M h(x), \quad \text{where } h(p) \neq 0.$$

As an example, consider the function $f(x) = x^3 - 3x + 2$, which has a simple root at $p = -2$ and a double root at $p = 1$. This can be verified by considering the derivatives

$$f'(x) = 3x^2 - 3 \quad \text{and} \quad f''(x) = 6x.$$

At the value $p = -2$ we have $f(-2) = 0$ and $f'(-2) = 9$, so that $M = 1$ in Definition 2.2; hence $p = -2$ is a simple root. For the value $p = 1$ we have $f(1) = 0, f'(1) = 0$, and $f''(1) = 6$, so that $M = 2$ in Definition 2.2; hence $p = 1$ is a double root. Also, notice that $f(x)$ has the factorization

$$f(x) = (x + 2)(x - 1)^2.$$

Speed of Convergence

The distinguishing property we seek is the following. If p is a simple root of $f(x) = 0$, then Newton's method will converge rapidly and the number of accurate decimal places (roughly) doubles with each iteration. On the other hand, if p is a multiple root, the error in each successive approximation is a fraction of the previous error. To make this precise, we define the order of convergence. This is a measure of how rapidly a sequence converges.

Definition 2.3 [Order of convergence] Let $\{p_k\}$ be a sequence that converges to p, and set $e_k = p - p_k$ for $k = 0, 1, \ldots$. If there exists a number R and a constant $A \neq 0$ such that

(19) $$\lim_{n \to \infty} \frac{|e_{n+1}|}{|e_n|^R} = A,$$

then R is called the *order of convergence* of the sequence.

If R is large, then the sequence $\{p_k\}$ converges rapidly to p, that is, relation

(19) implies that for large values of n we have the approximation $|e_{n+1}| \approx A|e_n|^R$. For example, suppose that $R = 2$ and $|e_n| \approx 10^{-2}$, then we could expect that $|e_{n+1}| \approx A \times 10^{-4}$.

Some sequences converge at a rate that is not an integer, and we will see that the order of convergence of the secant method is $R = (1 + \sqrt{5})/2 \approx 1.618033989$. When R takes on the integer values 1, 2, or 3 we say that the order of convergence is linear, quadratic, or cubic, respectively. It is worthwhile to emphasize this point:

(20)
$$\lim_{n \to \infty} \frac{|e_{n+1}|}{|e_n|^1} = A \quad \text{is linear convergence}$$

and

(21)
$$\lim_{n \to \infty} \frac{|e_{n+1}|}{|e_n|^2} = A \quad \text{is quadratic convergence.}$$

Example 2.7 [Quadratic Convergence at a Simple Root] Start with $p_0 = -2.4$ and use Newton–Raphson iteration to find the root $p = -2$ of the polynomial $f(x) = x^3 - 3x + 2$.

Solution. The iteration formula for computing $\{p_k\}$ is

(22)
$$p_k = g(p_{k-1}) = \frac{2p_{k-1}^3 - 2}{3p_{k-1}^2 - 3}.$$

Using formula (21) to check for quadratic convergence, we get the values in Table 2.5.

Table 2.5 Newton's method converges quadratically at a simple root

k	p_k	$p_{k+1} - p_k$	$e_k = p - p_k$	$\dfrac{|e_{k+1}|}{|e_k|^2}$
0	−2.400000000	0.323809524	0.400000000	0.476190475
1	−2.076190476	0.072594465	0.076190476	0.619469086
2	−2.003596011	0.003587422	0.003596011	0.664202613
3	−2.000008589	0.000008589	0.000008589	
4	−2.000000000	0.000000000	0.000000000	

A detailed look at the rate of convergence in Example 2.7 will reveal that the error in each successive iteration is proportional to the square of the error in the previous iteration. That is,

$$|p - p_{k+1}| \approx A|p - p_k|^2,$$

where $A \approx \frac{2}{3}$. To check this, we use

$$|p - p_3| = 0.000008589 \quad \text{and} \quad |p - p_2|^2 = |0.003596011|^2 = 0.000012931$$

and it is easy to see that

$$|p - p_3| = 0.000008589 \approx 0.000008621 = \tfrac{2}{3}|p - p_2|^2.$$

Remark. Theorem 2.2 will show that the constant A can be found by the calculation

$$A = \frac{1}{2}\frac{|f''(-2)|}{|f'(-2)|} = \frac{1}{2}\frac{|-12|}{|9|} = \frac{2}{3}.$$

Example 2.8 [Linear Convergence at a Double Root] Start with $p_0 = 1.2$ and use Newton–Raphson iteration to find the double root $p = 1$ of the polynomial $f(x) = x^3 - 3x + 2$.

Solution. Using formula (20) to check for linear convergence, we get the values in Table 2.6.

Table 2.6 Newton's method converges linear at a double root

k	p_k	$p_{k+1} - p_k$	$e_k = p - p_k$	$\dfrac{|e_{k+1}|}{|e_k|}$
0	1.200000000	−0.096969697	−0.200000000	0.515151515
1	1.103030303	−0.050673883	−0.103030303	0.508165253
2	1.052356420	−0.025955609	−0.052356420	0.496751115
3	1.026400811	−0.013143081	−0.026400811	0.509753688
4	1.013257730	−0.006614311	−0.013257730	0.501097775
5	1.006643419	−0.003318055	−0.006643419	0.500550093

Notice that the Newton–Raphson method is converging to the double root, but at a slow rate. The values $f(p_k)$ in Example 2.8 go to zero faster than the values of $f'(p_k)$, so that the quotient $f(p_k)/f'(p_k)$ in formula (5) is defined when $p_k \neq p$. The sequence is converging linearly and the error is decreasing by a factor of approximately $\tfrac{1}{2}$ with each successive iteration.

Theorem 2.2 [Order of convergence of Newton–Raphson iteration] Suppose that the sequence $\{p_k\}$ in (5) converges to the root p. If p is a simple root, then convergence is quadratic:

(23)
$$|e_{k+1}| \approx \frac{1}{2}\frac{|f''(p)|}{|f'(p)|}|e_k|^2 \qquad \text{for } k \text{ sufficiently large.}$$

If p is a multiple root of order M, convergence is linear:

(24)
$$|e_{k+1}| \approx \frac{M-1}{M}|e_k| \qquad \text{for } k \text{ sufficiently large.}$$

Pitfalls

The division-by-zero error was easy to anticipate, but there are other difficulties that are not so easy to spot. Suppose that the function is $f(x) = x^2 - 4x + 5$; then the sequence $\{p_k\}$ of real numbers generated by formula (5) will wander back and forth from left to right and *not* converge. A simple analysis of the situation reveals that $f(x) > 0$, and that the roots are complex numbers.

Sometimes the initial approximation p_0 is too far away from the desired root and the sequence $\{p_k\}$ converges to some other root. This usually happens when the slope $f'(p_0)$ is small and the tangent line to the curve $y = f(x)$ is nearly horizontal. For example, if $f(x) = \cos(x)$ and we seek the root $p = \pi/2$ and start with $p_0 = 3$, calculation reveals that $p_1 = -4.01525255$, $p_2 = -4.85265757, \ldots$ and $\{p_k\}$ will converge to a different root $-3\pi/2 \approx -4.71238898$.

Suppose that $f(x)$ is positive and monotone decreasing on the unbounded interval $[a, \infty)$ and $p_0 > a$; then the sequence $\{p_k\}$ might diverge to $+\infty$. For example, if $f(x) = x \exp(-x)$ and $p_0 = 10$, then

$$p_1 \approx 11.11111111, \quad p_2 \approx 12.21001221, \quad \ldots, \quad p_{15} \approx 25.98908679, \quad \ldots$$

and $\{p_k\}$ diverges slowly to $+\infty$ [see Figure 2.10(a)]. This particular function has another surprising problem. The value of $f(x)$ goes to zero rapidly as x gets large; for example, $f(p_{15}) \approx 0.00000000013424$ and it is possible that p_{15} could be mistaken for a root. For this reason we designed the stopping criterion in Algorithm 2.4 to involve the relative error $2|p_{k+1} - p_k|/(|p_k| + 10^{-6})$, and when $k = 15$ this value is 0.07695538, so that the tolerance $\delta = 10^{-6}$ will help guard against reporting a false root.

Another phenomenon, *cycling*, occurs when the terms in the sequence $\{p_k\}$ tend to repeat, or almost repeat. For example, if $f(x) = x^3 - x - 3$ and the initial approximation is $p_0 = 0$, then the sequence is

$$p_1 = -3.000000, \quad p_2 = -1.961538, \quad p_3 = -1.147176, \quad p_4 = -0.006579,$$

$$p_5 = -3.000389, \quad p_6 = -1.961818, \quad p_7 = -1.147430, \quad \ldots$$

and we are stuck in a cycle where $p_{k+4} \approx p_k$ for $k = 0, 1, 2, \ldots$ [see Figure 2.10(b)]. But if the starting value p_0 is sufficiently close to the root $p \approx 1.671699881$, then $\{p_k\}$ converges. If $p_0 = 2$, then the sequence converges,

$$p_1 = 1.727272727, \quad p_2 = 1.673691173, \quad p_3 = 1.671702570, \quad p_4 = 1.671699881.$$

When $f''(p) = 0$ at the root p, there is a chance of divergent oscillation. For example, if $f(x) = \arctan(x)$ and $p_0 = 1.4$, then

$$p_1 = -1.413618649, \quad p_2 = 1.450129315, \quad p_3 = -1.550625977, \quad \text{etc.}$$

[see Figure 2.10(c)]. But if the starting value is sufficiently close to the root $p = 0$, then a convergent sequence results. If $p_0 = 0.5$, then

$$p_1 = -0.079559511, \quad p_2 = 0.000335302, \quad p_3 = 0.000000000.$$

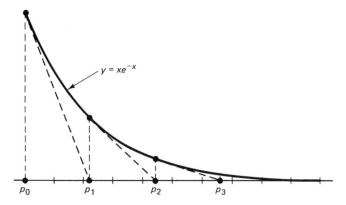

Figure 2.10(a). $f(x) = x \exp(-x)$ produces a divergent sequence.

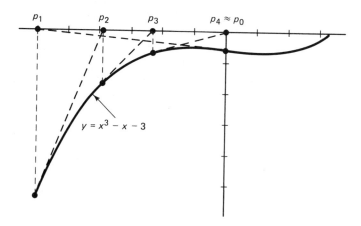

Figure 2.10(b). $f(x) = x^3 - x - 3$ produces a cyclic sequence.

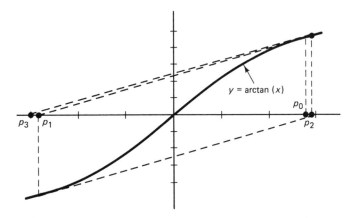

Figure 2.10(c). $f(x) = \arctan(x)$ produces divergent oscillations.

The situations above point to the fact that we must be honest in reporting an answer. Sometimes the sequence does not converge. It is not always the case that after N iterations a solution is found. The user of a root-finding algorithm needs to be warned of the situation when a root is not found. If there is other information concerning the context of the problem, then it is less likely that an erroneous root will be found. Sometimes $f(x)$ has a definite interval in which a root is meaningful. If knowledge of the behavior of the function or a graph is available, then it is easier to choose p_0.

The Secant Method

The Newton–Raphson algorithm requires the evaluation of two functions per iteration, $f(p_{k-1})$ and $f'(p_{k-1})$. If they are not complicated expressions, the method is desirable. In some cases it may require a considerable amount of effort to use the rules of calculus and derive the formula for $f'(x)$ from $f(x)$. Hence it is desirable to have a method that converges almost as fast as Newton's method yet involves only evaluations of $f(x)$. The secant method will require only one evaluation of $f(x)$ per step and at a simple root has an order of convergence $R \approx 1.618033989$. It is almost as fast as Newton's method, which has order 2.

The formula involved in the secant method is the same one that was used in the regula falsi method, except that the logical decisions regarding how to define each succeeding term is different. Two initial points $(p_0, f(p_0))$ and $(p_1, f(p_1))$ near the point $(p, 0)$ are needed, as shown in Figure 2.11. Define p_2 to be the point of intersection of the line through these two points; then Figure 2.11 shows that p_2 will be closer to p than to either p_0 or p_1. The equation relating $p_2, p_1,$ and p_0 is found by considering the slope

(25)
$$m = \frac{f(p_1) - f(p_0)}{p_1 - p_0} = \frac{0 - f(p_1)}{p_2 - p_1}.$$

The values of m in (25) are the slope of the secant line through the first two approximations and the slope of the line through $(p_1, f(p_1))$ and $(p_2, 0)$, respectively. Solve for $p_2 = g(p_1, p_0)$ and get

(26)
$$p_2 = g(p_1, p_0) = p_1 - \frac{f(p_1)(p_1 - p_0)}{f(p_1) - f(p_0)}.$$

The general term is given by the two-point iteration formula

(27)
$$p_{k+1} = g(p_k, p_{k-1}) = p_k - \frac{f(p_k)(p_k - p_{k-1})}{f(p_k) - f(p_{k-1})}.$$

Example 2.9 [Secant Method at a Simple Root]   Start with $p_0 = -2.6$ and $p_1 = -2.4$ and use the secant method to find the root $p = -2$ of the polynomial function $f(x) = x^3 - 3x + 2$.

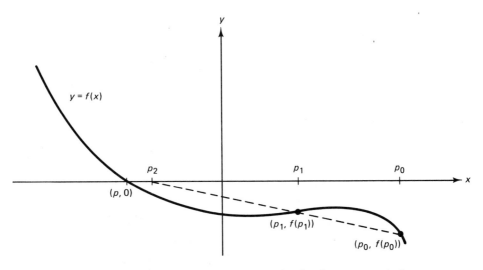

Figure 2.11. The geometric construction for the secant method.

Solution. In this case the iteration formula (27) is

$$(28) \qquad p_{k+1} = g(p_k, p_{k-1}) = p_k - \frac{(p_k^3 - 3p_k + 2)(p_k - p_{k-1})}{p_k^3 - p_{k-1}^3 - 3p_k + 3p_{k-1}}.$$

This can be algebraically manipulated to obtain

$$(29) \qquad p_{k+1} = g(p_k, p_{k-1}) = \frac{p_k^2 p_{k-1} + p_k p_{k-1}^2 - 2}{p_k^2 + p_k p_{k-1} + p_{k-1}^2 - 3}.$$

and the sequence of iterates is given in Table 2.7.

Table 2.7   Convergence of the secant method at a simple root

k	p_k	$p_{k+1} - p_k$	$e_k = p - p_k$	$\dfrac{\lvert e_{k+1}\rvert}{\lvert e_k\rvert^{1.618}}$
0	−2.600000000	0.200000000	0.600000000	0.914152831
1	−2.400000000	0.293401015	0.400000000	0.469497765
2	−2.106598985	0.083957573	0.106598985	0.847290012
3	−2.022641412	0.021130314	0.022641412	0.693608922
4	−2.001511098	0.001488561	0.001511098	0.825841116
5	−2.000022537	0.000022515	0.000022537	0.727100987
6	−2.000000022	0.000000022	0.000000022	
7	−2.000000000	0.000000000	0.000000000	

There is a relationship between the secant method and Newton's method. For a polynomial function $f(x)$, the secant method two-point formula $g(p_k, p_{k-1})$ will reduce to Newton's one-point formula $g(p_k)$ if p_k is replaced by

p_{k-1}. Indeed, if we replace p_k by p_{k-1} in (29), then the right side becomes the same as the right side of (22).

Proofs about the rate of convergence of the secant method can be found in advanced texts on numerical analysis. Let us state that the error terms satisfy the relationship

$$(30) \qquad |e_{k+1}| = |e_k|^{1.618} \left| \frac{f''(p)}{2f'(p)} \right|^{0.618}$$

where the order of convergence is $R = (1 + 5^{1/2})/2 \approx 1.618$ and the relation in (30) is valid only at simple roots.

To check this out we use Example 2.9 and the specific values $|p - p_5| = 0.000022537$, $|p - p_4|^{1.618} = 0.001511098^{1.618} = 0.000027296$, and $A = |f''(-2)/2f'(-2)|^{0.618} = (2/3)^{0.618} = 0.778351205$. Combine these and it is easy to see that

$$|p - p_5| = 0.000022537 \approx 0.000021246 = A|p - p_4|^{1.618}.$$

Accelerated Convergence

We would hope that there are root-finding techniques which converge faster than linearly when p is a root of order M. Our final result shows how a modification can be made to Newton's method so that convergence becomes quadratic at a multiple root.

Theorem 2.3 [Accelerated Newton–Raphson iteration] Suppose that p is a root of order $M > 1$. If the starting value p_0 is close to p, then the modified Newton–Raphson formula

$$(31) \qquad p_k = p_{k-1} - M \frac{f(p_{k-1})}{f'(p_{k-1})}.$$

will produce a sequence $\{p_k\}$ that converges quadratically to p.

Example 2.10 [Acceleration of Convergence at a Double Root] Start with $p_0 = 1.2$ and use accelerated Newton–Raphson iteration to find the double root $p = 1$ of $f(x) = x^3 - 3x + 2$.

Solution. Since $M = 2$, the acceleration formula (30) becomes

$$p_k = p_{k-1} - 2 \frac{f(p_{k-1})}{f'(p_{k-1})} = \frac{p_{k-1}^3 + 3p_{k-1} - 4}{3p_{k-1}^2 - 3},$$

and we obtain the values in Table 2.8.

Table 2.8 Acceleration of convergence at a double root

k	p_k	$p_{k+1} - p_k$	$e_k = p - p_k$	$\dfrac{\|e_{k+1}\|}{\|e_k\|^2}$
0	1.200000000	−0.193939394	−0.200000000	0.151515150
1	1.006060606	−0.006054519	−0.006060606	0.165718578
2	1.000006087	−0.000006087	−0.000006087	
3	1.000000000	0.000000000	0.000000000	

Table 2.9 compares speed of convergence of the various root-finding methods that we have studied so far. The value of the constant A is different for each method.

Table 2.9 Comparison of the speed of convergence

Method	Special Considerations	Relation between Successive Error Terms
Bisection		$\|e_{k+1}\| \approx \frac{1}{2}\|e_k\|$
Regula falsi		$\|e_{k+1}\| \approx A\|e_k\|$
Secant method	Multiple root	$\|e_{k+1}\| \approx A\|e_k\|$
Newton–Raphson	Multiple root	$\|e_{k+1}\| \approx A\|e_k\|$
Secant method	Simple root	$\|e_{k+1}\| \approx A\|e_k\|^{1.618}$
Newton–Raphson	Simple root	$\|e_{k+1}\| \approx A\|e_k\|^2$
Accelerated Newton–Raphson	Multiple root	$\|e_{k+1}\| \approx A\|e_k\|^2$

Algorithm 2.5 [*Secant Method*] Find a root of $f(x) = 0$ given two initial approximations p_0 and p_1 and using the iteration

$$p_{k+1} = p_k - \frac{f(p_k)(p_k - p_{k-1})}{f(p_k) - f(p_{k-1})} \quad \text{for } k = 1, 2, \ldots, n.$$

Delta := 10^{-6}, Epsilon := 10^{-6}, Small := 10^{-6} {Tolerances}

Max := 149   {Maximum number of iterations}

Cond := 0   {Condition for loop termination}

INPUT P0, P1   {These values must be close to the root}

Y0 := F(P0), Y1 := F(P1)   {Compute the function values}

```
DO   FOR K := 1 TO Max UNTIL Cond ≠ 0
     │  Df := [Y1 − Y0]/[P1 − P0]                    {Compute the slope}
     │     IF   Df = 0   THEN                        {Check division
     │     │        Cond := 1                                by zero}
     │     │        Dp := 0
     │     ELSE
     │     │        Dp := Y1/Df
     │     ENDIF
     │  P2 := P1 − Dp                                {New iterate}
     │  Y2 := F(P2)                                  {New function value}
     │  RelErr := 2*|Dp|/[|P2|+Small]                {Relative error}
     │  IF  RelErr<Delta AND |Y2|<Epsilon   THEN     {Check for
     │  └  IF Cond ≠ 1 THEN Cond := 2                  convergence}
     └──── P0 := P1, P1 := P2, Y0 := Y1, Y1 := Y2    {Update values}

         PRINT 'The current k-th iterate is' P2
         PRINT 'Consecutive iterates differ by' Dp
         PRINT 'The value of f(x) is' Y2
IF    Cond = 0   THEN
         PRINT 'The maximum number of iterations was exceeded.'
IF    Cond = 1   THEN
         PRINT 'Division by zero was encountered.'
IF    Cond = 2   THEN
         PRINT 'The root was found with the desired tolerances'
```

Exercises for Newton–Raphson and Secant Method

For problems involving calculations, you can use either a *calculator* or a *computer*.

1. Use Newton's square-root algorithm.
   (a) Start with $p_0 = 3$ and find $8^{1/2}$.
   (b) Start with $p_0 = 7$ and find $50^{1/2}$.
   (c) Start with $p_0 = 10$ and find $91^{1/2}$.
   (d) Start with $p_0 = -3$ and find $-(8)^{1/2}$.

2. *Cube-Root Algorithm.* Start with $f(x) = x^3 - A$, where A is any real number, and derive the recursive formula

$$p_k = \frac{2p_{k-1} + A/p_{k-1}^2}{3} \qquad \text{for } k = 1, 2, \ldots$$

for finding the cube root of A.

3. Use the iteration in Exercise 2 for finding cube roots.
   (a) Start with $p_0 = 2$ and find $7^{1/3}$.
   (b) Start with $p_0 = 3$ and find $30^{1/3}$.
   (c) Start with $p_0 = 6$ and find $200^{1/3}$.
   (d) Start with $p_0 = -2$ and find $(-7)^{1/3}$.

4. Consider $f(x) = x^N - A$, where N is a positive integer.
   (a) What real values are the solution to $f(x) = 0$ for the various choices of N and A that can arise?
   (b) Derive the recursive formula

   $$p_k = \frac{(N-1)p_{k-1} + A/p_{k-1}^{N-1}}{N} \qquad \text{for } k = 1, 2, \ldots$$

   for finding the nth root of A.

5. Establish the limit in equation (11).

6. Write an algorithm that will use Newton's method to find $A^{1/N}$. Input the real number A and integer N. Return the value 0 for the case of $0^{1/N}$. Have the algorithm print an error message in the case when A is negative and N is an even integer. For the other cases start with $p_0 = A$ and use the stopping criterion

   $$|p_k - p_{k-1}| < |p_k| \times 10^{-6}.$$

7. Let $f(x) = x^2 - 2x - 1$.
   (a) Find the Newton–Raphson formula $g(p_{k-1})$.
   (b) Start with $p_0 = 2.5$ and find p_1, p_2, and p_3.
   (c) Start with $p_0 = -0.5$ and find p_1, p_2, and p_3.

8. Let $f(x) = x^3 - x - 3$.
   (a) Find the Newton–Raphson formula $g(p_{k-1})$.
   (b) Start with $p_0 = 1.6$ and find p_1, p_2, and p_3.
   (c) Start with $p_0 = 0.0$ and find p_1, p_2, p_3, p_4, and p_5. What do you conjecture about this sequence?

9. Let $f(x) = x^3 - x + 2$.
   (a) Find the Newton–Raphson formula $g(p_{k-1})$.
   (b) Start with $p_0 = -1.5$ and find p_1, p_2, and p_3.

10. Suppose that the equations of motion for a projectile are

    $$y = f(t) = 1{,}600 \left[1 - \exp\left(-\frac{t}{5}\right) \right] - 160t,$$

    $$x = r(t) = 800 \left[1 - \exp\left(-\frac{t}{5}\right) \right].$$

    (a) Start with $p_0 = 8$ and find the elapsed time until impact.
    (b) Find the range.

11. Suppose that the equations of motion for a projectile are

    $$y = f(t) = 9{,}600 \left[1 - \exp\left(-\frac{t}{15}\right) \right] - 480t,$$

    $$x = r(t) = 2{,}400 \left[1 - \exp\left(-\frac{t}{15}\right) \right].$$

    (a) Start with $p_0 = 9$ and find the elapsed time until impact.
    (b) Find the range.

12. Let $f(x) = (x - 2)^4$.
    (a) Find the Newton–Raphson formula $g(p_{k-1})$.
    (b) Start with $p_0 = 2.1$ and compute $p_1, p_2, p_3,$ and p_4.
    (c) Is the sequence converging quadratically or linearly?

13. Let $f(x) = x^3 - 3x - 2$.
    (a) Find the Newton–Raphson formula $g(p_{k-1})$.
    (b) Start with $p_0 = 2.1$ and compute $p_1, p_2, p_3,$ and p_4.
    (c) Is the sequence converging quadratically or linearly?

14. Consider the function $f(x) = \cos(x)$.
    (a) Find the Newton–Raphson formula $g(p_{k-1})$.
    (b) We want to find the root $p = 3\pi/2$. Can we use $p_0 = 3$? Why?
    (c) We want to find the root $p = 3\pi/2$. Can we use $p_0 = 5$? Why?

15. Find the point on the parabola $y = x^2$ that is closest to the point $(3, 1)$. Using the following steps.
    (a) Show that $d(x) = (x - 3)^2 + (x^2 - 1)^2$ is the distance squared between $(3, 1)$ and the point (x, y) on the parabola.
    (b) Show that when the derivative of $d(x)$ is set equal to zero we obtain the equation $f(x) = 4x^3 - 2x - 6 = 0$.
    (c) Start with $p_0 = 1.0$ and find the root of $f(x) = 0$.

16. Find the point on the parabola $y = x^2$ closest to $(1, 3)$.
    (a) Show that $d(x) = (x - 1)^2 + (x^2 - 3)^2$.
    (b) Start with $p_0 = 1.5$ and find the root of $d'(x) = f(x) = 0$.

17. Can Newton–Raphson iteration be used to solve $f(x) = 0$ if $f(x) = x^2 - 14x + 50$? Why?

18. Can Newton–Raphson iteration be used to solve $f(x) = 0$ if $f(x) = x^{1/3}$? Why?

19. Can Newton–Raphson iteration be used to solve $f(x) = 0$ if $f(x) = (x - 3)^{1/2}$ and the starting value is $p_0 = 4$?

20. An open-top box is constructed from a rectangular piece of sheet metal measuring 10 by 16 inches. Squares of what size should be cut from the corners if the volume of the box is 100 cubic inches? *Hint.* If the removed squares measure x inches on a side, the volume of the box will be height $\times$ width $\times$ length $= x(10 - 2x)(16 - 2x)$, so we need to find the roots of $f(x) = 4x^3 - 52x^2 + 160x - 100$.

21. An open-top box is constructed from a rectangular piece of sheet metal measuring 10 by 14 inches. Squares of what size should be cut from the corners if the volume of the box is 100 cubic inches?

22. Consider the function $f(x) = x \exp(-x)$.
    (a) Find the Newton–Raphson formula $g(p_{k-1})$.
    (b) If $p_0 = 0.2$, then find $p_1, \ldots, p_4$. What is $\lim p_k$?
    (c) If $p_0 = 20$, then find $p_1, \ldots, p_4$. What is $\lim p_k$?
    (d) What is the value of $f(p_4)$ in part (c)?

23. Consider the function $f(x) = \arctan(x)$.
    (a) Find the Newton–Raphson formula $g(p_{k-1})$.
    (b) If $p_0 = 1.0$, then find $p_1, \ldots, p_4$. What is $\lim p_k$?
    (c) If $p_0 = 2.0$, then find $p_1, \ldots, p_4$. What is $\lim p_k$?

24. In celestial mechanics, Kepler's equation is $y = x - e \sin(x)$, where y is the planet's mean anomaly, x its eccentric anomaly, and e the eccentricity of its orbit. Let $y = 1$ and $e = 0.5$ and find the root of $2x - 2 - \sin(x) = 0$ in the interval $[0, \pi]$.

25. The natural frequency P for an elastic beam of length L that is fixed at one end and pinned at the other end obeys the relation $\tan(KL) = \tanh(KL)$, where $P = K^2 C$ (C depends on certain measurements of the beam).
   (a) Find the solution to $\tan(x) = \tanh(x)$ near $p_0 = 4$.
   (b) Find the solution to $\tan(x) = \tanh(x)$ near $p_0 = 7$.

26. Prove that the sequence $\{p_k\}$ in equation (5) of Theorem 2.1 converges to p. Use the following steps.
   (a) Show that if p is a fixed point of $g(x)$ in equation (4), then p is a zero of $f(x)$.
   (b) If p is zero of $f(x)$ and $f'(p) \neq 0$, then show that $g'(p) = 0$.
   (c) Use part (b) and Theorem 1.7 to show that the sequence $\{p_k\}$ in equation (5) converges to p.

27. If $h'(x)$ exists at x_0, then prove that $h(x)$ is continuous at x_0.

28. Prove equation (23) of Theorem 2.2. Use the following steps.
   By Theorem 1.2 we can expand $f(x)$ about $x = p_k$ to get

$$f(x) = f(p_k) + f'(p_k)(x - p_k) + \tfrac{1}{2}f''(c_k)(x - p_k)^2.$$

Since p is a zero of $f(x)$, we set $x = p$ and obtain

$$0 = f(p_k) + f'(p_k)(p - p_k) + \tfrac{1}{2}f''(c_k)(p - p_k)^2.$$

   (a) Now assume that $f'(x) \neq 0$ for all x near the root p.
   Use the facts given above and $f'(p_k) \neq 0$ to show that

$$p - p_k + \frac{f(p_k)}{f'(p_k)} = \frac{-f''(c_k)}{2f'(p_k)}(p - p_k)^2.$$

   (b) Assume that $f'(x)$ and $f''(x)$ do not change too rapidly, so that we can use the approximations $f'(p_k) \approx f'(p)$ and $f''(c_k) \approx f''(p)$. Now use part (a) to get

$$e_{k+1} \approx \frac{-f''(p)}{2f'(p)}(e_k)^2.$$

29. Suppose that A is a positive real number.
   (a) Show that A has the representation

$$A = q \times 2^{2m}, \qquad \text{where } \tfrac{1}{4} \leq q < 1 \quad \text{and} \quad m \text{ is an integer.}$$

   (b) Use part (a) to show that the square root is

$$A^{1/2} = q^{1/2} \times 2^m.$$

   Remark. Let $p_0 = (2q + 1)/3$, where $\tfrac{1}{4} \leq q < 1$, and use Newton's formula (10). After three iterations p_3 will be an approximation to $q^{1/2}$ with a precision of 24 binary digits. This is the algorithm that is often used in the computer's hardware to compute square roots.

30. (a) Show that formula (27) for the secant method is algebraically equivalent to

$$p_{k+1} = \frac{p_{k-1}f(p_k) - p_k f(p_{k-1})}{f(p_k) - f(p_{k-1})}.$$

(b) Explain why loss of significance in subtraction makes this formula inferior for computational purposes to the one given in formula (27).

In Exercises 31–33, use the secant method and formula (27) and compute the next two iterates p_2 and p_3.

31. Let $f(x) = x^2 - 2x - 1$.
Start with $p_0 = 2.6$, $p_1 = 2.5$.

32. Let $f(x) = x^3 - x - 3$.
Start with $p_0 = 1.7$, $p_1 = 1.67$.

33. Let $f(x) = x^3 - x + 2$.
Start with $p_0 = -1.5$, $p_1 = -1.52$.

34. Use the accelerated Newton–Raphson iteration in Theorem 2.3 to find the root p of order M for the following;
(a) $f(x) = (x - 2)^5$, $M = 5$, $p = 2$; start with $p_0 = 1$.
(b) $f(x) = \sin(x^3)$, $M = 3$, $p = 0$; start with $p_0 = 1$.
(c) $f(x) = (x^2 - 3)^4$, $M = 4$, $p = 3^{1/2}$; start with $p_0 = 2$.
(d) $f(x) = (x - 1)\ln(x)$, $M = 2$, $p = 1$; start with $p_0 = 2$.

35. Suppose that p is a root of order $M = 2$ for $f(x) = 0$. Prove that the accelerated Newton–Raphson iteration

$$p_k = p_{k-1} - 2\frac{f(p_{k-1})}{f'(p_{k-1})}$$

converges quadratically (cf. Exercise 28).

36. *Olver's method* for finding a zero of $f(x)$ uses the formula

$$p_{k+1} = p_k - \frac{f(p_k)}{f'(p_k)} - \frac{f''(p_k)[f(p_k)]^2}{2[f'(p_k)]^3}.$$

Oliver's method will generally yield cubic $(R = 3)$ convergence. Write a program and experiment to see if this is the case.

37. A Modified Newton–Raphson method for multiple roots.
Fact. If P is a root of multiplicity M, then $f(x) = (x - p)^M q(x)$, where $q(P) \neq 0$.
(a) Show that $h(x) = f(x)/f'(x)$ has a simple root at $x = P$.
(b) Show that when the Newton–Raphson method is applied to finding the simple root P of $h(x)$, we get $g(x) = x - h(x)/h'(x)$ which becomes

$$g(x) = x - \frac{f(x)f'(x)}{[f'(x)]^2 - f(x)f''(x)}.$$

(c) The iteration using $g(x)$ in part (b) converges quadratically to P. Tell why this happens.

2.4 Synthetic Division and Bairstow's Method

Polynomials are used in nearly all areas of numerical analysis and applied mathematics and science. The familiar form for a polynomial $P(x)$ of degree n

involves powers of x,

(1) $$P(x) = a_n x^n + a_{n-1} x^{n-1} + \ldots + a_2 x^2 + a_1 x + a_0,$$

where the numbers a_k are constants and $a_n \neq 0$. One goal of this section is to develop an efficient method for evaluating $P(x)$. The algorithm is known as Horner's method, and can easily be extended to evaluate $P'(x), P''(x), \ldots,$ $P^{(n)}(x)$.

The roots of certain polynomials arise frequently in physical applications. For example, the stability of electrical or mechanical systems is related to the real part of one of the complex roots of a certain polynomial whose degree can sometimes be higher than 5. If $n = 2$, the roots of $P(x)$ can be found by using the quadratic formula, and there exist formulas for degree 3 and 4. But for polynomials of degree $n \geq 5$ there is no explicit formula. The roots can be found using the Newton–Raphson method if complex variables and complex arithmetic is carried out in the calculations. If complex variables are available on the computer, then these changes are easily made. However, many problems in engineering and science involve polynomials with real coefficients, and the necessary calculations can be done using real arithmetic. The method for these cases is called the Newton–Bairstow or Bairstow method and is based on quadratic synthetic division.

The following results are readily available and are a basis for our work with polynomials. The proof of the theorem relies on the study of complex variables, and it is not given.

Theorem 2.4 [Fundamental theorem of algebra] Let $P(x)$ be a polynomial of degree $n \geq 1$ given by (1), with $a_0, a_1, \ldots, a_n$ real or complex numbers and $a_n \neq 0$. Then $P(x)$ has at least one zero; that is, there exists a complex number z such that $P(z) = 0$.

Corollary 2.2 [Factorization involving the roots of $P(x) = 0$] A polynomial $P(x)$ of degree n has n complex roots $x_1, x_2, \ldots, x_n$ (not necessarily distinct). If all these roots are used to form the factors $(x - x_j)$, then $P(x)$ can be expressed as their product:

(2) $$P(x) = a_n(x - x_1)(x - x_2) \ldots (x - x_n).$$

Theorem 2.5 [Case of real coefficients a_i] If the polynomial $P(x)$ of degree n has all real coefficients a_i and $a + bi$ is a root, then the conjugate complex number $a - bi$ is also a root of $P(x)$. Hence the complex roots occur in pairs and the associated quadratic factor for these two roots is

(3) $$x^2 - rx - s = x^2 - 2ax + a^2 + b^2 = (x - a - bi)(x - a + bi).$$

Thus $P(x)$ can be expressed as the product of j linear factors and k quadratic factors of the form (3), where $n = j + 2k$.

(4) $$P(x) = a_n(x - x_1) \ldots (x - x_j)(x^2 - r_1 x - s_1) \ldots (x^2 - r_k x - s_k),$$

where all the numbers $\{x_i\}_{i=1}^{i=j}$ and $\{r_i, s_i\}_{i=1}^{i=k}$ are real numbers.

Example 2.11 Discuss the factorizations for the polynomials

$$P_4(x) = x^4 - 6x^3 + 9x^2 + 4x - 12$$

and

$$P_5(x) = x^5 - x^4 + 3x^3 - 9x^2 + 16x - 10.$$

Solution. The first polynomial has real roots $x_1 = -1$, $x_2 = 2$, $x_3 = 2$, and $x_4 = 3$ and can be expressed as the product

$$P_4(x) = (x + 1)(x - 2)^2(x - 3).$$

The second polynomial has four complex roots $x_1 = -1 + 2i$, $x_2 = -1 - 2i$, $x_3 = 1 + i$, and $x_4 = 1 - i$ and one real root $x_5 = 1$. The quadratic factors $(x^2 + 2x + 5)$ and $(x^2 - 2x + 2)$ correspond to the pairs of roots x_1, x_2 and x_3, x_4, respectively. Hence $P_5(x)$ has the factorization

$$P_5(x) = (x - 1)(x^2 + 2x + 5)(x^2 - 2x + 2).$$

Evaluation of a Polynomial

Let the polynomial $P(x)$ of degree n have the form

(5)
$$P(x) = a_n x^n + a_{n-1} x^{n-1} + \ldots + a_2 x^2 + a_1 x + a_0.$$

We now develop an algorithm known as *Horner's method* or *synthetic division* for evaluating a polynomial. It can be thought of as nested multiplication, that is, $P_5(x)$ can be written in the form

$$P_5(x) = ((((a_5 x + a_4)x + a_3)x + a_2)x + a_1)x + a_0.$$

Since we do not want to enter the parentheses into lines of code in a computer program, we must develop an algorithm that will carry out this intention. The key is to consider how to evaluate $P(z)$, where z is considered a constant. Then the linear term $(x - z)$ can be divided into $P(x)$ and the result will be

(6)
$$P(x) = (x - z)Q_0(x) + R_0.$$

where $Q_0(x)$ is a polynomial of degree $n - 1$ and R_0 is a constant.

If we let $Q_0(x)$ be represented by

(7)
$$Q_0(x) = b_n x^{n-1} + b_{n-1} x^{n-2} + \ldots + b_3 x^2 + b_2 x + b_1$$

and set $R_0 = b_0$, equation (6) can be written

(8)
$$P(x) = (x - z)(b_n x^{n-1} + b_{n-1} x^{n-2} + \ldots + b_3 x^2 + b_2 x + b_1) + b_0.$$

When (8) is expressed in powers of x, we have

(9)
$$P(x) = b_n x^n + (b_{n-1} - z b_n)x^{n-1} + \ldots + (b_2 - z b_3)x^2 + (b_1 - z b_2)x + b_0 - z b_1.$$

The numbers b_k are found by comparing the coefficients of x^k in equations (5) and (9), as shown in Table 2.10.

Table 2.10 Coefficients b_k for Horner's method

x^k	Comparing (5) and (9)	Solving for b_k
x^n	$a_n = b_n$	$b_n = a_n$
x^{n-1}	$a_{n-1} = b_{n-1} - zb_n$	$b_{n-1} = a_{n-1} + zb_n$
$\vdots$	$\vdots$	$\vdots$
x^k	$a_k = b_k - zb_{k+1}$	$b_k = a_k + zb_{k+1}$
$\vdots$	$\vdots$	$\vdots$
x^0	$a_0 = b_0 - zb_1$	$b_0 = a_0 + zb_1$

The value $P(z)$ is b_0 and this can be easily seen from equation (6) and the fact that $R_0 = b_0$:

(10) $$P(z) = (z - z)Q_0(z) + R_0 = 0Q_0(z) + b_0 = b_0.$$

The recursive formulas for b_k are succinctly given by

(11) $$b_n = a_n \quad \text{and then} \quad b_k = a_k + zb_{k+1} \quad \text{for } k = n - 1, \ldots, 2, 1, 0$$

and are easy to use in a computer. A very simple algorithm is:

```
B(N) := A(N)
DO   FOR K=N−1 DOWNTO 0
    B(K) := A(K)+X*B(K+1)
```

The variable z was introduced in the equations above so that the term $(x - z)$ could be used. If the formulas in (11) are used with z replaced by x, the result will be $P(x) = b_0$. When hand calculations are done, it is easier to write the coefficients of $P(x)$ on a line and perform the calculation $b_k = a_k + xb_{k+1}$ below a_k in a column (see Table 2.11).

Table 2.11 Horner's table for the synthetic division process

Input	a_n	a_{n-1}	a_{n-2}	$\cdots$	a_k	$\cdots$	a_2	a_1	a_0
x		xb_n	xb_{n-1}	$\cdots$	xb_{k+1}	$\cdots$	xb_3	xb_2	xb_1
	b_n	b_{n-1}	b_{n-2}	$\cdots$	b_k	$\cdots$	b_2	b_1	$b_0 = P(x)$
									Output

Example 2.12 Evaluate $P(3)$ for the polynomial

$$P(x) = x^5 - 6x^4 + 8x^3 + 8x^2 + 4x - 40.$$

Solution. The coefficients $a_5 = 1, a_4 = -6, a_3 = 8, a_2 = 8, a_1 = 4$, and $a_0 = -40$ and the value $x = 3$ are used in Table 2.11 to compute the coefficients b_k and the value $P(3) = b_0$.

	a_5	a_4	a_3	a_2	a_1		a_0
Input	1	-6	8	8	4		-40
$x = 3$		3	-9	-3	15		57
	1	-3	-1	5	19		$17 = P(3) = b_0$
	b_5	b_4	b_3	b_2	b_1		Output

Therefore, $P(3) = 17$.

Calculation of the Derivative

Let the polynomial $P(x)$ of degree n have the form (5) and let z be fixed; then differentiation of both sides of the equation in (6) with respect to x results in

(12) $$P'(x) = Q_0(x) + (x - z)Q_0'(x).$$

Evaluating $P'(z)$, we have

(13) $$P'(z) = Q_0(z) + (z - z)Q_0'(z) = Q_0(z).$$

Synthetic division can be used to compute $Q_0(z)$. Using the same approach as before, we divide the term $(x - z)$ into $Q_0(x)$ to get

(14) $$Q_0(x) = (x - z)Q_1(x) + R_1.$$

If we let $Q_1(x)$ have the representation

(15) $$Q_1(x) = c_n x^{n-2} + c_{n-1} x^{n-3} + \ldots + c_3 x + c_2$$

and set $R_1 = c_1$, then equation (14) can be written

(16) $$Q_0(x) = (x - z)(c_n x^{n-2} + c_{n-1} x^{n-3} + \ldots + c_3 x + c_2) + c_1.$$

The numbers c_k are found by expanding (16) into powers of x and then comparing the coefficients of x^k with the ones given in equation (7). The recursive formulas for computing the coefficients of $Q_1(z)$ are

(17) $$c_n = b_n \quad \text{and then} \quad c_k = b_k + z c_{k+1} \quad \text{for } k = n - 1, \ldots, 2, 1.$$

The value $P'(z)$ is c_1 and this can be seen from equations (13) and (14) and the fact that $R_1 = c_1$:

(18) $$P'(z) = Q_0(z) = (z - z)Q_1(z) + R_1 = 0 + R_1 = c_1.$$

The variable z was introduced so that we could use the factor $(x - z)$. If the formulas in (17) are used with x replaced for z, the result will be $P'(x) = c_1$. When hand calculations are done, it is easier to write the coefficients a_k on a line and compute the sums $b_k = a_k + x b_{k+1}$ and $c_k = b_k + x c_{k+1}$ in a column (see Table 2.12).

Table 2.12 Horner's table for finding $P(x)$ and $P'(x)$

Input	a_n	a_{n-1}	a_{n-2}	$\cdots$	a_k	$\cdots$	a_2	a_1	a_0
x		xb_n	xb_{n-1}	$\cdots$	xb_{k+1}	$\cdots$	xb_3	xb_2	xb_1
	b_n	b_{n-1}	b_{n-2}	$\cdots$	b_k	$\cdots$	b_2	b_1	$b_0 = P(x)$
		xc_n	xc_{n-1}	$\cdots$	xc_{k+1}	$\cdots$	xc_3	xc_2	
	c_n	c_{n-1}	c_{n-2}	$\cdots$	c_k	$\cdots$	c_2	$c_1 = P'(x)$	

Example 2.13 Evaluate $P'(3)$ for the polynomial

$$P(x) = x^5 - 6x^4 + 8x^3 + 8x^2 + 4x - 40.$$

Solution. The coefficients $b_5 = 1$, $b_4 = -3$, $b_3 = -1$, $b_2 = 5$, $b_1 = 19$, and $b_0 = 17$ were found in Example 2.12. Using $x = 3$ in Table 2.12 to compute the coefficients c_k, we have

Input	1	-6	8	8	4	-40
$x = 3$		3	-9	-3	15	57
	1	-3	-1	5	19	$17 = P(3) = b_0$
		3	0	-3	6	
	1	0	-1	2	$25 = P'(3) = c_1$	

Therefore, $P'(3) = 25$.

Remark. Horner's method can be extended to find higher derivatives. Exercises 7–11 show how to make and use these extensions.

Our immediate goal is to incorporate Horner's method in Newton's method, so that we can find the real roots of a polynomial. Start with an initial approximation p_0 and use formulas (11) and (17) to compute $b_0 = P(p_0)$ and $c_1 = P'(p_0)$; then use the Newton-Raphson formula $p_1 = p_0 - b_0/c_1$ to find the next approximation.

Example 2.14 Start with the initial approximation $p_0 = 3$ for the real root of the polynomial

$$P(x) = x^5 - 6x^4 + 8x^3 + 8x^2 + 4x - 40.$$

Use Horner's method with Newton's method and compute the next two approximations, p_1 and p_2.

Solution. Using Examples 2.12 and 2.13, we have

$$p_1 = 3 - \frac{P(3)}{P'(3)} = 3 - \frac{17}{25} = 2.32.$$

Now use $x = 2.32$ in Table 2.12 to compute $P(2.32)$ and $P'(2.32)$.

Input	1	−6.00	8.0000	8.0000	4.0000	−40.0000
2.32		2.32	−8.5376	−1.2472	15.6664	45.6261
	1	−3.68	−0.5376	6.7528	19.6664	$5.6261 = P(2.32)$
		2.32	−3.1552	−8.5673	−4.2096	
	1	−1.36	−3.6928	−1.8145	$15.4568 = P'(2.32)$	

We find the next approximation p_2 with the calculation

$$p_2 = 2.32 - \frac{P(2.32)}{P'(2.32)} = 2.32 - \frac{5.6261}{15.45682} = 1.9560.$$

If more decimal places were used, $p_2 = 1.9560095$ and further calculations would produce $p_3 = 1.9992543$ and $p_4 = 1.9999998$. This sequence is converging to the root $x = 2$.

The Deflated Polynomial

When the iteration $p_1, p_2, \ldots$ in Example 2.14 is continued, it will converge to the zero $z = 2$ of $P(x)$. Synthetic division can be used to factor $P(x)$ as the product $P(x) = (x - 2)Q(x)$, that is,

Input	1	−6	8	8	4	−40
$x = 2$			2	−8	0	16	40
	1	−4	0	8	20	$0 = P(2)$

If the other roots of $P(x)$ are to be found, then they must be roots of quotient polynomial $Q(x) = x^4 - 4x^3 + 0x^2 + 8x + 20$. The polynomial $Q(x)$ is referred to as the *deflated polynomial*, because its degree is one less than that of $P(x)$. In Example 2.14 it is possible to factor $Q(x)$ as the product of two quadratic polynomials, that is, $Q(x) = (x^2 + 2x + 2)(x^2 - 6x + 10)$. Therefore, the polynomial $P(x)$ has the factorization

$$P(x) = (x - 2)(x^2 + 2x + 2)(x^2 - 6x + 10),$$

and the five roots of $P(x)$ are $x = 2, -1 \pm i, 3 \pm i$.

The Pitfalls for Finding Roots

The reader should be aware that sometimes the calculation of roots of a polynomial is an ill-conditioned problem. Higher-degree polynomials can be unstable in the sense that small changes in the coefficients will result in large changes in the computed roots. Also, at a multiple root there is a loss of significant digits due to round-off error. In this case the function value $P(x_k)$ and the correction term $P(x_k)/P'(x_k)$ in Newton's method can fail to have enough accuracy to make x_{k+1} closer to the root than x_k. Generally speaking, when we calculate a multiple root, we obtain only about half the accuracy of which the computer is capable.

Example 2.15 Root finding near $x = 1$ is ill conditioned for

$$P(x) = x^5 - 6x^4 + 8x^3 + 2x^2 - 9x + 4 = (x + 1)(x - 1)^3(x - 4).$$

Solution. Start with the initial guess $x_0 = 1.005$, carry only five decimal places in the calculations, and compute x_1.

1.00000	−6.00000	8.00000	2.00000	−9.00000	4.00000
	1.00500	−5.01998	2.99492	5.01989	−4.00001
1.00000	−4.99500	2.98002	4.99492	−3.98011	$-0.00001 = P(x_0)$
	1.00500	−4.00995	−1.03508	3.97964	
1.00000	−3.99000	−1.02993	3.95984	$-0.00047 = P'(x_0)$	

Hence the next approximation is

$$x_1 = 1.005 - \frac{-0.00001}{-0.00047} = 0.98372.$$

The errors in the approximations are seen to be

$$e_0 = 1.00500 - 1.00000 = 0.00500 \quad \text{and} \quad e_1 = 0.98372 - 1.00000 = -0.01628.$$

Therefore, the correction step in Newton's method seems to have failed. Actually it is a case of round-off error. If the factored form of $P(x)$ is used, $P(1.005) = -0.0000007506218$. It was the synthetic division process that failed, because only five decimal places were used and the answer was quantized in multiples of 0.00001 and hence the computed value of $P(1.005)$ was -0.00001. Unfortunately, there is no way to guarantee that a small function value is accurate if there are additions or subtractions involved.

Algorithm 2.6 [*Find the Real Roots of a Polynomial*] Synthetic division is used to compute $P(x)$ and $P'(x)$ for the polynomial

$$P(x) = a_N x^N + a_{N-1} x^{N-1} + \ldots + a_2 x^2 + a_1 x + a_0.$$

Newton's method is used to find real roots. If a root z is found, then $P(x) = (x - z)Q(x)$ and the quotient polynomial $Q(x)$ is used to find the next real root. This is called *deflation*.

```
Delta := 10⁻⁶, Epsilon := 10⁻⁶, Small := 10⁻⁶          {Tolerances}
Satisfied := "False"                                {Condition for loop termination}
Max := 99                                           {Maximum number of iterations}


INPUT N ; A(0), A(1), . . . , A(N)                     {Input coefficients}
INPUT Pold                                             {Input approximation}
B(N+1) := 0, C(N+1) := 0                                {Initialize variables}


DO  FOR J = 1 TO Max UNTIL Satisfied = "True"
    FOR  K = N DOWN TO 0 DO                             {Compute the values
        B(K) := A(K) + Pold*B(K+1)                         B(0) = P(X)
        C(K) := B(K) + Pold*C(K+1)                         C(1) = P'(X)}
    DX := B(0)/C(1)                                     {Change in X}
    Pnew := Pold − DX                                  {Newton–Raphson}
    RelErr := 2*|DX|/[|Pnew|+|Pold|]                   {Relative error}
    IF  RelErr≤Delta AND |B(0)|<Epsilon  THEN          {Check for
        Satisfied := "True" , Xroot := Pnew            convergence}
    Pold := Pnew                                       {Update iterate}


IF  Satisfied = "True"  THEN                            {Output}
    PRINT "The computed root of P(x) is" Xroot
    FOR  K = 0 TO N−1 DO                               {Deflate the polynomial,
        A(K) := B(K+1)                                 i.e., replace P(x) with Q(x);
    N := N−1                                           the degree of Q(x) is N−1}
ENDIF


IF  Satisfied ≠ "True"  THEN
    PRINT "Newton's method failed to find a root."
```

Quadratic Synthetic Division (Optional)

Let the polynomial $P(x)$ of degree n have the form (5) and let $x^2 - rx - s$ be a fixed quadratic term. Then $P(x)$ can be expressed

$$(19) \qquad P(x) = (x^2 - rx - s)Q(x) + u(x - r) + v,$$

where the terms $u(x - r) + v$ form the remainder when $P(x)$ is divided by $(x^2 - rx - s)$. Here $Q(x)$ is a polynomial of degree $n - 2$ and can be represented by

$$(20) \qquad Q(x) = b_n x^{n-2} + b_{n-1} x^{n-3} + \ldots + b_4 x^2 + b_3 x + b_2.$$

If we set $b_1 = u$ and $b_0 = v$, then equation (19) can be written

(21) $P(x) = (x^2 - rx - s)(b_n x^{n-2} + b_{n-1}x^{n-3} + \ldots + b_3 x + b_2) + b_1(x - r) + b_0.$

The terms can be expanded so that $P(x)$ is represented in powers of x:

$$P(x) = b_n x^n + (b_{n-1} - rb_n)x^{n-1} + (b_{n-2} - rb_{n-1} - sb_n)x^{n-2}$$

(22) $$+ \ldots + (b_k - rb_{k+1} - sb_{k+2})x^k + \ldots + (b_1 - rb_2 - sb_3)x$$

$$+ b_0 - rb_1 - sb_2.$$

The numbers b_k are found by comparing the coefficients of x^k in equations (22) and (5). The recursive formulas for computing the coefficients b_k of $Q(z)$ are

(23)
$$b_n = a_n, \quad b_{n-1} = a_{n-1} + rb_n \quad \text{and then}$$

$$b_k = a_k + rb_{k+1} + sb_{k+2} \qquad \text{for } k = n - 2, n - 3, \ldots, 1, 0.$$

When hand calculations are done, it is easier to write the coefficients a_k on a line and compute the sum $b_k = a_k + sb_{k+2} + rb_{k+1}$ below it. Notice that each new number b_k is multiplied by s and r and entered in the next two columns, except for b_1, which is entered only once, and b_0, which is the last number we compute (see Table 2.13).

Table 2.13   Table for quadratic synthetic division

Inputs	a_n	a_{n-1}	a_{n-2}	a_{n-3}	$\ldots$	a_k	$\ldots$	a_2	a_1	a_0
s			sb_n	sb_{n-1}	$\ldots$	sb_{k+2}	$\ldots$	sb_4	sb_3	sb_2
r		rb_n	rb_{n-1}	rb_{n-2}	$\ldots$	rb_{k+1}	$\ldots$	rb_3	rb_2	rb_1
	b_n	b_{n-1}	b_{n-2}	b_{n-3}	$\ldots$	b_k	$\ldots$	b_2	b_1	b_0
									Outputs	

Example 2.16 Use quadratic synthetic division to show how to divide $P(x) = x^5 + 6x^4 - 20x^2 + 22x + 8$ by $(x^2 + 2x - 3)$.

Solution. In this case we use $r = -2$ and $s = 3$ in Table 2.13.

Inputs	1	6	0	−20	22	8
$s = 3$			3	12	−15	6
$r = -2$		−2	−8	10	−4	−6
	1	4	−5	2	3	8
					b_1	b_0
					Outputs	

Therefore, the polynomial $P(x)$ can be factored

$$P(x) = (x^2 + 2x - 3)(x^3 + 4x^2 - 5x + 2) + 3(x + 2) + 8.$$

Now let us build on the previous idea and develop Bairstow's method for finding a quadratic factor $x^2 - rx - s$ of $P(x)$. Suppose that we start with an initial guess

(24)
$$x^2 - r_0 x - s_0$$

and that $P(x)$ can be expressed

(25)
$$P(x) = (x^2 - r_0 x - s_0)Q(x) + u(x - r_0) + v.$$

When u and v are small, the quadratic (24) is close to a factor of $P(x)$. We want to find new values r_1 and s_1 so that

(26)
$$x^2 - r_1 x - s_1$$

is closer to a factor of $P(x)$ than the quadratic in (24). Observe that u and v in (25) are functions of r and s, that is,

(27)
$$u = u(r, s) \quad \text{and} \quad v = v(r, s).$$

The new values r_1 and s_1 satisfy the relations

(28)
$$r_1 = r_0 + \Delta r \quad \text{and} \quad s_1 = s_0 + \Delta s.$$

The differentials of the functions u and v give the approximations

(29)
$$v(r_1, s_1) \approx v(r_0, s_0) + v_r(r_0, s_0)\,\Delta r + v_s(r_0, s_0)\,\Delta s,$$
$$u(r_1, s_1) \approx u(r_0, s_0) + u_r(r_0, s_0)\,\Delta r + u_s(r_0, s_0)\,\Delta s.$$

The new values r_1 and s_1 are to satisfy

(30)
$$u(r_1, s_1) = 0 \quad \text{and} \quad v(r_1, s_1) = 0.$$

When the quantities Δr and Δs are small, the approximations in (29) together with (30) are used to show that the quantities Δr and Δs should be chosen to be solutions of the linear system

(31)
$$0 = v(r_0, s_0) + v_r(r_0, s_0)\,\Delta r + v_s(r_0, s_0)\,\Delta s,$$
$$0 = u(r_0, s_0) + u_r(r_0, s_0)\,\Delta r + u_s(r_0, s_0)\,\Delta s.$$

If the values of the partial derivatives in (31) are known, Δr and Δs can be found by using Cramer's rule, and then the new values r_1 and s_1 are computed using the equations in (28). Let us announce that the partial derivatives in (31) are

(32)
$$v_r = c_1, \quad v_s = c_2, \quad u_r = c_2, \quad u_s = c_3,$$

where the coefficients c_k are given by the recursive formulas

(33)
$$c_n = b_n, \quad c_{n-1} = b_{n-1} + rc_n \quad \text{and then}$$
$$c_k = b_k + rc_{k+1} + sc_{k+2} \qquad \text{for } k = n - 2, n - 3, \ldots, 2, 1.$$

The formulas in (33) use the coefficients b_k that were computed with equations (23). Since $u(r_0, s_0) = b_1$ and $v(r_0, s_0) = b_0$, the linear system (31) can be rewritten as

(34)
$$c_1 \, \Delta r + c_2 \, \Delta s = -b_0,$$
$$c_2 \, \Delta r + c_3 \, \Delta s = -b_1.$$

Cramer's rule can be used to solve the linear system in (34) (see Section 3.2). The required determinants are

$$D = \begin{vmatrix} c_1 & c_2 \\ c_2 & c_3 \end{vmatrix} \qquad D_1 = \begin{vmatrix} -b_0 & c_2 \\ -b_1 & c_3 \end{vmatrix} \qquad D_2 = \begin{vmatrix} c_1 & -b_0 \\ c_2 & -b_1 \end{vmatrix},$$

and the new values r_1 and s_1 are computed using

(35)
$$r_1 = r_0 + \frac{D_1}{D} \quad \text{and} \quad s_1 = s_0 + \frac{D_2}{D}.$$

The iterative process is continued until good approximations to r and s have been found. If the initial guesses r_0 and s_0 are chosen small, the iteration does not tend to wander for a long time before converging. When $x \approx 0$, the larger powers of x can be neglected in equation (5) and we have the approximation

(36)
$$0 \approx P(x) \approx a_2 x^2 + a_1 x + a_0.$$

Hence the initial guesses could be

(37)
$$r_0 = -\frac{a_1}{a_2} \quad \text{and} \quad s_0 = -\frac{a_0}{a_2} \qquad \text{provided that } a_2 \neq 0.$$

If hand calculations are done, Table 2.13 can be extended to form an easy way to the calculate c_k's (see Table 2.14). Bairstow's method is a special case of Newton's method in two dimensions. In Chapter 8 we will develop Newton's method in two dimensions.

Table 2.14 Table for Bairstow's method

Inputs	a_n	a_{n-1}	a_{n-2}	a_{n-3}	$\cdots$	a_3	a_2	a_1	a_0
s			sb_n	sb_{n-1}	$\cdots$	sb_5	sb_4	sb_3	sb_2
r		rb_n	rb_{n-1}	rb_{n-2}	$\cdots$	rb_4	rb_3	rb_2	rb_1
	b_n	b_{n-1}	b_{n-2}	b_{n-3}	$\cdots$	b_3	b_2	b_1	b_0
			sc_n	sc_{n-1}	$\cdots$	sc_5	sc_4	sc_3	
		rc_n	rc_{n-1}	rc_{n-2}	$\cdots$	rc_4	rc_3	rc_2	
	c_n	c_{n-1}	c_{n-2}	c_{n-3}	$\cdots$	c_3	c_2	c_1	

Example 2.17 Given $P(x) = x^4 + x^3 + 3x^2 + 4x + 6$. Start with $r_0 = -2.1$, $s_0 = -1.9$ and use Bairstow's method to find $r_1, s_1, r_2, s_2, \ldots$, and the quadratic factors and roots of $P(x)$.

Solution. The table for calculating the r_1 and s_1 is

Inputs	1.0000	1.0000	3.0000	4.0000	6.0000
$s = -1.9$			−1.9000	2.0900	−6.4790
$r = -2.1$		−2.1000	2.3100	−7.1610	2.2491
	1.0000	−1.1000	3.4100 $\quad b_1 = -1.0710$	$b_0 = 1.7701$	
			−1.9000	6.0800	
		−2.1000	6.7200	−17.2830	
	1.0000 $\quad c_3 = -3.2000$	$c_2 = 8.2300$	$c_1 = -12.2740$		

The resulting linear system involving Δr and Δs is

$$-12.2740\ \Delta r + 8.2300\ \Delta s = -1.7701,$$
$$8.2300\ \Delta r + 3.2000\ \Delta s = 1.0710.$$

The determinants used in Cramer's rule are

(38)

$$D = \begin{vmatrix} -12.2740 & 8.2300 \\ 8.2300 & -3.2000 \end{vmatrix} = -28.4561000,$$

$$D_1 = \begin{vmatrix} -1.7701 & 8.2300 \\ 1.0710 & -3.2000 \end{vmatrix} = -3.15001000,$$

$$D_2 = \begin{vmatrix} -12.2740 & -1.7701 \\ 8.2300 & 1.0710 \end{vmatrix} = 1.42246900.$$

Finally, the new values r_1 and s_1 are found by using (35) and (38):

$$r_1 = -2.1 + \frac{-3.150010}{-28.4561} = -1.98930282,$$

$$s_1 = -1.9 + \frac{1.422469}{-28.4561} = -1.94998819.$$

Another iteration will produce $r_2 = -1.99999277$, $s_2 = -2.00015098$. The sequences are converging to $r = -2$, $s = -2$ and $P(x)$ has the factorization $P(x) = (x^2 + 2x + 2)(x^2 - x + 3)$. The four complex roots are $1 + i$, $1 - i$, $0.5 + i \times 1.65831239$, and $0.5 - i \times 1.65831239$.

Derivation of Bairstow's Recursive Formulas

We now derive the formulas that were announced in (33). The trick is to differentiate the equations in (23) with respect to r and s. To start, we notice that $b_n = a_n$ is a constant, so that its partial derivatives are zero. Continue down the list and get

$$\frac{\partial}{\partial r} b_n = 0, \qquad\qquad \frac{\partial}{\partial s} b_{n-1} = 0,$$

(39)
$$\frac{\partial}{\partial r} b_{n-1} = b_n, \qquad\qquad \frac{\partial}{\partial s} b_{n-2} = b_n + r \frac{\partial}{\partial s} b_{n-1} = b_n,$$

$$\frac{\partial}{\partial r} b_{n-2} = b_{n-1} + r \frac{\partial}{\partial r} b_{n-1}, \qquad \frac{\partial}{\partial s} b_{n-3} = b_{n-1} + r \frac{\partial}{\partial s} b_{n-2} + s \frac{\partial}{\partial s} b_{n-1}$$

$$= b_{n-1} + r b_n, \qquad\qquad = b_{n-1} + r b_n.$$

Using the product rule and differentiating the general term in (23) with respect to r and s, we find that

(40)
$$\frac{\partial}{\partial r} b_k = 0 + b_{k+1} + r \frac{\partial}{\partial r} b_{k+1} + 0 + s \frac{\partial}{\partial r} b_{k+2}$$

and

(41)
$$\frac{\partial}{\partial s} b_{k-1} = 0 + 0 + r \frac{\partial}{\partial s} b_k + b_{k+1} + s \frac{\partial}{\partial s} b_{k+1}.$$

Starting with (39) and then using (40) and (41), it follows that

(42)
$$\frac{\partial}{\partial r} b_k = \frac{\partial}{\partial s} b_{k-1} \qquad \text{for } k = n, n-1, \ldots, 2, 1.$$

If we define c_{k+1} to be the common term in (42), then (40) can be used to show that

(43)
$$c_{k+1} = \frac{\partial}{\partial r} b_k = b_{k+1} + r \frac{\partial}{\partial r} b_{k+1} + s \frac{\partial}{\partial r} b_{k+2} = b_{k+1} + r c_{k+2} + s c_{k+3}.$$

A compact method for calculating is obtained if we set

(44)
$$b_{k+1} = 0, \quad b_{k+2} = 0, \quad c_{k+1} = 0, \quad c_{k+2} = 0,$$

and use (23) and (43) to get

(45)
$$b_{k+1} = a_k + r b_{k+1} + s b_{k+2} \quad \text{and then}$$

$$c_{k+1} = b_k + r c_{k+1} + s c_{k+2} \quad \text{for } k = n, n-1, \ldots, 1, 0.$$

When roots of polynomials are desired, it is best first to find all the real roots by using Algorithm 2.6, and deflate the polynomial each time a root is found. Then Bairstow's method can be used to find the irreducible quadratic factors.

Algorithm 2.7 [*Bairstow's Method*] Find a quadratic factor of
$$P(x) = a_N x^N + a_{N-1} x^{N-1} + \ldots + a_2 x^2 + a_1 x + a_0.$$
A quadratic $x^2 - rx - s$ is found so that $P(x) = [x^2 - rx - s]Q(x)$, where $Q(x)$ is a polynomial of degree $N - 2$.

```
Delta := 10⁻⁵                                              {Tolerance}
INPUT N ; A(0), A(1), ... , A(N)                      {Input coefficients}
INPUT R, S                                       {Initial approximations}
B(N+1) := B(N+2) := 0, C(N+1) := 0, C(N+2) := 0           {Initialize}

DO   FOR  J = 1   TO   199   UNTIL   Error ≤ Delta
     FOR  K = N   DOWNTO  0   DO                          {Bairstow's
         B(K) := A(K) + R*B(K+1) + S*B(K+2)                recursive
         C(K) := B(K) + R*C(K+1) + S*C(K+2)               formulas}
     Det  := C(3)*C(1) − C(2)*C(2)                        {Compute
     DetR := C(2)*B(1) − C(3)*B(0)                 the determinants for
     DetS := C(2)*B(0) − C(1)*B(1)                    Cramer's rule}
        IF   Det = 0   THEN                            {Check division
               Dr := 0.05*R, Ds := 0.05*S                   by zero}
           ELSE   Dr := DetR/Det                     {Compute changes
               Dr := DetR/Det, Ds := DetS/Det           in R and S}
        ENDIF
     Error := |Dr| + |Ds|                              {Absolute error}
     R := R + Dr, S := S + Ds                           {Update iterates}

IF   Error ≤ Delta   THEN                                     {Output}
     PRINT "A quadratic factor is" X² − RX − S
     FOR  K = 0   TO   N−2                        {Deflate the polynomial,
          A(K) := B(K+2)                       i.e., replace P(X) with Q(X);
     N := N−2                                  the degree of Q(X) is N−2}

IF   Error > Delta   THEN                                     {Output}
     PRINT "Bairstow's method did not find a quadratic factor."
```

Exercises for Synthetic Division and Bairstow's Method

In Exercises 1–6:
- (a) Use Horner's method to show that z_0 is a root of $P(x)$, and find the polynomial $Q_0(x)$ so that $P(x) = (x - z_0)Q_0(x)$.
- (b) Show that $Q_0(z_1) = 0$ and find $Q_1(x)$ so that $Q_0(x) = (x - z_1)Q_1(x)$.

(c) From parts (a) and (b) observe that $P(x) = (x - z_0)(x - z_1)Q_1(x)$. Use the quadratic formula on $Q_1(x)$ to find the other roots.

(d) Start with p_0 and use Newton's method to find p_1 and p_2.

(e) Use a computer to find the roots of $P(x)$.

1. $P(x) = x^4 + x^3 - 13x^2 - x + 12$.
(a) $z_0 = 3$; (b) $z_1 = -1$; (c) find the remaining roots; (d) $p_0 = 1.1$.

2. $P(x) = x^4 + x^3 - 21x^2 - x + 20$.
(a) $z_0 = 4$; (b) $z_1 = -1$; (c) find the remaining roots; (d) $p_0 = 1.1$.

3. $P(x) = x^4 - 2x^3 - 7x^2 + 8x + 12$.
(a) $z_0 = 3$; (b) $z_1 = 2$; (c) find the remaining roots; (d) $p_0 = 2.1$.

4. $P(x) = x^4 - 4x^3 - 3x^2 + 14x - 8$.
(a) $z_0 = -2$; (b) $z_1 = 4$; (c) find the remaining roots; (d) $p_0 = 1.1$ (near a double root).

5. $P(x) = x^4 - 6x^3 + 9x^2 + 4x - 12$.
(a) $z_0 = 2$; (b) $z_1 = -1$; (c) find the remaining roots; (d) $p_0 = 2.1$ (near a double root).

6. $P(x) = x^4 - 5x^3 + 6x^2 + 4x - 8$.
(a) $z_0 = -1$; (b) $z_1 = 2$; (c) find the remaining roots; (d) $p_0 = 2.1$ (near a triple root).

7. *Higher Derivatives.* Let $P(x)$ be a polynomial of degree n. Let z be a fixed real number. Show that repeated synthetic division can be used to compute $P^{(k)}(z)/k!$. *Hint.* Show that

$$P(x) = Q_0(x)(x - z) + R_0, \qquad R_0 = \frac{P(z)}{0!},$$

$$Q_0(x) = Q_1(x)(x - z) + R_1, \qquad R_1 = \frac{P'(z)}{1!},$$

$$Q_1(x) = Q_2(x)(x - z) + R_2, \qquad R_2 = \frac{P''(z)}{2!},$$

$$\vdots$$

$$Q_{n-2}(x) = Q_{n-1}(x)(x - z) + R_{n-1}, \qquad R_{n-1} = \frac{P^{(n-1)}(z)}{(n-1)!},$$

$$Q_{n-1}(x) = Q_n(x)(x - z) + R_n, \qquad R_n = \frac{P^{(n)}(z)}{n!}.$$

In Exercises 8–11:

(a) Use the synthetic division method in Exercise 7 to find $P(2)$, $P'(2)$, $P''(2)$, $P^{(3)}(2)$, and $P^{(4)}(2)$.

(b) Find the Taylor polynomial expansion for $P(x)$ in powers of $(x - 2)$.

8. $P(x) = x^4 - 4x^3 - 3x^2 + 14x - 8$

9. $P(x) = x^4 - 2x^3 - 7x^2 + 8x + 12$

10. $P(x) = x^4 - 6x^3 + 9x^2 + 4x - 12$

11. $P(x) = x^4 - 5x^3 + 6x^2 + 4x - 8$

12. Given that the roots of $x^4 + x^3 - 13x^2 - x + 12 = 0$ are $-4, -1, 1$, and 3, find the roots of $1 + y - 13y^2 - y^3 + 12y^4 = 0$. *Hint.* Set $y = 1/x$.

13. Given that the roots of $x^4 + x^3 - 21x^2 - x + 20 = 0$ are $-5, -1, 1$, and 4, find the roots of $1 + y - 21y^2 - y^3 + 20y^4 = 0$.

14. In Chapter 7, Gauss–Legendre integration will be discussed. The method relies on knowing the zeros of the Legendre polynomials. Use a computer to find the zeros of the following Legendre polynomials.
    (a) $P_4(x) = (35x^4 - 30x^2 + 3)/8$
    (b) $P_5(x) = (63x^5 - 70x^3 + 15x)/8$
    (c) $P_6(x) = (693x^6 - 945x^4 + 315x^2 - 15)/48$

15. The Chebyshev polynomials are used in approximation theory; their zeros are nodes for certain approximating polynomials. Use a computer to find the zeros of the following Chebyshev polynomials.
    (a) $P_4(x) = 8x^4 - 8x^2 + 1$
    (b) $P_5(x) = 16x^5 - 20x^3 + 5x$
    (c) $P_6(x) = 32x^6 - 48x^4 + 18x^2 - 1$

16. *Modulus of the Roots of a Polynomial.* Prove the theorem.

    Theorem. Let $P(x)$ be a polynomial of degree n whose leading coefficient is 1, that is,

    $$P(x) = x^n + a_{n-1}x^{n-1} + a_{n-2}x^{n-2} + \ldots + a_2x^2 + a_1x + a_0,$$

    and set $T = 1 + \max\{|a_{n-1}|, |a_{n-2}|, \ldots, |a_1|, |a_0|\}$. If z is a root of $P(x)$, then $|z| < T$.
    Hint. Assume that $|z| \geq T$ and use synthetic division and the formulas in (11) to show that $|P(z)| = |b_0| \geq 1$.

17. Use the theorem in Exercise 16 to find a bound for the magnitude of the roots of the polynomial

    $$P(x) = x^5 - 6x^4 + 8x^3 + 2x^2 - 9x + 4.$$

    In Exercises 18–24:
    (a) Use quadratic synthetic division to find the quotient polynomial and the remainder term when $P(x)$ is divided by the given quadratic.
    (b) Start with r_0, s_0 and use Bairstow's method to find r_1, s_1 and r_2, s_2.
    (c) Use a computer to find a quadratic factor of $P(x)$.

18. $P(x) = x^4 + 2x^3 - x^2 - 2x + 10$
    (a) $x^2 - x + 1$ (b) $r_0 = 1, s_0 = -1$

19. $P(x) = x^4 - 2x^3 + 5x^2 - 4x + 40$
    (a) $x^2 + x + 4$ (b) $r_0 = -1, s_0 = -4$

20. $P(x) = x^4 + 2x^3 + 2x^2 + 10x + 25$
    (a) $x^2 + 4x + 4$ (b) $r_0 = -4, s_0 = -4$

21. $P(x) = x^4 + 4x^3 + 3x^2 + 10x + 50$
    (a) $x^2 - x + 4$ (b) $r_0 = 1, s_0 = -4$

22. $P(x) = x^5 - 3x^4 + 3x^3 - 3x^2 - 8x - 30$
    (a) $x^2 + 2x + 1$ (b) $r_0 = -2, s_0 = -1$

23. $P(x) = x^6 + 2x^5 + 8x^4 + 12x^3 + 37x^2 + 50x + 50$
    (a) $x^2 + 2x + 3$ (b) $r_0 = -2, s_0 = -3$

24. $P(x) = x^6 + 4x^5 + 8x^4 + 6x^3 + x^2 + 10x + 50$
    (a) $x^2 + 2x + 3$ (b) $r_0 = -2, s_0 = -3$

25. Write a report on the quotient-difference algorithm.

26. Assume that $a \neq 0$ and consider the quadratic equation

$$ax^2 + bx + c = 0.$$

The quadratic formula states that the roots are given by

$$x_1 = \frac{-b + (b^2 - 4ac)^{1/2}}{2a} \qquad x_2 = \frac{-b - (b^2 - 4ac)^{1/2}}{2a}.$$

Show that an equivalent way to calculate the roots is

$$x_1 = \frac{-2c}{b + (b^2 - 4ac)^{1/2}} \qquad x_2 = \frac{-2c}{b - (b^2 - 4ac)^{1/2}}.$$

Remark. If $b > 0$ then the second formula for x_1 should be used, and x_2 is computed with the first formula. However, if $b < 0$ then the first formula for x_1 and second formula for x_2 should be used.

27. Use the quadratic formula and its equivalent version given in Exercise 26 to find the roots of the quadratic equations.
    (a) $x^2 - 1,000.001x + 1 = 0$
    (b) $x^2 - 10,000.0001x + 1 = 0$
    (c) $x^2 - 100,000.00001x + 1 = 0$
    (d) $x^2 - 1,000,000.000001x + 1 = 0$

28. Study Exercises 26 and 27 and write an algorithm that will accurately compute the roots of a quadratic equation for the cases when $|b| \approx (b^2 - 4ac)^{1/2}$.

2.5 Aitken's Process and Steffensen's and Muller's Methods (Optional)

In Section 2.3 we saw that Newton's method converged slowly at a multiple root and the sequence of iterates $\{p_n\}$ exhibited linear convergence. Theorem 2.3 showed how to speed up convergence, but it depends on knowing the order of the root in advance.

Aitken's Process

An accelerating technique called *Aitken's Δ^2 process* can be used to speed up the convergence of any sequence that is linearly convergent. Assume that $\{p_n\}$ is *linearly convergent*, that is,

(1) $$\lim_{n \to \infty} p_n = p,$$

and the error terms $e_n = p - p_n$ satisfy the relation

(2)
$$\lim_{n \to \infty} \frac{e_{n+1}}{e_n} = A, \qquad \text{where } 0 < |A| < 1.$$

We shall construct a sequence $\{q_n\}$ that converges to p faster than the original sequence. First form the differences

(3)
$$\frac{p - p_{n+1}}{p - p_n} = A_n \quad \text{and} \quad \frac{p - p_{n+2}}{p - p_{n+1}} = A_{n+1}.$$

Suppose that n is large enough so that $A_n \approx A$ and $A_{n+1} \approx A$ in (3). Then $A_n \approx A_{n+1}$, so that formally we assume that $A_n = A_{n+1}$ and equate the two expressions in (3) to get

(4)
$$(p - p_{n+1})^2 = (p - p_{n+2})(p - p_n).$$

Now expand the terms in (4). The terms p^2 will cancel out on each side of the equation and we can solve for p and obtain

(5)
$$p = \frac{p_{n+2} p_n - p_{n+1}^2}{p_{n+2} - 2p_{n+1} + p_n}.$$

The terms p_n and $-p_n$ can be added to the right side of (5) to get

(6)
$$p = p_n - \frac{(p_{n+1} - p_n)^2}{p_{n+2} - 2p_{n+1} + p_n}.$$

Formula (6) is preferred over (5) because the values p_n are known and there is less round-off error due to the subtractions. We define the new sequence $\{q_n\}$ using the formula in (6):

(7)
$$q_n = p_n - \frac{(p_{n+1} - p_n)^2}{p_{n+2} - 2p_{n+1} + p_n} \qquad \text{for } n = 0, 1, \ldots .$$

Theorem 2.6   [Aitken's accelerated convergence] Let $\{p_n\}$ be any sequence that converges linearly to its limit p. Then the sequence given by formula (7) converges to p faster in the sense that

(8)
$$\lim_{n \to \infty} \frac{p - q_n}{p - p_n} = 0.$$

Proof. The proof is left as an exercise.

Example 2.18 Show that the sequence $\{p_n\}$ in Example 1.10 exhibits linear convergence, and show that the sequence $\{q_n\}$ obtained by Aitken's Δ^2 process converges faster.

Solution. The sequence $\{p_n\}$ was obtained by fixed-point iteration using the function $g(x) = \exp(-x)$ and starting with $p_0 = 0.5$. After convergence has been achieved, the limit is $P \approx 0.567143290$. The values of p_n and q_n are given in the Tables 2.15 and 2.16. For illustration, the value of q_0 is given by the calculation

$$q_0 = p_0 - \frac{(p_1 - p_0)^2}{p_2 - 2p_1 + p_0}$$

$$= 0.606530660 - \frac{(-0.061291448)^2}{0.095755331} = 0.567298989.$$

Table 2.15 Linearly convergent sequence $\{p_n\}$

n	p_n	$e_n = p_n - p$	$A_n = \dfrac{e_{n+1}}{e_{n-1}}$
1	0.606530660	0.039387369	−0.586616609
2	0.545239212	−0.021904079	−0.556119357
3	0.579703095	0.012559805	−0.573400269
4	0.560064628	−0.007078663	−0.563596551
5	0.571172149	0.004028859	−0.569155345
6	0.564862947	−0.002280343	−0.566002341

Table 2.16 Derived sequence $\{q_n\}$ using Aitken's process

n	q_n	$q_n - p$
0	0.567298989	0.000155699
1	0.567193142	0.000049852
2	0.567159364	0.000016074
3	0.567148453	0.000005163
4	0.567144952	0.000001662
5	0.567143825	0.000000534

Although the sequence $\{q_n\}$ in Table 2.16 converges linearly, it converges faster than $\{p_n\}$ in the sense of Theorem 2.6. Often Aitken's method gives a better improvement than this.

Steffensen's Method

We now combine Aitken's process with the fixed-point iteration algorithm of Chapter 1.

Algorithm 2.8 [*Steffensen's Acceleration*] Find a solution p of the equation $x = g(x)$ given an initial approximation p_0. It is assumed that $g(x)$ and $g'(x)$ are continuous and $|g'(p)| < 1$ and that ordinary fixed-point iteration converges to p.

Delta := 10^{-6}, Small := 10^{-6} {Tolerances}
Cond := 0   {Condition for loop termination}

INPUT   P0   {Input initial approximation}

```
DO  FOR  K = 1  TO  99  UNTIL  Cond ≠ 0
      P1 := g(P0)                              {First new iterate}
      P2 := g(P1)                              {Second new iterate}
      D1 := [P1 − P0]²                         {Form the
      D2 := P2 − 2*P1 + P0                      differences}
        IF  D2 = 0  THEN                       {Check division
             Cond := 1                              by zero}
             DP := P2 − P1
             P3 := P2
        ELSE
             DP := D1/D2
             P3 := P0 − DP                     {Aitken's improvement}
        ENDIF
      Relerr := 2*|DP|/[|P3|+Small]            {Relative error}
      IF  RelErr < Delta  THEN                 {Check for
       └─ IF Cond ≠ 1  THEN Cond := 2           convergence}
      P0 := P3                                 {Update iterate}
```

PRINT 'The computed fixed point of g(x) is' P0   {Output}
PRINT 'Consecutive iterates are closer than' DP
IF   Cond = 0   THEN
PRINT 'The maximum number of iterations was exceeded.'
IF   Cond = 1   THEN
PRINT 'Division by zero was encountered.'
IF   Cond = 2   THEN
PRINT 'The solution was found with the desired tolerance.'

Muller's Method

Muller's method is a generalization of the secant method, in the sense that it does not require the derivative of the function. It is an iterative method that requires three starting points: $(p_0, f(p_0))$, $(p_1, f(p_1))$, and $(p_2, f(p_2))$. A parabola is constructed that passes through the three points; then the quadratic formula is used to find a root of the quadratic for the next approximation. It has been proven that near a simple root Muller's method converges faster than the secant method and almost as fast as Newton's method. The method can be used to find real or complex zeros of a function and can be programmed to use complex arithmetic.

Without loss of generality, we assume that p_2 is the best approximation to the root and consider the parabola through the three starting values, as shown in Figure 2.12. Make the change of variable

(9)
$$t = x - p_2,$$

and use the differences

(10)
$$h_0 = p_0 - p_2 \quad \text{and} \quad h_1 = p_1 - p_2.$$

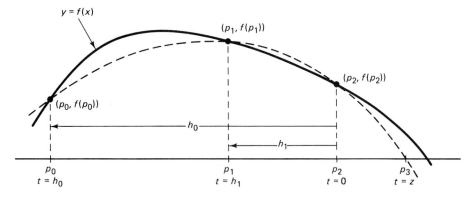

Figure 2.12. The starting approximations p_0, p_1, p_2 for Muller's method and the differences h_0 and h_1.

Consider the quadratic polynomial, involving the variable t,

(11)
$$y = at^2 + bt + c.$$

Each point is used to obtain an equation involving $a, b,$ and c:

$$\text{At } t = h_0: \quad ah_0^2 + bh_0 + c = f_0.$$

(12)
$$\text{At } t = h_1: \quad ah_1^2 + bh_1 + c = f_1.$$

$$\text{At } t = 0: \quad a0^2 + b0 + c = f_2.$$

From the third equation in (12) we see that

(13)
$$c = f_2.$$

Substituting $c = f_2$ into the first two equations in (12) and using the definitions $e_0 = f_0 - c$ and $e_1 = f_1 - c$ results in the system

(14)
$$ah_0^2 + bh_0 = f_0 - c = e_0,$$
$$ah_1^2 + bh_1 = f_1 - c = e_1.$$

Cramer's rule can be used to solve the linear system in (14) (see Section 3.2).

The required determinants are

$$D = \begin{vmatrix} h_0^2 & h_0 \\ h_1^2 & h_1 \end{vmatrix} = h_0 h_1 (h_0 - h_1),$$

(15)
$$D_1 = \begin{vmatrix} e_0 & h_0 \\ e_1 & h_1 \end{vmatrix} = e_0 h_1 - e_1 h_0,$$

$$D_2 = \begin{vmatrix} h_0^2 & e_0 \\ h_1^2 & e_1 \end{vmatrix} = e_1 h_0^2 - e_0 h_1^2.$$

The coefficients a and b are given by

(16)
$$a = \frac{D_1}{D} \quad \text{and} \quad b = \frac{D_2}{D}.$$

The quadratic formula is used to find the roots $t = z_1, z_2$ of (11).

(17)
$$z = \frac{-2c}{b \pm (b^2 - 4ac)^{1/2}}$$

Formula (17) is equivalent to the standard formula for the roots of a quadratic and is better in this case because we know that $c = f_2$.

To ensure stability of the method, we choose the root in (17) that has the smallest absolute value. If $b > 0$, use the positive sign with the square root, and if $b < 0$, use the negative sign. Then p_3 is shown in Figure 2.12 and is given by

(18)
$$p_3 = p_2 + z.$$

To update the iterates, choose p_0 and p_1 to be the two values selected from $\{p_0, p_1, p_2\}$ that lie closest to p_3 (i.e., throw out the one that is farthest away). Then replace p_2 with p_3. Although a lot of auxiliary calculations are done in Muller's method, it only requires one function evaluation per iteration.

If Muller's method is used to find the real roots of $f(x)$, then it is possible that one may encounter complex approximations because the roots of the quadratic in (17) might be complex. In these cases the complex component will have a small magnitude and can be set equal to zero so that the calculations proceed with real numbers.

Algorithm 2.9 [*Muller's Method*] Find a root of $f(x) = 0$. Three distinct initial approximations p_0, p_1, and p_2 are needed. (Eliminate the step that suppresses complex roots if complex variables are used.)

Delta $:= 10^{-6}$, Epsilon $:= 10^{-6}$, Small $:= 10^{-6}$ {Tolerances}
Max $:= 99$ {Maximum number of iterations}
Satisfied $:=$ "False"   {Condition for loop termination}

```
    INPUT P0, P1, P2                                  {Input initial approximations}
    Y0 := F(P0), Y1 := F(P1), Y2 := F(P2)                      {Function values}

DO  FOR K = 1 TO Max UNTIL Satisfied = "True"
    │ H0 := P0 − P2, H1 := P1 − P2                             {Form differences}
    │ C := Y2, E0 := Y0 − C, E1 := Y1 − C
    │ Det := H0∗H1∗[H0 − H1]                          {Compute determinants
    │ A := [E0∗H1 − H0∗E1]/Det                                 and solve the
    │ B := [H0∗H0∗E1 − H1∗H1∗E0]/Det                            linear system}
    │ IF  B∗B>4∗A∗C THEN Disc := SQRT(B∗B−4∗A∗C)
    │ └ ELSE Disc := 0                           {This suppresses complex roots}
    │ IF  B<0  THEN  Disc := −Disc                        {Find the smallest
    │ Z := −2∗C/[B + Disc]                                root of the quadratic}
    │ P3 := P2 + Z                                     {Newest approximation}
    │ IF  |P3−P1| < |P3−P0|  THEN                           {Sort to make
    │ └ U := P1, P1 := P0, P0 := U, V := Y1, Y1 := Y0, Y0 := V    the values
    │ IF  |P3−P2|<|P3−P1|  THEN                                 P0, P1 closest
    │ └ U := P2, P2 := P1, P1 := U, V := Y2, Y2 := Y1, Y1 := V    to P3}
    │ P2 := P3, Y2 := F(P2)                                 {Update iterate}
    │ RelErr := 2∗|Z|/[|P2|+Small]                         {Relative error}
    │ IF  RelErr < Delta  AND  |Y2| < Epsilon  THEN        {Check for
    │ └ Satisfied := "True"                                 convergence}
END

IF  Satisfied = "True"  THEN                                           {Output}
    │ PRINT "The computed root of f(x) is" P2
    │ PRINT "Consecutive iterates are closer than" |Z|
    └ PRINT "The current function value is" Y2
IF  Satisfied ≠ "True"  THEN
    PRINT "Muller's method did not find a root of f(x) = 0."
```

Comparison of Methods

Steffensen's method can be used with the Newton–Raphson fixed-point function $g(x) = x - f(x)/f'(x)$. In the next two examples we look at the roots of the polynomial $f(x) = x^3 - 3x + 2$. The Newton–Raphson function is $g(x) = (2x^3 - 2)/(3x^2 - 3)$. When this function is used in Algorithm 2.8, we get the calculations under the heading Steffensen with Newton in Tables 2.17 and 2.18. For example, starting with $p_0 = -2.4$, we would compute

$$p_1 = g(p_0) = -2.076190476,$$

$$p_2 = g(p_1) = -2.003596011.$$

Then Aitken's improvement will give $p_3 = -1.982618143$.

Example 2.19 [Convergence near a Simple Root] This is a comparison of methods for the function $f(x) = x^3 - 3x + 2$ near the simple root at $p = -2$.

Solution. Newton's method and the secant method for this function were given in Examples 2.7 and 2.9, respectively. The summary of calculations for the methods is shown in Table 2.17.

Table 2.17 Comparison of convergence near a simple root

k	Secant Method	Muller's Method	Newton's Method	Steffensen with Newton
0	−2.600000000	−2.600000000	−2.400000000	−2.400000000
1	−2.400000000	−2.500000000	−2.076190476	−2.076190476
2	−2.106598985	−2.400000000	−2.003596011	−2.003596011
3	−2.022641412	−1.985275287	−2.000008589	−1.982618143
4	−2.001511098	−2.000334062	−2.000000000	−2.000204982
5	−2.000022537	−2.000000218		−2.000000028
6	−2.000000022	−2.000000000		−2.000002389
7	−2.000000000			−2.000000000

Example 2.20 [Convergence near a Double Root] This is a comparison of methods for the function $f(x) = x^3 - 3x + 2$ near the double root at $p = 1$.

Solution. The summary of calculations is shown in Table 2.18.

Table 2.18 Comparison of convergence near a double root

k	Secant Method	Muller's Method	Newton's Method	Steffensen with Newton
0	1.400000000	1.400000000	1.200000000	1.200000000
1	1.200000000	1.300000000	1.103030303	1.103030303
2	1.138461538	1.200000000	1.052356417	1.052356417
3	1.083873738	1.003076923	1.026400814	0.996890433
4	1.053093854	1.003838922	1.013257734	0.998446023
5	1.032853156	1.000027140	1.006643418	0.999223213
6	1.020429426	0.999997914	1.003325375	0.999999193
7	1.012648627	0.999999747	1.001663607	0.999999597
8	1.007832124	1.000000000	1.000832034	0.999999798
9	1.004844757		1.000416075	0.999999999
	⋮		⋮	

Newton's method is the best choice for finding a simple root (see Table 2.17). At a double root, either Muller's method or Steffensen's method with the Newton–Raphson formula are good choices (see Table 2.18).

Exercises for Aitken's, Steffensen's, and Muller's Methods

1. Let $p_n = n$. Show that $p_{n+2} - 2p_{n+1} + p_n = 0$ for all n.

2. Let $p_n = 1/2^n$. Show that $q_n = 0$ for all n, where q_n is given by formula (7).

3. Let $p_n = 1/n$. Show that $q_n = 1/(2n + 2)$ for all n, hence there is little acceleration of convergence. Does $\{p_n\}$ converge to 0 linearly? Why?

4. The sequence $p_n = 1/(4^n + 4^{-n})$ converges linearly to 0. Use Aitken's formula (7) to find $q_1, q_2,$ and q_3 and hence speed up the convergence.

n	p_n	q_n
0	0.5	−0.26437542
1	0.23529412	_____
2	0.06225681	_____
3	0.01562119	_____
4	0.00390619	
5	0.00097656	

5. The sequence $\{p_n\}$ generated by fixed-point iteration, starting with $p_0 = 2.5$, and using the function $g(x) = (6 + x)^{1/2}$ converges linearly to $p = 3$. Use Aitken's formula (7) to find $q_1, q_2,$ and q_3 and hence speed up the convergence.

n	p_n	q_n
0	2.5	3.00024351
1	2.91547595	_____
2	2.98587943	_____
3	2.99764565	_____
4	2.99960758	
5	2.99993460	

6. The sequence $\{p_n\}$ generated by fixed-point iteration, starting with $p_0 = 3.14$, and using the function $g(x) = \ln(x) + 2$ converges linearly to $p \approx 3.14619322$. Use Aitken's formula (7) to find $q_1, q_2,$ and q_3 and hence speed up the convergence.

n	p_n	q_n
0	3.14	3.14619413
1	3.14422280	_____
2	3.14556674	_____
3	3.14599408	_____
4	3.14612992	
5	3.14617310	

7. For the equation $\cos(x) - 1 = 0$, the Newton–Raphson iteration function is $g(x) = x - [1 - \cos(x)]/\sin(x) = x - \tan(x/2)$. Use Steffensen's algorithm with $g(x)$ and start with $p_0 = 0.5$ and find $p_1, p_2,$ and p_3, then find the next three values.

8. For the equation $x - \sin(x) = 0$, the Newton–Raphson iteration function is $g(x) = x - [x - \sin(x)]/[1 - \cos(x)]$. Use Steffensen's algorithm with $g(x)$ and start with $p_0 = 0.5$ and find $p_1, p_2,$ and p_3, then find the next three values.

9. For the equation $\sin(x^3) = 0$, the Newton–Raphson iteration function is $g(x) = x - \sin(x^3)/[3x^2 \cos(x^3)] = x - \tan(x^3)/[3x^2]$. Use Steffensen's algorithm with $g(x)$ and start with $p_0 = 0.5$ and find $p_1, p_2,$ and p_3, then find the next three values.

10. *Convergence of Series.* Aitken's method can be used to speed up the convergence of a series. If the nth term of the series is

$$S_n = \sum_{k=1}^{n} A_k,$$

show that the derived series using Aitken's method is

$$T_n = S_n + \frac{A_{n+1}^2}{A_{n+1} - A_{n+2}}.$$

In Exercises 11–14, apply Aitken's method and the result of Exercise 10 to speed up convergence of the series.

11. $S_n = \sum_{k=1}^{n} (0.99)^k$

12. $S_n = \sum_{k=1}^{n} \frac{1}{4^k + 4^{-k}}$

13. $S_n = \sum_{k=1}^{n} \frac{k}{2^{k-1}}$

14. $S_n = \sum_{k=1}^{n} \frac{1}{k \times 2^k}$

15. Use Muller's method to find the root of $f(x) = x^3 - x - 2$. Start with $p_0 = 1.0$, $p_1 = 1.2$, and $p_2 = 1.4$ and find $p_3, p_4,$ and p_5.

16. Use Muller's method to find a root of $f(x) = 4x^2 - \exp(x)$. Start with $p_0 = 4.0$, $p_1 = 4.1$, and $p_2 = 4.2$ and find $p_3, p_4,$ and p_5.

17. Use Muller's method to find a root of $f(x) = 1 + 2x - \tan(x)$. Start with $p_0 = 1.5$, $p_1 = 1.4$, and $p_2 = 1.3$ and find $p_3, p_4,$ and p_5.

18. Use Muller's method to find a root of $f(x) = 3 \cos(x) + 2 \sin(x)$. Start with $p_0 = 2.4$, $p_1 = 2.3$, and $p_2 = 2.2$ and find $p_3, p_4,$ and p_5.

19. Start with formula (5) and add the terms p_{n+2} and $-p_{n+2}$ to the right side and show that an equivalent formula for p is

$$p = p_{n+2} - \frac{(p_{n+2} - p_{n+1})^2}{p_{n+2} - 2p_{n+1} + p_n}.$$

20. Prove Theorem 2.5.

21. Assume that the error in an iteration process satisfies the relation $e_{n+1} = Ke_n$ for some constant K and $|K| < 1$.
    (a) Find an expression for e_n that involves $e_0, K,$ and n.
    (b) Find an expression for the smallest integer N so that

$$|e_N| < 10^{-8}.$$

3

Direct Methods for Solving Linear Systems

3.1 Introduction to Vectors and Matrices

Vectors

A real N-dimensional vector $\mathbf{X}$ is an ordered set of N real numbers and is usually written in the coordinate form

(1)
$$\mathbf{X} = (x_1, x_2, \ldots, x_N).$$

Here the numbers $x_1, x_2, \ldots, x_N$ are called the *components* of $\mathbf{X}$. The set consisting of all N-dimensional vectors is called *N-dimensional space*. When a vector is used to denote a point or position in space it is called a *position vector*. When it is used to denote a movement between two points in space it is called a *displacement vector*.

Let another vector be $\mathbf{Y} = (y_1, y_2, \ldots, y_N)$. The two vectors $\mathbf{X}$ and $\mathbf{Y}$ are said to be equal if and only if each corresponding coordinate is the same, that is,

(2)
$$\mathbf{X} = \mathbf{Y} \text{ if and only if } x_j = y_j \text{ for } j = 1, 2, \ldots, N.$$

The sum of the vectors $\mathbf{X}$ and $\mathbf{Y}$ is computed component by component, using the definition

(3)
$$\mathbf{X} + \mathbf{Y} = (x_1 + y_1, x_2 + y_2, \ldots, x_N + y_N).$$

The opposite of the vector $\mathbf{X}$ is obtained by replacing each coordinate with its opposite:

$$(4) \qquad -\mathbf{X} = (-x_1, -x_2, \ldots, -x_N).$$

The difference $\mathbf{Y} - \mathbf{X}$ is formed by taking the difference in each coordinate:

$$(5) \qquad \mathbf{Y} - \mathbf{X} = (y_1 - x_1, y_2 - x_2, \ldots, y_N - x_N).$$

Vectors in N-dimensional space obey the algebraic property

$$(6) \qquad \mathbf{Y} - \mathbf{X} = \mathbf{Y} + (-\mathbf{X}).$$

If c is a real number (*scalar*), we define *scalar multiplication* $c\mathbf{X}$ as follows:

$$(7) \qquad c\mathbf{X} = (cx_1, cx_2, \ldots, cx_N).$$

If c and d are scalars, then the weighted sum $c\mathbf{X} + d\mathbf{Y}$ is called a *linear combination* of $\mathbf{X}$ and $\mathbf{Y}$, and we write

$$(8) \qquad c\mathbf{X} + d\mathbf{Y} = (cx_1 + dy_1, cx_2 + dy_2, \ldots, cx_N + dy_N).$$

The *dot product* of the two vectors $\mathbf{X}$ and $\mathbf{Y}$ is a scalar quantity (real number) defined by the equation

$$(9) \qquad \mathbf{X} \cdot \mathbf{Y} = x_1 y_1 + x_2 y_2 + \ldots + x_N y_N.$$

The *length* (or *norm*) of the vector $\mathbf{X}$ is defined by

$$(10) \qquad |\mathbf{X}| = (x_1^2 + x_2^2 + \ldots + x_N^2)^{1/2}.$$

Formula (10) is referred to as the *Euclidean norm* (or *length*) of the vector $\mathbf{X}$.

Scalar multiplication $c\mathbf{X}$ stretches the vector $\mathbf{X}$ when $|c| > 1$ and shrinks the vector when $|c| < 1$. This is shown by using equation (10):

$$(11) \qquad |c\mathbf{X}| = (c^2 x_1^2 + c^2 x_2^2 + \ldots + c^2 x_N^2)^{1/2}$$
$$= |c|(x_1^2 + x_2^2 + \ldots + x_N^2)^{1/2} = |c||\mathbf{X}|.$$

An important relationship exists between the dot product and length of a vector. If both sides of equation (10) are squared and equation (9) is used with $\mathbf{Y}$ being replaced with $\mathbf{X}$, then we have

$$(12) \qquad |\mathbf{X}|^2 = x_1^2 + x_2^2 + \ldots + x_N^2 = \mathbf{X} \cdot \mathbf{X}.$$

If $\mathbf{X}$ and $\mathbf{Y}$ are position vectors that locate the points $(x_1, x_2, \ldots, x_N)$ and $(y_1, y_2, \ldots, y_N)$ in N-dimensional space, then the *displacement vector* from $\mathbf{X}$ to $\mathbf{Y}$ is given by the difference

$$(13) \qquad \mathbf{Y} - \mathbf{X} \qquad \text{(displacement from position } \mathbf{X} \text{ to position } \mathbf{Y}\text{)}.$$

Notice that if a particle starts at the position $\mathbf{X}$ and moves through the displacement $\mathbf{Y} - \mathbf{X}$, then its new position is $\mathbf{Y}$. This can be obtained by the following vector sum:

$$(14) \qquad \mathbf{Y} = \mathbf{X} + (\mathbf{Y} - \mathbf{X}).$$

Using equations (10) and (13), we can write down the formula for the distance between two points in *N*-space,

(15)
$$|\mathbf{Y} - \mathbf{X}| = [(y_1 - x_1)^2 + (y_2 - x_2)^2 + \ldots + (y_N - x_N)^2]^{1/2}.$$

When the distance between points is computed using formula (15), we say that the points lie in *N-dimensional Euclidean space.*

Example 3.1 Consider the two points $\mathbf{X} = (3, 4)$ and $\mathbf{Y} = (5, 2)$ that lie in two-dimensional space. Then 3 is the first component and 4 is the second component of the vector $\mathbf{X}$ [see Figure 3.1(a)]. Formula (10) for two-dimensional space becomes the familiar Pythagorean rule for finding the length of the hypotenuse of a right triangle. Indeed, $|\mathbf{X}| = (9 + 16)^{1/2} = 5$ [see Figure 3.1(b)]. The sum $(3, 4) + (5, 2) = (8, 6)$ can be visualized as follows. Let $\mathbf{X}$ be a position vector with its initial point at the origin and its terminal end at $(3, 4)$; then let $\mathbf{Y}$ represent a displacement vector with its initial point placed at the position $(3, 4)$. The point in space located at the terminal end of $\mathbf{Y}$ is the position vector $\mathbf{X} + \mathbf{Y}$ [see Figure 3.1(c)]. The difference $(5, 2) - (3, 4) = (2, -2)$ can be visualized by letting $\mathbf{Y} = (5, 2)$ denote a position vector and then adding the vector $-\mathbf{X} = (-3, -4)$, as shown in Figure 3.1(d). The displacement $\mathbf{Y} - \mathbf{X} = (2, -2)$ is a vector that points from the position $\mathbf{X}$ to the position $\mathbf{Y}$ and its length is $2 \times 2^{1/2}$, as shown in Figure 3.1(e) and (f), respectively.

Example 3.2 Let $\mathbf{X} = (2, -3, 5, -1)$ and $\mathbf{Y} = (6, 1, 2, -4)$. The concepts mentioned above are now illustrated for vectors in 4-space.

Sum	$\mathbf{X} + \mathbf{Y} = (8, -2, 7, -5)$
Difference	$\mathbf{X} - \mathbf{Y} = (-4, -4, 3, 3)$
Scalar multiple	$3\mathbf{X} = (6, -9, 15, -3)$
Length	$|\mathbf{X}| = (4 + 9 + 25 + 1)^{1/2} = 39^{1/2}$
Dot product	$\mathbf{X} \cdot \mathbf{Y} = 12 - 3 + 10 + 4 = 23$

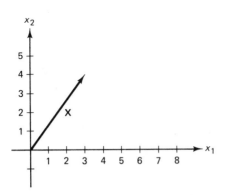

Figure 3.1(a). The components of the position vector $\mathbf{X} = (3, 4)$.

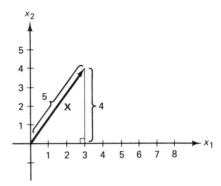

Figure 3.1(b). The length of the vector $\mathbf{X} = (3, 4)$.

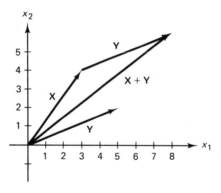

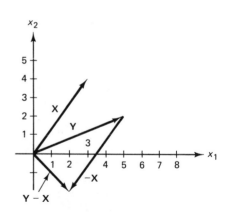

Figure 3.1(c). The sum of the vectors $\mathbf{X} = (3, 4)$ and $\mathbf{Y} = (5, 2)$.

Figure 3.1(d). The difference $\mathbf{Y} - \mathbf{X}$.

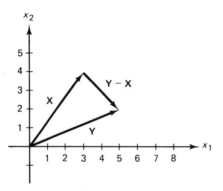

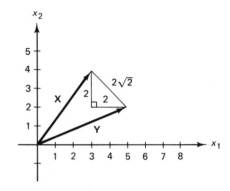

Figure 3.1(e). The displacement $\mathbf{Y} - \mathbf{X}$ from $\mathbf{X}$ to $\mathbf{Y}$.

Figure 3.1(f). The distance between the points $\mathbf{X}$ and $\mathbf{Y}$.

Displacement
from $\mathbf{X}$ to $\mathbf{Y}$ $\qquad\qquad \mathbf{Y} - \mathbf{X} = (4, 4, -3, -3)$

Distance
from $\mathbf{X}$ to $\mathbf{Y}$ $\qquad\qquad |\mathbf{Y} - \mathbf{X}| = (16 + 16 + 9 + 9)^{1/2} = 50^{1/2}$

It is sometimes useful to write vectors as columns instead of rows. For example

(16)
$$\mathbf{X} = \begin{pmatrix} x_1 \\ x_2 \\ \cdot \\ \cdot \\ \cdot \\ x_N \end{pmatrix} \quad \text{and} \quad \mathbf{Y} = \begin{pmatrix} y_1 \\ y_2 \\ \cdot \\ \cdot \\ \cdot \\ y_N \end{pmatrix}.$$

Then the linear combination $c\mathbf{X} + d\mathbf{Y}$ is

(17)
$$c\mathbf{X} + d\mathbf{Y} = \begin{pmatrix} cx_1 + dy_1 \\ cx_2 + dy_2 \\ \vdots \quad \vdots \\ cx_N + dy_N \end{pmatrix}.$$

By choosing c and d appropriately in equation (17) we have the sum $1\mathbf{X} + 1\mathbf{Y}$, the difference $1\mathbf{X} - 1\mathbf{Y}$, and scalar multiple $c\mathbf{X} + 0\mathbf{Y}$. We use the superscript T, for transpose, to indicate that a row vector should be converted to a column vector, and vice versa.

(18)
$$(x_1, x_2, \ldots, x_N)^T = \begin{pmatrix} x_1 \\ x_2 \\ \vdots \\ x_N \end{pmatrix}$$

and

$$\begin{pmatrix} x_1 \\ x_2 \\ \vdots \\ x_N \end{pmatrix}^T = (x_1, x_2, \ldots, x_N).$$

The set of vectors has a zero element $\mathbf{0}$ which is defined by

(19)
$$\mathbf{0} = (0, 0, \ldots, 0).$$

Theorem 3.1 [Vector algebra] Let $\mathbf{X}$, $\mathbf{Y}$, and $\mathbf{Z}$ be N-dimensional vectors, and let a and b be scalars (real numbers). Vectors in N-dimensional space have the following properties:

(20)	$\mathbf{Y} + \mathbf{X} = \mathbf{X} + \mathbf{Y}$	commutative property
(21)	$\mathbf{0} + \mathbf{X} = \mathbf{X} + \mathbf{0} = \mathbf{X}$	zero vector
(22)	$\mathbf{X} - \mathbf{X} = \mathbf{X} + (-\mathbf{X}) = \mathbf{0}$	opposite of a vector
(23)	$(\mathbf{X} + \mathbf{Y}) + \mathbf{Z} = \mathbf{X} + (\mathbf{Y} + \mathbf{Z})$	associative property
(24)	$(a + b)\mathbf{X} = a\mathbf{X} + b\mathbf{X}$	distributive property for scalars
(25)	$a(\mathbf{X} + \mathbf{Y}) = a\mathbf{X} + a\mathbf{Y}$	distributive properties for vectors
(26)	$a(b\mathbf{X}) = (ab)\mathbf{X}$	associative property for scalars

Matrices and Two-Dimensional Arrays

A *matrix* is a rectangular array of numbers that is arranged systematically in rows and columns. A matrix having M rows and N columns is called an $M \times N$ (read "M by N") matrix. The capital letter A denotes a matrix and the small subscripted letter $a_{i,j}$ denotes one of the numbers forming the matrix. We write

$$(27) \qquad A = (a_{i,j})_{M \times N} \qquad \text{for } 1 \leqq i \leqq M, \quad 1 \leqq j \leqq N,$$

where $a_{i,j}$ is the number in location (i, j) (i.e., stored in the ith row and jth column of the array). We refer to $a_{i,j}$ as the *element in location* (i, j). In expanded form we write

$$(28) \qquad \text{row } i \longrightarrow \begin{pmatrix} a_{1,1} & a_{1,2} & \cdots & a_{1,j} & \cdots & a_{1,N} \\ a_{2,1} & a_{2,2} & \cdots & a_{2,j} & \cdots & a_{2,N} \\ \vdots & \vdots & & \vdots & & \vdots \\ a_{i,1} & a_{i,2} & \cdots & a_{i,j} & \cdots & a_{i,N} \\ \vdots & \vdots & & \vdots & & \vdots \\ a_{M,1} & a_{M,2} & \cdots & a_{M,j} & \cdots & a_{M,N} \end{pmatrix} = A.$$

$$\text{column } j$$

The rows of the M by N matrix A are N-dimensional vectors:

$$(29) \qquad \begin{aligned} \mathbf{V}_1 &= (a_{1,1}, a_{1,2}, \ldots, a_{1,j}, \ldots, a_{1,N}) \\ \mathbf{V}_2 &= (a_{2,1}, a_{2,2}, \ldots, a_{2,j}, \ldots, a_{2,N}) \\ &\vdots \\ \mathbf{V}_i &= (a_{i,1}, a_{i,2}, \ldots, a_{i,j}, \ldots, a_{i,N}) \\ &\vdots \\ \mathbf{V}_M &= (a_{M,1}, a_{M,2}, \ldots, a_{M,j}, \ldots, a_{M,N}). \end{aligned}$$

The row vectors in (29) can also be viewed as 1 by N matrices. Here we have sliced the M by N matrix A into M pieces which are 1 by N matrices.

In this case we could express A as a column vector consisting of the row vectors $\mathbf{V}_i$, that is,

(30)
$$A = \begin{pmatrix} \mathbf{V}_1 \\ \mathbf{V}_2 \\ \vdots \\ \mathbf{V}_i \\ \vdots \\ \mathbf{V}_M \end{pmatrix}.$$

The columns of the M by N matrix A are M-dimensional vectors:

(31)
$$\mathbf{C}_1 = \begin{pmatrix} a_{1,1} \\ a_{2,1} \\ \vdots \\ a_{i,1} \\ \vdots \\ a_{M,1} \end{pmatrix}, \ldots, \mathbf{C}_j = \begin{pmatrix} a_{1,j} \\ a_{2,j} \\ \vdots \\ a_{i,j} \\ \vdots \\ a_{M,j} \end{pmatrix}, \ldots, \mathbf{C}_N = \begin{pmatrix} a_{1,N} \\ a_{2,N} \\ \vdots \\ a_{i,N} \\ \vdots \\ a_{M,N} \end{pmatrix}.$$

Each column vector in (31) can be viewed as an M by 1 matrix. In this case we could express A as a row vector consisting of the column vectors $\mathbf{C}_j$:

(32)
$$A = [\mathbf{C}_1, \mathbf{C}_2, \ldots, \mathbf{C}_j, \ldots, \mathbf{C}_N].$$

Example 3.3 Identify the row and column vectors associated with the 4 by 3 matrix

$$A = \begin{pmatrix} -2 & 4 & 9 \\ 5 & -7 & 1 \\ 0 & -3 & 8 \\ -4 & 6 & -5 \end{pmatrix}.$$

Solution. The four row vectors are $\mathbf{V}_1 = (-2, 4, 9)$, $\mathbf{V}_2 = (5, -7, 1)$, $\mathbf{V}_3 = (0, -3, 8)$, and $\mathbf{V}_4 = (-4, 6, -5)$. The three column vectors are

$$\mathbf{C}_1 = \begin{pmatrix} -2 \\ 5 \\ 0 \\ -4 \end{pmatrix} \quad \mathbf{C}_2 = \begin{pmatrix} 4 \\ -7 \\ -3 \\ 6 \end{pmatrix} \quad \mathbf{C}_3 = \begin{pmatrix} 9 \\ 1 \\ 8 \\ -5 \end{pmatrix}.$$

Notice how A can be represented with these vectors:

$$A = \begin{pmatrix} \mathbf{V}_1 \\ \mathbf{V}_2 \\ \mathbf{V}_3 \\ \mathbf{V}_4 \end{pmatrix} = (\mathbf{C}_1, \mathbf{C}_2, \mathbf{C}_3).$$

Let $A = (a_{i,j})_{M \times N}$ and $B = (b_{i,j})_{M \times N}$ be two matrices of the same dimensions M by N. The two matrices A and B are said to be equal if and only if each corresponding element is the same, that is,

(33) $\qquad A = B \quad$ if and only if $a_{i,j} = b_{i,j} \qquad$ for $1 \leq i \leq M, \quad 1 \leq j \leq N$.

The sum of the two M by N matrices A and B is computed element by element, using the definition

(34) $\qquad A + B = (a_{i,j} + b_{i,j})_{M \times N} \qquad$ for $1 \leq i \leq M, \quad 1 \leq j \leq N$.

The opposite of the matrix A is obtained by replacing each element with its opposite:

(35) $\qquad -A = (-a_{i,j})_{M \times N} \qquad$ for $1 \leq i \leq M, \quad 1 \leq j \leq N$.

The difference $A - B$ is formed by taking the difference in each coordinate:

(36) $\qquad A - B = (a_{i,j} - b_{i,j})_{M \times N} \qquad$ for $1 \leq i \leq M, \quad 1 \leq j \leq N$.

Example 3.4 Find the sum and difference of the matrices

$$A = \begin{pmatrix} -1 & 2 \\ 7 & 5 \\ 3 & -4 \end{pmatrix} \qquad B = \begin{pmatrix} -2 & 3 \\ 1 & -4 \\ -9 & 7 \end{pmatrix}.$$

Solution. Using formulas (34) and (36), we obtain

$$A + B = \begin{pmatrix} -3 & 5 \\ 8 & 1 \\ -6 & 3 \end{pmatrix} \qquad A - B = \begin{pmatrix} 1 & -1 \\ 6 & 9 \\ 12 & -11 \end{pmatrix}.$$

Notice that the sum of two matrices of different sizes is not defined; for example, the following matrices cannot be added!

$$\begin{pmatrix} -1 & 2 \\ 7 & 5 \\ 3 & -4 \end{pmatrix} \overset{?\ ?}{+} \begin{pmatrix} -2 & 3 & -9 \\ 1 & -4 & 7 \end{pmatrix}.$$

If c is a real number (scalar), we define scalar multiplication cA as follows:

(37) $\qquad cA = (ca_{i,j})_{M \times N} \qquad$ for $1 \leq i \leq M, \quad 1 \leq j \leq N$.

If p and q are scalars, then the weighted sum $pA + qB$ is called a linear combination of the matrices A and B, and we write

(38) $$pA + qB = (pa_{i,j} + qb_{i,j})_{M \times N} \qquad \text{for } 1 \leq i \leq M, \quad 1 \leq j \leq N.$$

The zero matrix of order M by N consists of all zeros:

(39) $$O = (0)_{M \times N}.$$

Example 3.5 Find the scalar multiples $2A$ and $3B$ and the linear combination $2A - 3B$ for the matrices

$$A = \begin{pmatrix} -1 & 2 \\ 7 & 5 \\ 3 & -4 \end{pmatrix} \qquad B = \begin{pmatrix} -2 & 3 \\ 1 & -4 \\ -9 & 7 \end{pmatrix}.$$

Solution. Using formula (37), we obtain

$$2A = \begin{pmatrix} -2 & 4 \\ 14 & 10 \\ 6 & -8 \end{pmatrix} \qquad 3B = \begin{pmatrix} -6 & 9 \\ 3 & -12 \\ -27 & 21 \end{pmatrix}.$$

The linear combination $2A - 3B$ is now found:

$$2A - 3B = \begin{pmatrix} -2+6 & 4-9 \\ 14-3 & 10+12 \\ 6+27 & -8-21 \end{pmatrix} = \begin{pmatrix} 4 & -5 \\ 11 & 22 \\ 33 & -29 \end{pmatrix}.$$

Theorem 3.2 [Matrix addition] Let A, B, and C be M by N matrices, and let c and d be scalars (real numbers); then

(40)	$B + A = A + B$	commutative property
(41)	$O + A = A + O = A$	identity for addition
(42)	$A - A = A + (-A) = O$	the opposite matrix
(43)	$(A + B) + C = A + (B + C)$	associative property
(44)	$(p + q)A = pA + qA$	distributive property for scalars
(45)	$p(A + B) = pA + pB$	distributive properties for matrices
(46)	$p(qA) = (pq)A$	associative property for scalars

Algorithm 3.1 [*Matrix Addition*] If $A = (a_{i,j})$ and $B = (b_{i,j})$ are matrices of dimension $M \times N$, then the sum $C = A + B$ is

$$C = (c_{i,j}) = (a_{i,j} + b_{i,j}) \qquad \text{for } 1 \leq i \leq M, \quad 1 \leq j \leq N.$$

```
INPUT  M                                          {Number of rows}
INPUT  N                                          {Number of columns}
VARiable declaration                              {Dimension the arrays}
        REAL A[1 .. M, 1 .. N] , B[1 .. M, 1 .. N] , C[1 .. M, 1 .. N]

FOR    Col = 1  TO  N  DO                          {Input columns of matrix A}
       FOR    Row = 1  TO  M  DO
              PRINT " A( " ; Row ; " ; " Col ; " ) = " ;    {Prompt for input}
              INPUT  A(Row, Col)

FOR    Col = 1  TO  N  DO                          {Input columns of matrix B}
       FOR    Row = 1  TO  M  DO
              PRINT " B( " ; Row ; " , " ; Col ; " ) = " ;    {Prompt for input}
              INPUT   B(Row, Col)

FOR    I = 1  TO  M  DO                            {Fix the row I}
       FOR    J = 1  TO  N  DO                     {Find the Jth element}
              C(I, J) := A(I, J) + B(I, J)         {Compute sum}

PRINT  'The sum of matrix A and matrix B is matrix C'

FOR    Row = 1  TO  M  DO                          {Output}
       FOR    Col = 1  TO  N  DO                   {Print rows of matrix C}
              PRINT   C(Row, Col),
              PRINT
```

Exercises for Introduction to Vectors and Matrices

In Exercises 1-4, given the vectors X and Y, find (a) $X + Y$, (b) $X - Y$, (c) $3X$, (d) $|X|$, (e) $X \cdot Y$, (f) $Y - X$, and (g) $|Y - X|$.

1. $X = (3, -4)$ and $Y = (-2, 8)$

2. $X = (-6, 3, 2)$ and $Y = (-8, 5, 1)$

3. $X = (4, -8, 1)$ and $Y = (1, -12, -11)$

4. $X = (1, -2, 4, 2)$ and $Y = (3, -5, -4, 0)$

5. Prove the commutative property (20) for vector addition.

6. Prove the associative property (22) for vector addition.

7. Prove the distributive property (24) for scalars.

8. Prove the distributive property (25) for vectors.

9. The angle θ between two vectors **X** and **Y** is given by the relation

$$\cos(\theta) = \frac{\mathbf{X} \cdot \mathbf{Y}}{|\mathbf{X}||\mathbf{Y}|}.$$

Find the angle between the following vectors.
(a) $\mathbf{X} = (-6, 3, 2)$ and $\mathbf{Y} = (2, -2, 1)$
(b) $\mathbf{X} = (4, -8, 1)$ and $\mathbf{Y} = (3, 4, 12)$

10. Two vectors **X** and **Y** are said to be orthogonal (perpendicular) if the angle between them is $\pi/2$.
(a) Show that **X** and **Y** are orthogonal if and only if

$$\mathbf{X} \cdot \mathbf{Y} = 0.$$

Use part (a) to determine if the following vectors are orthogonal.
(b) $\mathbf{X} = (-6, 4, 2)$ and $\mathbf{Y} = (6, 5, 8)$
(c) $\mathbf{X} = (-4, 8, 3)$ and $\mathbf{Y} = (2, -5, 16)$
(d) $\mathbf{X} = (-5, 7, 2)$ and $\mathbf{Y} = (4, 1, 6)$

11. Identify the row and column vectors associated with the following matrices.

(a) $\begin{pmatrix} -2 & 5 & 12 \\ 1 & 4 & -1 \\ 7 & 0 & 6 \\ 11 & -3 & 8 \end{pmatrix}$ (b) $\begin{pmatrix} 4 & 9 & 2 \\ 3 & 5 & 7 \\ 8 & 1 & 6 \end{pmatrix}$.

12. Find (a) $A + B$, (b) $A - B$, and (c) $3A - 2B$ for the matrices

$$A = \begin{pmatrix} -1 & 9 & 4 \\ 2 & -3 & -6 \\ 0 & 5 & 7 \end{pmatrix} \quad B = \begin{pmatrix} -4 & 9 & 2 \\ 3 & -5 & 7 \\ 8 & 1 & -6 \end{pmatrix}.$$

13. Find the sum of any two of the matrices below for which the sum is defined.

$$A = \begin{pmatrix} 1 & 2 \\ 4 & 8 \end{pmatrix} \quad B = \begin{pmatrix} -4 & 5 \\ 1 & 0 \end{pmatrix} \quad C = \begin{pmatrix} 9 & 2 & 0 \\ 3 & -5 & 0 \end{pmatrix}$$

$$D = \begin{pmatrix} -3 & 4 \\ 9 & 2 \\ 1 & -1 \end{pmatrix} \quad E = \begin{pmatrix} 0 & 0 & 0 \\ 0 & 0 & 0 \\ 0 & 0 & 0 \end{pmatrix} \quad F = \begin{pmatrix} 3 & 2 & 5 \\ 8 & 4 & 6 \\ 0 & 0 & 0 \end{pmatrix}.$$

14. The *transpose* of an $M \times N$ matrix A, denoted A^T, is an $N \times M$ matrix obtained from A by converting the rows of A to columns of A^T. That is, if $A = (a_{i,j})_{M \times N}$ and $A^T = (b_{j,i})_{N \times M}$, the elements satisfy the relation

$$b_{j,i} = a_{i,j} \quad \text{for } 1 \leqq i \leqq M, \quad 1 \leqq j \leqq N.$$

Find the transpose of the following matrices.

(a) $\begin{pmatrix} -2 & 5 & 12 \\ 1 & 4 & -1 \\ 7 & 0 & 6 \\ 11 & -3 & 8 \end{pmatrix}$

(b) $\begin{pmatrix} 4 & 9 & 2 \\ 3 & 5 & 7 \\ 8 & 1 & 6 \end{pmatrix}$

15. Prove the commutative property (40) for matrix addition.

16. Prove the associative property (43) for matrix addition.

17. Prove the distributive property (44) for scalars.

18. Prove the distributive property (45) for matrices.

19. Are any two of the following matrices equal? Why?

(i) $\begin{pmatrix} 1 & 0 \\ 0 & 1 \end{pmatrix}$

(ii) $\begin{pmatrix} 1 & 0 & 0 \\ 0 & 1 & 0 \end{pmatrix}$

(iii) $\begin{pmatrix} 1 & 0 & 0 \\ 0 & 1 & 0 \\ 0 & 0 & 0 \end{pmatrix}$

(iv) $\begin{pmatrix} 1 & 0 \\ 0 & 1 \\ 0 & 0 \end{pmatrix}$

20. The square matrix A of dimension $N \times N$ is said to be symmetric if $A = A^T$. (See Exercise 14 for the definition of A^T.) Determine whether the following square matrices are symmetric.

(a) $\begin{pmatrix} 1 & -7 & 4 \\ -7 & 2 & 0 \\ 4 & 0 & 3 \end{pmatrix}$

(b) $\begin{pmatrix} 4 & -7 & 1 \\ 0 & 2 & -7 \\ 3 & 0 & 4 \end{pmatrix}$

21. Write a report about the Gram–Schmidt orthogonalization procedure.

3.2 Properties of Vectors and Matrices

A linear combination of the variables $x_1, x_2, \ldots, x_N$ is a sum

(1) $$a_1 x_1 + a_2 x_2 + \ldots + a_N x_N$$

where a_k is the coefficient of x_k (for $k = 1, 2, \ldots, N$).

A linear equation in $x_1, x_2, \ldots, x_N$ is obtained by requiring the linear combination in (1) to take on a prescribed value b, that is,

(2) $$a_1 x_1 + a_2 x_2 + \ldots + a_N x_N = b.$$

Systems of linear equations arise frequently and if M equations in N unknowns are given, we write

(3)

$$\text{Eq } \langle 1 \rangle: \quad a_{1,1}x_1 + a_{1,2}x_2 + a_{1,3}x_3 + \ldots + a_{1,N}x_N = b_1$$

$$\text{Eq } \langle 2 \rangle: \quad a_{2,1}x_1 + a_{2,2}x_2 + a_{2,3}x_3 + \ldots + a_{2,N}x_N = b_2$$

$$\vdots \qquad \vdots \qquad \vdots \qquad \vdots \qquad \vdots$$

$$\text{Eq } \langle k \rangle: \quad a_{k,1}x_1 + a_{k,2}x_2 + a_{k,3}x_3 + \ldots + a_{k,N}x_N = b_k$$

$$\vdots \qquad \vdots \qquad \vdots \qquad \vdots \qquad \vdots$$

$$\text{Eq } \langle M \rangle: \quad a_{M,1}x_1 + a_{M,2}x_2 + a_{M,3}x_3 + \ldots + a_{M,N}x_N = b_M.$$

To keep track of the different coefficients in each equation, it was necessary to use the two subscripts (k, j). The first subscript locates equation k and the second subscript locates the variable x_j.

A solution to (3) is a set of numerical values $x_1, x_2, \ldots, x_N$ that satisfies all the equations Eq $\langle 1 \rangle$, Eq $\langle 2 \rangle, \ldots,$ Eq $\langle M \rangle$. Hence a solution can be viewed as an N-dimensional vector:

(4)

$$\mathbf{X} = (x_1, x_2, \ldots, x_N).$$

Example 3.6 Concrete (used for sidewalks, etc.) is a mixture of portland cement, sand, and gravel. A distributor has three batches available for contractors. Batch 1 contains cement, sand, and gravel mixed in the proportions $\frac{1}{8} : \frac{3}{8} : \frac{4}{8}$. Batch 2 has the proportions $\frac{2}{10} : \frac{5}{10} : \frac{3}{10}$, and batch 3 has the proportions $\frac{2}{5} : \frac{3}{5} : \frac{0}{5}$.

Let $x_1, x_2,$ and x_3 denote the amount (in cubic yards) to be used from each batch to form a mixture of 10 cubic yards. Also, suppose that the mixture is to contain $b_1 = 2.3$, $b_2 = 4.8$, and $b_3 = 2.9$ cubic yards of portland cement, sand, and gravel, respectively. Then the system of linear equations for the ingredients is

(5)

$$\langle \text{cement} \rangle \quad 0.125x_1 + 0.200x_2 + 0.400x_3 = 2.3$$

$$\langle \text{sand} \rangle \quad 0.375x_1 + 0.500x_2 + 0.600x_3 = 4.8$$

$$\langle \text{gravel} \rangle \quad 0.500x_1 + 0.300x_2 + 0.000x_3 = 2.9.$$

The solution to the linear system (5) is $x_1 = 4$, $x_2 = 3$, and $x_3 = 3$, which can be verified by direct substitution into the equations:

$$\langle \text{cement} \rangle \quad 0.125 \times 4 + 0.200 \times 3 + 0.400 \times 3 = 2.3$$

$$\langle \text{sand} \rangle \quad 0.375 \times 4 + 0.500 \times 3 + 0.600 \times 3 = 4.8$$

$$\langle \text{gravel} \rangle \quad 0.500 \times 4 + 0.300 \times 3 + 0.000 \times 3 = 2.9.$$

We now discuss how matrices and vectors are used to represent a linear system of equations. Each equation in (3) can be written using the dot-product notation for vectors:

Eq $\langle 1 \rangle$: $(a_{1,1}, a_{1,2}, \ldots, a_{1,j}, \ldots, a_{1,N}) \cdot (x_1, x_2, \ldots, x_j, \ldots, x_N) = b_1$

Eq $\langle 2 \rangle$: $(a_{2,1}, a_{2,2}, \ldots, a_{2,j}, \ldots, a_{2,N}) \cdot (x_1, x_2, \ldots, x_j, \ldots, x_N) = b_2$

$$\vdots$$

Eq $\langle k \rangle$: $(a_{k,1}, a_{k,2}, \ldots, a_{k,j}, \ldots, a_{k,N}) \cdot (x_1, x_2, \ldots, x_j, \ldots, x_N) = b_k$

$$\vdots$$

Eq $\langle M \rangle$: $(a_{M,1}, a_{M,2}, \ldots, a_{M,j}, \ldots, a_{M,N}) \cdot (x_1, x_2, \ldots, x_j, \ldots, x_N) = b_M.$

One usually stores the coefficients $a_{k,j}$ in a matrix A of dimension M by N and the unknowns x_j are stored in an N-dimensional vector **X**. The constants b_k are stored in an M-dimensional vector **B**. It is conventional to use column vectors for both **X** and **B** and write

(6) $$A\mathbf{X} = \begin{pmatrix} a_{1,1} & a_{1,2} & \cdots & a_{1,j} & \cdots & a_{1,N} \\ a_{2,1} & a_{2,2} & \cdots & a_{2,j} & \cdots & a_{2,N} \\ \vdots & \vdots & & \vdots & & \vdots \\ a_{k,1} & a_{k,2} & \cdots & a_{k,j} & \cdots & a_{k,N} \\ \vdots & \vdots & & \vdots & & \vdots \\ a_{M,1} & a_{M,2} & \cdots & a_{M,j} & \cdots & a_{M,N} \end{pmatrix} \begin{pmatrix} x_1 \\ x_2 \\ \vdots \\ x_j \\ \vdots \\ x_N \end{pmatrix} = \begin{pmatrix} b_1 \\ b_2 \\ \vdots \\ b_k \\ \vdots \\ b_M \end{pmatrix} = B.$$

The matrix multiplication $A\mathbf{X} = \mathbf{B}$ in (6) is reminiscent of the dot-product multiplication for ordinary vectors, because each element b_k in **B** is the result obtained by taking the dot product of row k in matrix A with the column vector **X**. We will use $A\mathbf{X}$ to denote the product of A times **X**.

Example 3.7 Express the results of Example 3.6 with matrix notation.

Solution. Equations (5) can be written in the matrix form

(7) $$\begin{pmatrix} 0.125 & 0.200 & 0.400 \\ 0.375 & 0.500 & 0.600 \\ 0.500 & 0.300 & 0.000 \end{pmatrix} \begin{pmatrix} x_1 \\ x_2 \\ x_3 \end{pmatrix} = \begin{pmatrix} 2.3 \\ 4.8 \\ 2.9 \end{pmatrix}.$$

In vector notation, the column vector $(4, 3, 3)^T$ is shown to be the solution by performing the matrix calculation

$$\begin{pmatrix} 0.125 & 0.200 & 0.400 \\ 0.375 & 0.500 & 0.600 \\ 0.500 & 0.300 & 0.000 \end{pmatrix} \begin{pmatrix} 4 \\ 3 \\ 3 \end{pmatrix} = \begin{pmatrix} 0.5 + 0.6 + 1.2 \\ 1.5 + 1.5 + 1.8 \\ 2.0 + 0.9 + 0.0 \end{pmatrix} = \begin{pmatrix} 2.3 \\ 4.8 \\ 2.9 \end{pmatrix}.$$

Matrix Multiplication

Let $A = (a_{i,k})_{M \times N}$ and $B = (b_{k,j})_{N \times P}$ be two matrices with the property that A has as many columns as B has rows. Then the matrix product AB is defined to be the matrix C of dimension M by P

$$(8) \qquad AB = C = (c_{i,j})_{M \times P}$$

where the element $c_{i,j}$ of C is given by the dot product of the ith row of A and the jth column of B:

$$(9) \qquad c_{i,j} = (a_{i,1}, a_{i,2}, \ldots, a_{i,N}) \cdot (b_{1,j}, b_{2,j}, \ldots, b_{N,j}) = \sum_{k=1}^{N} a_{i,k} b_{k,j}.$$

Using the notation $c_{i,j} = (\text{row}_i \ A) \cdot (\text{col}_j \ B)$ in equation (9), the matrix product AB can be viewed as follows:

$$AB = \begin{pmatrix} \text{row}_1 \ A \cdot \text{col}_1 \ B & \text{row}_1 \ A \cdot \text{col}_2 \ B & \cdots & \text{row}_1 \ A \cdot \text{col}_j \ B & \cdots & \text{row}_1 \ A \cdot \text{col}_P \ B \\ \text{row}_2 \ A \cdot \text{col}_1 \ B & \text{row}_2 \ A \cdot \text{col}_2 \ B & \cdots & \text{row}_2 \ A \cdot \text{col}_j \ B & \cdots & \text{row}_2 \ A \cdot \text{col}_P \ B \\ \vdots & \vdots & & \vdots & & \vdots \\ \text{row}_i \ A \cdot \text{col}_1 \ B & \text{row}_i \ A \cdot \text{col}_2 \ B & \cdots & \text{row}_i \ A \cdot \text{col}_j \ B & \cdots & \text{row}_i \ A \cdot \text{col}_P \ B \\ \vdots & \vdots & & \vdots & & \vdots \\ \text{row}_M \ A \cdot \text{col}_1 \ B & \text{row}_M \ A \cdot \text{col}_2 \ B & \cdots & \text{row}_M \ A \cdot \text{col}_j \ B & \cdots & \text{row}_M \ A \cdot \text{col}_P \ B \end{pmatrix}$$

Example 3.8 Find the product $C = AB$ for the following matrices, and tell why BA is not defined.

$$A = \begin{pmatrix} 2 & 3 \\ -1 & 4 \end{pmatrix} \qquad B = \begin{pmatrix} 5 & -2 & 1 \\ 3 & 8 & -6 \end{pmatrix}.$$

Solution. The matrix A has two columns and B has two rows, so that the matrix product AB is defined. The product of a 2 by 2 matrix and a 2 by 3 matrix is a 2 by 3 matrix. Computation reveals that

$$\begin{pmatrix} 2 & 3 \\ -1 & 4 \end{pmatrix} \begin{pmatrix} 5 & -2 & 1 \\ 3 & 8 & -6 \end{pmatrix} = \begin{pmatrix} 10 + 9 & -4 + 24 & 2 - 18 \\ -5 + 12 & 2 + 32 & -1 - 24 \end{pmatrix}.$$

Thus

$$C = \begin{pmatrix} 19 & 20 & -16 \\ 7 & 34 & -25 \end{pmatrix}.$$

When an attempt is made to form the product BA we discover that the dimensions are not compatible in this order because $(\text{row}_i \ B)$ is a three-dimensional vector and $(\text{col}_j \ A)$ is a two-dimensional vector. Hence the dot product $(\text{row}_i \ B) \cdot (\text{col}_j \ A)$ is not defined, that is,

$$\begin{pmatrix} 5 & -2 & 1 \\ 3 & 8 & -6 \end{pmatrix} \begin{pmatrix} 2 & 3 \\ -1 & 4 \end{pmatrix} = ?$$

If it happens that $AB = BA$, then we say that A and B *commute.*

Example 3.9 Show that the following matrices commute.

$$A = \begin{pmatrix} 1 & 2 \\ 2 & 1 \end{pmatrix} \qquad B = \begin{pmatrix} 2 & 1 \\ 1 & 2 \end{pmatrix}.$$

Solution. Forming the products AB and BA, we see that

$$AB = \begin{pmatrix} 1 & 2 \\ 2 & 1 \end{pmatrix} \begin{pmatrix} 2 & 1 \\ 1 & 2 \end{pmatrix} = \begin{pmatrix} 4 & 5 \\ 5 & 4 \end{pmatrix}$$

$$BA = \begin{pmatrix} 2 & 1 \\ 1 & 2 \end{pmatrix} \begin{pmatrix} 1 & 2 \\ 2 & 1 \end{pmatrix} = \begin{pmatrix} 4 & 5 \\ 5 & 4 \end{pmatrix}.$$

The reader may suspect that the matrices in Example 3.9 were contrived to satisfy $AB = BA$. Most often, even when AB and BA are both defined, the products are not necessarily the same. Observe that if AB were to equal BA, then a necessary condition would be that both A and B are square matrices of the same size (i.e., both have dimension N by N). The next example gives two matrices that do not commute. To emphasize this point, it is often stated that matrix multiplication is not a commutative operation.

Example 3.10 Show that the following matrices do not commute.

$$A = \begin{pmatrix} 1 & -2 & 3 \\ 2 & 5 & -1 \\ 6 & -3 & 4 \end{pmatrix} \qquad B = \begin{pmatrix} 2 & 3 & 1 \\ 4 & -1 & 2 \\ -5 & 2 & -3 \end{pmatrix}.$$

Solution. Computation reveals that

$$AB = \begin{pmatrix} -21 & 11 & -12 \\ 29 & -1 & 15 \\ -20 & 29 & -12 \end{pmatrix} \qquad BA = \begin{pmatrix} 14 & 8 & 7 \\ 14 & -19 & 21 \\ -19 & 29 & -29 \end{pmatrix}.$$

Some Special Matrices

The M by N matrix whose elements are all zero is called the *zero matrix of dimension M by N* and is denoted by

(10)
$$O = (0)_{M \times N}.$$

When the dimension is clear we use O to denote the zero matrix.

The *identity matrix of order N* is the square matrix given by

(11)
$$I_N = \begin{pmatrix} 1 & 0 & 0 & \cdots & 0 & 0 \\ 0 & 1 & 0 & \cdots & 0 & 0 \\ 0 & 0 & 1 & \cdots & 0 & 0 \\ \vdots & \vdots & \vdots & & \vdots & \vdots \\ 0 & 0 & 0 & \cdots & 1 & 0 \\ 0 & 0 & 0 & \cdots & 0 & 1 \end{pmatrix}.$$

When the dimension is clear we use I to denote the identity matrix. The identity matrix is often denoted with the notation

(12)
$$I_N = (\delta_{i,j})_{N \times N}, \quad \text{where } \delta_{i,j} = \begin{cases} 1 & \text{when } i = j \\ 0 & \text{when } i \neq j. \end{cases}$$

It is the multiplicative identity, as illustrated in the next example.

Example 3.11 Let A be a 2 by 3 matrix. Then $I_2 A = A I_3 = A$.

Solution. Multiplication of A on the left by I_2 results in

$$\begin{pmatrix} 1 & 0 \\ 0 & 1 \end{pmatrix} \begin{pmatrix} a_{11} & a_{12} & a_{13} \\ a_{21} & a_{22} & a_{23} \end{pmatrix} = \begin{pmatrix} a_{11} + 0 & a_{12} + 0 & a_{13} + 0 \\ 0 + a_{21} & 0 + a_{22} & 0 + a_{23} \end{pmatrix} = A.$$

Multiplication of A on the right by I_3 results in

$$\begin{pmatrix} a_{11} & a_{12} & a_{13} \\ a_{21} & a_{22} & a_{23} \end{pmatrix} \begin{pmatrix} 1 & 0 & 0 \\ 0 & 1 & 0 \\ 0 & 0 & 1 \end{pmatrix} = \begin{pmatrix} a_{11} + 0 + 0 & 0 + a_{12} + 0 & 0 + 0 + a_{13} \\ a_{21} + 0 + 0 & 0 + a_{22} + 0 & 0 + 0 + a_{23} \end{pmatrix} = A.$$

Some properties of matrix multiplication are given in the following theorem.

Theorem 3.3 [Matrix multiplication] Let p be a scalar (real number). If A, B, C, and D are matrices such that the indicated sums and products are defined, then

(13) $\qquad A(BC) = (AB)C \qquad\qquad$ matrix associative property

(14) $\qquad\qquad IA = AI = A \qquad\qquad$ identity matrix

(15) $\qquad A(B + C) = AB + AC \qquad$ left distributive property

(16) $\qquad (A + B)C = AC + BC \qquad$ right distributive property

(17) $\qquad p(AB) = (pA)B = A(pB) \qquad$ scalar associative property

The Inverse of a Nonsingular Matrix

The concept of inverse applies to matrices, but special attention must be given. An N by N matrix A is called *nonsingular* or *invertible* if there exists an N by N matrix B such that

(18)
$$AB = BA = I.$$

If no such matrix B can be found, A is said to be *singular*. When B can be found and (18) holds, we usually write $B = A^{-1}$ and use the familiar relation

(19)
$$AA^{-1} = A^{-1}A = I \qquad \text{if } A \text{ is nonsingular.}$$

It is easy to show that at most one matrix B can be found that satisfies relation (18). For suppose that C is also an inverse (i.e., $AC = CA = I$). Then properties (13) and (14) can be used to obtain

$$C = IC = (BA)C = B(AC) = BI = B.$$

Theorem 3.4   [Inverse of a 2 by 2 matrix] A necessary and sufficient condition for the matrix

(20)
$$A = \begin{pmatrix} a & b \\ c & d \end{pmatrix}$$

to have an inverse is that $ad - bc \neq 0$. If $ad - bc \neq 0$, then

(21)
$$A^{-1} = \frac{1}{ad - bc} \begin{pmatrix} d & -b \\ -c & a \end{pmatrix}.$$

Proof. If $ad - bc \neq 0$, then the following calculation is valid:

$$\frac{1}{ad - bc} \begin{pmatrix} a & b \\ c & d \end{pmatrix} \begin{pmatrix} d & -b \\ -c & a \end{pmatrix} = \frac{1}{ad - bc} \begin{pmatrix} ad - bc & -ab + ba \\ cd - dc & -cb + da \end{pmatrix} = \begin{pmatrix} 1 & 0 \\ 0 & 1 \end{pmatrix}.$$

Hence A^{-1} given in formula (21) is the inverse of A.

Suppose that $ad - bc = 0$. If both $a = 0$ and $b = 0$, then the first row of the product AB, for any martix B, will contain only zero elements, so $AB \neq I$. Hence A does not have an inverse. On the other hand, if either $a \neq 0$ or $b \neq 0$, then

$$B \begin{pmatrix} a & b \\ c & d \end{pmatrix} \begin{pmatrix} b \\ -a \end{pmatrix} = B \begin{pmatrix} ab - ba \\ cb - da \end{pmatrix} = B \begin{pmatrix} 0 \\ 0 \end{pmatrix} = \begin{pmatrix} 0 \\ 0 \end{pmatrix}.$$

Thus for any matrix B, BA cannot be I, because

$$(BA) \begin{pmatrix} b \\ -a \end{pmatrix} = \begin{pmatrix} 0 \\ 0 \end{pmatrix} \neq \begin{pmatrix} b \\ -a \end{pmatrix}.$$

Hence A does not have an inverse.

We remark that if $ad - bc = 0$, then in Sections 3.4 and 3.5 we will see that the algorithm for constructing A^{-1} will encounter a division-by-zero error.

Example 3.12 Find the inverse of the matrix

$$A = \begin{pmatrix} 3 & 1 \\ 7 & 4 \end{pmatrix}.$$

Solution. Using formula (21), we find that

$$A^{-1} = \frac{1}{12 - 7} \begin{pmatrix} 4 & -1 \\ -7 & 3 \end{pmatrix} = \begin{pmatrix} 0.8 & -0.2 \\ -1.4 & 0.6 \end{pmatrix}.$$

Determinants

The determinant of a square matrix A is a scalar quantity (real number) and is denoted by $\det A$. If A is an N by N matrix

$$A = \begin{pmatrix} a_{1,1} & a_{1,2} & \cdots & a_{1,N} \\ a_{2,1} & a_{2,2} & \cdots & a_{2,N} \\ \vdots & \vdots & & \vdots \\ a_{N,1} & a_{N,2} & \cdots & a_{N,N} \end{pmatrix}$$

then it is customary to write

$$\det A = \begin{vmatrix} a_{1,1} & a_{1,2} & \cdots & a_{1,N} \\ a_{2,1} & a_{2,2} & \cdots & a_{2,N} \\ \vdots & \vdots & & \vdots \\ a_{N,1} & a_{N,2} & \cdots & a_{N,N} \end{vmatrix}.$$

Although the notation for a determinant may look like a matrix, its properties are completely different. For one, the determinant is a scalar quantity (real number). One goal of numerical methods is to develop an efficient algorithm for computing the determinant. The definition of $\det A$ found in most linear algebra textbooks is not tractable for computation when $N > 3$. We will review how to compute determinants of the first three orders. Evaluation of higher-order determinants is done using Gaussian elimination and is mentioned in the body of Algorithm 3.6.

If $A = (a_{1,1})$ is a 1 by 1 matrix, we define $\det A = a_{1,1}$. If A is a 2 by 2 matrix,

$$A = \begin{pmatrix} a_{1,1} & a_{1,2} \\ a_{2,1} & a_{2,2} \end{pmatrix}$$

then we define

(22) $$\det A = a_{1,1}a_{2,2} - a_{1,2}a_{2,1}.$$

If B is a 3 by 3 matrix,

$$B = \begin{pmatrix} b_{1,1} & b_{1,2} & b_{1,3} \\ b_{2,1} & b_{2,2} & b_{2,3} \\ b_{3,1} & b_{3,2} & b_{3,3} \end{pmatrix}$$

then we define

(23)
$$\det B = b_{1,1}b_{2,2}b_{3,3} + b_{1,2}b_{2,3}b_{3,1} + b_{1,3}b_{2,1}b_{3,2}$$
$$- b_{1,3}b_{2,2}b_{3,1} - b_{1,2}b_{2,1}b_{3,3} - b_{1,1}b_{2,3}b_{3,2}.$$

Before defining the determinant of the square matrix A of order N, we must introduce some preliminary concepts. Consider the set of integers $S = \{1, 2, \ldots, N\}$. An ordering of the elements of S is called a *permutation*. For example, if $S = \{1, 2, 3\}$, then there are six permutations:

$$(1, 2, 3), \quad (1, 3, 2), \quad (2, 1, 3), \quad (2, 3, 1), \quad (3, 1, 2), \quad (3, 2, 1).$$

In general, the number of permutations of a set of N integers is $N!$.

Let $P = (j_1, j_2, \ldots, j_N)$ be a permutation of S. Let α_k be the number of integers following j_k that are smaller than j_k for $k = 1, 2, \ldots, N - 1$. The sum $\alpha_1 + \alpha_2 + \ldots + \alpha_{N-1}$ is called the *number of inversions* in the permutation. A permutation P is said to be *even* (*odd*) if it has an even (odd) number of inversions. For example, if $P = (3, 2, 1)$, then $\alpha_1 = 2$ and $\alpha_2 = 1$ and $\alpha_1 + \alpha_2 = 3$, so that P has an odd number of inversions. For each permutation P of S, we define

$$\delta(P) = \begin{cases} 0 & \text{if } P \text{ has an even number of inversions,} \\ 1 & \text{if } P \text{ has an odd number of inversions.} \end{cases}$$

For example, $\delta(3, 2, 1) = 1$ and $\delta(3, 1, 2) = 0$.

Making use of the definitions above, we shall define the *determinant* of the N by N matrix A as follows:

$$\det A = \sum (-1)^{\delta(P)} a_{1,j_1} a_{2,j_2} \ldots a_{N,j_N}$$

where the sum is taken over all permutations $P = (j_1, j_2, \ldots, j_N)$ of S. This definition is not tractable to implement when N is large. For $N = 10$ it would require $10 \times 10! \approx 3.6 \times 10^7$ multiplications.

Example 3.13 Find the determinant of the matrices

$$A = \begin{pmatrix} 4 & 9 \\ 1 & 3 \end{pmatrix} \qquad B = \begin{pmatrix} 2 & 3 & 8 \\ -4 & 5 & -1 \\ 7 & -6 & 9 \end{pmatrix}.$$

Solution. Using formula (22), we obtain

$$\det A = (4)(3) - (9)(1) = 3.$$

Using formula (23), det B is computed as follows:

$$\det B = (2)(5)(9) + (3)(-1)(7) + (8)(-4)(-6)$$
$$- (8)(5)(7) - (3)(-4)(9) - (2)(-1)(-6)$$
$$= (90) + (-21) + (192) - (280) - (-108) - (12) = 77.$$

The following theorem gives sufficient conditions for the existence and uniqueness of solutions $A\mathbf{X} = \mathbf{B}$ for square matrices.

Theorem 3.5 Let A be an arbitrary N by N matrix. The following statements are equivalent.

(24)   The linear system $A\mathbf{X} = \mathbf{B}$ has a solution for any N-dimensional column vector $\mathbf{B}$.

(25)   The only solution of $A\mathbf{X} = 0$ is $\mathbf{X} = 0$.

(26)   The matrix A is nonsingular (i.e., A^{-1} exists).

(27) $\det A \neq 0$.

Theorems 3.3 and 3.5 help relate matrix algebra to ordinary algebra. If statement (24) is true, then statement (26) together with properties (12) and (13) give the following line of reasoning:

(28) $A\mathbf{X} = \mathbf{B}$ implies that $A^{-1}A\mathbf{X} = A^{-1}\mathbf{B}$, which implies that $\mathbf{X} = A^{-1}\mathbf{B}$.

Example 3.14 Use the inverse matrix A^{-1} in Example 3.12 and the reasoning in (28) to solve the linear system of equations $A\mathbf{X} = \mathbf{B}$:

$$A\mathbf{X} = \begin{pmatrix} 3 & 1 \\ 7 & 4 \end{pmatrix} \begin{pmatrix} x_1 \\ x_2 \end{pmatrix} = \begin{pmatrix} 2 \\ 5 \end{pmatrix} = B.$$

Solution. Using the last statement in (28), we get

$$\mathbf{X} = A^{-1}\mathbf{B} = \begin{pmatrix} 0.8 & -0.2 \\ -1.4 & 0.6 \end{pmatrix} \begin{pmatrix} 2 \\ 5 \end{pmatrix} = \begin{pmatrix} 0.6 \\ 0.2 \end{pmatrix}.$$

Cramer's Rule

A practical method for solving 2×2 systems of linear equations is *Cramer's rule*, which we mention for the sake of completeness. Although Cramer's rule can be used to solve 3 by 3 systems, it is intractable when $N > 3$. In Section 3.4 we develop an efficient method for solving a large systems of equations. Consider the two linear equations

(29)
$$ax_1 + bx_2 = e,$$
$$cx_1 + dx_2 = f,$$

with the condition that $ad - bc \neq 0$. We can solve for the variable x_1 by eliminating the variable x_2. This is accomplished by multiplying the top equation by d and the bottom equation by b, and subtracting:

$$
\begin{array}{c}
adx_1 + bdx_2 = ed \\
-(bcx_1 + bdx_2 = bf) \\
\hline
adx_1 - bcx_1 = ed - bf
\end{array}
$$

Hence $(ad - bc)x_1 = ed - bf$ and we can solve for x_1 and obtain

(30)
$$
x_1 = \frac{ed - bf}{ad - bc}; \quad \text{similarly,} \quad x_2 = \frac{af - ec}{ad - bc}.
$$

The quotients in (30) can be expressed using determinants:

(31)
$$
x_1 = \frac{\begin{vmatrix} e & b \\ f & d \end{vmatrix}}{\begin{vmatrix} a & b \\ c & d \end{vmatrix}} \quad \text{and} \quad x_2 = \frac{\begin{vmatrix} a & e \\ c & f \end{vmatrix}}{\begin{vmatrix} a & b \\ c & d \end{vmatrix}}.
$$

Example 3.15 Use Cramer's rule to solve the linear system

$$
A\mathbf{X} = \begin{pmatrix} 3 & 1 \\ 7 & 4 \end{pmatrix} \begin{pmatrix} x_1 \\ x_2 \end{pmatrix} = \begin{pmatrix} 2 \\ 5 \end{pmatrix} = \mathbf{B}.
$$

Solution. Using the equations in (31), we get

$$
x_1 = \frac{\begin{vmatrix} 2 & 1 \\ 5 & 4 \end{vmatrix}}{\begin{vmatrix} 3 & 1 \\ 7 & 4 \end{vmatrix}} = \frac{8 - 5}{12 - 7} = 0.6 \qquad x_2 = \frac{\begin{vmatrix} 3 & 2 \\ 7 & 5 \end{vmatrix}}{\begin{vmatrix} 3 & 1 \\ 7 & 4 \end{vmatrix}} = \frac{15 - 14}{12 - 7} = 0.2.
$$

Algorithm 3.2 [*Linear Transformation*] If $A = (a_{i,k})$ is a matrix of dimension M by N and $\mathbf{X} = (x_k)$ is an N-dimensional column vector, then the product $\mathbf{Y} = A\mathbf{X}$, where $\mathbf{Y} = (y_i)$, is an M-dimensional column vector and the elements y_i are given by

$$
y_i = \sum_{k=1}^{N} a_{i,k} x_k \qquad \text{for } 1 \leq i \leq M.
$$

Remark. Since the input vector $\mathbf{X}$ is of dimension N and the output vector $\mathbf{Y}$ is of dimension M, we say that $\mathbf{Y} = A\mathbf{X}$ is a *linear transformation* from N-dimensional space into M-dimensional space.

```
        INPUT   M , N
        VARiable declaration                          {Dimension the arrays}
            REAL   A[1 .. M, 1 .. N] , X[1 .. N] , Y[1 .. M]


FOR     K = 1   TO  N  DO                    {Input the columns of matrix A}
        FOR    I = 1  TO  M  DO
               PRINT " A( " ; I ; " , " ; K ; " ) = " ;     {Prompt for input}
               INPUT   A(I, K)                         {Get element aᵢ,ₖ}


FOR     I = 1  TO  N  DO                     {Input the column vector X}
        PRINT " X( " ; I ; " ) = " ;                  {Prompt for input}
        INPUT   X( I )                                {Get element xᵢ}


FOR     I = 1  TO  M  DO                               {Fix the row i}
        SUM := 0                             {Compute the dot product
        FOR    K = 1  TO  N  DO                        for element yᵢ}
               SUM := SUM + A(I, K)*X(K)
        Y( I ) := SUM
        PRINT  Y( I )                                 {Output}
```

Algorithm 3.3 [*Matrix Multiplication*] If $A = (a_{i,k})$ is a matrix of dimension M by N and $B = (b_{k,j})$ is a matrix of dimension N by P, then their product $C = AB$, where $C = (c_{i,j})$ is a matrix of dimension M by P and the elements $c_{i,j}$ are given by

$$c_{i,j} = \sum_{k=1}^{N} a_{i,k}b_{k,j} \qquad \text{for } 1 \leqq i \leqq M, \quad 1 \leqq j \leqq P.$$

```
        INPUT   M , N , P
        VARiable declaration                          {Dimension the arrays}
            REAL   A[1 .. M, 1 .. N] , B[1 .. N, 1 .. P] , C[1 .. M, 1 .. P]


FOR     K = 1  TO  N  DO                     {Input the columns of matrix A}
        FOR    I = 1  TO  M  DO
               PRINT " A( " ; I ; " , " ; K ; " ) = " ;     {Prompt for input}
               INPUT   A(I, K)                         {Get element aᵢ,ₖ}


FOR     J = 1  TO  P  DO                     {Input the columns of matrix B}
        FOR    K = 1  TO  N  DO
               PRINT " B( " ; K ; " , " ; J ; " ) = " ;     {Prompt for input}
               INPUT   B(K, J)                         {Get element bₖ,ⱼ}
```

```
FOR     I = 1  TO  M  DO                          {Fix the row i}
        FOR     J = 1  TO  P  DO                   {Find the jth element}
                SUM := 0                           {Compute the dot product
                FOR     K = 1  TO  N  DO               for element c_{i,j}}
                        SUM := SUM + A(I, K)*B(K, J)
                C(I, J) := SUM
                PRINT   C(I, J) ;                  {Output}
        PRINT                                      {Move on to the next row}
```

Exercises for Properties of Vectors and Matrices

1. Find AB and BA for the following matrices.

$$A = \begin{pmatrix} -3 & 2 \\ 1 & 4 \end{pmatrix} \qquad B = \begin{pmatrix} 5 & 0 \\ 2 & -6 \end{pmatrix}.$$

2. Find AB and BA for the following matrices.

$$A = \begin{pmatrix} 1 & -2 & 3 \\ 2 & 0 & 5 \end{pmatrix} \qquad B = \begin{pmatrix} 3 & 0 \\ -1 & 5 \\ 3 & -2 \end{pmatrix}.$$

3. Let A, B, and C be given by

$$A = \begin{pmatrix} 3 & 1 \\ 0 & 4 \end{pmatrix} \qquad B = \begin{pmatrix} 1 & 2 \\ -2 & -6 \end{pmatrix} \qquad C = \begin{pmatrix} 2 & -5 \\ 3 & 4 \end{pmatrix}.$$

(a) Find $(AB)C$ and $A(BC)$.
(b) Find $A(B + C)$ and $AB + AC$.
(c) Find $(A + B)C$ and $AC + BC$.

4. We use the notation convention $A^2 = AA$.
Find A^2 and B^2 for the following matrices.

$$A = \begin{pmatrix} -1 & -7 \\ 5 & 2 \end{pmatrix} \qquad B = \begin{pmatrix} 2 & 0 & 6 \\ -1 & 5 & -4 \\ 3 & -5 & 2 \end{pmatrix}.$$

5. Find the determinant of the following matrices if it exists.

(a) $\begin{pmatrix} -1 & -7 \\ 5 & 2 \end{pmatrix}$

(b) $\begin{pmatrix} 2 & 0 & 1 \\ -3 & 0 & 5 \end{pmatrix}$

(c) $\begin{pmatrix} 1 & 2 \\ 3 & 4 \\ 0 & 0 \end{pmatrix}$

(d) $\begin{pmatrix} 2 & 0 & 6 \\ -1 & 5 & -4 \\ 3 & -5 & 2 \end{pmatrix}$

6. Solve the linear system $AX = B$ by first using Theorem 3.4 to find A^{-1} and then computing $X = A^{-1}B$.

(a) $\begin{pmatrix} 3 & 1 \\ 7 & 4 \end{pmatrix} \begin{pmatrix} x_1 \\ x_2 \end{pmatrix} = \begin{pmatrix} 3 \\ -2 \end{pmatrix}$

(b) $\begin{pmatrix} 3 & 4 \\ 1 & 3 \end{pmatrix} \begin{pmatrix} x_1 \\ x_2 \end{pmatrix} = \begin{pmatrix} 3 \\ -2 \end{pmatrix}$

(c) $\begin{pmatrix} 3 & 4 \\ 1 & 3 \end{pmatrix} \begin{pmatrix} x_1 \\ x_2 \end{pmatrix} = \begin{pmatrix} 7 \\ 5 \end{pmatrix}$

(d) $\begin{pmatrix} 6 & 4 \\ 7 & 3 \end{pmatrix} \begin{pmatrix} x_1 \\ x_2 \end{pmatrix} = \begin{pmatrix} 3 \\ -2 \end{pmatrix}$

7. Solve the linear systems in Exercise 6 by using Cramer's rule.

8. Given the following expressions for A, B, A^{-1}, and B^{-1}.

$$A = \begin{pmatrix} 5 & 3 \\ 3 & 2 \end{pmatrix} \qquad B = \begin{pmatrix} 2 & 3 \\ 5 & 7 \end{pmatrix}$$

$$A^{-1} = \begin{pmatrix} 2 & -3 \\ -3 & 5 \end{pmatrix} \qquad B^{-1} = \begin{pmatrix} -7 & 3 \\ 5 & -2 \end{pmatrix}.$$

(a) Verify that $AA^{-1} = I$ and $BB^{-1} = I$.
(b) Find $C = A^{-1}B^{-1}$ and $D = B^{-1}A^{-1}$.
(c) Find AB; determine whether C or D is the inverse of AB.

9. Let A and B be 2 by 2 matrices. If $AB = 0$, then does it follow that $A = 0$ or $B = 0$?

10. If A and B are nonsingular N by N matrices and $C = AB$, then show that $C^{-1} = B^{-1}A^{-1}$. *Hint.* You must use the associative property of matrix multiplication.

11. Let A be an M by N matrix and X an N-dimensional vector.
(a) How many multiplications are needed to calculate AX?
(b) How many additions are needed to calculate AX?

12. Let A be an M by N matrix and B be an N by P matrix.
(a) How many multiplications are needed to calculate AB?
(b) How many additions are needed to calculate AB?

13. Let A be an M by N matrix, and let B and C be N by P matrices. Prove the left distributive law for matrix multiplication

$$A(B + C) = AB + AC.$$

14. Let A and B be M by N matrices, and let C be an N by P matrix. Prove the right distributive law for matrix multiplication

$$(A + B)C = AC + BC.$$

15. Find XX^T and X^TX, where $X = (1, -1, 2)$. *Note:* X^T is the transpose of X.

16. Let $A = \begin{pmatrix} a & b \\ c & d \end{pmatrix}$ and assume that $ad - bc \neq 0$.

Show that $(A^T)^{-1} = (A^{-1})^T$, where A^T is the transpose of A.

17. Let $A = \begin{pmatrix} -3 & 2 \\ 1 & 4 \end{pmatrix}$ and $B = \begin{pmatrix} 5 & 0 \\ 2 & -6 \end{pmatrix}$.

(a) Show that det $(AB) = (\det A)(\det B)$.

(b) Show that det $(A + B) \neq \det A + \det B$.

18. Let $A = \begin{pmatrix} a_{1,1} & a_{1,2} \\ a_{2,1} & a_{2,2} \end{pmatrix}$ and $B = \begin{pmatrix} b_{1,1} & b_{1,2} \\ b_{2,1} & b_{2,2} \end{pmatrix}$, and let p be a scalar (real number).

(a) Show that det $(pA) = p^2(\det A)$.

(b) Show that det $(AB) = (\det A)(\det B)$.

(c) Show that det $(A^T) = \det A$.

(d) Show that det $(A + B) \neq \det A + \det B$ in general.

(e) Show that $(AB)^T = B^T A^T$.

19. Let $(x_1, y_1), (x_2, y_2), (x_3, y_3)$ be the vertices of a triangle in the xy-plane. It is a fact that the area of the triangle is given by

$$\text{Area} = \left| \frac{1}{2} \begin{vmatrix} x_1 & y_1 & 1 \\ x_2 & y_2 & 1 \\ x_3 & y_3 & 1 \end{vmatrix} \right|.$$

Find the areas of the triangles with the vertices:

(a) $(1, 1), (5, 0), (2, 4)$

(b) $(-1, 2), (3, 4), (7, -6)$

3.3 Upper–Triangular Linear Systems

We will now develop the *back-substitution algorithm*, which is useful for solving a linear system of equations that has an upper-triangular coefficient matrix. This algorithm will be incorporated in the algorithm for solving a general system in Section 3.4.

Definition 3.1 [Upper-triangular system] The linear system $AX = B$ is said to be an upper-triangular system and A is called an upper-triangular matrix provided that $a_{i,j} = 0$ for $i > j$, and we write

$$
\begin{array}{r}
a_{1,1}x_1 + a_{1,2}x_2 + a_{1,3}x_3 + \ldots + a_{1,N-1}x_{N-1} + \quad a_{1,N}x_N = b_1 \\
a_{2,2}x_2 + a_{2,3}x_3 + \ldots + a_{2,N-1}x_{N-1} + \quad a_{2,N}x_N = b_2 \\
a_{3,3}x_3 + \ldots + a_{3,N-1}x_{N-1} + \quad a_{3,N}x_N = b_3 \\
\vdots \qquad\qquad \vdots \\
a_{N-1,N-1}x_{N-1} + a_{N-1,N}x_N = b_{N-1} \\
a_{N,N}x_N = b_N.
\end{array}
$$

(1)

Theorem 3.6 [Back substitution] If $AX = B$ is an upper-triangular system with form given by (1) and the diagonal elements are nonzero,

(2) $$a_{k,k} \neq 0 \quad \text{for } k = 1, 2, \ldots, N,$$

there exists a unique solution to the linear system (1).

Constructive Proof. The solution is easy to find. The last equation involves only x_N, so we solve it first:

(3)
$$x_N = \frac{b_N}{a_{N,N}}.$$

Now x_N is known and it can be used in the next-to-last equation:

(4)
$$x_{N-1} = \frac{b_{N-1} - a_{N-1,N}x_N}{a_{N-1,N-1}}.$$

Now x_N and x_{N-1} are used to find x_{N-2}.

(5)
$$x_{N-2} = \frac{b_{N-2} - a_{N-2,N-1}x_{N-1} - a_{N-2,N}x_N}{a_{N-2,N-2}}.$$

Once the values $x_N, x_{N-1}, \ldots, x_{k+1}$ are known, the general step is

(6)
$$x_k = \frac{b_k - \sum_{j=k+1}^{N} a_{k,j}x_j}{a_{k,k}} \quad \text{for } k = N-1, N-2, \ldots, 1.$$

The uniqueness of the solution is easy to see. The Nth equation implies that $b_N/a_{N,N}$ is the only possible value of x_N. Then finite induction is used to establish that $x_{N-1}, x_{N-2}, \ldots, x_1$ are unique.

Example 3.16 Use back substitution to solve the linear system

$$4x_1 - x_2 + 2x_3 + 3x_4 = 20$$
$$-2x_2 + 7x_3 - 4x_4 = -7$$
$$6x_3 + 5x_4 = 4$$
$$3x_4 = 6.$$

Solution. Solving for x_4 in the last equation yields

$$x_4 = \tfrac{6}{3} = 2.$$

Using x_4 to solve for x_3 in the third equation, we obtain

$$x_3 = \frac{4 - 5(2)}{6} = -1.$$

Now x_3 and x_4 are used to find x_2 in the second equation:

$$x_2 = \frac{-7 - 7(-1) + 4(2)}{-2} = -4.$$

Finally, x_1 is obtained using the first equation:

$$x_1 = \frac{20 + 1(-4) - 2(-1) - 3(2)}{4} = 3.$$

The condition that $a_{k,k} \neq 0$ is essential because equation (6) involves division by $a_{k,k}$. If this requirement is not fulfilled, then either no solution exists or infinitely many solutions exist.

Example 3.17 Show that there is no solution to the linear system

(7)
$$
\begin{aligned}
4x_1 - x_2 + 2x_3 + 3x_4 &= 20 \\
0x_2 + 7x_3 - 4x_4 &= -7 \\
6x_3 + 5x_4 &= 4 \\
3x_4 &= 6.
\end{aligned}
$$

Solution. Using the last equation in (7), we must have $x_4 = 2$, which is substituted into the second and third equations to obtain

(8)
$$
\begin{aligned}
7x_3 - 8 &= -7 \\
6x_3 + 10 &= 4.
\end{aligned}
$$

The first equation in (8) implies that $x_3 = \frac{1}{7}$ and the second equation implies that $x_3 = -1$. This inconsistency leads to the conclusion that there is no solution to the linear system (7).

Example 3.18 Show that there are infinitely many solutions to

(9)
$$
\begin{aligned}
4x_1 - x_2 + 2x_3 + 3x_4 &= 20 \\
0x_2 + 7x_3 + 0x_4 &= -7 \\
6x_3 + 5x_4 &= 4 \\
3x_4 &= 6.
\end{aligned}
$$

Solution. Using the last equation in (9), we must have $x_4 = 2$, which is substituted into the second and third equations to get $x_3 = -1$, which checks out in both equations. But only the two values x_3 and x_4 have been obtained from the second through fourth equations, and when they are substituted into the first equation of (9) the result is

(10)
$$
x_2 = 4x_1 - 14,
$$

which has infinitely many solutions; hence (9) has infinitely many solutions. If we choose a value of x_1 in (10), then the value of x_2 is uniquely determined. For example, if we include the equation $x_1 = 2$ in the system (9), then from (10) we compute $x_2 = -6$.

Theorem 3.7 [Determinant of an upper-triangular matrix] If A is an upper-triangular matrix, then det (A), the determinant of A, is given by the product of the diagonal elements

(11)
$$
\det(A) = a_{1,1}a_{2,2} \ldots a_{N,N}.
$$

Proof. See any texts on linear algebra, such as [6] or [80].

The value of the determinant for the matrix in Example 3.16 is

$$\det (A) = 4(-2)(6)(3) = -144.$$

Algorithm 3.4 [*Back Substitution*] Consider the following upper-triangular system of equations where $a_{k,k} \neq 0$ for $k = 1, 2, \ldots, N$:

$$a_{1,1}x_1 + a_{1,2}x_2 + \ldots + a_{1,N-1}x_{N-1} + \quad a_{1,N}x_N = b_1$$
$$a_{2,2}x_2 + \ldots + a_{2,N-1}x_{N-1} + \quad a_{2,N}x_N = b_2$$
$$\vdots \qquad \qquad \vdots \qquad \vdots$$
$$a_{N-1,N-1}x_{N-1} + a_{N-1,N}x_N = b_{N-1}$$
$$a_{N,N}x_N = b_N.$$

Computing the solution, we have $x_N = b_N/a_{N,N}$ and then

$$x_R = \frac{b_R - \displaystyle\sum_{J=R+1}^{N} a_{R,J}x_J}{a_{R,R}} \qquad \text{for } R = N-1, N-2, \ldots, 1.$$

Remark. The value det (A) is not necessary for the solution of the linear system. However, its value is simply the product of the diagonal elements and is included in the algorithm for completeness.

```
INPUT N                                          {Number of equations}
VARiable declaration                             {Dimension the arrays}
     REAL   A[1 .. N, 1 .. N] , B[1 .. N] , X[1 .. N]
INPUT   A[1 .. N, 1 .. N] , B[1 .. N]

DET := A(N, N)                                   {Initialize the variable}
X(N) := B(N)/A(N, N)                             {Start the back substitution}

FOR    R = N−1  DOWNTO  1  DO
       DET := DET*A(R, R)                        {Multiply the diagonal elements}
       SUM := 0                                  {Solve for Xᵣ in row r}
            FOR    J = R+1  TO  N  DO
                   SUM := SUM + A(R, J)*X(J)
       X(R) := [B(R) − SUM]/A(R, R)              {End back substitution}

PRINT  "The value of  Det (A)  is "  DET          {Output}
PRINT  "The solution to the upper-triangular system is:"
       FOR    R = 1  TO  N  DO
              PRINT  "X(" ; R; ") = " ; X(R)
```

Exercises for Upper–Triangular Linear Systems

In Exercises 1–3, solve the upper-triangular system and find the value of the determinant of the matrix of coefficients.

1. $\begin{aligned} 3x_1 - 2x_2 + x_3 - x_4 &= 8 \\ 4x_2 - x_3 + 2x_4 &= -3 \\ 2x_3 + 3x_4 &= 11 \\ 5x_4 &= 15 \end{aligned}$

2. $\begin{aligned} 5x_1 - 3x_2 - 7x_3 + x_4 &= -14 \\ 11x_2 + 9x_3 + 5x_4 &= 22 \\ 3x_3 - 13x_4 &= -11 \\ 7x_4 &= 14 \end{aligned}$

3. $\begin{aligned} 4x_1 - x_2 + 2x_3 + 2x_4 - x_5 &= 4 \\ -2x_2 + 6x_3 + 2x_4 + 7x_5 &= 0 \\ x_3 - x_4 - 2x_5 &= 3 \\ -2x_4 - x_5 &= 10 \\ 3x_5 &= 6 \end{aligned}$

4. Write a computer program that uses Algorithm 3.4. Use some of Exercises 1–3 as test cases.

5. Consider the two upper-triangular matrices

$$A = \begin{pmatrix} a_{11} & a_{12} & a_{13} \\ 0 & a_{22} & a_{23} \\ 0 & 0 & a_{33} \end{pmatrix} \quad \text{and} \quad B = \begin{pmatrix} b_{11} & b_{12} & b_{13} \\ 0 & b_{22} & b_{23} \\ 0 & 0 & b_{33} \end{pmatrix}.$$

Show that their product $C = AB$ is also upper-triangular.

6. Solve the lower-triangular system $A\mathbf{X} = \mathbf{B}$ and find det (A).

$$\begin{aligned} 2x_1 &= 6 \\ -x_1 + 4x_2 &= 5 \\ 3x_1 - 2x_2 - x_3 &= 4 \\ x_1 - 2x_2 + 6x_3 + 3x_4 &= 2 \end{aligned}$$

7. Solve the lower-triangular system $A\mathbf{X} = \mathbf{B}$ and find det (A).

$$\begin{aligned} 5x_1 &= -10 \\ x_1 + 3x_2 &= 4 \\ 3x_1 + 4x_2 + 2x_3 &= 2 \\ -x_1 + 3x_2 - 6x_3 - x_4 &= 5 \end{aligned}$$

8. *The Forward-Substitution Algorithm.* A linear system is called lower-triangular provided that $a_{i,j} = 0$ when $j > i$. Given

$$\begin{aligned} a_{1,1}x_1 \qquad\qquad\qquad\qquad\qquad\qquad\qquad &= b_1 \\ a_{2,1}x_1 + a_{2,2}x_2 \qquad\qquad\qquad\qquad\qquad &= b_2 \\ a_{3,1}x_1 + a_{3,2}x_2 + a_{3,3}x_3 \qquad\qquad\qquad &= b_3 \\ \vdots \qquad\quad \vdots \qquad\quad \vdots \qquad\qquad\qquad\qquad\quad &\quad \vdots \\ a_{N-1,1}x_1 + a_{N-1,2}x_2 + a_{N-1,3}x_3 + \ldots + a_{N-1,N-1}x_{N-1} &= b_{N-1} \\ a_{N,1}x_1 + a_{N,2}x_2 + a_{N,3}x_3 + \ldots + a_{N,N-1}x_{N-1} + a_{N,N}x_N &= b_N, \end{aligned}$$

develop an algorithm for solving a lower-triangular system when the diagonal elements are nonzero (i.e., $a_{k,k} \neq 0$ for $k = 1, 2, \ldots, N$).

9. Show that back-substitution requires N divisions, $(N^2 - N)/2$ multiplications, and $(N^2 - N)/2$ additions or subtractions. *Hint.* You can use the formula

$$\sum_{k=1}^{M} k = (M^2 + M)/2.$$

3.4 Gaussian Elimination and Pivoting

In this section we develop an efficient scheme for solving a general system $A\mathbf{X} = \mathbf{B}$ of N equations and N unknowns. The crucial step is to construct an equivalent upper-triangular system $U\mathbf{X} = \mathbf{Y}$ that can be solved by the method of Section 3.3.

Two linear systems of dimension N by N are said to be *equivalent* provided that their solutions sets are the same. Theorems from linear algebra show that when certain transformations are applied to a given system, the solution sets do not change.

Definition 3.2 [Elementary transformations] The following operations applied to a linear system yield an equivalent system

(1) Interchanges: The order of two equations can be changed.

(2) Scaling: Multiplying an equation by a nonzero constant.

(3) Replacement: An equation can be replaced by the sum of that equation and a multiple of any other equation.

It is common to use (3) by replacing an equation with the difference of that equation and a multiple of another equation. These concepts are illustrated in the next example.

Example 3.19 Find the parabola $y = A + Bx + Cx^2$ that passes through the three points $(1, 1)$, $(2, -1)$, and $(3, 1)$.

Solution. For each point we obtain an equation relating the value of x to the value of y. The result is the system

$$\begin{aligned} A + B + C &= 1 \quad \text{at } (1, 1) \\ A + 2B + 4C &= -1 \quad \text{at } (2, -1) \\ A + 3B + 9C &= 1 \quad \text{at } (3, 1). \end{aligned}$$

(4)

The variable A is eliminated from the second and third equations by subtracting the first equation from them. This is an application of the replacement transformation (3) and the resulting system is

$$\begin{aligned} A + B + C &= 1 \\ B + 3C &= -2 \\ 2B + 8C &= 0. \end{aligned}$$

(5)

The variable B is eliminated from the third equation in (5) by subtracting from it two times the second equation. We arrive at the equivalent upper-triangular system

(6)
$$
\begin{aligned}
A + B + C &= 1 \\
B + 3C &= -2 \\
2C &= 4.
\end{aligned}
$$

The back-substitution algorithm is now used to find the coefficients $C = 4/2 = 2$, $B = -2 - 3(2) = -8$, and $A = 1 - (-8) - 2 = 7$, and the equation of the parabola is $y = 7 - 8x + 2x^2$.

It is efficient to store all the coefficients of the linear system $AX = \mathbf{B}$ in an array of dimension N by $N + 1$. The coefficients of $\mathbf{B}$ are stored in column $N + 1$ of the array (i.e., $a_{k,N+1} = b_k$). Each row contains all the coefficients necessary to represent one equation in the linear system. The augmented matrix is denoted by $[A, B]$ and the linear system is represented as follows:

(7).
$$
[A, B] = \begin{pmatrix}
a_{1,1} & a_{1,2} & \cdots & a_{1,N} & b_1 \\
a_{2,1} & a_{2,2} & \cdots & a_{2,N} & b_2 \\
\vdots & \vdots & & \vdots & \vdots \\
a_{N,1} & a_{N,2} & \cdots & a_{N,N} & b_N
\end{pmatrix}.
$$

The system $AX = \mathbf{B}$, with augmented matrix given in (7), can be solved by performing row operations on the augmented matrix $[A, B]$. The variables x_k are placeholders for the coefficients and can be omitted until the end of the calculation.

Definition 3.3 [Row operations] The following operations applied to the augmented matrix (7) yield an equivalent system.

(8) Interchanges: The order of two rows can be changed.

(9) Scaling: Multiplying a row by a nonzero constant.

(10) Replacement: A row can be replaced by the sum of that row and a multiple of any other row.

It is common to use (10) by replacing a row with the difference of that row and a multiple of another row.

Definition 3.4 [Pivot] The number $a_{r,r}$ in position (r, r) that is used to eliminate x_r in rows $r + 1, r + 2, \ldots, N$ is called the rth *pivotal element*, and r is called the *pivotal row*.

Example 3.20 Express the following system in augmented matrix form and find an equivalent upper-triangular system and the solution.

$$x_1 + 2x_2 + x_3 + 4x_4 = 13$$
$$2x_1 + 0x_2 + 4x_3 + 3x_4 = 28$$
$$4x_1 + 2x_2 + 2x_3 + x_4 = 20$$
$$-3x_1 + x_2 + 3x_3 + 2x_4 = 6.$$

Solution. The augmented matrix is

$$
\begin{matrix}
\text{pivot} \longrightarrow \\
m_{2,1} = 2 \\
m_{3,1} = 4 \\
m_{4,1} = -3
\end{matrix}
\left(
\begin{array}{cccc|c}
1 & 2 & 1 & 4 & 13 \\
2 & 0 & 4 & 3 & 28 \\
4 & 2 & 2 & 1 & 20 \\
-3 & 1 & 3 & 2 & 6
\end{array}
\right).
$$

The first row is used to eliminate elements in the first column below the diagonal. We refer to the first row as the pivotal row and the element $a_{1,1}$ is called the pivotal element. The values $m_{k,1}$ are the multiples of row 1 that are to be subtracted from row k for $k = 2, 3, 4$. The result after elimination is

$$
\begin{matrix}
\\
\text{pivot} \quad \longrightarrow \\
m_{3,2} = 1.5 \\
m_{4,2} = -1.75
\end{matrix}
\left(
\begin{array}{cccc|c}
1 & 2 & 1 & 4 & 13 \\
0 & -4 & 2 & -5 & 2 \\
0 & -6 & -2 & -15 & -32 \\
0 & 7 & 6 & 14 & 45
\end{array}
\right).
$$

The second row is used to eliminate elements in the second column that lie below the diagonal. The second row is the pivotal row and the values $m_{k,2}$ are the multiples of row 2 that are to be subtracted from row k for $k = 3, 4$. The result after elimination is

$$
\begin{matrix}
\\
\\
\text{pivot} \quad \longrightarrow \\
m_{4,3} = -1.9
\end{matrix}
\left(
\begin{array}{cccc|c}
1 & 2 & 1 & 4 & 13 \\
0 & -4 & 2 & -5 & 2 \\
0 & 0 & -5 & -7.5 & -35 \\
0 & 0 & 9.5 & 5.25 & 48.5
\end{array}
\right).
$$

Finally, the multiple $m_{4,3} = -1.9$ of the third row is subtracted from the fourth row and the result is the upper-triangular system

(11)
$$
\left(
\begin{array}{cccc|c}
1 & 2 & 1 & 4 & 13 \\
0 & -4 & 2 & -5 & 2 \\
0 & 0 & -5 & -7.5 & -35 \\
0 & 0 & 0 & -9 & -18
\end{array}
\right).
$$

The back-substitution algorithm can be used to solve (11) and we get

$$x_4 = 2, \quad x_3 = 4, \quad x_2 = -1, \quad x_1 = 3.$$

The process described above is called *Gaussian elimination* and must be modified so that it can be used in most circumstances. If $a_{k,k} = 0$, then row k cannot be used to eliminate the elements in column k, and row k must be changed with some row below the diagonal to obtain a nonzero pivot element. If this cannot be done, then the system of equations does not have a unique solution.

Theorem 3.8   [Gaussian elimination with back substitution] If A is an N by N nonsingular matrix, then an equivalent system $U\mathbf{X} = \mathbf{Y}$ exists where U is an upper-triangular matrix with $u_{k,k} \neq 0$, and back substitution can be used to solve for $\mathbf{X}$.

Proof. We will use the augmented matrix with $\mathbf{B}$ stored in column $N + 1$:

$$A\mathbf{X} = \begin{pmatrix} a_{1,1}^{(1)} & a_{1,2}^{(1)} & a_{1,3}^{(1)} & \cdots & a_{1,n}^{(1)} \\ a_{2,1}^{(1)} & a_{2,2}^{(1)} & a_{2,3}^{(1)} & \cdots & a_{2,n}^{(1)} \\ a_{3,1}^{(1)} & a_{3,2}^{(1)} & a_{3,3}^{(1)} & \cdots & a_{3,n}^{(1)} \\ \vdots & \vdots & \vdots & & \vdots \\ a_{n,1}^{(1)} & a_{n,2}^{(1)} & a_{n,3}^{(1)} & \cdots & a_{n,n}^{(1)} \end{pmatrix} \begin{pmatrix} x_1 \\ x_2 \\ x_3 \\ \vdots \\ x_n \end{pmatrix} = \begin{pmatrix} a_{1,n+1}^{(1)} \\ a_{2,n+1}^{(1)} \\ a_{3,n+1}^{(1)} \\ \vdots \\ a_{n,n+1}^{(1)} \end{pmatrix} = \mathbf{B}.$$

Then we will construct an equivalent upper-triangular system $U\mathbf{X} = \mathbf{Y}$:

$$U\mathbf{X} = \begin{pmatrix} a_{1,1}^{(1)} & a_{1,2}^{(1)} & a_{1,3}^{(1)} & \cdots & a_{1,n}^{(1)} \\ 0 & a_{2,2}^{(2)} & a_{2,3}^{(2)} & \cdots & a_{2,n}^{(2)} \\ 0 & 0 & a_{3,3}^{(3)} & \cdots & a_{3,n}^{(3)} \\ \vdots & \vdots & \vdots & & \vdots \\ 0 & 0 & 0 & \cdots & a_{n,n}^{(n)} \end{pmatrix} \begin{pmatrix} x_1 \\ x_2 \\ x_3 \\ \vdots \\ x_n \end{pmatrix} = \begin{pmatrix} a_{1,n+1}^{(1)} \\ a_{2,n+1}^{(2)} \\ a_{3,n+1}^{(3)} \\ \vdots \\ a_{n,n+1}^{(n)} \end{pmatrix} = \mathbf{Y}.$$

Step 1. Store the coefficients in the array. The superscript on $a_{r,c}^{(1)}$ means that this is the first time a number is stored in location (r, c).

$$\begin{matrix} a_{1,1}^{(1)} & a_{1,2}^{(1)} & a_{1,3}^{(1)} & \cdots & a_{1,n}^{(1)} & a_{1,n+1}^{(1)} \\ a_{2,1}^{(1)} & a_{2,2}^{(1)} & a_{2,3}^{(1)} & \cdots & a_{2,n}^{(1)} & a_{2,n+1}^{(1)} \\ a_{3,1}^{(1)} & a_{3,2}^{(1)} & a_{3,3}^{(1)} & \cdots & a_{3,n}^{(1)} & a_{3,n+1}^{(1)} \\ \vdots & \vdots & \vdots & & \vdots & \vdots \\ a_{n,1}^{(1)} & a_{n,2}^{(1)} & a_{n,3}^{(1)} & \cdots & a_{n,n}^{(1)} & a_{n,n+1}^{(1)}. \end{matrix}$$

Step 2. If necessary, switch rows so that $a_{1,1}^{(1)} \neq 0$; then eliminate x_1 in rows 2 through N. In this process $m_{r,1}$ is the multiple of row 1 that is subtracted from row r.

```
FOR      r = 2  TO  N  DO
     Set  m_r,1 := a_r,1^(1)/a_1,1^(1)  and set  a_r,1^(1) := 0
     FOR      c = 2  TO  N+1  DO
              a_r,c^(2) := a_r,c^(1) − m_r,1*a_1,c^(1)
```

The new elements are written $a_{r,c}^{(2)}$ to indicate that this is the second time that a number has been stored in the array at location (r, c). The result after step 2 is

$$
\begin{array}{cccccc}
a_{1,1}^{(1)} & a_{1,2}^{(1)} & a_{1,3}^{(1)} & \cdots & a_{1,n}^{(1)} & a_{1,n+1}^{(1)} \\
0 & a_{2,2}^{(2)} & a_{2,3}^{(2)} & \cdots & a_{2,n}^{(2)} & a_{2,n+1}^{(2)} \\
0 & a_{3,2}^{(2)} & a_{3,3}^{(2)} & \cdots & a_{3,n}^{(2)} & a_{3,n+1}^{(2)} \\
\vdots & \vdots & \vdots & & \vdots & \vdots \\
0 & a_{n,2}^{(2)} & a_{n,3}^{(2)} & \cdots & a_{n,n}^{(2)} & a_{n,n+1}^{(2)}.
\end{array}
$$

Step 3. If necessary, switch the second row with some row below it so that $a_{2,2}^{(2)} \neq 0$, then eliminate x_2 in rows 3 through N. In this process $m_{r,2}$ is the multiple of row 2 that is subtracted from row r.

```
FOR    r = 3  TO  N  DO
       Set  m_r,2 := a⁽²⁾_r,2 / a⁽²⁾_2,2 and set a⁽²⁾_r,2 := 0
       FOR     c = 3  TO  N+1  DO
               a⁽³⁾_r,c := a⁽²⁾_r,c − m_r,2 * a⁽²⁾_2,c
```

The new elements are written $a_{r,c}^{(3)}$ to indicate that this is the third time that a number has been stored in the array at location (r, c). The result after step 3 is

$$
\begin{array}{cccccc}
a_{1,1}^{(1)} & a_{1,2}^{(1)} & a_{1,3}^{(1)} & \cdots & a_{1,n}^{(1)} & a_{1,n+1}^{(1)} \\
0 & a_{2,2}^{(2)} & a_{2,3}^{(2)} & \cdots & a_{2,n}^{(2)} & a_{2,n+1}^{(2)} \\
0 & 0 & a_{3,3}^{(3)} & \cdots & a_{3,n}^{(3)} & a_{3,n+1}^{(3)} \\
\vdots & \vdots & \vdots & & \vdots & \vdots \\
0 & 0 & a_{n,3}^{(3)} & \cdots & a_{n,n}^{(3)} & a_{n,n+1}^{(3)}.
\end{array}
$$

Step $p + 1$. *This is the general step.* If necessay, switch row p with some row beneath it so that $a_{p,p}^{(p)} \neq 0$, then eliminate x_p in rows $p + 1$ through N. Here $m_{r,p}$ is the multiple of row p that is subtracted from row r.

```
FOR    r = p+1  TO  N  DO
       Set  m_r,p := a⁽ᵖ⁾_r,p / a⁽ᵖ⁾_p,p and set a⁽ᵖ⁾_r,p := 0
       FOR     c = p+1  TO  N+1  DO
               a⁽ᵖ⁺¹⁾_r,c := a⁽ᵖ⁾_r,c − m_r,p * a⁽ᵖ⁾_p,c
```

The final result after x_{N-1} has been eliminated from row N is

$$
\begin{array}{cccccc}
a_{1,1}^{(1)} & a_{1,2}^{(1)} & a_{1,3}^{(1)} & \cdots & a_{1,n}^{(1)} & a_{1,n+1}^{(1)} \\
0 & a_{2,2}^{(2)} & a_{2,3}^{(2)} & \cdots & a_{2,n}^{(2)} & a_{2,n+1}^{(2)} \\
0 & 0 & a_{3,3}^{(3)} & \cdots & a_{3,n}^{(3)} & a_{3,n+1}^{(3)} \\
\vdots & \vdots & \vdots & & \vdots & \vdots \\
0 & 0 & 0 & \cdots & a_{n,n}^{(n)} & a_{n,n+1}^{(n)}.
\end{array}
$$

The upper-triangularization process is now complete.

Since A is nonsingular, when row operations are performed the successive matrices are also nonsingular. This guarantees that $a_{k,k}^{(k)} \neq 0$ for all k in the construction process. Hence back substitution can be used to solve $U\mathbf{X} = \mathbf{Y}$ for $\mathbf{X}$, and the theorem is proven.

Pivoting to Avoid $a_{p,p}^{(p)} = 0$

If $a_{p,p}^{(p)} = 0$, then row p cannot be used to eliminate the elements in column p below the diagonal. It is necessary to find row k, where $a_{k,p}^{(p)} \neq 0$ and $k > p$ and then interchange row p and row k so that a nonzero pivot element is obtained. This process is called *pivoting*, and the criterion for deciding which row to choose is called a *pivoting strategy*. The trivial pivoting strategy is as follows. If $a_{p,p}^{(p)} \neq 0$, then do not switch rows. If $a_{p,p}^{(p)} = 0$, then locate the first row below row p in which $a_{k,p}^{(p)} \neq 0$ and switch rows k and p. This will result in a new element $a_{p,p}^{(p)} \neq 0$, which is a nonzero pivot element.

Pivoting to Reduce Error

If there is more than one nonzero element in column p that lies on or below the diagonal, then there is a choice to determine which rows to interchange. We must determine ahead of time a good pivoting strategy. The partial pivoting strategy is the most common one and is used in Algorithm 3.5. Because the computer uses fixed-precision arithmetic, it is possible that a small error is introduced each time an arithmetic operation is performed. To reduce the propagation of error, it is suggested that one check the magnitude of all the elements in column p that lie on or below the diagonal, and locate row k in which the element has the largest absolute value, that is,

$$|a_{k,p}| = \max\{|a_{p,p}|, |a_{p+1,p}|, \ldots, |a_{N-1,p}|, |a_{N,p}|\},$$

and then switch row p with row k if $k > p$. Usually the larger pivot element will result in a smaller error being propagated.

In Sections 3.5 and 3.6 we will find that it takes a total of $(4N^3 + 9N^2 - 7N)/6$ arithmetic operations to solve an $N \times N$ system. When $N = 20$ the total number of arithmetic operations that must be performed is 5910, and the propagation of error in the computations could result in an erroneous answer. The technique of scaled partial pivoting or equilibrating can be used to further reduce the effect of error propagation. The reader can find an excellent discussion about these pivoting refinements in Reference 29.

Example 3.21 The values $x_1 = x_2 = 1.000$ are the solution to

$$1.133x_1 + 5.281x_2 = 6.414$$
$$24.14x_1 - 1.210x_2 = 22.93.$$

Use four-digit arithmetic and Gaussian elimination with trivial pivoting to find a computed approximate solution to the system.

Solution. The multiple $m_{2,1} = 24.14/1.133 = 21.31$ of row 1 is to be subtracted from row 2 to obtain the upper-triangular system. Using four digits in the calculations we obtain the new coefficients

$$a_{2,2}^{(2)} = -1.210 - 21.31 \times 5.281 = -1.210 - 112.5 = -113.7$$
$$a_{2,3}^{(2)} = 22.93 - 21.31 \times 6.414 = 22.93 - 136.7 = -113.8.$$

The computed upper-triangular system is

$$1.133x_1 + 5.281x_2 = 6.414$$
$$-113.7x_2 = -113.8.$$

Back substitution is used to compute $x_2 = -113.8/-113.7 = 1.001$, and $x_1 = (6.414 - 5.281 \times 1.001)/1.133 = (6.414 - 5.286)/1.133 = 0.9956.$

Example 3.22 Use four-digit arithmetic and Gaussian elimination with partial pivoting to solve the linear system in Example 3.21.

Solution. Interchanging the rows, we have

$$24.14x_1 - 1.210x_2 = 22.93$$
$$1.133x_1 + 5.281x_2 = 6.414.$$

This time $m_{2,1} = 1.133/24.14 = 0.04693$ is the multiple of row 1 that is to be subtracted from row 2. The new coefficients are

$$a_{2,2}^{(2)} = 5.281 - 0.04693(-1.210) = 5.281 + 0.05679 = 5.338,$$
$$a_{2,3}^{(2)} = 6.414 - 0.04693(22.93) = 6.414 - 1.076 = 5.338.$$

The computed upper-triangular system is

$$24.14x_1 + 1.210x_2 = 22.93$$
$$5.338x_2 = 5.338.$$

Back substitution is used to compute $x_2 = 5.338/5.338 = 1.000$, and $x_1 = (22.93 + 1.210 \times 1.000)/(24.14) = 24.14/24.14 = 1.000.$

Ill-Conditioning

A matrix A is called *ill-conditioned* if there exists a vector $\mathbf{B}$ for which small changes in the coefficients of A or $\mathbf{B}$ will produce large changes in $\mathbf{X} = A^{-1}\mathbf{B}$. The system $A\mathbf{X} = \mathbf{B}$ is ill-conditioned when A is ill-conditioned, and numerical methods for computing the solution are prone to have more error. For example, consider the linear system

(12)
$$\begin{pmatrix} 20{,}514 & 4{,}424 & 978 & 224 \\ 4{,}424 & 978 & 224 & 54 \\ 978 & 224 & 54 & 14 \\ 224 & 54 & 14 & 4 \end{pmatrix} \begin{pmatrix} x_1 \\ x_2 \\ x_3 \\ x_4 \end{pmatrix} = \begin{pmatrix} 20{,}514 \\ 4{,}424 \\ 978 \\ 224 \end{pmatrix}.$$

This matrix and column vector occur when the method of "least squares" is used to find the cubic equation that passes through the points $(2, 8)$, $(3, 27)$, $(4, 64)$, and $(5, 125)$. The solution vector is easily seen to be $\mathbf{X} = (1, 0, 0, 0)^T$, but a solution obtained by a computer was $\mathbf{X} = (1.000004, -0.000038, 0.000126, -0.000131)^T$. Suppose that the coefficient $a_{1,1}$ in the matrix is changed to $a_{1,1} = 20,515$. This change is 1 part in 20,514. However, the solution to the resulting perturbed linear system $A\mathbf{X} = \mathbf{B}$ is

$$\mathbf{X} = (0.642857, 3.75000, -12.3928, 12.7500)^T.$$

Hence the linear system in (12) is ill-conditioned.

Algorithm 3.5 [*Upper-Triangularization, Then Back Substitution*] Let A be a nonsingular matrix of dimension N by N, and consider

$$A\mathbf{X} = \mathbf{B}.$$

Let the coefficients of A and the constants of $\mathbf{B}$ be stored in the augmented matrix $[A, B] = (a_{i,j})$, which has N rows and $N + 1$ columns; that is, the column vector $\mathbf{B} = (a_{i,N+1})$ is stored in column $N + 1$ of the augmented matrix $[A, B]$. The system can be written

$$
\begin{aligned}
a_{1,1}x_1 + \quad a_{1,2}x_2 + \ldots + \quad a_{1,N-1}x_{N-1} + \quad a_{1,N}x_N &= a_{1,N+1} \\
a_{2,1}x_1 + \quad a_{2,2}x_2 + \ldots + \quad a_{2,N-1}x_{N-1} + \quad a_{2,N}x_N &= a_{2,N+1} \\
\vdots \qquad\qquad \vdots \qquad\qquad\qquad \vdots \qquad\qquad\qquad \vdots \qquad\qquad \vdots \\
a_{N-1,1}x_1 + a_{N-1,2}x_2 + \ldots + a_{N-1,N-1}x_{N-1} + a_{N-1,N}x_N &= a_{N-1,N+1} \\
a_{N,1}x_1 + \quad a_{N,2}x_2 + \ldots + \quad a_{N,N-1}x_{N-1} + \quad a_{N,N}x_N &= a_{N,N+1}
\end{aligned}
$$

Row operations will be used to eliminate x_P in column P.

The array will be overwritten and the portion on and above the main diagonal will be used to store an equivalent upper-triangular system:

$$
\begin{aligned}
u_{1,1}x_1 + u_{1,2}x_2 + \ldots + u_{1,N-1}x_{N-1} + \quad u_{1,N}x_N &= u_{1,N+1} \\
u_{2,2}x_2 + \ldots + u_{2,N-1}x_{N-1} + \quad u_{2,N}x_N &= u_{2,N+1} \\
\vdots \qquad\qquad\qquad \vdots \qquad\qquad \vdots \\
u_{N-1,N-1}x_{N-1} + u_{N-1,N}x_N &= u_{N-1,N+1} \\
u_{N,N}x_N &= u_{N,N+1}.
\end{aligned}
$$

Back substitution is then used to solve the upper-triangular system.

Remarks. Rather than actually interchange rows, row interchanges are recorded in the pointer vector row (P). If either the matrix A or vector $\mathbf{B}$ is to be used later, then a copy must be stored someplace besides the augmented matrix.

```
         INPUT N , A[1 .. N, 1 .. N+1]                    {Input augmented matrix}
         FOR    J = 1  TO  N  DO                          {Initialize the pointer vector}
              └──── Row(J) := J
FOR      P = 1  TO  N−1  DO                               {Start upper-triangularization}
         FOR    K = P+1  TO  N  DO                         {Find pivot element}
                IF  |A(Row(K), P)| > |A(Row(P), P)|  THEN
                           T := Row(P)                    {Switch the index for
                           Row(P) := Row(K)                 the pth pivot row
                           Row(K) := T                      if necessary}
              └──── ENDIF                                 {End of simulated row interchange}
                IF     A(Row(P), P) = 0  THEN
                       PRINT "The matrix is singular."
                       ENDIF
         FOR    K = P+1  TO  N  DO
                       M := A(Row(K), P) / A(Row(P), P)   {Form multiplier}
                FOR    C = P+1  TO  N+1  DO               {Eliminate Xₚ}
              └──── └──── A(Row(K), C) := A(Row(K), C) − M*A(Row(P), C)
END                                                      {End of the upper-triangularization routine}
                IF     A(Row(N), N) = 0  THEN
                       PRINT "The matrix is singular."
                       TERMINATE ALGORITHM
                       ENDIF

         X(N) := A(Row(N), N+1) / A(Row(N), N)           {Start the back substitution}
FOR      K = N−1  DOWNTO  1  DO
         SUM := 0
                FOR    C = K+1  TO  N  DO
              └──── SUM := SUM + A(Row(K), C)*X(C)
         X(K) := [A(Row(K), N+1) − SUM] / A(Row(K), K)   {End back substitution}

FOR      K = 1  TO  N  DO
      └──── PRINT X(K)                                   {Output}
```

Exercises for Gaussian Elimination and Pivoting

In Exercises 1–4, show that $AX = B$ is equivalent to the upper-triangular system $UX = Y$ and find the solution.

1.

$$2x_1 + 4x_2 - 6x_3 = -4 \qquad 2x_1 + 4x_2 - 6x_3 = -4$$
$$x_1 + 5x_2 + 3x_3 = 10 \qquad 3x_2 + 6x_3 = 12$$
$$x_1 + 3x_2 + 2x_3 = 5 \qquad 3x_3 = 3$$

2.

$$x_1 + x_2 + 6x_3 = 7$$
$$-x_1 + 2x_2 + 9x_3 = 2$$
$$x_1 - 2x_2 + 3x_3 = 10$$

$$x_1 + x_2 + 6x_3 = 7$$
$$3x_2 + 15x_3 = 9$$
$$12x_3 = 12$$

3.

$$2x_1 - 2x_2 + 5x_3 = 6$$
$$2x_1 + 3x_2 + x_3 = 13$$
$$-x_1 + 4x_2 - 4x_3 = 3$$

$$2x_1 - 2x_2 + 5x_3 = 6$$
$$5x_2 - 4x_3 = 7$$
$$0.9x_3 = 1.8$$

4.

$$-5x_1 + 2x_2 - x_3 = -1$$
$$x_1 + 0x_2 + 3x_3 = 5$$
$$3x_1 + x_2 + 6x_3 = 17$$

$$-5x_1 + 2x_2 - x_3 = -1$$
$$0.4x_2 + 2.8x_3 = 4.8$$
$$-10x_3 = -10$$

5. Find the parabola $y = A + Bx + Cx^2$ that passes through the three points $(1, 4)$, $(2, 7)$, and $(3, 14)$.

6. Find the parabola $y = A + Bx + Cx^2$ that passes through the three points $(1, 6)$, $(2, 5)$, and $(3, 2)$.

7. Find the parabola $y = A + Bx + Cx^2$ that passes through the three points $(1, 2)$, $(2, 2)$, and $(4, 8)$.

In Exercises 8–10, show that $AX = B$ is equivalent to the upper-triangular system $UX = Y$ and find the solution.

8.
$$4x_1 + 8x_2 + 4x_3 + 0x_4 = 8$$
$$x_1 + 5x_2 + 4x_3 - 3x_4 = -4$$
$$x_1 + 4x_2 + 7x_3 + 2x_4 = 10$$
$$x_1 + 3x_2 + 0x_3 - 2x_4 = -4$$

$$4x_1 + 8x_2 + 4x_3 + 0x_4 = 8$$
$$3x_2 + 3x_3 - 3x_4 = -6$$
$$4x_3 + 4x_4 = 12$$
$$x_4 = 2$$

9.
$$2x_1 + 4x_2 - 4x_3 + 0x_4 = 12$$
$$x_1 + 5x_2 - 5x_3 - 3x_4 = 18$$
$$2x_1 + 3x_2 + x_3 + 3x_4 = 8$$
$$x_1 + 4x_2 - 2x_3 + 2x_4 = 8$$

$$2x_1 + 4x_2 - 4x_3 + 0x_4 = 12$$
$$3x_2 - 3x_3 - 3x_4 = 12$$
$$4x_3 + 2x_4 = 0$$
$$3x_4 = -6$$

10.
$$x_1 + 2x_2 + 0x_3 - x_4 = 9$$
$$2x_1 + 3x_2 - x_3 + 0x_4 = 9$$
$$0x_1 + 4x_2 + 2x_3 - 5x_4 = 26$$
$$5x_1 + 5x_2 + 2x_3 - 4x_4 = 32$$

$$x_1 + 2x_2 + 0x_3 - x_4 = 9$$
$$-x_2 - x_3 + 2x_4 = -9$$
$$-2x_3 + 3x_4 = -10$$
$$1.5x_4 = -3$$

11. Find the solution to the following linear system.

$$x_1 + x_2 + 0x_3 + 4x_4 = 3$$
$$2x_1 - x_2 + 5x_3 + 0x_4 = 2$$
$$5x_1 + 2x_2 + x_3 + 2x_4 = 5$$
$$-3x_1 + 0x_2 + 2x_3 + 6x_4 = -2$$

12. (a) Write a computer program that uses Algorithm 3.5. Use some of Exercises 8–11, and 19 as test cases.

(b) Extend the program so that it will compute $R = B - AX$. This quantity is called the *residual* and can be used for a check to see that the correct solution was obtained.

13. Find the solution to the following linear system.

$$
\begin{aligned}
x_1 + 2x_2 &= 7 \\
2x_1 + 3x_2 - x_3 &= 9 \\
4x_2 + 2x_3 + 3x_4 &= 10 \\
2x_3 - 4x_4 &= 12
\end{aligned}
$$

14. Find the solution to the following linear system.

$$
\begin{aligned}
x_1 + x_2 &= 5 \\
2x_1 - x_2 + 5x_3 &= -9 \\
3x_2 - 4x_3 + 2x_4 &= 19 \\
2x_3 + 6x_4 &= 2
\end{aligned}
$$

15. Many applications involve matrices with many zeros. Of practical importance are *tridiagonal systems* of the form

$$
\begin{aligned}
d_1 x_1 + c_1 x_2 &= b_1 \\
a_1 x_1 + d_2 x_2 + c_2 x_3 &= b_2 \\
a_2 x_2 + d_3 x_3 + c_3 x_4 &= b_3 \\
& \vdots \\
a_{N-2} x_{N-2} + d_{N-1} x_{N-1} + c_{N-1} x_N &= b_{N-1} \\
a_{N-1} x_{N-1} + d_N x_N &= b_N.
\end{aligned}
$$

Write an algorithm that will solve a tridiagonal system. You may assume that row interchanges are not needed and that row k can be used to eliminate x_k in row $k + 1$.

16. How could you modify Algorithm 3.5 so that it will efficiently solve M linear systems with the same matrix A but different column vectors **B**? The M linear systems look like

$$
A\mathbf{X}_1 = \mathbf{B}_1, \quad A\mathbf{X}_2 = \mathbf{B}_2, \quad \ldots, \quad A\mathbf{X}_M = \mathbf{B}_M.
$$

17. Write a report about the Gauss–Jordan method.

18. Write a report about the topic of pivoting. Include a discussion about full pivoting and scaled partial pivoting.

19. Write a report about band systems of equations. Include a discussion about their use in solving differential equations.

20. Write a report about ill-conditioned matrices. Be sure to include a discussion of the condition number of a matrix.

21. The Hilbert matrix is a classical ill-conditioned matrix, and small changes in its coefficients will produce a large change in the solution to the perturbed system.

(a) Solve $A\mathbf{X} = \mathbf{B}$ using the Hilbert matrix of dimension 5 by 5:

$$A = \begin{pmatrix} \frac{1}{1} & \frac{1}{2} & \frac{1}{3} & \frac{1}{4} & \frac{1}{5} \\ \frac{1}{2} & \frac{1}{3} & \frac{1}{4} & \frac{1}{5} & \frac{1}{6} \\ \frac{1}{3} & \frac{1}{4} & \frac{1}{5} & \frac{1}{6} & \frac{1}{7} \\ \frac{1}{4} & \frac{1}{5} & \frac{1}{6} & \frac{1}{7} & \frac{1}{8} \\ \frac{1}{5} & \frac{1}{6} & \frac{1}{7} & \frac{1}{8} & \frac{1}{9} \end{pmatrix} \qquad B = \begin{pmatrix} 1 \\ 0 \\ 0 \\ 0 \\ 0 \end{pmatrix}.$$

(b) Solve $A\mathbf{X} = \mathbf{B}$ using

$$A = \begin{pmatrix} 1.0 & 0.5 & 0.33333 & 0.25 & 0.2 \\ 0.5 & 0.33333 & 0.25 & 0.2 & 0.16667 \\ 0.33333 & 0.25 & 0.2 & 0.16667 & 0.14286 \\ 0.25 & 0.2 & 0.16667 & 0.14286 & 0.125 \\ 0.2 & 0.16667 & 0.14286 & 0.125 & 0.11111 \end{pmatrix} \qquad B = \begin{pmatrix} 1 \\ 0 \\ 0 \\ 0 \\ 0 \end{pmatrix}.$$

Note: The matrices in parts (a) and (b) are different.

22. The Rockmore Corp. is considering the purchase of a new computer, and will choose either the DoGood 174 or the MightDo 11. They test both computers' ability to solve the linear system

$$34x + 55y - 21 = 0,$$
$$55x + 89y - 34 = 0.$$

The DoGood 174 computer gives $x = -0.11$ and $y = 0.45$ and its check for accuracy is found by substitution:

$$34(-0.11) + 55(0.45) - 21 = 0.01,$$
$$55(-0.11) + 89(0.45) - 34 = 0.00.$$

The MightDo 11 computer gives $x = -0.99$ and $y = 1.01$ and its check for accuracy is found by substitution:

$$34(-0.99) + 55(1.01) - 21 = 0.89,$$
$$55(-0.99) + 89(1.01) - 34 = 1.44.$$

which computer gave the better answer? Why?

23. Write a report about iterative refinement, which is sometimes referred to as "residual correction."

3.5 Matrix Inversion

The Vector Decomposition of A^{-1}

The following discussion is presented for matrices of dimension 3 by 3, but the concepts apply to matrices of dimension N by N. Consider the three linear systems $AC_j = E_j$ for $j = 1, 2, 3$, where E_1, E_2, and E_3 are the standard base vectors

$$AC_1 = \begin{pmatrix} 5 & 6 & -3 \\ 1 & 3 & 1 \\ 4 & 2 & -6 \end{pmatrix} \begin{pmatrix} 2.0 \\ -1.0 \\ 1.0 \end{pmatrix} = \begin{pmatrix} 1 \\ 0 \\ 0 \end{pmatrix} = E_1,$$

$$AC_2 = \begin{pmatrix} 5 & 6 & -3 \\ 1 & 3 & 1 \\ 4 & 2 & -6 \end{pmatrix} \begin{pmatrix} -3.0 \\ 1.8 \\ -1.4 \end{pmatrix} = \begin{pmatrix} 0 \\ 1 \\ 0 \end{pmatrix} = E_2,$$

$$AC_3 = \begin{pmatrix} 5 & 6 & -3 \\ 1 & 3 & 1 \\ 4 & 2 & -6 \end{pmatrix} \begin{pmatrix} -1.5 \\ 0.8 \\ -0.9 \end{pmatrix} = \begin{pmatrix} 0 \\ 0 \\ 1 \end{pmatrix} = E_3.$$

This can be compactly written with the matrix notation

$$AC_1 = E_1, \quad AC_2 = E_2, \quad AC_3 = E_3.$$

The column vectors C_1, C_2, C_3 and E_1, E_2, E_3 are used to form the matrices $C = [C_1, C_2, C_3]$ and $I = [E_1, E_2, E_3]$. Using the row-by-column rule for matrix multiplication, we obtain

$$A[C_1, C_2, C_3] = [E_1, E_2, E_3] = I.$$

Hence, we have found that $A^{-1} = [C_1, C_2, C_3]$. When the matrices are displayed this is easy to see:

$$AC = \begin{pmatrix} 5 & 6 & -3 \\ 1 & 3 & 1 \\ 4 & 2 & -6 \end{pmatrix} \begin{pmatrix} 2.0 & -3.0 & -1.5 \\ -1.0 & 1.8 & 0.8 \\ 1.0 & -1.4 & -0.9 \end{pmatrix} = \begin{pmatrix} 1 & 0 & 0 \\ 0 & 1 & 0 \\ 0 & 0 & 1 \end{pmatrix} = I.$$

Conversely, if we want to find the inverse matrix, then this is accomplished by first representing A^{-1} by its three columns

(1)
$$A^{-1} = [C_1, C_2, C_3]$$

Then the identity matrix $I = [E_1, E_2, E_3]$ is used to write

(2)
$$A[C_1, C_2, C_3] = [E_1, E_2, E_3] = I.$$

Hence the solution of (2) is equivalent to

(3)
$$AC_1 = E_1, \quad AC_2 = E_2, \quad AC_3 = E_3.$$

When displayed, we see that we must solve three linear systems,

$$AC_1 = \begin{pmatrix} a_{1,1} & a_{1,2} & a_{1,3} \\ a_{2,1} & a_{2,2} & a_{2,3} \\ a_{3,1} & a_{3,2} & a_{3,3} \end{pmatrix} \begin{pmatrix} c_{1,1} \\ c_{2,1} \\ c_{3,1} \end{pmatrix} = \begin{pmatrix} 1 \\ 0 \\ 0 \end{pmatrix} = E_1,$$

(4)
$$AC_2 = \begin{pmatrix} a_{1,1} & a_{1,2} & a_{1,3} \\ a_{2,1} & a_{2,2} & a_{2,3} \\ a_{3,1} & a_{3,2} & a_{3,3} \end{pmatrix} \begin{pmatrix} c_{1,2} \\ c_{2,2} \\ c_{3,2} \end{pmatrix} = \begin{pmatrix} 0 \\ 1 \\ 0 \end{pmatrix} = E_2,$$

$$AC_3 = \begin{pmatrix} a_{1,1} & a_{1,2} & a_{1,3} \\ a_{2,1} & a_{2,2} & a_{2,3} \\ a_{3,1} & a_{3,2} & a_{3,3} \end{pmatrix} \begin{pmatrix} c_{1,3} \\ c_{2,3} \\ c_{3,3} \end{pmatrix} = \begin{pmatrix} 0 \\ 0 \\ 1 \end{pmatrix} = E_3.$$

When the solution vectors in (4) are combined to form $C = [C_1, C_2, C_3]$, then it is easy to see that C is the inverse of A:

$$AC = \begin{pmatrix} a_{1,1} & a_{1,2} & a_{1,3} \\ a_{2,1} & a_{2,2} & a_{2,3} \\ a_{3,1} & a_{3,2} & a_{3,3} \end{pmatrix} \begin{pmatrix} c_{1,1} & c_{1,2} & c_{1,3} \\ c_{2,1} & c_{2,2} & c_{2,3} \\ c_{3,1} & c_{3,2} & c_{3,3} \end{pmatrix} = \begin{pmatrix} 1 & 0 & 0 \\ 0 & 1 & 0 \\ 0 & 0 & 1 \end{pmatrix}.$$

A slight modification of the Gaussian elimination process will give an algorithm for finding the inverse matrix. In Section 3.4 the augmented matrix contained one extra column. In the case of matrix inversion we add three column vectors E_1, E_2, E_3 to the right of A to form the augmented matrix $[A, E_1, E_2, E_3] = [A, I]$. A succession of row operations are used to eliminate the elements above as well as below the diagonal in the augmented matrix so that the identity matrix appears on the left side and the three solution vectors appear the right side, that is, we reduce the augmented matrix to the form $[I, C_1, C_2, C_3] = [I, C]$. Since the elements in column P that lie above and below the diagonal are eliminated, it is not necessary to use back substitution.

Example 3.23 Find the inverse of the matrix.

$$\begin{pmatrix} 2 & 0 & 1 \\ 3 & 2 & 5 \\ 1 & -1 & 0 \end{pmatrix}.$$

Solution. Start with the augmented matrix $[A, I]$.

$$\left(\begin{array}{ccc|ccc} 2 & 0 & 1 & 1 & 0 & 0 \\ 3 & 2 & 5 & 0 & 1 & 0 \\ 1 & -1 & 0 & 0 & 0 & 1 \end{array} \right).$$

Interchange rows 1 and 3 so that $a_{1,1} = 1$.

$$\left(\begin{array}{ccc|ccc} 1 & -1 & 0 & 0 & 0 & 1 \\ 3 & 2 & 5 & 0 & 1 & 0 \\ 2 & 0 & 1 & 1 & 0 & 0 \end{array}\right).$$

Eliminate elements in column 1 that lie below the diagonal.

$$\left(\begin{array}{ccc|ccc} 1 & -1 & 0 & 0 & 0 & 1 \\ 0 & 5 & 5 & 0 & 1 & -3 \\ 0 & 2 & 1 & 1 & 0 & -2 \end{array}\right).$$

Divide row 2 by 5 so that $a_{2,2} = 1$.

$$\left(\begin{array}{ccc|ccc} 1 & -1 & 0 & 0 & 0 & 1 \\ 0 & 1 & 1 & 0 & 0.2 & -0.6 \\ 0 & 2 & 1 & 1 & 0 & -2 \end{array}\right).$$

Eliminate elements in column 2 that lie above and below the diagonal.

$$\left(\begin{array}{ccc|ccc} 1 & 0 & 1 & 0 & 0.2 & 0.4 \\ 0 & 1 & 1 & 0 & 0.2 & -0.6 \\ 0 & 0 & -1 & 1 & -0.4 & -0.8 \end{array}\right).$$

Change the sign in row 3 so that $a_{3,3} = 1$ and eliminate the elements in column 3 that lie above the diagonal.

$$\left(\begin{array}{ccc|ccc} 1 & 0 & 0 & 1 & -0.2 & -0.4 \\ 0 & 1 & 0 & 1 & -0.2 & -1.4 \\ 0 & 0 & 1 & -1 & 0.4 & 0.8 \end{array}\right).$$

The identity matrix now appears on the left side of the augmented matrix and the inverse on the right side. Thus

$$A^{-1} = \left(\begin{array}{ccc} 1 & -0.2 & -0.4 \\ 1 & -0.2 & -1.4 \\ -1 & 0.4 & 0.8 \end{array}\right).$$

The solution of an N by N system of linear equations can be accomplished by inverting the matrix A. The solution to $A\mathbf{X} = \mathbf{B}$ is given by $\mathbf{X} = A^{-1}\mathbf{B}$. However, this is not a computationally efficient way to solve the system and the reader is encouraged to work out Exercise 8 to determine how inefficient it is. Sometimes the inverse is useful in its own right. For example, in the statistical treatment of the fitting of a function to observed data by the method of least squares, the entries of A^{-1} give information about the magnitude of errors in the data.

Exercises for Matrix Inversion

In Exercises 1–7:
 (a) Find the inverse of the given matrix.
 (b) Check your answer by computing the product AA^{-1}.

1. $\begin{pmatrix} 1 & 1 & 2 \\ 1 & 2 & 4 \\ 2 & 4 & 7 \end{pmatrix}$

2. $\begin{pmatrix} 1 & -3 & 3 \\ -2 & 4 & -5 \\ 1 & -5 & 3 \end{pmatrix}$

3. $\begin{pmatrix} 1 & -2 & 3 \\ -2 & 4 & -5 \\ 1 & -5 & 3 \end{pmatrix}$

4. $\begin{pmatrix} 1 & 3 & -2 \\ -2 & -4 & 6 \\ 1 & 5 & 2 \end{pmatrix}$

5. $\begin{pmatrix} 2 & -3 & -5 & 2 \\ 1 & -4 & 7 & 4 \\ 0 & 2 & 0 & -1 \\ 2 & 1 & 4 & 1 \end{pmatrix}$

6. $\begin{pmatrix} 3 & -9 & 27 & -81 \\ -4 & 16 & -64 & 256 \\ 5 & -25 & 125 & -625 \\ -6 & 36 & -216 & 1,296 \end{pmatrix}$

7. $\begin{pmatrix} 16 & -120 & 240 & -140 \\ -120 & 1,200 & -2,700 & 1,680 \\ 240 & -2,700 & 6,480 & -4,200 \\ -140 & 1,680 & -4,200 & 2,800 \end{pmatrix}$

8. Formulas for the arithmetic operations count for solving $AX = B$ for finding A^{-1} and for computing the product $A^{-1}B$ are readily available and are given in the table below.
 (a) Complete the table.
 (b) Discuss the amount of work required to solve $AX = B$ directly using Gaussian elimination, and that required to find A^{-1} and then compute the product $X = A^{-1}B$.
 (c) If the goal is to solve $AX = B$, then which way is best?

N	Operations Needed to Solve $AX = B$: $(4N^3 + 9N^2 - 7N)/6$ (see Algorithm 3.5)	Operations Needed to Find A^{-1}: $(16N^3 - 9N^2 - N)/6$ (see Problem 11)		Operations Needed to Multiply $A^{-1}B$: $2N^2 - N$ (see Algorithm 3.3)
3	28	58	+	15
5	–	295	+	45
8	428	–	+	120
10	–	–	+	190
15	–	–	+	–
20	–	–	+	–

9. Let $A = (a_{i,j})_{N \times N}$ be a nonsingular upper-triangular matrix, $a_{i,j} = 0$ when $i > j$. Then $C = A^{-1}$ is also upper-triangular. Let $I_N = (d_{i,j})_{N \times N}$, where $d_{i,i} = 1$ and $d_{i,j} = 0$ when $i \neq j$. An extension of the back-substitution algorithm can be made to find $A^{-1} = (c_{i,j})$. Show that the elements of A^{-1} can be computed using the following algorithm:

$$
\begin{aligned}
&\text{For} \quad j = N \quad \text{downto} \quad 1 \quad \text{do} \\
&\quad\quad \text{For} \quad i = j \quad \text{downto} \quad 1 \quad \text{do} \\
&\quad\quad\quad\quad c_{i,j} = [d_{i,j} - \sum_{k=i+1}^{j} a_{i,k} c_{k,j}]/a_{i,i}
\end{aligned}
$$

Remark. When $i = j$ the lower index for the summation is larger than the upper index. This is interpreted to mean that no terms are to be added, that is,

$$
0 = \sum_{k=j+1}^{j} a_{j,k} c_{k,j}.
$$

10. Let $A = (a_{i,j})_{N \times N}$ be a nonsingular lower-triangular matrix, $a_{i,j} = 0$ when $j > i$. Then $C = A^{-1}$ is also lower-triangular. Let $I_N = (d_{i,j})_{N \times N}$, where $d_{i,i} = 1$ and $d_{i,j} = 0$ when $i \neq j$. An extension of the forward-substitution algorithm in Exercise 8 of Section 3.3 can be made to find $A^{-1} = (c_{i,j})$. Show that the elements of A^{-1} can be computed using the following algorithm:

$$
\begin{aligned}
&\text{For} \quad j = 1 \quad \text{to} \quad N \quad \text{do} \\
&\quad\quad \text{For} \quad i = j \quad \text{to} \quad N \quad \text{do} \\
&\quad\quad\quad\quad c_{i,j} = [d_{i,j} - \sum_{k=j}^{i-1} a_{i,k} c_{k,j}]/a_{i,i}
\end{aligned}
$$

Remark. When $i = j$ the lower index for the summation is larger than the upper index. This is interpreted to mean that no terms are to be added, that is,

$$
0 = \sum_{k=j}^{i-1} a_{j,k} c_{k,j}.
$$

11. Modify Algorithm 3.5 to find A^{-1}. Use the special augmented matrix $[A, \mathbf{E}_1, \mathbf{E}_2, \ldots, \mathbf{E}_N]$, where $\mathbf{E}_J$ is the standard basis vector and is stored in column $N + J$ of the augmented matrix.

12. Use a computer to find the inverse of the Hilbert matrix of dimension 5 by 5 given in Exercise 19 of Section 3.4.

3.6 Triangular Factorization

In Section 3.3 we saw how easy it is to solve an upper-triangular system. Now we introduce the concept of factorization of a given matrix A into the product of an upper-triangular matrix U with nonzero diagonal elements and a lower-triangular matrix L that has 1's along the diagonal. For ease of notation we illustrate the concepts with matrices of dimension 4 by 4, but they apply to an arbitrary system of dimension N by N.

Definition 3.5 $[A = LU$ factorization] The nonsingular matrix A has a triangular factorization if it can be expressed as the product

(1)
$$A = LU.$$

In matrix form this is written as

$$
\begin{pmatrix}
a_{1,1} & a_{1,2} & a_{1,3} & a_{1,4} \\
a_{2,1} & a_{2,2} & a_{2,3} & a_{2,4} \\
a_{3,1} & a_{3,2} & a_{3,3} & a_{3,4} \\
a_{4,1} & a_{4,2} & a_{4,3} & a_{4,4}
\end{pmatrix}
=
\begin{pmatrix}
1 & 0 & 0 & 0 \\
m_{2,1} & 1 & 0 & 0 \\
m_{3,1} & m_{3,2} & 1 & 0 \\
m_{4,1} & m_{4,2} & m_{4,3} & 1
\end{pmatrix}
\begin{pmatrix}
u_{1,1} & u_{1,2} & u_{1,3} & u_{1,4} \\
0 & u_{2,2} & u_{2,3} & u_{2,4} \\
0 & 0 & u_{3,3} & u_{3,4} \\
0 & 0 & 0 & u_{4,4}
\end{pmatrix}.
$$

The condition that A is nonsingular implies that $u_{k,k} \neq 0$ for all k. The notation for the entries in L are $m_{i,j}$, and the reason for the choice of $m_{i,j}$ instead of $l_{i,j}$ will be pointed out soon.

Solution of a Linear System

Suppose that the coefficient matrix A for the linear system $AX = \mathbf{B}$ has a triangular factorization (1); then the solution to

(2)
$$LUX = \mathbf{B}$$

can be obtained by defining $\mathbf{Y} = U\mathbf{X}$ and then solving two systems

(3)   first solve $L\mathbf{Y} = \mathbf{B}$ for $\mathbf{Y}$, then solve $U\mathbf{X} = \mathbf{Y}$ for $\mathbf{X}$.

In equation form, we must first solve the lower-triangular system

(4)
$$
\begin{aligned}
y_1 &&&&&&= b_1 \\
m_{2,1}y_1 &+& y_2 &&&&= b_2 \\
m_{3,1}y_1 &+& m_{3,2}y_2 &+& y_3 &&= b_3 \\
m_{4,1}y_1 &+& m_{4,2}y_2 &+& m_{4,3}y_3 &+& y_4 = b_4
\end{aligned}
$$

to obtain $y_1, y_2, y_3,$ and y_4 and use them in solving the system

(5)
$$
\begin{aligned}
u_{1,1}x_1 + u_{1,2}x_2 + u_{1,3}x_3 + u_{1,4}x_4 &= y_1 \\
u_{2,2}x_2 + u_{2,3}x_3 + u_{2,4}x_4 &= y_2 \\
u_{3,3}x_3 + u_{3,4}x_4 &= y_3 \\
u_{4,4}x_4 &= y_4.
\end{aligned}
$$

Example 3.24 Solve

$$
\begin{aligned}
4x_1 + 3x_2 - x_3 &= -2 \\
-2x_1 - 4x_2 + 5x_3 &= 20 \\
x_1 + 2x_2 + 6x_3 &= 7.
\end{aligned}
$$

Use the triangular factorization method and the fact that

$$\begin{pmatrix} 4 & 3 & -1 \\ -2 & -4 & 5 \\ 1 & 2 & 6 \end{pmatrix} = \begin{pmatrix} 1 & 0 & 0 \\ -0.5 & 1 & 0 \\ 0.25 & -0.5 & 1 \end{pmatrix} \begin{pmatrix} 4 & 3 & -1 \\ 0 & -2.5 & 4.5 \\ 0 & 0 & 8.5 \end{pmatrix}.$$

Solution. Use the forward substitution method to solve

(6)
$$\begin{aligned} y_1 \qquad\qquad &= -2 \\ -0.5y_1 + \quad y_2 \quad &= 20 \\ 0.25y_1 - 0.5y_2 + y_3 &= \quad 7. \end{aligned}$$

Compute the values $y_1 = -2$, $y_2 = 20 + 0.5(2) = 19$, and $y_3 = 7 - 0.25(-2) + 0.5(19) = 17$. Next write the system $U\mathbf{X} = \mathbf{Y}$:

(7)
$$\begin{aligned} 4x_1 + \quad 3x_2 - \quad x_3 &= -2 \\ -2.5x_2 + 4.5x_3 &= 19 \\ 8.5x_3 &= 17. \end{aligned}$$

Now use back substitution and compute the solution $x_3 = 17/8.5 = 2$, $x_2 = [19 - 4.5(2)]/(-2.5) = -4$, and $x_1 = [-2 - 3(-4) + 2]/4 = 3$.

Triangular Factorization

We now discuss how to obtain the triangular factorization. If row interchanges are not necessary when using Gaussian elimination, the multipliers $m_{i,j}$ are the subdiagonal entries in L.

Example 3.25 Use Gaussian elimination to construct the triangular factorization of the matrix

$$A = \begin{pmatrix} 4 & 3 & -1 \\ -2 & -4 & 5 \\ 1 & 2 & 6 \end{pmatrix}.$$

Solution. The matrix L will be constructed from an identity matrix placed at the left. For each row operation used to construct the upper-triangular matrix, the multipliers $m_{i,j}$ will be put in their proper place at the left. Start with

$$A = \begin{pmatrix} 1 & 0 & 0 \\ 0 & 1 & 0 \\ 0 & 0 & 1 \end{pmatrix} \begin{pmatrix} 4 & 3 & -1 \\ -2 & -4 & 5 \\ 1 & 2 & 6 \end{pmatrix}.$$

Row 1 is used to eliminate the elements of A in column 1 below $a_{1,1}$. The multiples $m_{2,1} = -0.5$ and $m_{3,1} = 0.25$ of row 1 are subtracted from rows 2 and 3, respectively.

These multipliers are put in the matrix at the left and the result is

$$A = \begin{pmatrix} 1 & 0 & 0 \\ -0.5 & 1 & 0 \\ 0.25 & 0 & 1 \end{pmatrix} \begin{pmatrix} 4 & 3 & -1 \\ 0 & -2.5 & 4.5 \\ 0 & 1.25 & 6.25 \end{pmatrix}.$$

Row 2 is used to eliminate the elements of A in column 2 below $a_{2,2}$. The multiple $m_{3,2} = -0.5$ of the second row is subtracted from row 3, and the multiplier is entered in the matrix at the left and we have the desired triangular factorization of A:

(8)
$$A = \begin{pmatrix} 1 & 0 & 0 \\ -0.5 & 1 & 0 \\ 0.25 & -0.5 & 1 \end{pmatrix} \begin{pmatrix} 4 & 3 & -1 \\ 0 & -2.5 & 4.5 \\ 0 & 0 & 8.5 \end{pmatrix}.$$

Theorem 3.9   [Direct factorization $A = LU$, no row interchanges] If row interchanges are not necessary in solving the system $A\mathbf{X} = \mathbf{B}$ of N equations and N unknowns, the upper-triangularization step of Gaussian elimination produces the factorization $A = LU$, where L is a lower-triangular matrix with $l_{kk} = 1$ and U is an upper-triangular matrix with $u_{kk} \neq 0$ for $k = 1, 2, \ldots, N$. After finding the factorization, $\mathbf{X}$ can be found in two steps.

   1. Solve $L\mathbf{Y} = \mathbf{B}$ for $\mathbf{Y}$ using forward substitution.
   2. Solve $U\mathbf{X} = \mathbf{Y}$ for $\mathbf{X}$ using back substitution.

   Proof. We will show that when the Gaussian elimination process is followed and $\mathbf{B}$ is stored in column $N + 1$ of the augmented matrix, the result after the upper-triangularization step is the equivalent upper-triangular system $U\mathbf{X} = \mathbf{Y}$. The matrices L, U and vectors $\mathbf{B}, \mathbf{Y}$ will have the form

$$L = \begin{pmatrix} 1 & 0 & 0 & & 0 \\ m_{2,1} & 1 & 0 & \cdots & 0 \\ m_{3,1} & m_{3,2} & 1 & \cdots & 0 \\ \vdots & \vdots & \vdots & & \vdots \\ m_{n,1} & m_{n,2} & m_{n,3} & \cdots & 1 \end{pmatrix} \quad \mathbf{B} = \begin{pmatrix} a_{1,n+1}^{(1)} \\ a_{2,n+1}^{(1)} \\ a_{3,n+1}^{(1)} \\ \vdots \\ a_{n,n+1}^{(1)} \end{pmatrix}$$

$$U = \begin{pmatrix} a_{1,1}^{(1)} & a_{1,2}^{(1)} & a_{1,3}^{(1)} & \cdots & a_{1,n}^{(1)} \\ 0 & a_{2,2}^{(2)} & a_{2,3}^{(2)} & \cdots & a_{2,n}^{(2)} \\ 0 & 0 & a_{3,3}^{(3)} & \cdots & a_{3,n}^{(3)} \\ \vdots & \vdots & \vdots & & \vdots \\ 0 & 0 & 0 & \cdots & a_{n,n}^{(n)} \end{pmatrix} \quad \mathbf{Y} = \begin{pmatrix} a_{1,n+1}^{(1)} \\ a_{2,n+1}^{(2)} \\ a_{3,n+1}^{(3)} \\ \vdots \\ a_{n,n+1}^{(n)} \end{pmatrix}.$$

Remark. To find just L and U, the $(n + 1)$st column is not needed.

Step 1. Store the coefficients in the array. The superscript on $a_{r,c}^{(1)}$ means that this is the first time a number is stored in location (r, c).

$$
\begin{array}{ccccc}
a_{1,1}^{(1)} & a_{1,2}^{(1)} & a_{1,3}^{(1)} & \cdots & a_{1,n}^{(1)} \quad a_{1,n+1}^{(1)} \\
a_{2,1}^{(1)} & a_{2,2}^{(1)} & a_{2,3}^{(1)} & \cdots & a_{2,n}^{(1)} \quad a_{2,n+1}^{(1)} \\
a_{3,1}^{(1)} & a_{3,2}^{(1)} & a_{3,3}^{(1)} & \cdots & a_{3,n}^{(1)} \quad a_{3,n+1}^{(1)} \\
\cdot & \cdot & \cdot & & \cdot \quad \cdot \\
\cdot & \cdot & \cdot & & \cdot \quad \cdot \\
a_{n,1}^{(1)} & a_{n,2}^{(1)} & a_{n,3}^{(1)} & \cdots & a_{n,n}^{(1)} \quad a_{n,n+1}^{(1)} .
\end{array}
$$

Step 2. Eliminate x_1 in rows 2 through N and store the multiplier $m_{r,1}$ used to eliminate x_1 in row r in the array at location $(r, 1)$.

```
FOR    r = 2  TO  N  DO
       m_r,1 := a_r,1^(1)/a_1,1^(1)  and  a_r,1 := m_r,1
       FOR    c = 2  TO N+1  DO
              a_r,c^(2) := a_r,c^(1) − m_r,1*a_1,c^(1)
```

The new elements are written $a_{r,c}^{(2)}$ to indicate that this is the second time that a number has been stored in the array at location (r, c). The result after step 2 is

$$
\begin{array}{ccccc}
a_{1,1}^{(1)} & a_{1,2}^{(1)} & a_{1,3}^{(1)} & \cdots & a_{1,n}^{(1)} \quad a_{1,n+1}^{(1)} \\
m_{2,1} & a_{2,2}^{(2)} & a_{2,3}^{(2)} & \cdots & a_{2,n}^{(2)} \quad a_{2,n+1}^{(2)} \\
m_{3,1} & a_{3,2}^{(2)} & a_{3,3}^{(2)} & \cdots & a_{3,n}^{(2)} \quad a_{3,n+1}^{(2)} \\
\cdot & \cdot & \cdot & & \cdot \quad \cdot \\
\cdot & \cdot & \cdot & & \cdot \quad \cdot \\
m_{n,1} & a_{n,2}^{(2)} & a_{n,3}^{(2)} & \cdots & a_{n,n}^{(2)} \quad a_{n,n+1}^{(2)} .
\end{array}
$$

Step 3. Eliminate x_2 in rows 3 through N and store the multiplier $m_{r,2}$ used to eliminate x_2 in row r in the array at location $(r, 2)$.

```
FOR    r = 3  TO  N  DO
       m_r,2 := a_r,2^(2)/a_2,2^(2)  and  a_r,2 := m_r,2
       FOR    c = 3  TO N+1  DO
              a_r,c^(3) := a_r,c^(2) − m_r,2*a_2,c^(2)
```

The new elements are written $a_{r,c}^{(3)}$ to indicate that this is the third time that a number has been stored in the array at location (r, c). The result after step 3 is

$$
\begin{array}{ccccc}
a_{1,1}^{(1)} & a_{1,2}^{(1)} & a_{1,3}^{(1)} & \cdots & a_{1,n}^{(1)} \quad a_{1,n+1}^{(1)} \\
m_{2,1} & a_{2,2}^{(2)} & a_{2,3}^{(2)} & \cdots & a_{2,n}^{(2)} \quad a_{2,n+1}^{(2)} \\
m_{3,1} & m_{3,2} & a_{3,3}^{(3)} & \cdots & a_{3,n}^{(3)} \quad a_{3,n+1}^{(3)} \\
\cdot & \cdot & \cdot & & \cdot \quad \cdot \\
\cdot & \cdot & \cdot & & \cdot \quad \cdot \\
m_{n,1} & m_{n,2} & a_{n,3}^{(3)} & \cdots & a_{n,n}^{(3)} \quad a_{n,n+1}^{(3)}
\end{array}
$$

Step $p + 1$. This is the general step. Eliminate x_p in rows $p + 1$ through N and store the multipliers $m_{r,p}$ at locations (r, p).

```
FOR      r = p+1  TO  N  DO
    m_{r,p} := a_{r,p}^{(p)}/a_{p,p}^{(p)}  and   a_{r,p} := m_{r,p}
    FOR      c = p+1  TO  N+1   DO
        a_{r,c}^{(p+1)} := a_{r,c}^{(p)} - m_{r,p}*a_{p,c}^{(p)}
```

The final result after x_{N-1} has been eliminated from row N is

$$
\begin{array}{cccccc}
a_{1,1}^{(1)} & a_{1,2}^{(1)} & a_{1,3}^{(1)} & \cdots & a_{1,n}^{(1)} & a_{1,n+1}^{(1)} \\
m_{2,1} & a_{2,2}^{(2)} & a_{2,3}^{(2)} & \cdots & a_{2,n}^{(2)} & a_{2,n+1}^{(2)} \\
m_{3,1} & m_{3,2} & a_{3,3}^{(3)} & \cdots & a_{3,n}^{(3)} & a_{3,n+1}^{(3)} \\
\vdots & \vdots & \vdots & & \vdots & \vdots \\
m_{n,1} & m_{n,2} & m_{n,3} & \cdots & a_{n,n}^{(n)} & a_{n,n+1}^{(n)}.
\end{array}
$$

The upper-triangularization process is now complete. Notice that one array is used to store the elements of both L and U. The 1's of L are not stored, nor are the 0's of L and U that lie above and below the diagonal, respectively. Only the essential coefficients needed to reconstruct L and U are stored!

We must now verify that the product is $LU = A$. Suppose that $C = LU$; and consider the case when $r \leq c$. Then $c_{r,c}$ is:

(9)
$$
c_{r,c} = m_{r,1}a_{1,c}^{(1)} + m_{r,2}a_{2,c}^{(2)} + m_{r,r-1}a_{r-1,c}^{(r-1)} + a_{r,c}^{(r)}.
$$

Using the replacement equations in steps 1 through $p + 1 = r$, we obtain the following substitutions:

(10)
$$
\begin{aligned}
m_{r,1}a_{1,c}^{(1)} &= a_{r,c}^{(1)} - a_{r,c}^{(2)}, \\
m_{r,2}a_{2,c}^{(2)} &= a_{r,c}^{(2)} - a_{r,c}^{(3)}, \\
&\vdots \\
m_{r,r-1}a_{r-1,c}^{(r-1)} &= a_{r,c}^{(r-1)} - a_{r,c}^{(r)}.
\end{aligned}
$$

When the substitutions in (10) are used in (9), the result is

$$
c_{r,c} = a_{r,c}^{(1)} - a_{r,c}^{(2)} + a_{r,c}^{(2)} - a_{r,c}^{(3)} + \ldots + a_{r,c}^{(r-1)} - a_{r,c}^{(r)} + a_{r,c}^{(r)} = a_{r,c}^{(1)}.
$$

The other case $r > c$ is similar to prove.

Counting Arithmetic Operations

The process for triangularizing is the same for both the Gaussian elimination and triangular factorization methods. We can count the operations if we look at the first N columns of the augmented matrix in Theorem 3.9. The outer loop of step $p + 1$ requires $N - p = N - (p + 1) + 1$ divisions to compute the multipliers $m_{r,p}$. Inside the loops, but for the first N columns only, a total of $(N - p)(N - p)$ multiplications and the same number of subtractions are required to compute the new row elements $a_{r,c}^{(p+1)}$. This process is carried out for $p = 1, 2, \ldots, N - 1$. Thus the triangular factorization portion $A = LU$ requires

(11) $$\sum_{p=1}^{N-1} (N - p)(N - p + 1) = \frac{N^3 - N}{3} \quad \text{multiplications and divisions}$$

and

(12) $$\sum_{p=1}^{N-1} (N - p)(N - p) = \frac{2N^3 - 3N^2 + N}{6} \quad \text{subtractions.}$$

To establish (11) we use the summation formulas

$$\sum_{k=1}^{M} k = \frac{M(M + 1)}{2} \quad \text{and} \quad \sum_{k=1}^{M} k^2 = \frac{M(M + 1)(2M + 1)}{6}$$

Using the change of variable $k = N - p$, we rewrite (11) as

$$\sum_{p=1}^{N-1} (N - p)(N - p + 1) = \sum_{p=1}^{N-1} (N - p) + \sum_{p=1}^{N-1} (N - p)^2$$

$$= \sum_{k=1}^{N-1} k + \sum_{k=1}^{N-1} k^2$$

$$= \frac{(N - 1)(N)}{2} + \frac{(N - 1)(N)(2N - 1)}{6}$$

$$= \frac{N^3 - N}{3}$$

Once the triangular factorization $A = LU$ has been obtained, the solution to the lower-triangular system $LY = B$ will require $0 + 1 + \ldots + N - 1 = (N^2 - N)/2$ multiplications and subtractions; no divisions are required because the diagonal elements of L are 1's. Then the solution of the upper-triangular system $UX = Y$ requires $1 + 2 + \ldots + N = (N^2 + N)/2$ multiplications and divisions and $(N^2 - N)/2$ subtractions. Therefore, finding the solution to $LUX = B$ requires

$$N^2 \quad \text{multiplications and divisions} \quad \text{and} \quad N^2 - N \quad \text{subtractions.}$$

We see that the bulk of the calculations lie in the triangularization portion of the solution. If the linear system is to be solved many times, with the same coefficient matrix A but with different column vectors $\mathbf{B}$, then it is not necessary to triangularize the matrix each time if the factors are saved. This is the reason the triangular factorization method is usually chosen over the elimination method. However, if only one linear system is solved, then the two methods are the same, except that the triangular factorization method stores the multipliers.

Permutation Matrices

The $A = LU$ factorization in Theorem 3.9 assumes that there are no row interchanges. It is possible that a nonsingular matrix A cannot be directly factored as $A = LU$.

Example 3.26 Show that the following matrix cannot be directly factored $A = LU$.

$$A = \begin{pmatrix} 1 & 2 & 6 \\ 4 & 8 & -1 \\ -2 & 3 & 5 \end{pmatrix}.$$

Solution. Suppose that A has a direct factorization LU; then

(13)
$$\begin{pmatrix} 1 & 2 & 6 \\ 4 & 8 & -1 \\ -2 & 3 & 5 \end{pmatrix} = \begin{pmatrix} 1 & 0 & 0 \\ m_{21} & 1 & 0 \\ m_{31} & m_{32} & 1 \end{pmatrix} \begin{pmatrix} u_{11} & u_{12} & u_{13} \\ 0 & u_{22} & u_{23} \\ 0 & 0 & u_{33} \end{pmatrix}.$$

The matrices L and U on the right-hand side of (13) can be multiplied and each element of the product compared with the corresponding element of matrix A. In the first column $1 = 1u_{11}$, then $4 = m_{21}u_{11} = m_{21}$, and finally $-2 = m_{31}u_{11} = m_{31}$. In the second column $2 = 1u_{12}$, then $8 = m_{21}u_{12} + 1u_{22} = 4(2) + u_{22}$ implies that $u_{22} = 0$, and finally $3 = m_{31}u_{12} + m_{32}u_{22} = (-2)(2) + m_{32}(0) = -4$, which is a contradiction. Therefore, A does not have an LU factorization.

A permutation of the first N positive integers $1, 2, \ldots, N$ is an arrangement $j_1, j_2, \ldots, j_N$ of these integers in a definite order. For example, $1, 4, 2, 3, 5$ is a permutation of the five integers 1, 2, 3, 4, 5. The standard base vectors $\mathbf{E}_J = (0, 0, \ldots, 1, \ldots, 0)^T$ for $J = 1, 2, \ldots, N$ of N-dimensional space are used in the next definition.

Definition 3.6 An N by N permutation matrix P is a matrix whose rows are the vectors $\mathbf{E}_1, \mathbf{E}_2, \ldots, \mathbf{E}_N$ in some arrangement, that is,

(14)
$$P = [\mathbf{E}_{j_1}, \mathbf{E}_{j_2}, \ldots, \mathbf{E}_{j_N}]^T,$$

where $j_1, j_2, \ldots, j_N$ is a permutation of the integers $1, 2, \ldots, N$.

For example, the following 4 by 4 matrix is a permutation matrix.

(15)
$$P = \begin{pmatrix} 0 & 1 & 0 & 0 \\ 1 & 0 & 0 & 0 \\ 0 & 0 & 0 & 1 \\ 0 & 0 & 1 & 0 \end{pmatrix} = [\mathbf{E}_2, \mathbf{E}_1, \mathbf{E}_4, \mathbf{E}_3]^T.$$

If A is an N by N matrix and P is the N by N permutation matrix given in (14), the product PA is a matrix whose rows consist of the rows of A rearranged in the order $j_1, j_2, \ldots, j_N$.

Example 3.27 Let A be a 4 by 4 matrix and let P be the permutation matrix given in (15); then PA is the matrix whose rows consist of the rows of A rearranged in the order row$_2$ A, row$_1$ A, row$_4$ A, row$_3$ A.

Solution. Computing the product, we have

$$
\begin{pmatrix} 0 & 1 & 0 & 0 \\ 1 & 0 & 0 & 0 \\ 0 & 0 & 0 & 1 \\ 0 & 0 & 1 & 0 \end{pmatrix}
\begin{pmatrix} a_{1,1} & a_{1,2} & a_{1,3} & a_{1,4} \\ a_{2,1} & a_{2,2} & a_{2,3} & a_{2,4} \\ a_{3,1} & a_{3,2} & a_{3,3} & a_{3,4} \\ a_{4,1} & a_{4,2} & a_{4,3} & a_{4,4} \end{pmatrix}
=
\begin{pmatrix} a_{2,1} & a_{2,2} & a_{2,3} & a_{2,4} \\ a_{1,1} & a_{1,2} & a_{1,3} & a_{1,4} \\ a_{4,1} & a_{4,2} & a_{4,3} & a_{4,4} \\ a_{3,1} & a_{3,2} & a_{3,3} & a_{3,4} \end{pmatrix}.
$$

Theorem 3.10 If A is a nonsingular matrix, then there exists a permutation matrix P so that PA has a triangular factorization

(16)
$$
PA = LU.
$$

The proof can be found in advanced texts; see Reference [94].

Example 3.28 If rows 2 and 3 of the matrix in Example 3.26 are interchanged, the resulting matrix PA can be factored.

Solution. The permutation matrix that switches rows 2 and 3 is $P = [\mathbf{E}_1, \mathbf{E}_3, \mathbf{E}_2]^T$. Computing the product PA, we obtain

$$
PA =
\begin{pmatrix} 1 & 0 & 0 \\ 0 & 0 & 1 \\ 0 & 1 & 0 \end{pmatrix}
\begin{pmatrix} 1 & 2 & 6 \\ 4 & 8 & -1 \\ -2 & 3 & 5 \end{pmatrix}
=
\begin{pmatrix} 1 & 2 & 6 \\ -2 & 3 & 5 \\ 4 & 8 & -1 \end{pmatrix}.
$$

Now Gaussian elimination without row interchanges can be used:

$$
\begin{array}{c}
\text{pivot} \longrightarrow \\
m_{2,1} = -2 \\
m_{3,1} = 4
\end{array}
\begin{pmatrix} 1 & 2 & 6 \\ -2 & 3 & 5 \\ 4 & 8 & -1 \end{pmatrix}.
$$

After x_2 has been eliminated from column 2, row 3, we have

$$
\begin{array}{c}
\text{pivot} \longrightarrow \\
m_{3,2} = 0
\end{array}
\begin{pmatrix} 1 & 2 & 6 \\ 0 & 7 & 17 \\ 0 & 0 & -25 \end{pmatrix} = U.
$$

The lower-triangular matrix is formed with the multipliers

$$L = \begin{pmatrix} 1 & 0 & 0 \\ m_{2,1} & 1 & 0 \\ m_{3,1} & m_{3,2} & 1 \end{pmatrix} = \begin{pmatrix} 1 & 0 & 0 \\ -2 & 1 & 0 \\ 4 & 0 & 1 \end{pmatrix}.$$

Hence we have found the factorization

$$PA = \begin{pmatrix} 1 & 2 & 6 \\ -2 & 3 & 5 \\ 4 & 8 & -1 \end{pmatrix} = \begin{pmatrix} 1 & 0 & 0 \\ -2 & 1 & 0 \\ 4 & 0 & 1 \end{pmatrix} \begin{pmatrix} 1 & 2 & 6 \\ 0 & 7 & 17 \\ 0 & 0 & -25 \end{pmatrix} = LU.$$

Modifying the Gaussian Elimination Algorithm

The only difference between Gaussian elimination and triangular factorization is the delay in performing calculations involving the column vector **B** until the factorization of A is accomplished. If row interchanges are required, then we will keep track of them in the pointer vector Row (P). The elements of **B** are not changed during the triangularization portion of the algorithm, and they are rearranged correctly using Row (P) during forward-substitution step. The value of the determinant det (A) is the product of the diagonal elements of U multiplied by $(-1)^Q$, where Q is the number of row interchanges that are required. By Theorem 3.10, we can find a factorization of a PA. Since the permutation matrix keeps track of row interchanges, they can be remembered in a pointer vector Row (P), and this will save space.

Algorithm 3.6 [$PA = LU$ *Factorization with Row Interchanges*] Let A be an N by N nonsingular matrix. Consider the linear system

$$AX = B.$$

We can construct an upper-triangular matrix U with $u_{kk} \neq 0$ and a lower-triangular matrix L with $l_{kk} = 1$ (for $k = 1, 2, \ldots, N$) and a permutation matrix P that rearranges the rows of A so that $PA = LU$. The solution **X** is constructed using the four steps:

1. Compute the matrices L, U, and P.
2. Find the column vector P**B**.
3. Solve the lower-triangular system $LY = P$**B** for **Y**.
4. Solve the upper-triangular system $UX = Y$ for **X**.

Remarks. This algorithm is an extension of Algorithm 3.5.

Since $l_{kk} = 1$ these values do not need to be stored. The coefficients of L below the main diagonal and the nonzero coefficients of U overwrite the matrix A.

The permutation matrix P is not actually needed. The pointer vector Row (P) serves in its place to keep track of row interchanges.

If matrix A is to be used later, a copy must be stored someplace else. The value DET is the determinant of A.

```
VARiable declaration
        REAL   A[1 .. N, 1 .. N], X[1 .. N], Y[1 .. N], B[1 .. N]
        INTEGER   Row [1 .. N]
        INPUT N , A[1 .. N, 1 .. N]              {Input the matrix A at this time}
        DET := 1                                {Initialize the variable}

        FOR    J = 1  TO  N  DO                  {Initialize the pointer vector}
        └──── Row (J) := J

FOR    P = 1  TO  N−1  DO                        {Start LU factorization}

        FOR    K = P+1  TO  N  DO                {Find the pivot element}
        │   IF  |A(Row(K), P)| > |A(Row(P), P)|   THEN
        │           T := Row(P)                  {Switch the index for
        │           Row(P) := Row(K)                 the pth pivot row
        │           Row(K) := T                      if necessary}
        │              DET := − DET              {Change the sign of DET}
        └──── ENDIF                              {End of simulated row interchange}

        IF    A(Row(P), P) = 0   THEN
        │   PRINT "The matrix is singular."
        │   TERMINATE  ALGORITHM
        ENDIF

        DET := DET∗A(Row(P), P)                  {Multiply the diagonal elements}

        FOR    K = P+1  TO  N  DO
        │   A(Row(K), P) := A(Row(K), P)/A(Row(P), P)  {Find the multipliers}
        │   FOR  C = P+1  TO  N  DO              {Eliminate X_P}
        └────└──── A(Row(K), C) := A(Row(K),C) − A(Row(K),P)∗A(Row(P),C)

END                                             {End of the L∗U factorization routine}

        DET := DET∗A(Row(N), N)                  {Multiply the diagonal elements}

        INPUT   B[1 .. N]                        {Input the column vector now}

        Y(1) := B(Row(1))                        {Start the forward substitution}
```

```
FOR    K = 2  TO  N  DO
  │    SUM := 0
  │    FOR  C = 1  TO  K−1  DO
  │    └── SUM := SUM + A(Row(K), C)*Y(C)      {End forward substitution}
  └────  Y(K)  := B(Row(K))  −  SUM


       IF    A(Row(N), N) = 0  THEN
  │          PRINT "The matrix is singular."
  │          TERMINATE  ALGORITHM
       ENDIF


       X(N) := Y(N)/A(Row(N), N)                {Start the back substitution}

FOR    K = N−1  DOWNTO  1  DO
  │    SUM := 0
  │        FOR    C = K+1  TO  N  DO
  │        └── SUM := SUM+A(Row(K), C)*X(C)
  └──  X(K) := [Y(K)−SUM]/A(Row(K), K)         {End back substitution}

FOR    K = 1  TO  N  DO
  └── PRINT   X(K)                               {Output}
```

Exercises for Triangular Factorization

1. Solve $LY = \mathbf{B}$, $UX = \mathbf{Y}$ and verify that $\mathbf{B} = A\mathbf{X}$ for (a) $B^T = (-4, 10, 5)$ and (b) $B^T = (20, 49, 32)$, where $A = LU$ is

$$
\begin{pmatrix} 2 & 4 & -6 \\ 1 & 5 & 3 \\ 1 & 3 & 2 \end{pmatrix} = \begin{pmatrix} 1 & 0 & 0 \\ \frac{1}{2} & 1 & 0 \\ \frac{1}{2} & \frac{1}{3} & 1 \end{pmatrix} \begin{pmatrix} 2 & 4 & -6 \\ 0 & 3 & 6 \\ 0 & 0 & 3 \end{pmatrix}.
$$

2. Solve $LY = \mathbf{B}$, $UX = \mathbf{Y}$ and verify that $\mathbf{B} = A\mathbf{X}$ for (a) $B^T = (7, 2, 10)$ and (b) $B^T = (25, 35, 7)$, where $A = LU$ is

$$
\begin{pmatrix} 1 & 1 & 6 \\ -1 & 2 & 9 \\ 1 & -2 & 3 \end{pmatrix} = \begin{pmatrix} 1 & 0 & 0 \\ -1 & 1 & 0 \\ 1 & -1 & 1 \end{pmatrix} \begin{pmatrix} 1 & 1 & 6 \\ 0 & 3 & 15 \\ 0 & 0 & 12 \end{pmatrix}.
$$

3. Solve $LY = \mathbf{B}$, $UX = \mathbf{Y}$ and verify that $\mathbf{B} = A\mathbf{X}$ for (a) $B^T = (6, 13, 3)$ and (b) $B^T = (3.0, 1.5, -1.5)$, where $A = LU$ is

$$
\begin{pmatrix} 2 & -2 & 5 \\ 2 & 3 & 1 \\ -1 & 4 & -4 \end{pmatrix} = \begin{pmatrix} 1 & 0 & 0 \\ 1 & 1 & 0 \\ -0.5 & 0.6 & 1 \end{pmatrix} \begin{pmatrix} 2 & -2 & 5 \\ 0 & 5 & -4 \\ 0 & 0 & 0.9 \end{pmatrix}.
$$

4. Find the triangular factorization $A = LU$ for the matrices

(a) $\begin{pmatrix} -5 & 2 & -1 \\ 1 & 0 & 3 \\ 3 & 1 & 6 \end{pmatrix}$ (b) $\begin{pmatrix} 1 & 0 & 3 \\ 3 & 1 & 6 \\ -5 & 2 & -1 \end{pmatrix}$

5. Find the triangular factorization $A = LU$ for the matrices

(a) $\begin{pmatrix} 4 & 2 & 1 \\ 2 & 5 & -2 \\ 1 & -2 & 7 \end{pmatrix}$ (b) $\begin{pmatrix} 1 & -2 & 7 \\ 4 & 2 & 1 \\ 2 & 5 & -2 \end{pmatrix}$

6. Solve $LY = B$, $UX = Y$ and verify that $B = AX$ for (a) $B^T = (8, -4, 10, -4)$ and (b) $B^T = (28, 13, 23, 4)$, where $A = LU$ is

$$\begin{pmatrix} 4 & 8 & 4 & 0 \\ 1 & 5 & 4 & -3 \\ 1 & 4 & 7 & 2 \\ 1 & 3 & 0 & -2 \end{pmatrix} = \begin{pmatrix} 1 & 0 & 0 & 0 \\ \frac{1}{4} & 1 & 0 & 0 \\ \frac{1}{4} & \frac{2}{3} & 1 & 0 \\ \frac{1}{4} & \frac{1}{3} & -\frac{1}{2} & 1 \end{pmatrix} \begin{pmatrix} 4 & 8 & 4 & 0 \\ 0 & 3 & 3 & -3 \\ 0 & 0 & 4 & 4 \\ 0 & 0 & 0 & 1 \end{pmatrix}.$$

7. Solve $LY = B$, $UX = Y$ and verify that $B = AX$ for (a) $B^T = (12, 18, 8, 8)$ and (b) $B^T = (-2, -10, 11, 4)$, where $A = LU$ is

$$\begin{pmatrix} 2 & 4 & -4 & 0 \\ 1 & 5 & -5 & -3 \\ 2 & 3 & 1 & 3 \\ 1 & 4 & -2 & 2 \end{pmatrix} = \begin{pmatrix} 1 & 0 & 0 & 0 \\ \frac{1}{2} & 1 & 0 & 0 \\ 1 & -\frac{1}{3} & 1 & 0 \\ \frac{1}{2} & \frac{2}{3} & \frac{1}{2} & 1 \end{pmatrix} \begin{pmatrix} 2 & 4 & -4 & 0 \\ 0 & 3 & -3 & -3 \\ 0 & 0 & 4 & 2 \\ 0 & 0 & 0 & 3 \end{pmatrix}.$$

8. Solve $LY = B$, $UX = Y$ and verify that $B = AX$ for (a) $B^T = (9, 9, 26, 32)$ and (b) $B^T = (2.0, 7.5, -7.0, 7.5)$, where $A = LU$ is

$$\begin{pmatrix} 1 & 2 & 0 & -1 \\ 2 & 3 & -1 & 0 \\ 0 & 4 & 2 & -5 \\ 5 & 5 & 2 & -4 \end{pmatrix} = \begin{pmatrix} 1 & 0 & 0 & 0 \\ 2 & 1 & 0 & 0 \\ 0 & -4 & 1 & 0 \\ 5 & 5 & -3.5 & 1 \end{pmatrix} \begin{pmatrix} 1 & 2 & 0 & -1 \\ 0 & -1 & -1 & 2 \\ 0 & 0 & -2 & 3 \\ 0 & 0 & 0 & 1.5 \end{pmatrix}.$$

9. Find the triangular factorization $A = LU$ for the matrix

$$\begin{pmatrix} 1 & 1 & 0 & 4 \\ 2 & -1 & 5 & 0 \\ 5 & 2 & 1 & 2 \\ -3 & 0 & 2 & 6 \end{pmatrix}.$$

10. Establish the formula in (12).

11. Write a computer program that uses Algorithm 3.6 to solve a linear system of equations $A\mathbf{X} = \mathbf{B}$. Use some of Exercises 1–3 and 6–8 as test cases.

12. Write an output routine for Algorithm 3.6 so that it will print out the matrices $A, P, L,$ and U. Use some of the exercises above as test cases; switch rows of your test case to see what happens.

13. Write a report on LU factorization of matrices. Include a discussion of Doolittle's method, Choleski's method, and symmetric matrices.

14. Write a report on determinants. Include a discussion and examples of row interchanges, column interchanges, row replacement, column replacement, and scalar multiplication.

15. Modify Algorithm 3.6 so that it will compute A^{-1} by repeatedly solving N linear systems

$$A\mathbf{C}_J = \mathbf{E}_J, \ J = 1, 2, \ldots, N.$$

Then

$$A[\mathbf{C}_1, \mathbf{C}_2, \ldots, \mathbf{C}_N] = [\mathbf{E}_1, \mathbf{E}_2, \ldots, \mathbf{E}_N] = I, \quad \text{and} \quad A^{-1} = [\mathbf{C}_1, \mathbf{C}_2, \ldots, \mathbf{C}_N].$$

Make sure that you compute the LU factorization only once!

16. Prove that a triangular factorization is unique in the following sense: If A is nonsingular and $L_1U_1 = A = L_2U_2$, then $L_1 = L_2$ and $U_1 = U_2$.

17. Prove that the product of two N by N upper-triangular matrices is an upper-triangular matrix.

18. Prove that the inverse of a nonsingular N by N upper-triangular matrix is an upper-triangular matrix.

19. Kirchhoff's voltage law says that the sum of the voltage drops around any closed path in the network in a given direction is zero. When this principle is applied to the figure below, we obtain the following linear system of equations:

$$\begin{aligned}
(R_1 + R_3 + R_4)I_1 + \quad & R_3I_2 + \quad & R_4I_3 = E_1 \\
R_3I_1 + (R_2 + R_3 + R_5)I_2 - \quad & & R_5I_3 = E_2 \\
R_4I_1 - \quad & R_5I_2 + (R_4 + R_5 + R_6)I_3 = 0. &
\end{aligned}$$

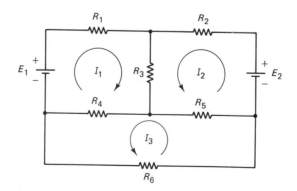

Solve for the currents I_1, I_2, and I_3 if

(a) $R_1 = 1$, $R_2 = 1$, $R_3 = 2$, $R_4 = 1$, $R_5 = 2$, $R_6 = 4$, and $E_1 = 23$, $E_2 = 29$

(b) $R_1 = 1$, $R_2 = 0.75$, $R_3 = 1$, $R_4 = 2$, $R_5 = 1$, $R_6 = 4$, and $E_1 = 12$, $E_2 = 21.5$

(c) $R_1 = 1$, $R_2 = 2$, $R_3 = 4$, $R_4 = 3$, $R_5 = 1$, $R_6 = 5$, and $E_1 = 41$, $E_2 = 38$

4

Numerical Interpolation
and Extrapolation

4.1 Introduction to Interpolation

In Chapter 1 we saw how a Taylor polynomial can be used to approximate the function $f(x)$. The information needed to construct the Taylor polynomial is the value of f and its derivatives at x_0. A shortcoming is that the higher-order derivatives must be known, and often they are either not available or they are hard to compute.

Suppose that the function $y = f(x)$ is known at the $N + 1$ points $(x_0, y_0), \ldots, (x_N, y_N)$, where the values x_k are spread out over the interval $[a, b]$ and satisfy

$$a \leqq x_0 < x_1 < \ldots < x_N \leqq b \quad \text{and} \quad y_k = f(x_k).$$

A polynomial $P(x)$ of degree N shall be constructed which passes through these $N + 1$ points. In the construction, only the numerical values x_k and y_k are needed. Hence the higher-order derivatives are not necessary. The polynomial $P(x)$ can be used to approximate $f(x)$ over the entire interval $[a, b]$. However, if the error function $E(x) = f(x) - P(x)$ is required, then we will need to know $f^{(N+1)}(x)$ and a bound for its magnitude, that is,

$$M = \max \{|f^{(N+1)}(x)|: \quad \text{for } a \leqq x \leqq b\}.$$

Situations in statistical and scientific analysis arise where the function $y = f(x)$ is available only at $N + 1$ tabulated points (x_k, y_k) and a method is needed to approximate $f(x)$ at nontabulated abscissas. If there is a significant amount of error in the tabulated values, then the methods of curve fitting in Chapter 5 should be considered. On the other hand, if the points (x_k, y_k) are known to a high degree of accuracy, then the polynomial curve $y = P(x)$ that passes through them can be considered. When $x_0 < x < x_N$ the approximation $P(x)$ is called an *interpolated value*. If either $x < x_0$ or $x_N < x$, then $P(x)$ is called an *extrapolated value*. Polynomials are used to design software algorithms to approximate functions, for numerical differentiation, for numerical integration, and for making computer-drawn curves that must pass through specified points.

Let us briefly mention how to evaluate the polynomial $P(x)$:

(1) $$P(x) = a_N x^N + a_{N-1} x^{N-1} + \ldots + a_2 x^2 + a_1 x + a_0.$$

Horner's method of synthetic division is an efficient way to evaluate $P(x)$. The derivative $P'(x)$ is

(2) $$P'(x) = N a_N x^{N-1} + (N-1) a_{N-1} x^{N-2} + \ldots + 2 a_2 x + a_1$$

and the indefinite integral $I(x)$, which satisfies $I'(x) = P(x)$, is

(3) $$I(x) = \frac{a_N x^{N+1}}{N+1} + \frac{a_{N-1} x^N}{N} + \ldots + \frac{a_2 x^3}{3} + \frac{a_1 x^2}{2} + a_0 x + C,$$

where C is the constant of integration. Algorithm 4.1 shows how to adapt Horner's method to evaluate $P'(x)$ and $I(x)$.

Algorithm 4.1 [*Polynomial Calculus*] Let $P(x)$ be a polynomial of degree N and have the form (1). The following three algorithms are used to evaluate $P(x)$, $P'(x)$, and $I(x)$, respectively.

```
INPUT   N                            {Degree of P(x)}
INPUT   A(0), A(1), ..., A(N)        {Coefficients of P(x)}
INPUT   C                            {Constant of integration}
INPUT   X                            {Independent variable}
```

(i) Algorithm to evaluate P(x)	Space-saving version:
`B(N) := A(N)`	`Poly := A(N)`
`FOR  K = N−1  DOWNTO  0  DO`	`FOR  K = N−1  DOWNTO  0  DO`
`    B(K) := A(K) + B(K+1)*X`	`    Poly := A(K) + Poly*X`
`PRINT "The value P(x) is", B(0)`	`"The value P(x) is", Poly`

(ii) Algorithm to evaluate P'(x)	Space-saving version:
`D(N−1) := N*A(N)`	`Deriv := N*A(N)`
`FOR  K = N−1  DOWNTO  1  DO`	`FOR  K = N−1  DOWNTO  1  DO`
`    D(K−1) := K*A(K) + D(K)*X`	`    Deriv := K*A(K) + Deriv*X`
`PRINT "The value P'(x) is", D(0)`	`"The value P'(x) is", Deriv`

(iii) Algorithm to evaluate I(x)	Space-saving version:
I(N+1) := A(N)/[N+1] FOR K = N DOWNTO 1 DO └── I(K) := A(K−1)/K + I(K+1)∗X I(0) := C + I(1)∗X PRINT "The value of I(x) is", I(0)	Integ := A(N)/[N+1] FOR K = N DOWNTO 1 DO └── Integ := A(K−1)/K + Integ∗X Integ := C + Integ∗X "The value of I(x) is", Integ

Example 4.1 The polynomial $P(x) = -0.02x^3 + 0.2x^2 - 0.4x + 1.28$ passes through the four points $(1, 1.06)$, $(2, 1.12)$, $(3, 1.34)$, and $(5, 1.78)$. Find (a) $P(4)$, (b) $P'(4)$, (c) $\int_1^4 P(x)\,dx$, and (d) $P(5.5)$.

(e) Show how to find the coefficients of $P(x)$.

    Solution. Use Algorithm 4.1(i)–(iii) with $x = 4$.
    (a) $b_3 = a_3 = -0.02$
        $b_2 = a_2 + b_3x = 0.2 + (-0.02)(4) = 0.12$
        $b_1 = a_1 + b_2x = -0.4 + (0.12)(4) = 0.08$
        $b_0 = a_0 + b_1x = 1.28 + (0.08)(4) = 1.60.$
The interpolated value is $P(4) = 1.60$ [see Figure 4.1(a)].
    (b) $d_2 = 3a_3 = -0.06$
        $d_1 = 2a_2 + d_2x = 0.4 + (-0.06)(4) = 0.16$
        $d_0 = a_1 + d_1x = -0.4 + (0.16)(4) = 0.24.$
The numerical derivative is $P'(4) = 0.24$ [see Figure 4.1(b)].

    (c) $i_4 = \dfrac{a_3}{4} = -0.005$

        $i_3 = \dfrac{a_2}{3} + i_4x = 0.06666667 + (-0.005)(4) = 0.04666667$

        $i_2 = \dfrac{a_1}{2} + i_3x = -0.2 + (0.04666667)(4) = -0.01333333$

        $i_1 = a_0 + i_2x = 1.28 + (-0.01333333)(4) = 1.22666667$

        $i_0 = 0 + i_1x = 0 + (1.22666667)(4) = 4.90666667.$

Hence $I(4) = 4.90666667$. Similarly, $I(1) = 1.14166667$. Therefore, $\int_1^4 P(x)\,dx = I(4) - I(1) = 3.765$ [see Figure 4.1(c)].
    (d) Use Algorithm 4.1(i) with $x = 5.5$.

$$b_3 = a_3 = -0.02$$

$$b_2 = a_2 + b_3x = 0.2 + (-0.02)(5.5) = 0.09$$

$$b_1 = a_1 + b_2x = -0.4 + (0.09)(5.5) = 0.095$$

$$b_0 = a_0 + b_1x = 1.28 + (0.095)(5.5) = 1.8025.$$

The extrapolated value is $P(5.5) = 1.8025$ [see Figure 4.1(a)].

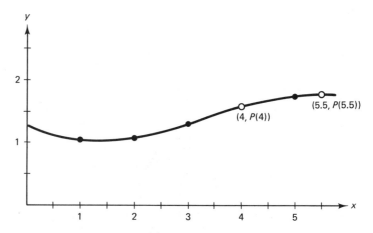

Figure 4.1(a). The graph $y = P(x)$ is used for interpolation at $(4, P(4))$ and extrapolation at $(5.5, P(5.5))$.

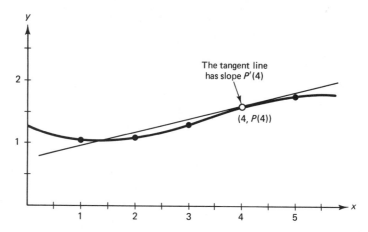

Figure 4.1(b). The graph $y = P(x)$ with slope $P'(x)$ is used to find the slope at the interpolation point $(4, P(4))$.

(e) The methods of Chapter 3 can be used to find the coefficients. Assume that $P(x) = A + Bx + Cx^2 + Dx^3$; then at each value $x = 1, 2, 3, 5$ we get a linear equation involving A, B, C, and D.

$$
\begin{aligned}
\text{At } x = 1: & \quad A + 1B + 1C + 1D = 1.06 \\
\text{At } x = 2: & \quad A + 2B + 4C + 8D = 1.12 \\
\text{At } x = 3: & \quad A + 3B + 9C + 27D = 1.34 \\
\text{At } x = 5: & \quad A + 5B + 25C + 125D = 1.78.
\end{aligned}
$$

(4)

The solution to (4) is $A = 1.28$, $B = -0.4$, $C = 0.2$, and $D = -0.02$.

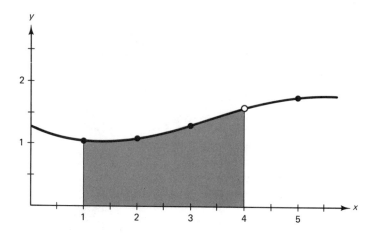

Figure 4.1(c). The graph $y = P(x)$ of the interpolation polynomial is used to estimate the integral over $[1, 4]$.

This method for finding the coefficients is mathematically sound, but sometimes the matrix is difficult to solve accurately. In this chapter we design algorithms specifically for polynomials.

Let us return to the topic of using a polynomial to calculate approximations to a known function. In Chapter 1 we saw that the fifth-degree Taylor polynomial for $f(x) = \ln(1 + x)$ is

(5)
$$T(x) = x - \frac{x^2}{2} + \frac{x^3}{3} - \frac{x^4}{4} + \frac{x^5}{5}.$$

If $T(x)$ is used to approximate $\ln(1 + x)$ on the interval $[0, 1]$, then the error is 0 at $x = 0$ and is largest when $x = 1$ (see Table 4.1). Indeed, the error between $T(1)$ and the correct value $\ln(2)$ is 13%. We seek a polynomial of degree 5 that will approximate $\ln(1 + x)$ better over the interval $[0, 1]$. The polynomial $P(x)$ in Example 4.2 is an interpolating polynomial and will approximate $\ln(1 + x)$ with an error no bigger than 0.00002385 over the interval $[0, 1]$.

Table 4.1   Values of the Taylor polynomial $T(x)$ of degree 5, and the function $\ln(1 + x)$ and error $\ln(1 + x) - T(x)$ on $[0, 1]$

x	Taylor Polynomial, $T(x)$	Function, $\ln(1 + x)$	Error, $\ln(1 + x) - T(x)$
0.0	0.00000000	0.00000000	0.00000000
0.2	0.18233067	0.18232156	−0.00000911
0.4	0.33698133	0.33647224	−0.00050909
0.6	0.47515200	0.47000363	−0.00514837
0.8	0.61380267	0.58778666	−0.02601601
1.0	0.78333333	0.69314718	−0.09018615

Example 4.2 Consider the function $f(x) = \ln(1 + x)$ and the polynomial

$$P(x) = 0.02957206x^5 - 0.12895295x^4 + 0.28249626x^3$$
$$- 0.48907554x^2 + 0.99910735x$$

based on the six nodes $x_k = k/5$ for $k = 0, 1, \ldots, 5$. The following are empirical descriptions of the approximation $P(x) \approx \ln(1 + x)$.

1. $P(x_k)$ agrees with $f(x_k)$ at each node [see Table 4.2].
2. The maximum error on the interval $[-0.1, 1.1]$ occurs at $x = -0.1$ and $|\text{error}| \leq 0.00026334$ for $-0.1 \leq x \leq 1.1$ (see Figure 4.3). Hence the graph of $y = P(x)$ would appear identical to that of $y = \ln(1 + x)$ (see Figure 4.2).

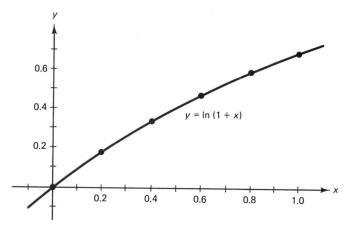

Figure 4.2. The graph of $y = P(x)$ which appears to agree with the graph of the curve $y = \ln(1 + x)$.

3. The maximum error on the interval $[0, 1]$ occurs at $x = 0.06472456$ and $|\text{error}| \leq 0.00002385$ for $0 \leq x \leq 1$ (see Figure 4.3).

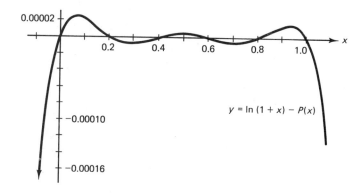

Figure 4.3. The graph of the error $y = E(x) = \ln(1 + x) - P(x)$.

Remark. At a node x_k we have $f(x_k) = P(x_k)$. Hence $E(x_k) = 0$ at a node. The graph of $E(x) = f(x) - P(x)$ looks like a vibrating string, with the nodes being the abscissa where there is no displacement.

Table 4.2   Values of the approximating polynomial $P(x)$ of Example 4.2, and the function $f(x) = \ln (1 + x)$ and error $E(x)$ on $[-0.1, 1.1]$

x	Approximating Polynomial $P(x)$	Function, $f(x) = \ln (1 + x)$	Error Function, $E(x) = f(x) - P(x)$
−0.1	−0.10509718	−0.10536052	−0.00026334
0.0	0.00000000	0.00000000	0.00000000
0.1	0.09528988	0.09531018	0.00002030
0.2	0.18232156	0.18232156	0.00000000
0.3	0.26237015	0.26236426	−0.00000589
0.4	0.33647224	0.33647224	0.00000000
0.5	0.40546139	0.40546511	0.00000372
0.6	0.47000363	0.47000363	0.00000000
0.7	0.53063292	0.53062825	−0.00000467
0.8	0.58778666	0.58778666	0.00000000
0.9	0.64184118	0.64185389	0.00001271
1.0	0.69314718	0.69314718	0.00000000
1.1	0.74206529	0.74193734	−0.00012795

Exercises for Introduction to Interpolation

1. Consider $P(x) = -0.02x^3 + 0.1x^2 - 0.2x + 1.66$, which passes through the four points $(1, 1.54)$, $(2, 1.5)$, $(3, 1.42)$, and $(5, 0.66)$.
   (a) Find $P(4)$.
   (b) Find $P'(4)$.
   (c) Find the integral of $P(x)$ taken over $[1, 4]$.
   (d) Find the extrapolated value $P(5.5)$.
   (e) Show how to find the coefficients of $P(x)$.

2. Consider $P(x) = -0.04x^3 + 0.14x^2 - 0.16x + 2.08$, which passes through the four points $(0, 2.08)$, $(1, 2.02)$, $(2, 2.00)$, and $(4, 1.12)$.
   (a) Find $P(3)$.
   (b) Find $P'(3)$.
   (c) Find the integral of $P(x)$ taken over $[0, 3]$.
   (d) Find the extrapolated value $P(4.5)$.
   (e) Show how to find the coefficients of $P(x)$.

3. Consider $P(x) = -0.029166667x^3 + 0.275x^2 - 0.570833333x + 1.375,$ which passes through $(1, 1.05)$, $(2, 1.10)$, $(3, 1.35)$, and $(5, 1.75)$.
   (a) Show that the ordinates 1.05, 1.1, 1.35, and 1.75 differ from those of Example 4.1 by less than 1.8%, yet the coefficients of x^3 and x differ by more than 42%.
   (b) Find $P(4)$ and compare with Example 4.1.

(c) Find $P'(4)$ and compare with Example 4.1.

(d) Find the integral of $P(x)$ taken over $[1, 4]$ and compare with Example 4.1.

(e) Find the extrapolated value $P(5.5)$ and compare with Example 4.1.

Remark. Part (a) shows that the computation of the coefficients of an interpolating polynomial is an ill-conditioned problem.

In Exercises 4–6, for the given function $f(x)$, the fifth-degree polynomial $P(x)$ passes through the six points $(0, f(0))$, $(0.2, f(0.2))$, $(0.4, f(0.4))$, $(0.6, f(0.6))$, $(0.8, f(0.8))$, and $(1, f(1))$. The six coefficients of $P(x)$ are $a_0, a_1, \ldots, a_5$, where

$$P(x) = a_5 x^5 + a_4 x^4 + a_3 x^3 + a_2 x^2 + a_1 x + a_0.$$

(a) Use Algorithm 4.1(i) to compute the interpolated values $P(0.3)$, $P(0.4)$, and $P(0.5)$ and compare with $f(0.3), f(0.4)$, and $f(0.5)$.

(b) Use Algorithm 4.1(i) to compute the extrapolated values $P(-0.1)$ and $P(1.1)$ and compare with $f(-0.1)$ and $f(1.1)$.

(c) Use a computer and make a table of values for $P(x_k), f(x_k)$, and $E(x_k)$, where $x_k = k/100$ for $k = 0, 1, \ldots, 100$.

4. If $f(x) = \exp(x)$, then the six coefficients are

$$a_5 = 0.01385431, \qquad a_2 = 0.49906876,$$
$$a_4 = 0.03486637, \qquad a_1 = 1.00008255,$$
$$a_3 = 0.17040984, \qquad a_0 = 1.00000000.$$

5. If $f(x) = \sin(x)$, then the six coefficients are

$$a_5 = 0.00725244, \qquad a_2 = 0.00024394,$$
$$a_4 = 0.00161306, \qquad a_1 = 0.99997802,$$
$$a_3 = -0.16761647, \qquad a_0 = 0.00000000.$$

6. If $f(x) = \cos(x)$, then the six coefficients are

$$a_5 = -0.00396206, \qquad a_2 = -0.49944788,$$
$$a_4 = 0.04604659, \qquad a_1 = -0.00004812,$$
$$a_3 = -0.00228623, \qquad a_0 = 1.00000000.$$

4.2 Linear and Lagrange Approximation

Interpolation means to estimate a missing function value by taking a weighted average of known function values at neighboring points. Linear interpolation is the process of approximating a small portion of the curve with a line segment that passes through two points on the curve. Suppose that the points (x_0, y_0) and (x_1, y_1) are used. The slope between the points is $m = (y_1 - y_0)/(x_1 - x_0)$, and the point-slope formula for the line yields

(1) $$y = mx + B, \qquad \text{where } B = y_1 - mx_1.$$

The right side of (1) is a polynomial of degree $\leqq 1$. The French mathematician Joseph Louis Lagrange used a different method to find this polynomial. He noticed that it could be written as

(2)
$$P_1(x) = y_0 \frac{x - x_1}{x_0 - x_1} + y_1 \frac{x - x_0}{x_1 - x_0}.$$

Each term on the right side of (2) involves a linear factor, hence the sum is a polynomial of degree $\leqq 1$. Therefore, it will suffice to show that $P_1(x_0) = y_0$ and $P_1(x_1) = y_1$. Calculation reveals that

(3)
$$P_1(x_0) = y_0 \frac{x_0 - x_1}{x_0 - x_1} + y_1 \frac{x_0 - x_0}{x_1 - x_0} = y_0 \times 1 + y_1 \times 0 = y_0,$$

$$P_1(x_1) = y_0 \frac{x_1 - x_1}{x_0 - x_1} + y_1 \frac{x_1 - x_0}{x_1 - x_0} = y_0 \times 0 + y_1 \times 1 = y_1.$$

The terms $(x - x_1)/(x_0 - x_1)$ and $(x - x_0)/(x_1 - x_0)$ were constructed so that they produced the quotients $1/1, 0/1$ and $0/1, 1/1$ at x_0 and x_1, respectively. This forced the values $P_1(x_k)$ to be linear combinations of y_0 and y_1 which readily simplified to yield $P_1(x_k) = y_k$ (for $k = 1, 2$).

A common use for $P_1(x)$ is the approximation of a function $f(x)$ on the interval $[a, b]$ where the endpoints of the interval are used for the nodes, that is, $x_0 = a$ and $x_1 = b$ [see Figure 4.4(a)]. Sometimes the nodes are chosen inside the interval $[a, b]$. If $a < x_0$ and $x_1 < b$, both interpolation and extrapolation are involved, as shown in Figure 4.4(b). After comparing these graphs, we may speculate that a better fit between the line segment and curve will occur when the nodes lie inside $[a, b]$. It will be necessary to study the error term $E_1(x) = f(x) - P_1(x)$ to understand this matter fully.

Example 4.3 Consider the curve $y = f(x) = \sin(x)$ over the interval $[0.4, 1.0]$. Find the linear approximations using (a) the nodes 0.4 and 1.0 and (b) the nodes 0.5 and 0.9.

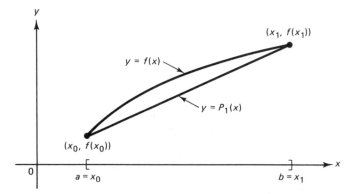

Figure 4.4(a). Linear approximation where the nodes are the endpoints of the interval $[a, b]$.

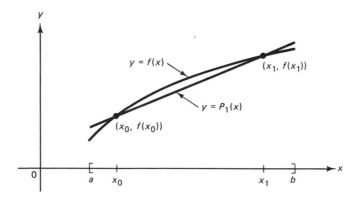

Figure 4.4(b). Linear approximation where the nodes lie inside the interval $[a, b]$.

Solution. (a) Using equation (2) with $x_0 = 0.4$ and $x_1 = 1.0$ yields

$$P_1(x) = \sin(0.4)\frac{x - 1.0}{-0.6} + \sin(1.0)\frac{x - 0.4}{0.6},$$

which can be simplified to obtain

$P_1(x) = 0.75342107x + 0.08804991$ based on nodes 0.4 and 1.0.

(b) When the nodes $x_0 = 0.5$ and $x_1 = 0.9$ are used, we get

$$Q_1(x) = \sin(0.5)\frac{x - 0.9}{-0.4} + \sin(0.9)\frac{x - 0.5}{0.4},$$

which can be simplified to obtain

$Q_1(x) = 0.75975343x + 0.09954882$ based on nodes 0.5 and 0.9.

Notice that the error in approximation at $x = 0.7$ is reduced from 0.02877302 to 0.01284146 when $Q_1(0.7)$ is used (see Table 4.3). However, error is introduced at the endpoints where $Q_1(0.4)$ and $Q_1(1.0)$ are extrapolated values. But the overall error in $Q_1(x)$ is less than the overall error in $P_1(x)$.

Table 4.3 Comparison of the linear approximations $P_1(x)$ and $Q_1(x)$ to sin (x) using the nodes 0.4, 1 and 0.5, 0.9, respectively

x	$P_1(x)$	$\sin(x) - P_1(x)$	$Q_1(x)$	$\sin(x) - Q_1(x)$
0.4	0.38941834	0.00000000	0.40345020	−0.01403185
0.5	0.46476045	0.01466509	0.47942554	0.00000000
0.6	0.54010256	0.02453992	0.55540088	0.00924159
0.7	0.61544466	0.02877302	0.63137622	0.01284146
0.8	0.69078677	0.02656932	0.70735157	0.01000452
0.9	0.76612888	0.01719803	0.78332691	0.00000000
1.0	0.84147098	0.00000000	0.85930225	−0.01783127

Quadratic Interpolation and Approximation

The quadratic curve $y = P_2(x)$ that passes through the three points (x_0, y_0), (x_1, y_1), and (x_2, y_2) where x_0, x_1, x_2 are distinct has the form

$$(4) \quad P_2(x) = y_0 \frac{(x - x_1)(x - x_2)}{(x_0 - x_1)(x_0 - x_2)} + y_1 \frac{(x - x_0)(x - x_2)}{(x_1 - x_0)(x_1 - x_2)} + y_2 \frac{(x - x_0)(x - x_1)}{(x_2 - x_0)(x_2 - x_1)}.$$

A note of caution must be observed when using the formula for $P_2(x)$ given in equation (4), namely that the abscissas $x_0, x_1,$ and x_2 must be distinct. To establish formula (4), it will suffice to show that $P_2(x)$ goes through the three points. The following calculations will suffice to prove it:

$$P_2(x_0) = y_0 \frac{(x_0 - x_1)(x_0 - x_2)}{(x_0 - x_1)(x_0 - x_2)} + y_1 \frac{(x_0 - x_0)(x_0 - x_2)}{(x_1 - x_0)(x_1 - x_2)} + y_2 \frac{(x_0 - x_0)(x_0 - x_1)}{(x_2 - x_0)(x_2 - x_1)}$$

$$= y_0 \times 1 + y_1 \times 0 + y_2 \times 0 = y_0,$$

$$P_2(x_1) = y_0 \frac{(x_1 - x_1)(x_1 - x_2)}{(x_0 - x_1)(x_0 - x_2)} + y_1 \frac{(x_1 - x_0)(x_1 - x_2)}{(x_1 - x_0)(x_1 - x_2)} + y_2 \frac{(x_1 - x_0)(x_1 - x_1)}{(x_2 - x_0)(x_2 - x_1)}$$

$$= y_0 \times 0 + y_1 \times 1 + y_2 \times 0 = y_1,$$

$$P_2(x_2) = y_0 \frac{(x_2 - x_1)(x_2 - x_2)}{(x_0 - x_1)(x_0 - x_2)} + y_1 \frac{(x_2 - x_0)(x_2 - x_2)}{(x_1 - x_0)(x_1 - x_2)} + y_2 \frac{(x_2 - x_0)(x_2 - x_1)}{(x_2 - x_0)(x_2 - x_1)}$$

$$= y_0 \times 0 + y_1 \times 0 + y_2 \times 1 = y_2.$$

The polynomial $P_2(x)$ is referred to as the Lagrange polynomial of degree $N = 2$ that is based on the nodes (x_0, y_0), (x_1, y_1), and (x_2, y_2). If $x_0 < x < x_2$, then $P_2(x)$ is said to be an interpolated value. If $x < x_0$ or $x_2 < x$, then $P_2(x)$ is called an extrapolated value.

Cubic Interpolation and Approximation

The cubic interpolating curve $y = P_3(x)$ that passes through the four points (x_0, y_0), (x_1, y_1), (x_2, y_2), and (x_3, y_3), where x_0, x_1, x_2, x_3 are distinct, has the form

$$(5) \quad P_3(x) = y_0 \frac{(x - x_1)(x - x_2)(x - x_3)}{(x_0 - x_1)(x_0 - x_2)(x_0 - x_3)} + y_1 \frac{(x - x_0)(x - x_2)(x - x_3)}{(x_1 - x_0)(x_1 - x_2)(x_1 - x_3)}$$

$$+ y_2 \frac{(x - x_0)(x - x_1)(x - x_3)}{(x_2 - x_0)(x_2 - x_1)(x_2 - x_3)} + y_3 \frac{(x - x_0)(x - x_1)(x - x_2)}{(x_3 - x_0)(x_3 - x_1)(x_3 - x_2)}.$$

To establish formula (5), it will suffice to observe that

$$P_3(x_0) = y_0 \times 1 + y_1 \times 0 + y_2 \times 0 + y_3 \times 0 = y_0,$$
$$P_3(x_1) = y_0 \times 0 + y_1 \times 1 + y_2 \times 0 + y_3 \times 0 = y_1,$$
$$P_3(x_2) = y_0 \times 0 + y_1 \times 0 + y_2 \times 1 + y_3 \times 0 = y_2,$$
$$P_3(x_3) = y_0 \times 0 + y_1 \times 0 + y_2 \times 0 + y_3 \times 1 = y_3.$$

Example 4.4 Consider $f(x) = \sin(x)$ over the interval $[0.4, 1.0]$.
(a) Find the quadratic approximation using the nodes $0.4, 0.7$, and 1.
(b) Find the cubic approximation using the nodes $0.4, 0.6, 0.8$, and 1.
(c) Compare the polynomials at the points $0.4, 0.5, 0.6, 0.7, 0.8, 0.9$, and 1.

Solution. (a) Using equation (4) with $x_0 = 0.4$, $x_1 = 0.7$, and $x_2 = 1$, we write the Lagrange form of the quadratic polynomial $P_2(x)$:

$$P_2(x) = \sin(0.4)\frac{(x - 0.7)(x - 1)}{(0.4 - 0.7)(0.4 - 1)} + \sin(0.7)\frac{(x - 0.4)(x - 1)}{(0.7 - 0.4)(0.7 - 1)}$$

$$+ \sin(1.0)\frac{(x - 0.4)(x - 0.7)}{(1.0 - 0.4)(1.0 - 0.7)}$$

which can be simplified to obtain the ordinary polynomial

(6)
$$P_2(x) = -0.31970026x^2 + 1.20100144x - 0.03983019.$$

(b) Using equation (5) with $x_0 = 0.4$, $x_1 = 0.6$, $x_2 = 0.8$, and $x_3 = 1$, we write the Lagrange form of the cubic polynomial $P_3(x)$:

$$P_3(x) = \sin(0.4)\frac{(x - 0.6)(x - 0.8)(x - 1)}{(0.4 - 0.6)(0.4 - 0.8)(0.4 - 1)} + \sin(0.6)\frac{(x - 0.4)(x - 0.8)(x - 1)}{(0.6 - 0.4)(0.6 - 0.8)(0.6 - 1)}$$

$$+ \sin(0.8)\frac{(x - 0.4)(x - 0.6)(x - 1)}{(0.8 - 0.4)(0.8 - 0.6)(0.8 - 1)} + \sin(1)\frac{(x - 0.4)(x - 0.6)(x - 0.8)}{(1 - 0.4)(1 - 0.6)(1 - 0.8)},$$

which can be simplified to obtain the ordinary polynomial

(7)
$$P_3(x) = -0.12683772x^3 - 0.05307353x^2 + 1.02559085x - 0.00420862.$$

(c) The arithmetic involved in changing the Lagrange forms to the simplified forms in (6) and (7) is quite tedious and can be done with a computer. Let us use the Lagrange form to evaluate $P_3(0.5)$.

$$P_3(0.5) = 0.38941834\frac{(-0.1)(-0.3)(-0.5)}{(-0.2)(-0.4)(-0.6)} + 0.56464247\frac{(+0.1)(-0.3)(-0.5)}{(+0.2)(-0.2)(-0.4)}$$

$$+ 0.71735609\frac{(+0.1)(-0.1)(-0.5)}{(+0.4)(+0.2)(-0.2)} + 0.84147098\frac{(+0.1)(-0.1)(-0.3)}{(+0.6)(+0.4)(+0.2)}$$

$$= 0.12169323 + 0.52935232 - 0.22417378 + 0.05259194$$

$$= 0.47946371.$$

A comparison of the other calculated values is given in Table 4.4.

Table 4.4 Comparison of the quadratic approximation $P_2(x)$ and cubic approximation $P_3(x)$ to sin (x) on $[0.4, 1.0]$

x	$P_2(x)$	sin $(x) - P_2(x)$	$P_3(x)$	sin $(x) - P_3(x)$
0.4	0.38941834	0.00000000	0.38941834	0.00000000
0.5	0.48074546	−0.00131992	0.47946371	−0.00003817
0.6	0.56567858	−0.00103610	0.56464247	0.00000000
0.7	0.64421769	0.00000000	0.64419361	0.00002408
0.8	0.71636279	0.00099330	0.71735609	0.00000000
0.9	0.78211389	0.00121302	0.78336889	−0.00004198
1.0	0.84147098	0.00000000	0.84147098	0.00000000

Application to Function Approximation

Theorem 4.1 [Error terms for approximations] Suppose that $f(x)$ is continuous on an interval $[a, b]$ containing the distinct values x_k. Consider the following approximations:

(8)   Linear: $f(x) = P_1(x) + E_1(x)$ with nodes x_0 and x_1.

(9)   Quadratic: $f(x) = P_2(x) + E_2(x)$ with nodes x_0, x_1, and x_2.

(10)   Cubic: $f(x) = P_3(x) + E_3(x)$ with nodes x_0, x_1, x_2, and x_3.

If the derivatives up to the order $N + 1$ are continuous ($N = 1, 2$, or 3), then there exists a value $c_N = c_N(x)$ in the interval (a, b) so that

$$(11) \qquad E_1(x) = (x - x_0)(x - x_1)\frac{f^{(2)}(c_1)}{2!},$$

$$(12) \qquad E_2(x) = (x - x_0)(x - x_1)(x - x_2)\frac{f^{(3)}(c_2)}{3!},$$

$$(13) \qquad E_3(x) = (x - x_0)(x - x_1)(x - x_2)(x - x_3)\frac{f^{(4)}(c_3)}{4!}.$$

Remark. The form of the error $E_N(x)$ involves a product of $N + 1$ linear factors $(x - x_0) \ldots (x - x_N)$ and makes $E_N(x_j) = 0$. If $P_N(x)$ is used to either interpolate or extrapolate, then the error terms (11)–(13) apply.

It will suffice to establish formula (11) since formulas (12) and (13) can be proven using a similar technique. We define the special function $g(t)$ as follows:

$$(14) \qquad g(t) = f(t) - P_1(t) - E_1(x)\frac{(t - x_0)(t - x_1)}{(x - x_0)(x - x_1)}.$$

Notice that x, x_0, x_1 are constants with respect to the variable t and that $g(t)$ is zero at these three values, that is,

$$g(x) = f(x) - P_1(x) - E_1(x) = E_1(x) - E_1(x) = 0,$$
$$g(x_0) = f(x_0) - P_1(x_0) - E_1(x_0)0 = 0,$$
$$g(x_1) = f(x_1) - P_1(x_1) - E_1(x_1)0 = 0.$$

For convenience we can consider that $x_0 < x < x_1$. Rolle's theorem* can be applied to $g(t)$ on the interval $[x_0, x]$ to find a value d_0 with $x_0 < d_0 < x$ so that

(15) $$g'(d_0) = 0.$$

A second application of Rolle's theorem to $g(t)$ on $[x, x_1]$ will produce a value d_1 with $x < d_1 < x_1$ so that

(16) $$g'(d_1) = 0.$$

Now observe that $g'(t)$ is zero at the two values $t = d_0, d_1$. A third use of Rolle's theorem, but this time applied to $g'(t)$ on the interval $[d_0, d_1]$ will find the value c_1 for which

(17) $$g^{(2)}(c_1) = 0.$$

Now let us use (14) and compute the derivatives of $g(t)$:

(18) $$g'(t) = f'(t) - P_1'(t) - E_1(x)\frac{(t - x_0) + (t - x_1)}{(x - x_0)(x - x_1)}.$$

(19) $$g^{(2)}(t) = f^{(2)}(t) - P_1^{(2)}(t) - E_1(x)\frac{2}{(x - x_0)(x - x_1)}.$$

Since $P_1(t)$ is a polynomial of degree $N = 1$, its second derivative is $P_1^{(2)}(t) \equiv 0$, hence $P_1^{(2)}(c_1) = 0$. When $g^{(2)}(c_1) = 0$ and $P_1^{(2)}(c_1) = 0$ are substituted into (19), the result is

(20) $$0 = f^{(2)}(c_1) - 0 - E_1(x)\frac{2}{(x - x_0)(x - x_1)}.$$

It is easy to solve for $E_1(x)$ in (20) and get equation (11), and the proof is complete.

If interpolation between equally spaced nodes is the only calculation performed, we may use the following simplified error bounds.

Theorem 4.2 [Error bounds for interpolation, equally spaced nodes] Let $f(x)$ be continuous on an interval containing the equally spaced nodes $x_k = x_0 + hk$. Suppose that the derivatives up to the order $N + 1$ are continuous $(N = 1, 2, 3)$ and that there exists constants M_{N+1} so that

(21) $$|f^{(N+1)}(c)| \leq M_{N+1} \quad \text{for } x_0 \leq c \leq x_N.$$

Rolle's Theorem. Let $g(t)$ be continuous on $[a, b]$ and differentiable on (a, b). If $g(a) = g(b) = 0$, then there exists at least one number c in (a, b) such that $g'(c) = 0$.

Then the error terms (11)–(13) for interpolation have the following useful bounds on their magnitude:

(22) $$|E_1(x)| \leq h^2 \frac{M_2}{8} \qquad \text{valid over } [x_0, x_1],$$

(23) $$|E_2(x)| \leq h^3 \frac{M_3}{9 \times 3^{1/2}} \qquad \text{valid over } [x_0, x_2],$$

(24) $$|E_3(x)| \leq h^4 \frac{M_4}{24} \qquad \text{valid over } [x_0, x_3].$$

Proof. We shall establish (22) and leave the others for the reader. Using the change of variables $x - x_0 = t$ and $x - x_1 = t - h$, for $0 \leq t \leq h$, the error term $E_1(x)$ can be written:

(25) $$E_1(x) = (x - x_0)(x - x_1) \frac{f^{(2)}(c_1)}{2} = (t^2 - th) \frac{f^{(2)}(c_1)}{2}.$$

The bound we use for the second derivative is

(26) $$|f^{(2)}(c_1)| \leq M_2 \qquad \text{for } x_0 \leq c_1 \leq x_1.$$

Now consider the special function

(27) $$\phi(t) = t^2 - th \qquad \text{for } 0 \leq t \leq h.$$

The derivative of $\phi(t)$ is $\phi'(t) = 2t - h$. The extreme values of $\phi(t)$ occur either at an endpoint $t = 0, h$, where $\phi(0) = 0$ and $\phi(h) = 0$, or at the critical point $t = h/2$, where $\phi'(h/2) = 0$ and $\phi(h/2) = -h^2/4$. Hence the maximum for the absolute value $|\phi(t)|$ occurs at $t = h/2$ and we have the bound

(28) $$|\phi(t)| \leq \frac{h^2}{4} \qquad \text{for } 0 \leq t \leq h.$$

Taking the absolute values in (25) and using the bounds given in (26) and (28) results in

(29) $$|E_1(x)| \leq |\phi(t)| \frac{|f^{(2)}(c_1)|}{2} \leq \frac{h^2 M_2}{8}.$$

Theorefore, (22) is established.

Comparison of Accuracy and $O(h^{N+1})$

The significance of Theorem 4.2 is to understand a simple relationship between the size of the error terms for linear, quadradic, and cubic interpolation. In each case the error bound $|E_N(x)|$ depends on h in two ways. First, h^{N+1} is explicitly present so that $|E_N(x)|$ is proportional to h^{N+1}. Second, the value c_N generally depend on h, and tends to x_0 as h goes to zero. Therefore, as h goes to zero, $E_N(x)$ converges to zero with the same rapidity that h^{N+1} converges to zero. The notation $O(h^{N+1})$ is frequently used to discuss this behavior. It is read "oh of h^{N+1}." For example, the error bound (22) can be expressed as

$$|E_1(x)| = O(h^2).$$

The notation $O(h^2)$ stands in place of $h^2 M_2/8$ in relation (29) and is meant to convey the idea that the error term is approximately a multiple of h^2, that is,

$$|E_1(x)| = O(h^2) \approx Ch^2.$$

As a consequence, if the derivatives of $f(x)$ are uniformly bounded on the interval and $|h| < 1$, then choosing N large will make h^{N+1} small and the higher-degree approximating polynomial will usually have less error.

Example 4.5 Compare the error bounds for the approximations $P_1(x)$, $P_2(x)$, and $P_3(x)$ to $f(x) = \sin(x)$ on the interval $[0.4, 1.0]$.

Solution. The bounds for the derivatives of $f(x)$ are

$$|f^{(2)}(c)| = |-\sin(c)| \leq \sin(1.0) = 0.84147098 = M_2,$$
$$|f^{(3)}(c)| = |-\cos(c)| \leq \cos(0.4) = 0.92106099 = M_3,$$
$$|f^{(4)}(c)| = |\sin(c)| \leq \sin(1.0) = 0.84147098 = M_4.$$

For $P_1(x)$ the spacing of the nodes is $h = 0.6$, so that

$$|E_1(x)| \leq h^2 \frac{M_2}{8} \leq (0.6)^2 \frac{0.84147098}{8} = 0.03786619.$$

For $P_2(x)$ the spacing of the nodes is $h = 0.3$, so that

$$|E_2(x)| \leq h^3 \frac{M_3}{9 \times 3^{1/2}} \leq (0.3)^3 \frac{0.92106099}{9 \times 3^{1/2}} = 0.00159532.$$

For $P_3(x)$ the spacing of the nodes is $h = 0.2$, so that

$$|E_3(x)| \leq h^4 \frac{M_4}{24} \leq (0.2)^4 \frac{0.84147098}{24} = 0.00005610$$

Using Examples 4.3 and 4.4, we see that $E_1(0.7) = 0.02877302$, $E_2(0.9) = 0.00121302$, and $E_3(0.9) = -0.00004198$. Although these values do not represent the maximum error over the interval, they do give some indication that the error bounds are reasonable.

Polynomial Approximation and Uniqueness

The Lagrange polynomial $P_N(x)$ of degree $\leq N$ that passes through the $N + 1$ points $(x_0, y_0), (x_1, y_1), \ldots, (x_N, y_N)$ has the form

(30)
$$P_N(x) = \sum_{k=0}^{N} y_k L_k(x)$$

where $L_k(x)$ is the *Lagrange coefficient polynomial*, defined by

(31)
$$L_k(x) = \frac{(x - x_0) \ldots (x - x_{k-1})(x - x_{k+1}) \ldots (x - x_N)}{(x_k - x_0) \ldots (x_k - x_{k-1})(x_k - x_{k+1}) \ldots (x_k - x_N)}.$$

It is understood that the terms $(x - x_k)$ and $(x_k - x_k)$ do not appear on the right side of equation (31). It is often convenient to introduce the product notation to express (31) and we write

(32)
$$L_k(x) = \frac{\prod\limits_{\substack{j=0 \\ j \neq k}}^{N} (x - x_j)}{\prod\limits_{\substack{j=0 \\ j \neq k}}^{N} (x_k - x_j)}$$

Here the notation indicates that products of the linear factors $(x - x_j)$ are to be formed, but the factor $(x - x_k)$ is to be left out (skipped). Then the nonzero product of terms $(x_k - x_j)$ is to be formed, where again the term $(x_k - x_k)$ is to be skipped. Finally, the quotient of these two products yields $L_k(x)$.

When $P_N(x)$ is used to approximate a continuous function $f(x)$ that has $N + 1$ continuous derivatives, we are permitted to write

(33)
$$f(x) = P_N(x) + E_N(x)$$

and there exists a value $c = c(x)$ so that

(34)
$$E_N(x) = (x - x_0)(x - x_1) \ldots (x - x_N) \frac{f^{(N+1)}(c)}{(N + 1)!}.$$

The proof of formula (34) is indicated in Exercise 13.

Proof of (30). It is easy to see that the Lagrange coefficient polynomial $L_k(x)$ has the property

(35)
$$L_k(x_j) = 1 \text{ when } j = k \quad \text{and} \quad L_k(x_j) = 0 \text{ when } j \neq k.$$

To show that $P_N(x)$ goes through the point (x_j, y_j), fix j and use the substitutions (35) in (30) to obtain

$$P_N(x_j) = y_0 L_0(x_j) + \ldots + y_j L_j(x_j) + \ldots + y_N L_N(x_j)$$
$$= y_0 \times 0 + \ldots + y_j \times 1 + \ldots + y_N \times 0$$
$$= y_j.$$

To show that $P_N(x)$ is unique, we can use the *Fundamental Theorem of Algebra*, which states that a polynomial

$$T(x) = a_N x^N + a_{N-1} x^{N-1} + \ldots + a_2 x^2 + a_1 x + a_0$$

of degree $\leq N$ has at most N roots. In other words, if $T(x)$ is zero at $N + 1$ distinct abscissae, then it is identically zero (i.e., $a_j = 0$ for $j = 0, 1, \ldots, N$). Now suppose that $P_N(x)$ is not unique and that there exists another polynomial $Q(x)$ of degree $\leqq N$ which passes through the $N + 1$ points. Then we form the difference polynomial

(36)
$$T(x) = P_N(x) - Q(x).$$

Now observe that $T(x)$ is of degree $\leq N$ and has $N + 1$ roots, that is,

$$T(x_j) = P_N(x_j) - Q(x_j) = y_j - y_j = 0 \qquad \text{for } j = 0, 1, \ldots, N.$$

Therefore, we conclude from the Fundamental Theorem of Algebra that $T(x) \equiv 0$ for all x. It follows from (36) that $Q(x) = P_N(x)$. Hence the uniqueness of $P_N(x)$ is established.

The Lagrange polynomial is important because we know that the error term has the form given in (34). In Section 4.3 we develop the Newton interpolation polynomial, which is algebraically equivalent to the Lagrange interpolation polynomial. However, the Newton polynomial is easier to use for computational purposes, and its error term is identical to (34).

Algorithm 4.2 [*Lagrange Approximation*] Given the $N + 1$ points (x_k, y_k) for $k = 0, 1, \ldots, N$, evaluate the Lagrange polynomial

$$P(x) = \sum_{k=0}^{N} y_k L_k(x)$$

where $L_k(x)$ is the Lagrange coefficient polynomial given in (32).

Remark. The nodes x_k must be distinct to preclude division by zero.

```
          INPUT   N                              {Degree of the polynomial}
          INPUT   X₀, X₁, ..., Xₙ                {The nodes of Pₙ(x)}
          INPUT   Y₀, Y₁, ..., Yₙ                {The ordinates of the points}

          INPUT   T                              {The independent variable}
          Sum := 0                               {Initialize variable}
     FOR  K = 0  TO  N  DO
          Term := Yₖ                             {Initialize variable}
          FOR  J = 0  TO  N  DO                  {Form terms to be added
               If  J ≠ K  THEN                    to avoid division by 0}
                    Term := Term * [T−Xⱼ]/[Xₖ−Xⱼ]
          Sum := Sum + Term

     PRINT 'The value of the Lagrange interpolating polynomial'
           'based on the nodes X₀, X₁, ..., Xₙ is' Sum
```

Exercises for Linear and Lagrange Approximation

1. Find Lagrange polynomials that approximate $f(x) = x^3$.
   (a) Find the linear interpolation polynomial $P_1(x)$ using the nodes $x_0 = -1$ and $x_1 = 0$.

(b) Find the quadratic interpolation polynomial $P_2(x)$ using the nodes $x_0 = -1$, $x_1 = 0$, and $x_2 = 1$.

(c) Find the cubic interpolation polynomial $P_3(x)$ using the nodes $x_0 = -1$, $x_1 = 0$, $x_2 = 1$, and $x_3 = 2$.

(d) Find the linear interpolation polynomial $P_1(x)$ using the nodes $x_0 = 1$ and $x_1 = 2$.

(e) Find the quadratic interpolation polynomial $P_2(x)$ using the nodes $x_0 = 0$, $x_1 = 1$, and $x_2 = 2$.

2. Let $f(x) = x + 2/x$.

(a) Use quadratic Lagrange interpolation based on the nodes $x_0 = 1$, $x_1 = 2$, and $x_2 = 2.5$ to approximate $f(1.5)$ and $f(1.2)$.

(b) Use cubic Lagrange interpolation based on the nodes $x_0 = 0.5$, $x_1 = 1$, $x_2 = 2$, and $x_3 = 2.5$ to approximate $f(1.5)$ and $f(1.2)$.

3. Let $f(x) = 8x/2^x$.

(a) Use quadratic Lagrange interpolation based on the nodes $x_0 = 0$, $x_1 = 1$, and $x_2 = 2$ to approximate $f(1.5)$ and $f(1.3)$.

(b) Use cubic Lagrange interpolation based on the nodes $x_0 = 0$, $x_1 = 1$, $x_2 = 2$, and $x_3 = 3$ to approximate $f(1.5)$ and $f(1.3)$.

4. Let $f(x) = 2 \sin(x\pi/6)$, where x is in radians.

(a) Use quadratic Lagrange interpolation based on the nodes $x_0 = 0$, $x_1 = 1$, and $x_2 = 3$ to approximate $f(2)$ and $f(2.4)$.

(b) Use cubic Lagrange interpolation based on the nodes $x_0 = 0$, $x_1 = 1$, $x_2 = 3$, and $x_3 = 5$ to approximate $f(2)$ and $f(2.4)$.

5. Let $f(x) = 2 \sin(x\pi/6)$, where x is in radians.

(a) Use quadratic Lagrange extrapolation based on the nodes $x_0 = 0$, $x_1 = 1$, and $x_2 = 3$ to approximate $f(4)$ and $f(3.5)$.

(b) Use cubic Lagrange interpolation based on the nodes $x_0 = 0$, $x_1 = 1$, $x_2 = 3$, and $x_3 = 5$ to approximate $f(4)$ and $f(3.5)$.

6. Write down the error term $E_3(x)$ for cubic Lagrange interpolation to $f(x)$, where interpolation is to be exact at the four nodes $x_0 = -1$, $x_1 = 0$, $x_2 = 3$, and $x_3 = 4$ and $f(x)$ is given by

(a) $f(x) = 4x^3 - 3x + 2$ (b) $f(x) = x^4 - 2x^3$

(c) $f(x) = x^5 - 5x^4$

7. Let $L_0(x), L_1(x), \ldots, L_N(x)$ be the Lagrange coefficient polynomials based on the $N + 1$ nodes $x_0, x_1, \ldots, x_N$. Show that $L_k(x_k) = 1$ for $k = 0, 1, \ldots, N$ and that $L_k(x_j) = 0$ whenever $j \neq k$.

8. Consider the Lagrange coefficient polynomials $L_k(x)$ that are used for quadratic interpolation at the nodes x_0, x_1, x_2.

Define $g(x) = L_0(x) + L_1(x) + L_2(x) - 1$.

(a) Show that g is a polynomial of degree ≤ 2.

(b) Show that $g(x_k) = 0$ for $k = 0, 1, 2$.

(c) Show that $g(x) = 0$ for all x.

Hint. Use the Fundamental Theorem of Algebra.

9. Let $f(x)$ be a polynomial of degree $\leq N$. Let $P_N(x)$ be the Lagrange interpolating polynomial of degree $\leq N$ based on the $N + 1$ nodes $x_0, x_1, \ldots, x_N$. Show that $f(x) = P_N(x)$ for all x. *Hint.* Show that the error term $E_N(x)$ is identically zero.

10. Consider the function $f(x) = \sin(x)$ on the interval $[0, 1]$. Use Theorem 4.2 to determine the step size h so that
    (a) linear Lagrange interpolation has an accuracy of 10^{-6} [i.e., find h so that $|E_1(x)| < 5 \times 10^{-7}$].
    (b) quadratic Lagrange interpolation has an accuracy of 10^{-6} [i.e., find h so that $|E_2(x)| < 5 \times 10^{-7}$].
    (c) cubic Lagrange interpolation has an accuracy of 10^{-6}.

11. Start with equation (12) and prove inequality (23). Let $x_1 = x_0 + h$, $x_2 = x_0 + 2h$. Prove that if $x_0 \leq x \leq x_2$, then

$$|x - x_0||x - x_1||x - x_2| \leq \frac{2h^3}{3 \times 3^{1/2}}.$$

Hint. Use the substitutions $t = x - x_1$, $t + h = x - x_0$, and $t - h = x - x_2$ and the function $v(t) = t^3 - th^2$ on the interval $-h \leq t \leq h$. Set $v'(t) = 0$ and solve for t in terms of h.

12. *Linear Interpolation in Two Dimensions.* Consider the polynomial $z = P(x, y) = A + Bx + Cy$ that passes through the three points (x_0, y_0, z_0), (x_1, y_1, z_1), and (x_2, y_2, z_2). Then $A, B,$ and C are the solution values for the linear system of equations

$$A + x_0 B + y_0 C = z_0$$
$$A + x_1 B + y_1 C = z_1$$
$$A + x_2 B + y_2 C = z_2.$$

    (a) Find $A, B,$ and C so that $z = P(x, y)$ passes through the points $(1, 1, 5)$, $(2, 1, 3)$, and $(1, 2, 9)$.
    (b) Find $A, B,$ and C so that $z = P(x, y)$ passes through the points $(1, 1, 2.5)$, $(2, 1, 0)$, and $(1, 2, 4)$.
    (c) Find $A, B,$ and C so that $z = P(x, y)$ passes through the points $(2, 1, 5)$, $(1, 3, 7)$, and $(3, 2, 4)$.
    (d) Can values $A, B,$ and C be found so that $z = P(x, y)$ passes through the points $(1, 2, 5)$, $(3, 2, 7)$, and $(1, 2, 0)$? Why?

13. Use the Generalized Rolle's theorem* and the special function

$$g(t) = f(t) - P_N(t) - E_N(x)\frac{(t - x_0)(t - x_1) \dots (t - x_N)}{(x - x_0)(x - x_1) \dots (x - x_N)},$$

where $P_N(x)$ is the Lagrange polynomial of degree N to prove that the error term $E_N(x) = f(x) - P_N(x)$ has the form

$$E_N(x) = (x - x_0)(x - x_1) \dots (x - x_N)\frac{f^{(N+1)}(c)}{(N + 1)!}.$$

Hint. Find $g^{(N+1)}(t)$ and then evaluate it at $t = c$.

Exercises 14–17 illustrate the concept of $O(h^N)$.

Generalized Rolle's Theorem. Let $g(t)$ be continuous on $[a, b]$ and suppose that its derivatives $g^{(k)}(t)$, for $k = 1, 2, \dots, N + 1$, are continuous on (a, b). If there are $N + 2$ distinct points $x, x_0, x_1, \dots, x_N$ such that $g(x) = 0$ and $g(x_k) = 0$ for $k = 0, 1, \dots, N$, then there exists a value c in (a, b) such that $g^{(N+1)}(c) = 0$.

14. Use the Taylor series for $\sin (x)$ expanded about $x = 0$, and establish the result $\sin (h) = h + O(h^3)$.

15. Use the Taylor series for $\cos (x)$ expanded about $x = 0$, and establish the result $\cos (h) = 1 + O(h^2)$.

16. Use the Taylor series for $\sin (x)$ expanded about $x = 0$, and establish the result $\sin (h)/h = 1 + O(h^2)$.

17. Use the Taylor series for $\cos (x)$ expanded about $x = 0$, and establish the result $[1 - \cos (h)]/h^2 = \frac{1}{2} + O(h^2)$.

4.3 Newton Polynomials

It is sometimes useful to find several approximating polynomials $P_1(x)$, $P_2(x), \ldots, P_N(x)$ and then choose the one that suits our needs. If the Lagrange polynomials are used, there is no constructive relationship between $P_{N-1}(x)$ and $P_N(x)$. Each polynomial has to be constructed individually, and the work required to compute the higher-degree polynomials involves many computations. We take a new approach and construct Newton polynomials that have the recursive pattern

(1) $\quad P_1(x) = a_0 + a_1(x - x_0),$

(2) $\quad P_2(x) = a_0 + a_1(x - x_0) + a_2(x - x_0)(x - x_1),$

(3) $\quad P_3(x) = a_0 + a_1(x - x_0) + a_2(x - x_0)(x - x_1) + a_3(x - x_0)(x - x_1)(x - x_2),$

$$\vdots$$

(4)
$$\begin{aligned} P_N(x) = {}& a_0 + a_1(x - x_0) + a_2(x - x_0)(x - x_1) + a_3(x - x_0)(x - x_1)(x - x_2) \\ & + a_4(x - x_0)(x - x_1)(x - x_2)(x - x_3) + \cdots \\ & + a_N(x - x_0) \ldots (x - x_{N-1}). \end{aligned}$$

Here the polynomial $P_N(x)$ is obtained from $P_{N-1}(x)$ using the recursive relationship

(5) $\quad P_N(x) = P_{N-1}(x) + a_N(x - x_0)(x - x_1)(x - x_2) \ldots (x - x_{N-1}).$

The polynomial (4) is said to be a Newton polynomial with the N *centers* $x_0, x_1, \ldots, x_{N-1}$. It involves sums of products of linear factors up to $a_N(x - x_0)(x - x_1)(x - x_2) \ldots (x - x_{N-1})$ so that $P_N(x)$ will simplify to be an ordinary polynomial of degree $\le N$.

Example 4.6 Given the centers $x_0 = 1$, $x_1 = 3$, $x_2 = 4$, and $x_3 = 4.5$ and the coefficients $a_0 = 5$, $a_1 = -2$, $a_2 = 0.5$, $a_3 = -0.1$, and $a_4 = 0.003$, find $P_1(x)$, $P_2(x)$, $P_3(x)$, and $P_4(x)$ and evaluate $P_k(2.5)$ for $k = 1, 2, 3, 4$.

Solution. Using the formulas (1)-(4), we have

$$P_1(x) = 5 - 2(x - 1),$$

$$P_2(x) = 5 - 2(x - 1) + 0.5(x - 1)(x - 3),$$

$$P_3(x) = 5 - 2(x - 1) + 0.5(x - 1)(x - 3) - 0.1(x - 1)(x - 3)(x - 4),$$

$$P_4(x) = P_3(x) + 0.003(x - 1)(x - 3)(x - 4)(x - 4.5).$$

Evaluating the polynomials at $x = 2.5$ results in

$$P_1(2.5) = 5 - 2[1.5] = 2,$$

$$P_2(2.5) = P_1(2.5) + 0.5(1.5)(-0.5) = 1.625,$$

$$P_3(2.5) = P_2(2.5) - 0.1(1.5)(-0.5)(-1.5) = 1.5125,$$

$$P_4(2.5) = P_3(2.5) + 0.03(1.5)(-0.5)(-1.5)(-2.0) = 1.50575.$$

Nested Multiplication

If N is fixed and the polynomial $P_N(x)$ is evaluated many times, then nested multiplication should be used. The process is similar to nested multiplication for ordinary polynomials, except that the centers x_k must be subtracted from the independent variable x. The nested multiplication form for $P_3(x)$ is

(6) $$P_3(x) = [[a_3(x - x_2) + a_2](x - x_1) + a_1](x - x_0) + a_0.$$

To evaluate $P_3(x)$ for a given value of x, start with the innermost grouping and form successively the quantities

(7)
$$S_3 = a_3,$$

$$S_2 = S_3(x - x_2) + a_2,$$

$$S_1 = S_2(x - x_1) + a_1,$$

$$S_0 = S_1(x - x_0) + a_0.$$

The quantity S_0 is now $P_3(x)$.

Example 4.7 Compute $P_3(2.5)$ in Example 4.6 using nested multiplication.

Solution. Using (6), we write

$$P_3(x) = [[-0.1(x - 4) + 0.5](x - 3) - 2](x - 1) + 5.$$

The values in (7) are

$$S_3 = -0.1,$$

$$S_2 = -0.1(2.5 - 4) + 0.5 = 0.65,$$

$$S_1 = 0.65(2.5 - 3) - 2 = -2.325,$$

$$S_0 = -2.325(2.5 - 1) + 5 = 1.5125.$$

Therefore, $P_3(2.5) = 1.5125.$

Algorithm 4.3 [*Nested Multiplication for Newton Polynomials*] Evaluate the Newton polynomial with centers $x_0, x_1, \ldots, x_{N-1}$.

$$P(x) = a_0 + a_1(x - x_0) + a_2(x - x_0)(x - x_1)$$
$$+ a_3(x - x_0)(x - x_1)(x - x_2)$$
$$+ \ldots + a_N(x - x_0)(x - x_1) \ldots (x - x_{N-1}).$$

```
        INPUT   N                          {Degree of P(x)}
        INPUT   A(0), A(1), ..., A(N)       {Coefficients of P(x)}
        INPUT   X(0), X(1), ..., X(N-1)     {Centers for P(x)}

        INPUT   X                           {Independent variable}
        Sum := A(N)                         {Initialize variable}
FOR   K = N-1   DOWNTO   0   DO             {Implement nested
        Sum := Sum*[X - X(K)] + A(K)             multiplication}
```

PRINT 'The value of the polynomial P(X) is' Sum

Polynomial Approximation, Nodes and Centers

Suppose that we want to find the coefficients a_k for all the polynomials $P_1(x), \ldots, P_N(x)$ that approximate a given function $f(x)$. Then $P_k(x)$ will be based on the centers $x_0, x_1, \ldots, x_k$ and have the nodes $x_0, x_1, \ldots, x_{k+1}$. For the polynomial $P_1(x)$ the coefficients a_0 and a_1 have a familiar meaning. In this case

(8) $$P_1(x_0) = f(x_0) \quad \text{and} \quad P_1(x_1) = f(x_1).$$

Using (1) and (8) to solve for a_0, we find that

(9) $$f(x_0) = P_1(x_0) = a_0 + a_1(x_0 - x_0) = a_0.$$

Hence $a_0 = f(x_0)$. Next, using (1), (8), and (9), we have

$$f(x_1) = P_1(x_1) = a_0 + a_1(x_1 - x_0) = f(x_0) + a_1(x_1 - x_0)$$

which can be solved for a_1 and we get

(10) $$a_1 = \frac{f(x_1) - f(x_0)}{x_1 - x_0}.$$

Hence a_1 is the slope of the secant line through the two points $(x_0, f(x_0))$ and $(x_1, f(x_1))$.

The coefficients a_0 and a_1 are the same for both $P_1(x)$ and $P_2(x)$. Evaluating (2) at the node x_2, we find that

(11) $$f(x_2) = P_2(x_2) = a_0 + a_1(x_2 - x_0) + a_2(x_2 - x_0)(x_2 - x_1).$$

The values for a_0 and a_1 in (9) and (10) can be used in (11) to obtain

$$a_2 = \frac{f(x_2) - a_0 - a_1(x_2 - x_0)}{(x_2 - x_0)(x_2 - x_1)}$$

$$= \left[\frac{f(x_2) - f(x_0)}{x_2 - x_0} - \frac{f(x_1) - f(x_0)}{x_1 - x_0}\right] \Big/ (x_2 - x_1).$$

For computational purposes we prefer to write this last quantity as

(12)
$$a_2 = \left[\frac{f(x_2) - f(x_1)}{x_2 - x_1} - \frac{f(x_1) - f(x_0)}{x_1 - x_0}\right] \Big/ (x_2 - x_0).$$

The two formulas for a_2 can be shown to be equivalent by writing the quotients over the common denominator $(x_2 - x_1)(x_2 - x_0)(x_1 - x_0)$. The details are left for the reader. The numerator in (12) is the difference between first order divided differences. In order to proceed, we need to introduce the idea of divided differences.

Definition 4.1 [Divided differences] The divided differences for a function $y = f(x)$ are defined as follows:

$$f[x_k] = f(x_k)$$

$$f[x_{k-1}, x_k] = \frac{f[x_k] - f[x_{k-1}]}{x_k - x_{k-1}},$$

(13)
$$f[x_{k-2}, x_{k-1}, x_k] = \frac{f[x_{k-1}, x_k] - f[x_{k-2}, x_{k-1}]}{x_k - x_{k-2}},$$

$$f[x_{k-3}, x_{k-2}, x_{k-1}, x_k] = \frac{f[x_{k-2}, x_{k-1}, x_k] - f[x_{k-3}, x_{k-2}, x_{k-1}]}{x_k - x_{k-3}}.$$

The recursive rule for constructing divided differences is

(14)
$$f[x_{k-j}, x_{k-j+1}, \ldots, x_k] = \frac{f[x_{k-j+1}, \ldots, x_k] - f[x_{k-j}, \ldots, x_{k-1}]}{x_k - x_{k-j}}$$

and is used to construct the divided differences in Table 4.5.

Table 4.5 Divided-difference table for $y = f(x)$

x_k	$f[x_k]$	$f[\ ,\ ]$	$f[\ ,\ ,\ ]$	$f[\ ,\ ,\ ,\ ]$	$f[\ ,\ ,\ ,\ ,\ ]$
x_0	$f[x_0]$				
x_1	$f[x_1]$	$f[x_0, x_1]$			
x_2	$f[x_2]$	$f[x_1, x_2]$	$f[x_0, x_1, x_2]$		
x_3	$f[x_3]$	$f[x_2, x_3]$	$f[x_1, x_2, x_3]$	$f[x_0, x_1, x_2, x_3]$	
x_4	$f[x_4]$	$f[x_3, x_4]$	$f[x_2, x_3, x_4]$	$f[x_1, x_2, x_3, x_4]$	$f[x_0, x_1, x_2, x_3, x_4]$

The coefficient a_k of $P_N(x)$ depends on the values $f(x_j)$ for $j = 0, 1, \ldots, k$. The next theorem shows that a_k can be computed using divided-difference

(15)
$$a_k = f[x_0, x_1, \ldots, x_k].$$

Theorem 4.3 [Newton polynomial interpolation] The interpolating polynomial $P_N(x)$ of degree N that agrees with the function $y = f(x)$ at the $N + 1$ points $(x_k, y_k) = (x_k, f(x_k))$ for $k = 0, 1, \ldots, N$ is

(16)
$$P_N(x) = a_0 + a_1(x - x_0) + \ldots + a_N(x - x_0)(x - x_1) \ldots (x - x_{N-1}),$$

where $a_k = f[x_0, x_1, \ldots, x_k]$. Also, if we define

(17)
$$f(x) = P_N(x) + E_N(x) \quad \text{and} \quad E_N(x) = f(x) - P_N(x),$$

and if $f(x)$ has $N + 1$ continuous derivatives, then the error term is

(18)
$$E_N(x) = (x - x_0)(x - x_1) \ldots (x - x_N)\frac{f^{(N+1)}(c)}{(N + 1)!},$$

for some c in the interval $[x_0, x_N]$. The error term $E_N(x)$ is the same as the one for Lagrange interpolation in (34) of Section 4.2.

It is of interest to start with a known function $f(x)$ that is a polynomial of degree N, and compute its divided-difference table. In this case we know that $f^{(N+1)}(x) = 0$ for all x, and calculation will reveal that the $(N + 1)$st divided difference is zero. This will happen because the divided difference (14) is proportional to a numerical approximation for the jth derivative. More details can be found in Reference 29.

Example 4.8 Let $f(x) = x^3 - 4x$. Construct the divided-difference table based on the nodes $x_0 = 1, x_1 = 2, \ldots, x_5 = 6$, and find the Newton polynomial $P_3(x)$ based on x_0, x_1, x_2, x_3.

Solution

x_k	$f[x_k]$	First Divided Difference	Second Divided Difference	Third Divided Difference	Fourth Divided Difference	Fifth Divided Difference
$x_0 = 1$	-3					
$x_1 = 2$	0	3				
$x_2 = 3$	15	15	6			
$x_3 = 4$	48	33	9	1		
$x_4 = 5$	105	57	12	1	0	
$x_5 = 6$	192	87	15	1	0	0

The coefficients $a_0 = -3$, $a_1 = 3$, $a_2 = 6$, and $a_3 = 1$ of $P_3(x)$ appear on the diagonal of the divided-difference table. The centers $x_0 = 1$, $x_1 = 2$, and $x_2 = 3$ are the values in the first column. Using formula (3), we write

$$P_3(x) = -3 + 3(x - 1) + 6(x - 1)(x - 2) + (x - 1)(x - 2)(x - 3).$$

Example 4.9 Construct a divided-difference table for $f(x) = \cos(x)$ based on the five points $(k, \cos(k))$ $k = 0, 1, 2, 3, 4$. Use it to find the coefficients a_k and all interpolating polynomials up to $P_4(x)$. Then compute the approximation $P_4(2.1)$.

Solution. For simplicity we round off the values to seven decimal places.

x_k	$f[x_k]$	$f[\ ,\ ]$	$f[\ ,\ ,\ ]$	$f[\ ,\ ,\ ,\ ]$	$f[\ ,\ ,\ ,\ ,\ ]$
0.0	1.0000000				
1.0	0.5403023	-0.4596977			
2.0	-0.4161468	-0.9564491	-0.2483757		
3.0	-0.9899925	-0.5738457	0.1913017	0.1465591	
4.0	-0.6536436	0.3363489	0.4550973	0.0879319	-0.0146568

The Newton interpolating polynomials are

$P_1(x) = 1 - 0.4596977(x)$,

$P_2(x) = 1 - 0.4596977(x) - 0.2483757(x)(x - 1)$,

$P_3(x) = 1 - 0.4596977(x) - 0.2483757(x)(x - 1) + 0.1465591(x)(x - 1)(x - 2)$,

$P_4(x) = P_3(x) - 0.0146568(x)(x - 1)(x - 2)(x - 3)$.

The graphs of $P_1(x), P_2(x)$, and $P_3(x)$ are shown in Figure 4.5.

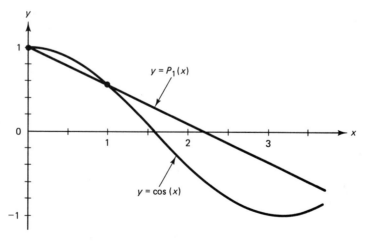

Figure 4.5(a). Graph of the Newton polynomial
$$P_1(x) = 1 - 0.459698(x)$$

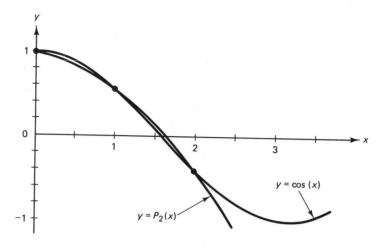

Figure 4.5(b). Graph of the Newton polynomial
$$P_2(x) = 1 - 0.459698(x) - 0.248376(x)(x - 1).$$

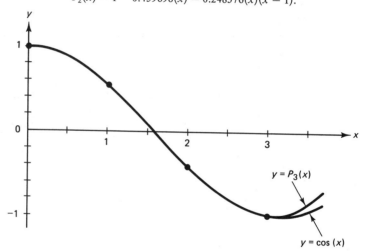

Figure 4.5(c). Graph of the Newton polynomial
$$P_3(x) = 1 - 0.459698(x) - 0.248376(x)(x - 1) + 0.146559(x)(x - 1)(x - 2).$$

Sample calculation for Example 4.9. Show how to find a_2. This will require the calculation of $f[x_0, x_1], f[x_1, x_2]$ and $a_2 = f[x_0, x_1, x_2]$.

$$f[x_0, x_1] = \frac{f[x_1] - f[x_0]}{x_1 - x_0} = \frac{0.5403023 - 1.0000000}{1.0 - 0.0} = -0.4596977,$$

$$f[x_1, x_2] = \frac{f[x_2] - f[x_1]}{x_2 - x_1} = \frac{-0.4161468 - 0.5403023}{2.0 - 1.0} = -0.9564491,$$

$$f[x_0, x_1, x_2] = \frac{f[x_1, x_2] - f[x_0, x_1]}{x_2 - x_0} = \frac{-0.9564491 + 0.4596977}{2.0 - 0.0} = -0.2483757.$$

The value $P_4(2.1)$ is obtained using the computation

$$P_4(2.1) = 1 - 0.4596977(2.1) - 0.2483757(2.1)(1.1)$$
$$+ 0.1465591(2.1)(1.1)(0.1) - 0.0146568(2.1)(1.1)(0.1)(-0.9)$$
$$= 1 - 0.9653652 - 0.5737479 + 0.0338552 + 0.0030471 = -0.5022107.$$

For computational purposes the divided differences in Table 4.5 need to be stored in an array. For convenience we use $D(k,j)$ and make the correspondence

(19) $$D(k,j) = f[x_{k-j}, x_{k-j+1}, \ldots, x_k] \qquad \text{(for } j \leq k\text{)}.$$

Then relation (14) implies that

(20) $$D(k,j) = \frac{D(k,j-1) - D(k-1,j-1)}{x_k - x_{k-j}}.$$

Notice that the value a_k in (15) is a diagonal element $a_k = D(k,k)$.

The algorithm for computing the divided differences and for evaluating $P_N(x)$ is now given. We remark that one of the exercises indicates how to modify the algorithm so that a_k is computed using a one-dimensional array.

Algorithm 4.4 [*Newton Polynomial Interpolation*] Suppose that the function $y = f(x)$ passes through the $N + 1$ points $(x_k, y_k) = (x_k, f(x_k))$ for $k = 0, 1, \ldots, N$. The Newton polynomial of degree $\leq N$ that passes through the $N + 1$ points is

$$P(x) = D_{0,0} + D_{1,1}(x - x_0) + D_{2,2}(x - x_0)(x - x_1)$$
$$+ \ldots + D_{N,N}(x - x_0)(x - x_1) \ldots (x - x_{N-1}),$$

where

$$D_{k,0} = f(x_k) = y_k \quad \text{and} \quad D_{k,j} = \frac{D_{k,j-1} - D_{k-1,j-1}}{x_k - x_{k-j}}.$$

Remark. The polynomial construction only requires $N + 1$ points with distinct abscissas.

```
INPUT   N                            {Degree of P(x)}
INPUT   X(0), X(1), ..., X(N)        {Abscissas of the points}
INPUT   Y(0), Y(1), ..., Y(N)        {Ordinates of the points}

FOR   K = 0  TO  N  DO               {Store ordinates in column 0
      D(K,0) := Y(K)                  of the array D(K,J)}

FOR   J = 1  TO  N  DO               {Compute the divided-
      FOR   K = J  TO  N  DO          difference table}
            D(K,J) := [D(K,J-1) - D(K-1,J-1)]/[X(K) - X(K-J)]
```

```
INPUT  X                                         {Independent variable}

     Sum := D(N,N)                               {Nested multiplication
FOR  K = N−1  DOWNTO  0  DO                              is used to
 └──── Sum := Sum∗[X − X(K)] + D(K,K)                 evaluate P(x)}
```

PRINT 'The value of the Newton polynomial P(x) is' Sum
       'Based on the interpolation nodes' $x_0, x_1, \ldots, x_N$
       'and the ordinates' $y_0, y_1, \ldots, y_N$

Exercises for Newton Polynomials

In Exercises 1–4, use the centers $x_0, x_1, x_2,$ and x_3 and coefficients $a_0, a_1, a_2, a_3,$ and a_4 to find the Newton polynomials $P_1(x), P_2(x), P_3(x),$ and $P_4(x)$ and evaluate them at the value $x = c$. *Hint*. Use equations (1)–(4) and the techniques of Example 4.6.

1. $a_0 = 4,$ $a_1 = -1,$ $a_2 = 0.4,$ $a_3 = 0.01,$ $a_4 = -0.002,$
   $x_0 = 1,$ $x_1 = 3,$ $x_2 = 4,$ $x_3 = 4.5,$ $c = 2.5$

2. $a_0 = 5,$ $a_1 = -2,$ $a_2 = 0.5,$ $a_3 = -0.1,$ $a_4 = 0.003,$
   $x_0 = 0,$ $x_1 = 1,$ $x_2 = 2,$ $x_3 = 3,$ $c = 2.5$

3. $a_0 = 7,$ $a_1 = 3,$ $a_2 = 0.1,$ $a_3 = 0.05,$ $a_4 = -0.04,$
   $x_0 = -1,$ $x_1 = 0,$ $x_2 = 1,$ $x_3 = 4,$ $c = 3$

4. $a_0 = -2,$ $a_1 = 4,$ $a_2 = -0.04,$ $a_3 = 0.06,$ $a_4 = 0.005,$
   $x_0 = -3,$ $x_1 = -1,$ $x_2 = 1,$ $x_3 = 4,$ $c = 2$

In Exercises 5–12:
(a) Compute the divided-difference table for the tabulated function.
(b) Write down the Newton polynomials $P_1(x), P_2(x), P_3(x),$ and $P_4(x)$.
(c) Evaluate the Newton polynomials in part (b) at the given values of x.
(d) Compare the values in part (c) with function value $f(x)$.

5. $f(x) = 3 \times 2^x$
    $x = 1.5, 2.5$

k	x_k	$f(x_k)$
0	−1.0	1.5
1	0.0	3.0
2	1.0	6.0
3	2.0	12.0
4	3.0	24.0

6. $f(x) = 3 \times 2^x$
    $x = 1.5, 2.5$

k	x_k	$f(x_k)$
0	1.0	6.0
1	2.0	12.0
2	0.0	3.0
3	3.0	24.0
4	−1.0	1.5

7. $f(x) = 3.6/x$
     $x = 2.5, 3.5$

k	x_k	$f(x_k)$
0	1.0	3.60
1	2.0	1.80
2	3.0	1.20
3	4.0	0.90
4	5.0	0.72

8. $f(x) = 3.6/x$
     $x = 2.5, 3.5$

k	x_k	$f(x_k)$
0	3.0	1.20
1	2.0	1.80
2	4.0	0.90
3	1.0	3.60
4	5.0	0.72

9. $f(x) = 3 \sin^2 (\pi x/6)$
     $x = 1.5, 3.5$

k	x_k	$f(x_k)$
0	0.0	0.00
1	1.0	0.75
2	2.0	2.25
3	3.0	3.00
4	4.0	2.25

10. $f(x) = 3x \times 2^{-x}$
     $x = 1.5, 3.5$

k	x_k	$f(x_k)$
0	0.0	0.000
1	1.0	1.500
2	2.0	1.500
3	3.0	1.125
4	4.0	0.750

11. $f(x) = x^{1/2}$
     $x = 4.5, 7.5$

k	x_k	$f(x_k)$
0	4.0	2.00000
1	5.0	2.23607
2	6.0	2.44949
3	7.0	2.64575
4	8.0	2.82843

12. $f(x) = \exp(-x)$
     $x = 0.5, 1.5$

k	x_k	$f(x_k)$
0	0.0	1.00000
1	1.0	0.36788
2	2.0	0.13534
3	3.0	0.04979
4	4.0	0.01832

13. Verify that the following modification of Algorithm 4.4 is an equivalent way to compute the Newton polynomial interpolation.

```
FOR  K = 0  TO  N  DO
└──── A(K) := Y(K)
FOR  J = 1  TO  N  DO
│   FOR  K = N  DOWNTO  J  DO
└── └──── A(K) := [A(K) − A(K−1)]/[X(K) − X(K−J)]
    INPUT  X
    Sum := A(N)
FOR  K = N−1  DOWNTO  0  DO
    Sum := Sum*[X − X(K)] + A(K)
```

14. Write a report on Aitken's iterated interpolation.

15. Write a report on Neville's iterated interpolation.

16. Write a report on inverse interpolation.

17. Write a report on Lozenge diagrams.

18. Write a report on Newton's backward-difference formula.

19. Consider the M points $(x_0, y_0), \ldots, (x_M, y_M)$.
    (a) If the $(N + 1)$st divided differences are zero, then show that the $(N + 2)$nd up to the Mth divided differences are zero.
    (b) If the $(N + 1)$st divided differences are zero, then show that there exists a polynomial $P_N(x)$ of degree N such that
    $$P_N(x_k) = y_k \qquad \text{for } k = 0, 1, \ldots, M.$$

In Exercises 20–23, use the result of Problem 19 to find the polynomial $P_N(x)$ that goes through the $M + 1$ points $(N < M)$.

20.

x_k	y_k
0	-2
1	2
2	4
3	4
4	2
5	-2

21.

x_k	y_k
1	8
2	17
3	24
4	29
5	32
6	33

22.

x_k	y_k
0	5
1	5
2	3
3	5
4	17
5	45
6	95

23.

x_k	y_k
0	1
1	0
2	9
3	34
4	81
5	156
6	265

4.4 Chebyshev Polynomials (Optional)

We now study polynomial approximations to $f(x)$ on $[-1, 1]$. Our goal is to minimize the maximum of the error term $|E_N(x)| = |f(x) - P_N(x)|$, where the maximum is taken over all x in the interval $[-1, 1]$. Equation (34) of Section 4.2 gave a form of $E_N(x)$ that involved the nodes x_k (for $k = 0, 1, \ldots, N$);

(1)
$$E_N(x) = Q(x)\frac{f^{(N+1)}(c)}{(N + 1)!},$$

where $Q(x)$ is the polynomial of degree $N + 1$:

(2)
$$Q(x) = (x - x_0)(x - x_1) \ldots (x - x_{N-1})(x - x_N).$$

If the situation permits us to choose freely the nodes where the function is to be evaluated, then the so-called Chebyshev nodes should be used, because they will minimize the maximum value of $|Q(x)|$. This leads us to a discussion of Chebyshev polynomials and some of their properties. To begin, the first eight Chebyshev polynomials are given in Table 4.6.

Table 4.6 Chebyshev polynomials $T_0(x)$ through $T_7(x)$

$$T_0(x) = 1$$
$$T_1(x) = x$$
$$T_2(x) = 2x^2 - 1$$
$$T_3(x) = 4x^3 - 3x$$
$$T_4(x) = 8x^4 - 8x^2 + 1$$
$$T_5(x) = 16x^5 - 20x^3 + 5x$$
$$T_6(x) = 32x^6 - 48x^4 + 18x^2 - 1$$
$$T_7(x) = 64x^7 - 112x^5 + 56x^3 - 7x$$

Properties of Chebyshev Polynomials

Several important properties about Chebyshev polynomials need to be discussed.

Property 1. Recurrence Relation

Chebyshev polynomials can be generated in the following way. Set $T_0(x) = 1$ and $T_1(x) = x$ and use the recurrence relation

(3)
$$T_k(x) = 2xT_{k-1}(x) - T_{k-2}(x) \qquad \text{for } k = 2, 3, \ldots.$$

Property 2. Leading Coefficient

The coefficient of x^N in $T_N(x)$ is 2^{N-1} when $N \geq 1$.

Property 3. Symmetry

When $N = 2M$, $T_{2M}(x)$ is an even function, that is,

(4)
$$T_{2M}(-x) = T_{2M}(x).$$

When $N = 2M + 1$, $T_{2M+1}(x)$ is an odd function, that is,

(5)
$$T_{2M+1}(-x) = -T_{2M+1}(x).$$

Property 4. Trigonometric Representation on $[1, 1]$

(6)
$$T_N(x) = \cos\left(N \arccos(x)\right) \qquad \text{for } -1 \leq x \leq 1.$$

Property 5. Distinct Zeros in $[-1, 1]$

$T_N(x)$ has N distinct zeros x_k that lie in the interval $[-1, 1]$ (see Figure 4.6):

(7)
$$x_k = \cos\left(\frac{(2k + 1)\pi}{2N}\right) \qquad \text{for } k = 0, 1, \ldots, N - 1.$$

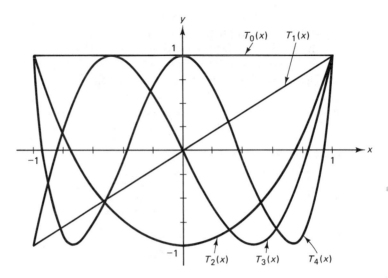

Figure 4.6. Graphs of the Chebyshev polynomials $T_0(x)$, $T_1(x)$, ..., $T_4(z)$ over $[-1, 1]$.

These values are called the *Chebyshev abscissas* (*nodes*).

Property 6. Extreme Values

(8)
$$|T_N(x)| \leq 1 \qquad \text{for } -1 \leq x \leq 1.$$

Property 1 is often used as the definition for higher-degree Chebyshev polynomials. Let us show that $T_3(x) = 2xT_2(x) - T_1(x)$. Using the expressions for $T_1(x)$ and $T_2(x)$ in Table 4.6, we obtain

$$2xT_2(x) - T_1(x) = 2x(2x^2 - 1) - x = 4x^3 - 3x = T_3(x).$$

Property 2 is proved by observing that the recurrence relation doubles the leading coefficient of $T_{N-1}(x)$ to get the leading coefficient of $T_N(x)$.

Property 3 is established by showing that $T_{2M}(x)$ involves only even powers of x and $T_{2M+1}(x)$ involves only odd powers of x. The details are left for the reader.

The proof of property 4 relies on the use of trigonometric identities and is sometimes used as an equivalent definition for the Chebyshev polynomials. We show the equivalence when $N = 2$:

$$\cos\left(2 \arccos\left(x\right)\right) = \cos^2\left(\arccos\left(x\right)\right) - \sin^2\left(\arccos\left(x\right)\right)$$

$$= x^2 - [(1 - x^2)^{1/2}]^2 = 2x^2 - 1.$$

Properties 5 and 6 are consequences of property 4.

Minimax

The Russian mathematician Chebyshev studied how to minimize the upper bound for $|E_N(x)|$. One upper bound can be formed by taking the product of the maximum value of $|Q(x)|$ over all x in $[-1, 1]$ and the maximum value $|f^{(N+1)}(x)/[(N + 1)!]|$ over all x in $[-1, 1]$. To minimize the factor $\max \{||Q(x)||\}$, Chebyshev discovered that $x_0, x_1, \ldots, x_N$ should be chosen so that $Q(x) = (1/2^N)T_{N+1}(x)$.

Theorem 4.4 Let N be fixed. Among all possible choices for $Q(x)$ in equation (2) (and thus among all possible choices of distinct nodes $\{x_k\}$ in $[-1, 1]$), the polynomial $T(x) = (1/2^N)T_{N+1}(x)$ is the unique choice which was the property

$$\max \{||T(x)||\} \leq \max \{||Q(x)||\},$$

where the maximum value of $|T(x)|$ and $|Q(x)|$ is taken over all values of x in the interval $-1 \leq x \leq 1$. Moreover,

(9)
$$\max\{||T(x)||\} = \frac{1}{2^N}.$$

Proof. The proof can be found in Reference 29.

The consequence of this result can be stated by saying that for Lagrange interpolation $f(x) = P_N(x) + E_N(x)$ on $[-1, 1]$, the minimum value of the error bound $\max \{||Q(x)||\} \max \{|f^{(N+1)}(x)/(N + 1)!|\}$ is achieved when the nodes $\{x_k\}$ are the Chebyshev abscissas of $T_{N+1}(x)$. As an illustration, we look at the Lagrange coefficient polynomials that are used in forming $P_3(x)$. First we use equally spaced nodes and then the Chebyshev nodes. Recall that the Lagrange polynomial of degree $N = 3$ has the form

(10)
$$P_3(x) = f(x_0)L_0(x) + f(x_1)L_1(x) + f(x_2)L_2(x) + f(x_3)L_3(x).$$

Equally Spaced Nodes

If $f(x)$ is approximated by a polynomial of degree at most $N = 3$ on $[-1, 1]$, the equally spaced nodes $x_0 = -1$, $x_1 = -\frac{1}{3}$, $x_2 = \frac{1}{3}$, and $x_3 = 1$ are easy to use for calculations. Substitution of these values into formula (31) of Section 4.2 and simplifying will produce the coefficient polynomials $L_k(x)$ in Table 4.7.

Table 4.7 Lagrange coefficient polynomials used to form $P_3(x)$ based on equally spaced nodes $x_k = -1 + 2k/3$

$$L_0(x) = -0.06250000 + 0.06250000x + 0.56250000x^2 - 0.56250000x^3$$
$$L_1(x) = 0.56250000 - 1.68750000x - 0.56250000x^2 + 1.68750000x^3$$
$$L_2(x) = 0.56250000 + 1.68750000x - 0.56250000x^2 - 1.68750000x^3$$
$$L_3(x) = -0.06250000 - 0.06250000x + 0.56250000x^2 + 0.56250000x^3$$

Chebyshev Nodes

When $f(x)$ is to be approximated by a polynomial of degree at most 3, using the Chebyshev nodes $x_0 = \cos\left(7\pi/8\right)$, $x_1 = \cos\left(5\pi/8\right)$, $x_2 = \cos\left(3\pi/8\right)$, and $x_3 = \cos\left(\pi/8\right)$, the coefficients polynomials are tedious to find (but this can be done by a computer). The results after simplification are shown in Table 4.8.

Table 4.8 Coefficient polynomials used to form $P_3(x)$ based on the Chebyshev nodes $x_k = \cos\left((7 - 2k)\pi/8\right)$

$$C_0(x) = -0.10355339 + 0.11208538x + 0.70710678x^2 - 0.76536686x^3$$
$$C_1(x) = 0.60355339 - 1.57716102x - 0.70710678x^2 + 1.84775906x^3$$
$$C_2(x) = 0.60355339 + 1.57716102x - 0.70710678x^2 - 1.84775906x^3$$
$$C_3(x) = -0.10355339 - 0.11208538x + 0.70710678x^2 + 0.76536686x^3$$

Example 4.10 Compare the Lagrange interpolating polynomials of degree $N = 3$ for $f(x) = \exp(x)$ that are obtained by using the coefficient polynomials in Tables 4.7 and 4.8, respectively.

Solution. Using equally spaced nodes we get the polynomial

$$P(x) = 0.99519577 + 0.99904923x + 0.54788486x^2 + 0.17615196x^3.$$

This is obtained by finding the function values

$$f(x_0) = \exp(-1) = 0.36787944, f(x_1) = \exp\left(\frac{-1}{3}\right) = 0.71653131,$$

$$f(x_2) = \exp\left(\frac{1}{3}\right) = 1.39561243, f(x_3) = \exp(1) = 2.71828183,$$

and using the coefficient polynomials $L_k(x)$ in Table 4.7, and forming the linear combination

$$P(x) = 0.36787944L_0(x) + 0.71653131L_1(x)$$
$$+ 1.39561243L_2(x) + 2.71828183L_3(x).$$

Similarly, when the Chebyshev nodes are used, we obtain

$$V(x) = 0.99461532 + 0.99893323x + 0.54290072x^2 + 0.17517569x^3.$$

Notice that the coefficients are different from those of $P(x)$. This is a consequence of using different nodes and function values:

$$f(x_0) = \exp(-0.92387953) = 0.39697597,$$
$$f(x_1) = \exp(-0.38268343) = 0.68202877,$$
$$f(x_2) = \exp(0.38268343) = 1.46621380,$$
$$f(x_3) = \exp(0.92387953) = 2.51904417.$$

Then the alternate set of coefficient polynomials $C_k(x)$ in Table 4.8 are used to form the linear combination:

$$V(x) = 0.39697597\,C_0(x) + 0.68202877\,C_1(x)$$
$$+ 1.46621380\,C_2(x) + 2.51904417\,C_3(x).$$

For a comparison of the accuracy of $P(x)$ and $V(x)$, the error functions are graphed in Figure 4.7(a) and (b), respectively. The maximum error $|\exp(x) - P(x)|$ occurs at $x = 0.75490129$, and

$$|\exp(x) - P(x)| \leqq 0.00998481 \qquad \text{for } -1 \leqq x \leqq 1.$$

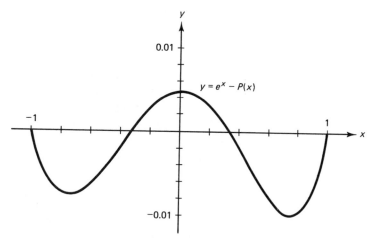

Figure 4.7(a). The error function for Lagrange approximation $y = \exp(x) - P(x)$ over $[-1, 1]$.

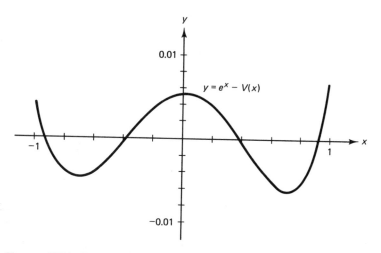

Figure 4.7(b). The error function for Chebyshev approximation $y = \exp(x) - V(x)$ over $[-1, 1]$.

The maximum error $|\exp(x) - V(x)|$ occurs at $x = 1$, and we get

$$|\exp(x) - V(x)| \leq 0.00665687 \qquad \text{for } -1 \leq x \leq 1.$$

Notice that the maximum error in $V(x)$ is about two-thirds the maximum error in $P(x)$. Also, the error is spread out more evenly over the interval.

Runge Phenomenon

We now look deeper to see the advantage of using the Chebyshev interpolation nodes. Consider Lagrange interpolating to $f(x)$ over the interval $[-1, 1]$ based on equally spaced nodes. Does the error $E_N(x) = f(x) - P_N(x)$ tend to zero as N increases? For functions like $\sin(x)$ or $\exp(x)$ where all the derivatives are bounded by the same constant M, the answer is yes. In general the answer to this question is no, and it is easy to find functions for which the sequence $\{P_N(x)\}$ does not converge. If $f(x) = 1/(1 + 12x^2)$, the maximum of the error term $E_N(x)$ grows when $N \longrightarrow \infty$. This nonconvergence is called the *Runge phenomenon*. The Lagrange polynomial of degree 10 based on 11 equally spaced nodes for this function is shown in Figure 4.8(a). Wild oscillations occur near the end of the interval. If the number of nodes is increased, then the oscillations become larger. This problem occurs because the nodes are equally spaced!

If the Chebyshev nodes are used to construct an interpolating polynomial of degree 10 to $f(x) = 1/(1 + 12x^2)$, the error is much smaller as seen in Figure 4.8(b). Under the condition that Chebyshev nodes are used, the error $E_N(x)$ will go to zero as $N \longrightarrow \infty$. In general, if $f(x)$ and $f'(x)$ are continuous on $[-1, 1]$, then it can be proven that Chebyshev interpolation will produce a sequence of polynomials $\{P_N(x)\}$ that converges uniformly to $f(x)$ over $[-1, 1]$.

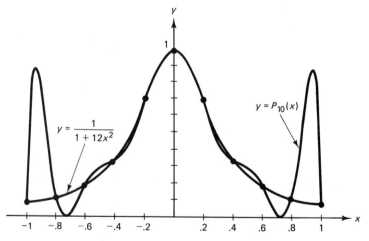

Figure 4.8(a). Polynomial approximation to $y = 1/[1 + 12x^2]$ using eleven equally spaced nodes over $[-1, 1]$.

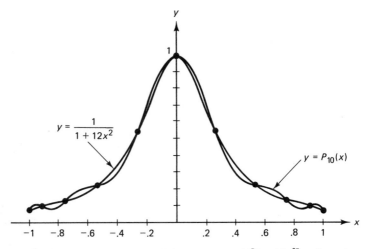

Figure 4.8(b). Polynomial approximation to $y = 1/[1 + 12x^2]$ using eleven Chebyshev nodes over $[-1, 1]$.

Transforming the Interval

Sometimes it is necessary to take a problem stated on an interval $[a, b]$ and reformulate the problem on the interval $[c, d]$ where the solution is known. If the approximation $P_N(x)$ to $f(x)$ is to be obtained on the interval $[a, b]$, then we change the variable so that the problem is reformulated on $[-1, 1]$:

(11) $$x = \left(\frac{b - a}{2}\right)t + \frac{a + b}{2} \quad \text{or} \quad t = 2\frac{x - a}{b - a} - 1,$$

where $a \leq x \leq b$ and $-1 \leq t \leq 1$.

The required Chebyshev nodes of $T_{N+1}(t)$ on $[-1, 1]$ are

(12) $$t_k = \cos\left((2N + 1 - 2k)\frac{\pi}{2N + 2}\right) \quad \text{for } k = 0, 1, \ldots, N.$$

and the interpolating nodes on $[a, b]$ are obtained by using (11):

(13) $$x_k = t_k\frac{b - a}{2} + \frac{a + b}{2} \quad \text{for } k = 0, 1, \ldots, N.$$

The next result gives the error bound for this approximation on $[a, b]$.

Theorem 4.5 [Lagrange–Chebyshev approximation] Let $f^{(N+1)}(x)$ be continuous on $[a, b]$. If the Lagrange polynomial $P_N(x)$ is based on the Chebyshev nodes (13), then

(14) $$|f(x) - P_N(x)| \leq \left(\frac{b - a}{4}\right)^{N+1}\frac{2M}{(N + 1)!},$$

where $M = \max\{|f^{(N+1)}(x)|\}$ is taken over $[a, b]$.

Example 4.11 For $f(x) = \sin(x)$ on $[0, \pi/4]$, find the Chebyshev nodes and the error bound (14) for the Lagrange polynomial $P_5(x)$.

Solution. Formulas (12) and (13) are used to find the nodes;

$$x_k = \cos\left(\frac{(11-2k)\pi}{12}\right)\frac{\pi}{8} + \frac{\pi}{8} \qquad \text{for } k = 0, 1, \ldots, 5.$$

Using the bound $|f^{(6)}(x)| \leq |-\sin(\pi/4)| = 2^{-1/2} = M$ in (14), we get

$$|f(x) - P_N(x)| \leq \left(\frac{\pi}{8}\right)^6\left(\frac{2}{6!}\right)2^{-1/2} \leq 0.00000720.$$

Orthogonal Property

In Example 4.10, the Chebyshev nodes were used to find the Lagrange interpolating polynomial. In general this implies that the Chebyshev polynomial of degree N can be obtained by Lagrange interpolation based on the $N + 1$ nodes that are the $N + 1$ zeros of $T_{N+1}(x)$. However, a direct approach to finding the approximation polynomial is to express $P_N(x)$ as a linear combination of the polynomials $T_k(x)$ which were given in Table 4.6. Therefore, the Chebyshev interpolating polynomial can be written in the form:

$$(15) \qquad P_N(x) = \sum_{k=0}^{N} c_k T_k(x) = c_0 T_0(x) + c_1 T_1(x) + \ldots + c_N T_N(x).$$

The coefficients $\{c_k\}$ in (15) are easy to find. The technical proof requires the use of following orthogonality properties. Let

$$(16) \qquad x_k = \cos\left(\pi \frac{2k+1}{2N+2}\right) \qquad \text{for } k = 0, 1, \ldots, N;$$

$$(17) \qquad \sum_{k=0}^{N} T_i(x_k)T_j(x_k) = 0 \qquad \text{when } i \neq j,$$

$$(18) \qquad \sum_{k=0}^{N} T_i(x_k)T_j(x_k) = \frac{N+1}{2} \qquad \text{when } i = j \neq 0,$$

$$(19) \qquad \sum_{k=0}^{N} T_0(x_k)T_0(x_k) = N + 1.$$

Property 4 and the identities (17)–(19) can be used to prove the following theorem.

Theorem 4.6 [Chebyshev polynomial interpolation] Let $f^{(N+1)}(x)$ be continuous on $[-1, 1]$, and let $\{x_k\}$ denote the zeros of $T_{N+1}(x)$. Then $f(x)$ can be expresed as

$$(20) \qquad f(x) = P_N(x) + E_N(x),$$

where $P_N(x)$ is given in (15) and $E_N(x)$ is given (1).

The coefficients for the Chebyshev interpolating polynomial in (15) are computed using the formulas

(21) $$c_0 = \frac{1}{N+1} \sum_{k=0}^{N} f(x_k) T_0(x_k) = \frac{1}{N+1} \sum_{k=0}^{N} f(x_k)$$

and

(22) $$c_j = \frac{2}{N+1} \sum_{k=0}^{N} f(x_k) T_j(x_k) = \frac{2}{N+1} \sum_{k=0}^{N} f(x_k) \cos\left(j\pi \frac{2k+1}{2N+2}\right)$$

$$\text{for } j = 1, 2, \ldots, N.$$

Example 4.12 Find the Chebyshev polynomial $P_3(x)$ that approximates the function $f(x) = \exp(x)$ over $[-1, 1]$.

Solution. The coefficients are calculated using (21) and (22), and the nodes $x_k = \cos(\pi(2k+1)/8)$ for $k = 0, 1, 2, 3$.

$$c_0 = \frac{1}{4} \sum_{k=0}^{3} \exp(x_k) T_0(x_k) = \frac{1}{4} \sum_{k=0}^{3} \exp(x_k) = 1.26606568,$$

$$c_1 = \frac{1}{2} \sum_{k=0}^{3} \exp(x_k) T_1(x_k) = \frac{1}{2} \sum_{k=0}^{3} \exp(x_k) x_k = 1.13031500,$$

$$c_2 = \frac{1}{2} \sum_{k=0}^{3} \exp(x_k) T_2(x_k) = \frac{1}{2} \sum_{k=0}^{3} \exp(x_k) \cos\left(2\pi \frac{2k+1}{8}\right) = 0.27145036,$$

$$c_3 = \frac{1}{2} \sum_{k=0}^{3} \exp(x_k) T_3(x_k) = \frac{1}{2} \sum_{k=0}^{3} \exp(x_k) \cos\left(3\pi \frac{2k+1}{8}\right) = 0.04379392.$$

Therefore, the Chebyshev polynomial $P_3(x)$ for $\exp(x)$ is

(23) $$P_3(x) = 1.26606568\,T_0(x) + 1.13031500\,T_1(x) + 0.27145036\,T_2(x)$$
$$+ 0.04379392\,T_3(x).$$

If the Chebyshev polynomial (23) is expanded in powers of x, the result is

$$P_3(x) = 0.99461532 + 0.99893324x + 0.54290072x^2 + 0.17517568x^3,$$

which is the same as the polynomial $V(x)$ in Example 4.10. If the goal is to find the Chebyshev polynomial then formulas (21) and (22) are preferred.

Algorithm 4.5 [*Chebyshev Interpolating Polynomial*] Let $f^{(N+1)}(x)$ be continuous on $[-1, 1]$. The nodes $x_k = \cos((2k+1)\pi/(2N+2))$ are used to generate the Chebyshev polynomial approximation:

$$P(x) = \sum_{j=0}^{N} c_j T_j(x),$$

where $\{T_j(x) : j = 0, 1, \ldots, N\}$ are the Chebyshev coefficient polynomials.

```
INPUT   N                                              {Degree of polynomial}
Pi := 3.1415926535
D := Pi/(2*N+2)


FOR  K = 0  TO  N  DO
      X(K) := COS((2*K+1)*D)                           {Make the nodes}
      Y(K) := F(X(K))                                  {Store the function values}
      C(K) := 0                                        {Initialize the array}


FOR  K = 0  DO  N  DO                                   {Calculate
      Z := (2*K+1)*D                                     the sums
      FOR  J = 0  TO  N  DO                               for the
            C(J) := C(J) + Y(K)*COS(J*Z)                coefficients}


      C(0) := C(0)/(N+1)                               {Multiple of
FOR  J = 1  DO  N  DO                                    sum forms
      C(J) := 2*C(J)/(N+1)                              coefficient}
```

{Procedure for evaluating the Chebyshev polynomial approximation}

```
INPUT   X                                              {Input independent variable}


      T(0) := 1, T(1) := X                             {Recursively generate
IF    N>1   THEN                                         the Chebyshev
      FOR  J = 1  TO  N−1  DO                            coefficient
            T(J+1) := 2*X*T(J) − T(J−1)                 polynomials}


      P := C(0)*T(0)                                    {Evaluate
FOR  J = 1  TO  N  DO                                    Chebyshev
      P := P + C(J)*T(J)                                 polynomial}


PRINT "The value of the Chebyshev poly. P(x) is" P     {Output}
```

Exercises for Chebyshev Polynomials

1. Use property (1) and
   (a) construct $T_4(x)$ from $T_3(x)$ and $T_2(x)$.
   (b) construct $T_5(x)$ from $T_4(x)$ and $T_3(x)$.

2. Use property (1) and
   (a) construct $T_6(x)$ from $T_5(x)$ and $T_4(x)$.
   (b) construct $T_7(x)$ from $T_6(x)$ and $T_5(x)$.

3. Use mathematical induction to prove property 2.

4. Use mathematical induction to prove property 3.

5. Find the maximum and minimum values of $T_2(x)$.

6. Find the maximum and minimum values of $T_3(x)$. *Hint.* $T_3'(1) = 0$ and $T_3'(-1) = 0$.

7. Find the maximum and minimum values of $T_4(x)$. *Hint.* $T_4'(0) = 0$, $T_4'(2^{-1/2}) = 0$, and $T_4'(-2^{-1/2}) = 0$.

8. Let $f(x) = \sin(x)$ on $[-1, 1]$.
   (a) Use the coefficient polynomials in Table 4.8 to obtain the Lagrange-Chebyshev polynomial approximation $P_3(x)$.
   (b) Find the error bound for $|\sin(x) - P_3(x)|$.

9. Let $f(x) = \ln(x + 2)$ on $[-1, 1]$.
   (a) Use the coefficient polynomials in Table 4.8 to obtain the Lagrange-Chebyshev polynomial approximation $P_3(x)$.
   (b) Find the error bound for $|\ln(x + 2) - P_3(x)|$.

10. The Lagrange polynomial of degree $N = 2$ has the form
$$f(x) = f(x_0)L_0(x) + f(x_1)L_1(x) + f(x_2)L_2(x).$$
If the Chebyshev nodes $x_0 = \cos(5\pi/6)$, $x_1 = 0$, and $x_2 = \cos(\pi/6)$ are used, show that the coefficient polynomials are
$$L_0(x) = -0.57735027x + \frac{2x^2}{3},$$
$$L_1(x) = 1.0000000 - \frac{4x^2}{3},$$
$$L_2(x) = 0.57735027x + \frac{2x^2}{3}.$$

11. Let $f(x) = \cos(x)$ on $[-1, 1]$.
    (a) Use the coefficient polynomials in Exercise 10 to get the the Lagrange-Chebyshev polynomial approximation $P_2(x)$.
    (b) Find the error bound for $|\cos(x) - P_2(x)|$.

12. Let $f(x) = \exp(x)$ on $[-1, 1]$.
    (a) Use the coefficient polynomials in Exercise 10 to get the the Lagrange-Chebyshev polynomial approximation $P_2(x)$.
    (b) Find the error bound for $|\exp(x) - P_2(x)|$.

In Exercises 13–15, compare the Taylor polynomial and the Lagrange-Chebyshev approximation to $f(x)$ on $[-1, 1]$. Find their error bounds.

13. $f(x) = \sin(x)$ and $N = 7$, the Lagrange-Chebyshev polynomial is
$$\sin(x) \approx 0.99999998x - 0.16666599x^3 + 0.00832995x^5 - 0.00019297x^7.$$

14. $f(x) = \cos(x)$ and $N = 6$, the Lagrange-Chebyshev polynomial is
$$\cos(x) \approx 1 - 0.49999734x^2 + 0.04164535x^4 - 0.00134608x^6.$$

15. $f(x) = \exp(x)$ and $N = 7$, the Lagrange–Chebyshev polynomial is

$$\exp(x) \approx 0.99999980 + 0.99999998x + 0.50000634x^2 + 0.16666737x^3$$
$$+ 0.04163504x^4 + 0.00832984x^5 + 0.00143925x^6 + 0.00020399x^7.$$

In Exercises 16–20, use formulas (21) and (22) to compute the coefficients $\{c_k\}$ for the Chebyshev polynomial approximation $P_N(x)$ to $f(x)$ over $[-1, 1]$, when (a) $N = 3$, (b) $N = 4$, (c) $N = 5$, and (d) $N = 6$. (e) Use a computer program to solve parts (a)–(d).

16. $f(x) = \exp(x)$ 17. $f(x) = \sin(x)$

18. $f(x) = \cos(x)$ 19. $f(x) = \ln(x + 2)$

20. $f(x) = (x + 2)^{1/2}$

21. Write a report on the economization of power series.

22. Write a report on the Legendre polynomials.

23. Write a report on the Gram polynomials.

24. Write a report on the Remes algorithm.

25. Prove equation (17).   26. Prove equation (18).

4.5 Padé Approximations

In this section we introduce the notion of rational approximations for functions. The function $f(x)$ will be approximated over a small portion of its domain. For example, if $f(x) = \cos(x)$, it is sufficient to have a formula to generate approximations on the interval $[0, \pi/2]$. Then trigonometric identities can be used to compute $\cos(x)$ for any value x that lies outside $[0, \pi/2]$.

A rational approximation to $f(x)$ on $[a, b]$ is the quotient of two polynomials $P_N(x)$ and $Q_M(x)$ of degrees N and M, respectively. We use the notation $R_{N, M}(x)$ to denote this quotient:

(1)
$$R_{N, M}(x) = \frac{P_N(x)}{Q_M(x)} \quad \text{for } a \leqq x \leqq b.$$

Our goal is to make the maximum error as small as possible. For a given amount of computational effort, one can usually construct a rational approximation that has a smaller overall error on $[a, b]$ than a polynomial approximation. Our development is an introduction and will be limited to Padé approximations.

The *method of Padé* requires that $f(x)$ and its derivatives are continuous at $x = 0$. There are two reasons for the arbitrary choice of $x = 0$. First, it makes the manipulations simpler. Second, a change of variable can be used to shift the calculations over to an interval that contains zero. The polynomials used in (1) are

(2)
$$P_N(x) = p_0 + p_1x + p_2x^2 + \ldots + p_Nx^N$$

and

(3)
$$Q_M(x) = 1 + q_1x + q_2x^2 + \ldots + q_Mx^M.$$

The polynomials in (2) and (3) are constructed so that $f(x)$ and $R_{N,M}(x)$ agree at $x = 0$ and their derivatives up to $N + M$ agree at $x = 0$. In the case $Q_0(x) = 1$, the approximation is just the Maclaurin expansion for $f(x)$. For a fixed value of $N + M$ the error is smallest when $P_N(x)$ and $Q_M(x)$ have the same degree or when $P_N(x)$ has degree one higher than $Q_M(x)$.

Notice that the constant coefficient of $Q_M(x)$ is $q_0 = 1$. This is permissible, because it cannot be 0 and $R_{N,M}(x)$ is not changed when both $P_N(x)$ and $Q_M(x)$ are divided by the same constant. Hence the rational function $R_{N,M}(x)$ has $N + M + 1$ unknown coefficients. Assume that $f(x)$ is analytic, and has the Maclaurin expansion

(4)
$$f(x) = a_0 + a_1x + a_2x^2 + \ldots + a_kx^k + \ldots,$$

and form the difference $f(x)Q_M(x) - P_N(x) = Z(x)$:

(5)
$$\left(\sum_{j=0}^{\infty} a_jx^j \right)\left(\sum_{j=0}^{M} q_jx^j \right) - \sum_{j=0}^{N} p_jx^j = \sum_{j=N+M+1}^{\infty} c_jx^j.$$

The lower index $j = M + N + 1$ in the summation on the right side of (5) is chosen because the first $N + M$ derivatives of $f(x)$ and $R_{N,M}(x)$ are to agree at $x = 0$.

When the left side of (5) is multiplied out and the coefficients of the powers of x^j are set equal to zero for $j = 0, 1, \ldots, N + M$, the result is a system of $N + M + 1$ linear equations:

(6)
$$a_0 - p_0 = 0$$
$$q_1a_0 + a_1 - p_1 = 0$$
$$q_2a_0 + q_1a_1 + a_2 - p_2 = 0$$
$$q_3a_0 + q_2a_1 + q_1a_2 + a_3 - p_3 = 0$$
$$q_Ma_{N-M} + q_{M-1}a_{N-M+1} + \ldots + a_N - p_N = 0$$

and

(7)
$$q_Ma_{N-M+1} + q_{M-1}a_{N-M+2} + \ldots + q_1a_N \qquad + a_{N+1} = 0$$
$$q_Ma_{N-M+2} + q_{M-1}a_{N-M+3} + \ldots + q_1a_{N+1} \qquad + a_{N+2} = 0$$
$$\vdots \qquad\qquad\qquad\qquad\qquad\qquad\qquad \vdots$$
$$q_Ma_N \qquad + q_{M-1}a_{N+1} \qquad + \ldots + q_1a_{N+M-1} + a_{N+M} = 0.$$

Notice that in each equation the sum of the subscripts on the factors of each product is the same, and this sum increases consecutively from 0 to $N + M$. The M equations in (7) involve only the unknowns $q_1, q_2, \ldots, q_M$ and must be solved first. Then the equations in (6) are used successively to find $p_0, p_1, \ldots, p_N$.

Example 4.13 Establish the Padé approximation

$$(8) \qquad \cos(x) \approx R_{4,4}(x) = \frac{15{,}120 - 6{,}900x^2 + 313x^4}{15{,}120 + 660x^2 + 13x^4}.$$

See Figure 4.9(a) for the graphs of $\cos(x)$ and $R_{4,4}(x)$ on $[-5, 5]$, and Figure 4.9(b) for the graph of $\cos(x) - R_{4,4}(x)$ on $[-1, 1]$.

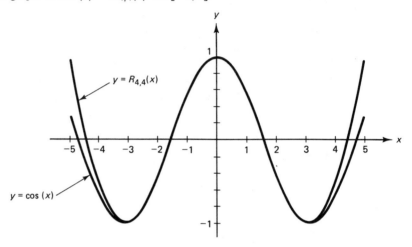

Figure 4.9(a). The graph of $y = \cos(x)$ and its Padé approximation $y = R_{4,4}(x)$ over $[-5, 5]$.

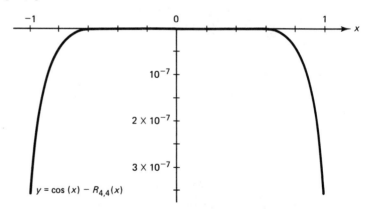

Figure 4.9(b). The graph of the error $y = \cos(x) - R_{4,4}(x)$ for the Padé approximation over $[-1, 1]$.

Solution. If the Maclaurin expansion for $\cos(x)$ was used, we would obtain nine equations in nine unknowns. Instead, notice that both $\cos(x)$ and $R_{4,4}(x)$ are even functions and involve powers of (x^2). We can simplify the computations if we start with $f(x) = \cos(x^{1/2})$:

$$(9) \qquad f(x) = 1 - \frac{1}{2}x + \frac{1}{24}x^2 - \frac{1}{720}x^3 + \frac{1}{40{,}320}x^4 - \cdots.$$

In this case equation (5) becomes

$$\left(1 - \frac{x}{2} + \frac{x^2}{24} - \frac{x^3}{720} + \frac{x^4}{40,320} + \dots\right)(1 + q_1 x + q_2 x^2) - p_0 - p_1 x - p_2 x^2$$

$$= 0 + 0x + 0x^2 + 0x^3 + 0x^4 + c_5 x^5 + c_6 x^6 + \dots.$$

When the coefficients of the first five powers of x are compared, we get the following system of linear equations:

$$1 - p_0 = 0$$

$$-\frac{1}{2} + q_1 - p_1 = 0$$

(10)

$$\frac{1}{24} - \frac{1}{2}q_1 + q_2 - p_2 = 0$$

$$-\frac{1}{720} + \frac{1}{24}q_1 - \frac{1}{2}q_2 = 0$$

$$\frac{1}{40,320} - \frac{1}{720}q_1 + \frac{1}{24}q_2 = 0.$$

The last two equations in (10) must be solved first. They can be rewritten in a form that is easy to solve:

$$q_1 - 12q_2 = \frac{1}{30} \quad \text{and} \quad -q_1 + 30q_2 = \frac{-1}{56}.$$

First find q_2 by adding the equations, then find q_1:

$$q_2 = \frac{1}{18}\left(\frac{1}{30} - \frac{1}{56}\right) = \frac{13}{15,120},$$

(11)

$$q_1 = \frac{1}{30} + \frac{156}{15,120} = \frac{11}{252}.$$

Now the first three equations of (10) are used. It is obvious that $p_0 = 1$, and we can use q_1 and q_2 in (11) to solve for p_1 and p_2:

$$p_1 = -\frac{1}{2} + \frac{11}{252} = -\frac{115}{252},$$

(12)

$$p_2 = \frac{1}{24} - \frac{11}{504} + \frac{13}{15,120} = \frac{313}{15,120}.$$

Now use the coefficients in (11) and (12) to form the rational approximation to $f(x)$:

(13)

$$f(x) \approx \frac{1 - 115x/252 + 313x^2/15,120}{1 + 11x/252 + 13x^2/15,120}.$$

Since $\cos(x) = f(x^2)$, we can substitute x^2 for x in equation (13) and the result is the formula for $R_{4,4}(x)$ in (8).

Exercises for Padé Approximations

1. Establish the Padé approximation:
$$\exp(x) \approx R_{1,1}(x) = \frac{2+x}{2-x}.$$

2. (a) Find the Padé approximation $R_{1,1}(x)$ for $f(x) = \ln(1+x)/x$. *Hint.* Start with the Maclaurin expansion:
$$f(x) = 1 - \frac{x}{2} + \frac{x^2}{3} - \dots.$$

   (b) Use the result in part (a) to establish the approximation
$$\ln(1+x) \approx R_{2,1}(x) = \frac{6x+x^2}{6+4x}.$$

3. (a) Find $R_{1,1}(x)$ for $f(x) = \tan(x^{1/2})/x^{1/2}$. *Hint.* Start with the Maclaurin expansion:
$$f(x) = 1 + \frac{x}{3} + \frac{2x^2}{15} + \dots.$$

   (b) Use the result in part (a) to establish the approximation
$$\tan(x) \approx R_{3,2}(x) = \frac{15x - x^3}{15 - 6x^2}.$$

4. (a) Find $R_{1,1}(x)$ for $f(x) = \arctan(x^{1/2})/x^{1/2}$. *Hint.* Start with the Maclaurin expansion:
$$f(x) = 1 - \frac{x}{3} + \frac{x^2}{5} - \dots.$$

   (b) Use the result in part (a) to establish the approximation
$$\arctan(x) \approx R_{3,2}(x) = \frac{15x - 4x^3}{15 + x^2}.$$

5. Establish the Padé approximation:
$$\exp(x) \approx R_{2,2}(x) = \frac{12 + 6x + x^2}{12 - 6x + x^2}.$$

6. (a) Find the Padé approximation $R_{2,2}(x)$ for $f(x) = \ln(1+x)/x$. *Hint.* Start with the Maclaurin expansion:
$$f(x) = 1 - \frac{x}{2} + \frac{x^2}{3} - \frac{x^3}{4} + \frac{x^4}{5} - \dots.$$

   (b) Use the result in part (a) to establish
$$\ln(1+x) \approx R_{3,2}(x) = \frac{30x + 21x^2 + x^3}{30 + 36x + 9x^2}.$$

7. (a) Find $R_{2,2}(x)$ for $f(x) = \tan(x^{1/2})/x^{1/2}$. *Hint.* Start with the Maclaurin expansion:
$$f(x) = 1 + \frac{x}{3} + \frac{2x^2}{15} + \frac{17x^3}{315} + \frac{62x^4}{2{,}835} + \dots.$$

(b) Use the result in part (a) to establish

$$\tan(x) \approx R_{5,4}(x) = \frac{945x - 105x^3 + x^5}{945 - 420x^2 + 15x^4}.$$

8. (a) Find $R_{2,2}(x)$ for $f(x) = \arctan(x^{1/2})/x^{1/2}$. *Hint.* Start with the Maclaurin expansion:

$$f(x) = 1 - \frac{x}{3} + \frac{x^2}{5} - \frac{x^3}{7} + \frac{x^4}{9} - \cdots.$$

(b) Use the result in part (a) to establish

$$\arctan(x) \approx R_{5,4}(x) = \frac{945x + 735x^3 + 64x^5}{945 + 1{,}050x^2 + 225x^4}.$$

9. Establish the Padé approximation:

$$\exp(x) \approx R_{2,2}(x) = \frac{120 + 60x + 12x^2 + x^3}{120 - 60x + 12x^2 - x^3}.$$

10. Establish the Padé approximation:

$$\exp(x) \approx R_{4,4}(x) = \frac{1{,}680 + 840x + 180x^2 + 20x^3 + x^4}{1{,}680 - 840x + 180x^2 - 20x^3 + x^4}.$$

11. Compare the following approximations to $f(x) = \exp(x)$.

$$\text{Taylor:} \quad T_4(x) = 1 + x + \frac{x^2}{2} + \frac{x^3}{6} + \frac{x^4}{24}$$

$$\text{Padé:} \quad R_{2,2}(x) = \frac{12 + 6x + x^2}{12 - 6x + x^2}$$

Evaluate $T_4(x)$ and $R_{2,2}(x)$ for $x = -0.8, -0.4, 0.4$, and 0.8.

12. Compare the following approximations to $f(x) = \ln(1 + x)$.

$$\text{Taylor:} \quad T_5(x) = x - \frac{x^2}{2} + \frac{x^3}{3} - \frac{x^4}{4} + \frac{x^5}{5}$$

$$\text{Padé:} \quad R_{3,2}(X) = \frac{30x + 21x^2 + x^3}{30 + 36x + 9x^2}$$

Evaluate $T_5(x)$ and $R_{3,2}(x)$ for $x = -0.6, -0.2, 0.2$, and 0.6.

13. Compare the following approximations to $f(x) = \tan(x)$.

$$\text{Taylor:} \quad T_9(x) = x + \frac{x^3}{3} + \frac{2x^5}{15} + \frac{17x^7}{315} + \frac{62x^9}{2{,}835}$$

$$\text{Padé:} \quad R_{5,4}(x) = \frac{945x - 105x^3 + x^5}{945 - 420x^2 + 15x^4}$$

Evaluate $T_9(x)$ and $R_{5,4}(x)$ for $x = 0.4, 0.8$, and 1.2.

5

Curve Fitting

5.1 Least-Squares Line

In science and engineering it is often the case that an experiment produces a set of data points $(x_1, y_1), \ldots, (x_N, y_N)$, where the abscissas $\{x_k\}$ are distinct. One goal of numerical methods is to determine a formula $y = f(x)$ that relates these variables. Usually, a class of allowable formulas is chosen and then coefficients must be determined. There are many possibilities for the type of function that can be used. Often there is an underlying mathematical model, based on the physical situation, that will determine the form of the function. In this section we emphasize the class of linear functions of the form

$$(1) \qquad y = f(x) = Ax + B.$$

In Chapter 4 we saw how to construct a polynomial that passes through a set of points. If all the numerical values $\{x_k\}, \{y_k\}$ are known to several significant digits of accuracy, then polynomial interpolation can be used successfully, otherwise it cannot. Some experiments are devised using specialized equipment so that the data points will have at least five digits of accuracy. However, many experiments are done with equipment that is reliable only to three or fewer digits of accuracy. Often there is an experimental error in the measurements, and although three digits are recorded for the values $\{x_k\}$ and $\{y_k\}$, it is realized that the true value $f(x_k)$ satisfies

(2) $$f(x_k) = y_k + e_k,$$

where e_k is the measurement error.

How do we find the best linear approximation of form (1) that goes near (not always through) the points? To answer this question we need to discuss the *errors* (also called *deviations* or *residuals*);

(3) $$e_k = f(x_k) - y_k \qquad \text{for } 1 \le k \le N.$$

There are several norms that can be used with the residuals in (3) to measure how far the curve $y = f(x)$ lies from the data.

(4) Maximum error: $\quad E_\infty(f) = \max_{1 \le k \le N} \{|f(x_k) - y_k|\}.$

(5) Average error: $\quad E_1(f) = \dfrac{1}{N} \sum_{k=1}^{N} |f(x_k) - y_k|.$

(6) Root-mean-square (rms) error: $\quad E_2(f) = \left[\dfrac{1}{N} \sum_{k=1}^{N} |f(x_k) - y_k|^2 \right]^{1/2}.$

The next example shows how to apply these norms when a function and a set of points are given.

Example 5.1 Compare the maximum error, average error, and rms error for the linear approximation $y = f(x) = 8.6 - 1.6x$ to the data points $(-1, 10)$, $(0, 9)$, $(1, 7)$, $(2, 5)$, $(3, 4)$, $(4, 3)$, $(5, 0)$, and $(6, -1)$.

Solution. The errors are found using the values for $f(x_k)$ and e_k given in Table 5.1.

TABLE 5.1   Calculations for finding $E_1(f)$, $E_2(f)$
for Example 5.1

x_k	y_k	$f(x_k) = 8.6 - 1.6x_k$	$|e_k|$	e_k^2
-1	10.0	10.2	0.2	0.04
0	9.0	8.6	0.4	0.16
1	7.0	7.0	0.0	0.00
2	5.0	5.4	0.4	0.16
3	4.0	3.8	0.2	0.04
4	3.0	2.2	0.8	0.64
5	0.0	0.6	0.6	0.36
6	-1.0	-1.0	0.0	0.00
			2.6	1.40

(7) $\quad E_\infty(f) = \max \{0.2, 0.4, 0.0, 0.4, 0.2, 0.8, 0.6, 0.0\} = 0.8,$

(8) $\quad E_1(f) = \dfrac{1}{8}(2.6) = 0.325,$

(9) $E_2(f) = \left(\dfrac{1.4}{8}\right)^{1/2} \approx 0.41833.$

We can see that the maximum error is largest, and if one point is badly in error, then its value determines $E_\infty(f)$. The average error $E_1(f)$ simply averages the absolute value of the error at the various points. It is often used because it is easy to compute. The error $E_2(f)$ is often used when the statistical nature of the errors is considered.

A "best-fitting" line is found by minimizing one of the quantities in equations (4)–(6). Hence there are three best-fitting lines that we could find. The third norm $E_2(f)$ is the traditional choice because it is much easier to minimize $E_2(f)$ computationally.

Definition 5.1 Let $(x_1, y_1), \ldots, (x_N, y_N)$ be a set of N points where the abscissas $\{x_k\}$ are distinct. The *least-squares line* $y = f(x) = Ax + B$ is the line for which $E_2(f)$ is a minimum.

Finding the Least-Squares Line

Given the N sample points $(x_1, x_2), \ldots, (x_N, y_N)$, we must find parameters A and B for the line $y = f(x) = Ax + B$ which minimize the quantity $E_2(f)$. This quantity, given in (6), will be minimal if and only if the sum of the squares of the residuals is minimal. Hence it will suffice to minimize the function $E(A, B)$ given by

(10) $$E(A, B) = \sum_{k=1}^{N} (Ax_k + B - y_k)^2 = N[E_2(f)]^2.$$

At a point that minimizes the value of $E(A, B)$, the partial derivatives $\partial E/\partial A$ and $\partial E/\partial B$ are both zero. Notice that $\{x_k\}$ and $\{y_k\}$ are constants in equation (10) and A and B are the variables! Hold B fixed and differentiate with respect to A to obtain

(11) $$\frac{\partial E}{\partial A} = \sum_{k=1}^{N} 2(Ax_k + B - y_k)^1(x_k + 0 - 0) = 2\sum_{k=1}^{N} (Ax_k^2 + Bx_k - x_k y_k).$$

Now hold A fixed and differentiate with respect to B to obtain

(12) $$\frac{\partial E}{\partial B} = \sum_{k=1}^{N} 2(Ax_k + B - y_k)^1(0 + 1 - 0) = 2\sum_{k=1}^{N} (Ax_k + B - y_k).$$

Setting the partial derivatives equal to zero in (11) and (12) and using the distributive properties of the summation yields

(13) $$0 = \sum_{k=1}^{N} (Ax_k^2 + Bx_k - x_k y_k) = A\sum_{k=1}^{N} x_k^2 + B\sum_{k=1}^{N} x_k - \sum_{k=1}^{N} x_k y_k,$$

(14) $$0 = \sum_{k=1}^{N} (Ax_k + B - y_k) = A\sum_{k=1}^{N} x_k + NB - \sum_{k=1}^{N} y_k.$$

Equations (13) and (14) can be arranged to form a 2 by 2 system that is often referred to as the *normal equations*:

(15)
$$\left(\sum_{k=1}^{N} x_k^2\right)A + \left(\sum_{k=1}^{N} x_k\right)B = \sum_{k=1}^{N} x_k y_k,$$

$$\left(\sum_{k=1}^{N} x_k\right)A + NB = \sum_{k=1}^{N} y_k.$$

The solution to the linear system (15) can be obtained by using Cramer's rule and is left as an exercise (see Exercise 6). However, the method employed in Algorithm 5.1 translates the data points so that a well conditioned matrix is used (see Problem 8).

Example 5.2 Find the least-squares line for the data points given in Example 5.1.

Solution. The sums required for the normal equations (15) are easily obtained using the values in Table 5.2.

TABLE 5.2 Obtaining the coefficients for normal equations

x_k	y_k	x_k^2	$x_k y_k$
-1	10	1	-10
0	9	0	0
1	7	1	7
2	5	4	10
3	4	9	12
4	3	16	12
5	0	25	0
6	-1	36	-6
20	37	92	25

The linear system involving A and B is

$$92A + 20B = 25$$

$$20A + 8B = 37.$$

The solution is obtained by the calculations

$$A = \frac{\begin{vmatrix} 25 & 20 \\ 37 & 8 \end{vmatrix}}{\begin{vmatrix} 92 & 20 \\ 20 & 8 \end{vmatrix}} = \frac{-540}{336} \approx -1.6071429$$

and

$$B = \frac{\begin{vmatrix} 92 & 25 \\ 20 & 37 \end{vmatrix}}{\begin{vmatrix} 92 & 20 \\ 20 & 8 \end{vmatrix}} = \frac{2{,}904}{336} \approx 8.6428571.$$

Therefore, the least-squares line is (see Figure 5.1)

$$y = -1.6071429x + 8.6428571.$$

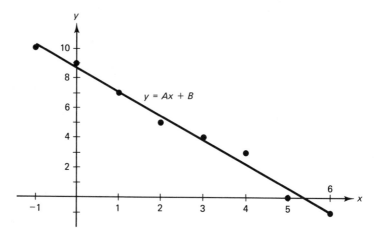

Figure 5.1. The "least-squares line" $y = -1.6071429x + 8.6428571$.

The Power Fit $y = Ax^M$

Some situations involve $f(x) = Ax^M$, where M is a known constant. In these cases there is only one parameter A to be found. Using the least-squares technique, we seek a minimum of the function $E(A)$:

(16)
$$E(A) = \sum_{k=1}^{N} (Ax_k^M - y_k)^2.$$

In this case it will suffice to solve $E'(A) = 0$. The derivative is

(17)
$$E'(A) = 2 \sum_{k=1}^{N} (Ax_k^M - y_k)(x_k^M) = 2 \sum_{k=1}^{N} (Ax_k^{2M} - x_k^M y_k).$$

Hence the coefficient A for the power fit $y = Ax^M$ is

(18)
$$A = \frac{\displaystyle\sum_{k=1}^{N} x_k^M y_k}{\displaystyle\sum_{k=1}^{N} x_k^{2M}}.$$

Example 5.3 Students collected the experimental data in Table 5.3. The relation is $d = \frac{1}{2}gt^2$, where d is distance in meters and t is time in seconds. Find the gravitational constant g.

Solution. The values in Table 5.3 are used to find the summations required in formula (18), where the power used is $M = 2$.

TABLE 5.3 Obtaining the coefficient for a power fit

Time, t_k	Distance, d_k	$d_k t_k^2$	t_k^4
0.200	0.1960	0.00784	0.0016
0.400	0.7850	0.12560	0.0256
0.600	1.7665	0.63594	0.1296
0.800	3.1405	2.00992	0.4096
1.000	4.9075	4.90750	1.0000
		7.68680	1.5664

The coefficient is $A = 7.68680/1.5664 = 4.9073$ and we get $d = 4.9073 t^2$ and $g = 2A = 9.8146$ m/sec^2.

Algorithm 5.1 [*Least-Squares Line*] Suppose that N points $(x_1, y_1), \ldots,$ (x_N, y_N) are given. Find the least-squares line $y = Ax + B$ that fits the data points.

Remark. The algorithm stated below is computationally efficient. It gives reliable results in cases where the linear system (15) is ill-conditioned. The points are first translated so that they have mean $(0, 0)$ and the slope is calculated. Then the y-intercept is computed. The proof is given in Exercises 5 and 8.

```
        READ N                              {Number of points}
FOR     K = 1  TO  N  DO
        READ X(K), Y(K)                     {Get the N points}


        Xmean := 0
FOR     K = 1  TO  N  DO                     {Find the mean x̄}
        Xmean := Xmean + X(K)
        Xmean := Xmean/N


        Ymean := 0
FOR     K = 1  TO  N  DO                     {Find the mean ȳ}
        Ymean := Ymean + Y(K)
        Ymean := Ymean/N
```

```
        SumX := 0                                    {Sum (xₖ – x̄)²}
FOR     K = 1   TO   N   DO
└────   SumX := SumX + [X(K) – Xmean]²

        SumXY := 0                                   {Sum (xₖ – x̄)(yₖ – ȳ)}
FOR     K = 1   TO   N   DO
└────   SumXY := SumXY + [X(K) – Xmean] * [Y(K) – Ymean]

        A := SumXY/SumX                              {Compute the slope}
        B := Ymean – A*Xmean                         {Compute the y-intercept}

PRINT "Least-squares line y = ax+b"                  {Output}
PRINT "The coefficients are"
PRINT "a = " , A
PRINT "b = " , B
```

Exercises for Least-Squares Line

In Exercises 1–3, find the least-squares line $y = f(x) = Ax + B$ for the data and calculate $E_2(f)$.

1. (a)

x_k	y_k	$f(x_k)$
−2	1	1.2
−1	2	1.9
0	3	2.6
1	3	3.3
2	4	4.0

(b)

x_k	y_k	$f(x_k)$
−6	7	7.0
−2	5	4.6
0	3	3.4
2	2	2.2
6	0	−0.2

(c)

x_k	y_k	$f(x_k)$
−4	−3	−3.0
−1	−1	−0.9
0	0	−0.2
2	1	1.2
3	2	1.9

2. (a)

x_k	y_k	$f(x_k)$
−4	1.2	0.44
−2	2.8	3.34
0	6.2	6.24
2	7.8	9.14
4	13.2	12.04

(b)

x_k	y_k	$f(x_k)$
−6	−5.3	−6.00
−2	−3.5	−2.84
0	−1.7	−1.26
2	0.2	0.32
6	4.0	3.48

(c)

x_k	y_k	$f(x_k)$
−8	6.8	7.32
−2	5.0	3.81
0	2.2	2.64
4	0.5	0.30
6	−1.3	−0.87

3. (a)

x_k	y_k	$f(x_k)$
−2	1	0.4
0	3	3.3
2	6	6.2
4	8	9.1
6	13	12.0

(b)

x_k	y_k	$f(x_k)$
−2	7	7.6
1	5	3.7
2	2	2.4
4	0	−0.2
5	−2	−1.5

(c)

x_k	y_k	$f(x_k)$
−3	−6	−6.7
1	−4	−3.3
3	−2	−1.6
5	0	0.1
9	4	3.5

4. Find the *power fit* $y = Ax$, where $M = 1$ which is a line through the origin, for the data and calculate $E_2(f)$.

(a)

x_k	y_k	$f(x_k)$
1	1.6	1.58
2	2.8	3.16
3	4.7	4.74
4	6.4	6.32
5	8.0	7.90

(b)

x_k	y_k	$f(x_k)$
3	1.6	1.722
4	2.4	2.296
5	2.9	2.870
6	3.4	3.444
8	4.6	4.592

(c)

x_k	y_k	$f(x_k)$
−4	−3	−2.8
−1	−1	−0.7
0	0	0.0
2	1	1.4
3	2	2.1

5. Define the data means $\bar{x}$ and $\bar{y}$ by

$$\bar{x} = \frac{1}{N} \sum_{k=1}^{N} x_k \quad \text{and} \quad \bar{y} = \frac{1}{N} \sum_{k=1}^{N} y_k.$$

Show that the point $(\bar{x}, \bar{y})$ lies on the least-squares line.

6. Show that the solution of the system in (15) is given by

$$A = \frac{1}{D}\left(N \sum_{k=1}^{N} x_k y_k - \sum_{k=1}^{N} x_k \sum_{k=1}^{N} y_k\right),$$

$$B = \frac{1}{D}\left(\sum_{k=1}^{N} x_k^2 \sum_{k=1}^{N} y_k - \sum_{k=1}^{N} x_k \sum_{k=1}^{N} x_k y_k\right),$$

$$D = N \sum_{k=1}^{N} x_k^2 - \left(\sum_{k=1}^{N} x_k\right)^2.$$

7. Show that the value D in Exercise 6 is nonzero.
   Hint. Show that $D = N \sum_{k=1}^{N} (x_k - \bar{x})^2$.

8. Show that the coefficients A and B for the least-squares line can be computed as follows. First compute the data means $\bar{x}$ and $\bar{y}$ in Exercise 5, then perform the calculations

$$C = \sum_{k=1}^{N} (x_k - \bar{x})^2 \qquad A = \frac{1}{C} \sum_{k=1}^{N} (x_k - \bar{x})(y_k - \bar{y}) \qquad B = \bar{y} - A\bar{x}.$$

Hint. Use $X_k = x_k - \bar{x}$, $Y_k = y_k - \bar{y}$ and first find the line $Y = AX$.

9. Compute the missing entries in the table and determine which curve fits the data better. Use $E_2(f)$ to check the fits.

x_k	y_k	$y_k \approx 2.069 x_k^2$	$y_k \approx 1.516 x_k^3$
0.3	0.1	0.19	0.04
0.6	0.4	.	.
0.9	1.2	$\overline{}$	$\overline{}$
1.2	2.8	2.98	2.62
1.5	5.0	.	.

10. Write a report on linear correlation of data. Be sure to include a discussion of the coefficient of correlation.

11. Find the power fits $y = Ax^2$ and $y = Bx^3$ for the following data and use $E_2(f)$ to determine which curve fits best.

(a)

x_k	y_k
2.0	5.1
2.3	7.5
2.6	10.6
2.9	14.4
3.2	19.0

(b)

x_k	y_k
2.0	5.9
2.3	8.3
2.6	10.7
2.9	13.7
3.2	17.0

12. Find the power fits $y = A/x$ and $y = B/x^2$ for the following data and use $E_2(f)$ to determine which curve fits best.

(a)

x_k	y_k
0.5	7.1
0.8	4.4
1.1	3.2
1.8	1.9
4.0	0.9

(b)

x_k	y_k
0.7	8.1
0.9	4.9
1.1	3.3
1.6	1.6
3.0	0.5

13. Hooke's law states that $F = kx$, where F is the force (ounces) used to stretch a spring and x is the increase in its length in (inches). Find an approximation to the spring constant k for the following data.

(a)

x_k	F_k
0.2	3.6
0.4	7.3
0.6	10.9
0.8	14.5
1.0	18.2

(b)

x_k	F_k
0.2	5.3
0.4	10.6
0.6	15.9
0.8	21.2
1.0	26.4

14. Find the gravitational constant g for the following sets of data. Use the power fit that was shown in Example 5.3.

(a)

Time, t_k	Distance, d_k
0.200	0.1960
0.400	0.7835
0.600	1.7630
0.800	3.1345
1.000	4.8975

(b)

Time, t_k	Distance, d_k
0.200	0.1965
0.400	0.7855
0.600	1.7675
0.800	3.1420
1.000	4.9095

15. Derive the normal equation for finding the least-squares linear fit through the origin $y = Ax$.

16. Derive the normal equation for finding the least-squares power fit $y = Ax^2$.

17. Derive the normal equations for finding the least-squares parabola $y = Ax^2 + B$.

5.2 Curve Fitting

Least-Squares Method

The method of least-squares curve fitting can be extended to many nonlinear cases. For example, start with the N data points $\{(x_k, y_k)\}$ and consider the exponential fit

(1)
$$y = C \exp(Ax).$$

The least-squares procedure requires that we find a minimum of

(2)
$$E(A, C) = \sum_{k=1}^{N} [C \exp(Ax_k) - y_k]^2.$$

The partial derivative of $E(A, C)$ with respect to A and C are

(3)
$$\frac{\partial E}{\partial A} = 2 \sum_{k=1}^{N} [C \exp(Ax_k) - y_k][Cx_k \exp(Ax_k)],$$

(4)
$$\frac{\partial E}{\partial C} = 2 \sum_{k=1}^{N} [C \exp(Ax_k) - y_k][\exp(Ax_k)].$$

When the derivatives in (3) and (4) are set equal to zero the resulting normal equations for the exponential fit are

(5)
$$C \sum_{k=1}^{N} x_k \exp(2Ax_k) - \sum_{k=1}^{N} x_k y_k \exp(Ax_k) = 0,$$
$$C \sum_{k=1}^{N} \exp(2Ax_k) - \sum_{k=1}^{N} y_k \exp(Ax_k) = 0.$$

The system of equations in (5) is nonlinear in the unknowns A and C, and can be solved using Newton's method for systems of equations. An alternative method is to find the minimum of $E(A, C)$ directly by using a method such as the Nelder–Mead simplex algorithm. Both of these methods are iterative and require a good initial approximation to the solution.

Data Linearization Method

Another common approach for the exponential fit is to take the logarithm of both sides of (1). We will end up with a linear relationship between transformed variables. The first step is

(6)
$$\ln(y) = Ax + \ln(C).$$

Now use the following change of variables (and constant):

(7)
$$Y = \ln(y) \qquad X = x \qquad B = \ln(C).$$

This results in a linear relation between the variables X and Y:

(8)
$$Y = AX + B.$$

The method for finding the least-squares line in Section 5.1 is applied to the transformed data points $\{(X_k, Y_k)\} = \{(x_k, \ln(y_k))\}$. The coefficients A and B are found by solving the linear system

(9)
$$A\left(\sum_{k=1}^{N} x_k^2\right) + B\left(\sum_{k=1}^{N} x_k\right) = \sum_{k=1}^{N} x_k \ln(y_k),$$
$$A\left(\sum_{k=1}^{N} x_k\right) + BN = \sum_{k=1}^{N} \ln(y_k).$$

The required coefficient C in (1) must be computed by the formula

(10)
$$C = \exp(B).$$

This technique is called *data linearization* and is a popular method that is used when one does not want to go to the effort of using an iterative procedure that is inherent in the least-squares method.

Example 5.4 Use the data linearization technique and the change of variables (and constant) in (7) to find the exponential fit $y = C \exp(Ax)$ for the five data points $(0, 1.5)$, $(1, 2.5)$, $(2, 3.5)$, $(3, 5.0)$, and $(4, 7.5)$.

Solution. Table 5.4 shows the transformed data points that must be used in computing the summations in the linear system (9).

TABLE 5.4 Obtaining coefficients for the exponential fit by the method of data linearization

x_k	y_k	$\ln(y_k)$	x_k^2	$x_k \ln(y_k)$
0	1.5	0.40547	0	0.00000
1	2.5	0.91629	1	0.91629
2	3.5	1.25276	4	2.50553
3	5.0	1.60944	9	4.82831
4	7.5	2.01490	16	8.05961
10		6.19886	30	16.30974

The linear system (9) for finding A and C is

$$30A + 10B = 16.30974$$
$$10A + 5B = 6.19886,$$

and the solution is $A = 0.391202$ and $B = 0.457367$. Then C is obtained by the calculation $C = \exp(0.457367) = 1.5799$, and the resulting exponential fit is (see Figure 5.2)

(11) $y = 1.5799 \exp(0.39120x)$ (fit by data linearization).

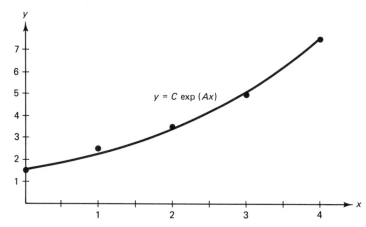

Figure 5.2. The exponential fit $y = 1.5799 \exp(0.39120x)$ obtained by using data linearization.

It is worthwhile to compare the solution (11) obtained by data linearization, with the solution found by minimizing $E(A, C)$. If a suitable iterative method is used, then the direct least-squares fit, obtained by minimizing (2) is

(12) $y = 1.6109 \exp(0.38357x)$ (fit by direct least squares).

A comparison of the solutions is given in Table 5.5. There is a slight difference in the coefficients and it is seen that the function values differ by no more than 2% over the interval $[0, 4]$. For purposes of interpolation there will not be much discrepancy between the curves when viewed by the eye. However, if there is a normal distribution of the errors in the data, then (12) is usually the better choice for extrapolating beyond the data. The least-squares prediction at $x = 10$ is 74.603, which differs by about 6% from the value 78.993 that is predicted by data linearization.

TABLE 5.5   Comparison of the two exponential fits
for Example 5.4

x_k	y_k	$1.5799e^{0.39120x}$	$1.6109e^{0.38357x}$
0	1.5	1.5799	1.6104
1	2.5	2.3363	2.3640
2	3.5	3.4548	3.4692
3	5.0	5.1088	5.0911
4	7.5	7.5548	7.4713

Transformations for Data Linearization

The technique of data linearization has been used by scientists to fit curves such as $y = C \exp(Ax)$, $y = Cx^A$, $y = A \ln(x) + B$ and $y = A/x + B$. Once the curve has been choosen, a suitable transformation of the variables must be found so that a linear relation is obtained. For example, the reader can verify that $y = D/(x + C)$ is transformed into a linear problem $Y = AX + B$ by using the change of variables (and constants) $X = xy$, $Y = y$, $C = -1/A$, and $D = -B/A$. Other useful transformations are given in Table 5.6.

TABLE 5.6 Change of variable(s) for linearization of data

Function, $y = f(x)$	Linearized Form, $Y = AX + B$	Change of Variable(s) and Constants
$y = \dfrac{A}{x} + B$	$y = A\dfrac{1}{x} + B$	$X = \dfrac{1}{x}, Y = y$
$y = \dfrac{D}{x + C}$	$y = \dfrac{-1}{C}(xy) + \dfrac{D}{C}$	$X = xy, Y = y$ $C = \dfrac{-1}{A}, D = \dfrac{-B}{A}$
$y = \dfrac{1}{Ax + B}$	$\dfrac{1}{y} = Ax + B$	$X = x, Y = \dfrac{1}{y}$
$y = \dfrac{x}{A + Bx}$	$\dfrac{1}{y} = A\dfrac{1}{x} + B$	$X = \dfrac{1}{x}, Y = \dfrac{1}{y}$
$y = A \ln(x) + B$	$y = A \ln(x) + B$	$X = \ln(x), Y = y$
$y = C \exp(Ax)$	$\ln(y) = Ax + \ln(C)$	$X = x, Y = \ln(y)$ $C = \exp(B)$
$y = Cx^A$	$\ln(y) = A \ln(x) + \ln(C)$	$X = \ln(x), Y = \ln(y)$ $C = \exp(B)$
$y = (Ax + B)^{-2}$	$y^{-1/2} = Ax + B$	$X = x, Y = y^{-1/2}$
$y = Cx \exp(-Dx)$	$\ln\left(\dfrac{y}{x}\right) = -Dx + \ln(C)$	$X = x, Y = \ln\left(\dfrac{y}{x}\right)$ $C = \exp(B), D = -A$
$y = \dfrac{L}{1 + C \exp(Ax)}$	$\ln\left(\dfrac{L}{y} - 1\right) = Ax + \ln(C)$	$X = x, Y = \ln\left(\dfrac{L}{y} - 1\right)$ $C = \exp(B)$ and L is a constant that must be given

Linear Least Squares

The linear least-squares problem is stated as follows. Suppose that N data points $\{(x_k, y_k)\}$ and a set of M linear independent functions $\{f_j(x)\}$ are given. We want to find M coefficients $\{c_j\}$ so that the function $f(x)$ given by the linear combination

$$(13) \qquad f(x) = \sum_{j=1}^{M} c_j f_j(x)$$

will minimize the sum of the squares of the errors

$$(14) \qquad E(c_1, c_2, \ldots, c_M) = \sum_{k=1}^{N} [f(x_k) - y_k]^2 = \sum_{k=1}^{N} \sum_{j=1}^{M} [c_j f_j(x_k) - y_k]^2.$$

For E to be minimized, it is necessary that each partial derivative be zero (i.e., $\partial E / \partial c_i = 0$ for $i = 1, 2, \ldots, M$), and this results in the system of equations;

$$(15) \qquad \sum_{k=1}^{N} \sum_{j=1}^{M} [c_j f_j(x_k) - y_k][f_i(x_k)] = 0 \qquad \text{for } i = 1, 2, \ldots, M.$$

Interchanging the order of the summations in (15) will produce an M by M system of linear equations where the unknowns are the coefficients $\{c_j\}$. They are called the normal equations:

$$(16) \qquad \sum_{j=1}^{M} \left[\sum_{k=1}^{N} f_i(x_k) f_j(x_k) \right] c_j = \sum_{k=1}^{N} f_i(x_k) y_k \qquad \text{for } i = 1, 2, \ldots, M.$$

The Matrix Formulation

Although (16) is easily recognized as a system of M equations in M unknowns, one must be clever so that wasted computations are not performed when writing the system in matrix notation. The key is to write down the matrices F and F^T as follows:

$$F = \begin{pmatrix} f_1(x_1) & f_2(x_1) & \cdots & f_M(x_1) \\ f_1(x_2) & f_2(x_2) & \cdots & f_M(x_2) \\ f_1(x_3) & f_2(x_3) & \cdots & f_M(x_3) \\ \vdots & \vdots & & \vdots \\ f_1(x_N) & f_2(x_N) & \cdots & f_M(x_N) \end{pmatrix} \qquad F^T = \begin{pmatrix} f_1(x_1) & f_1(x_2) & f_1(x_3) & \cdots & f_1(x_N) \\ f_2(x_1) & f_2(x_2) & f_2(x_3) & \cdots & f_2(x_N) \\ \vdots & \vdots & \vdots & & \vdots \\ f_M(x_1) & f_M(x_2) & f_M(x_3) & \cdots & f_M(x_N) \end{pmatrix}.$$

Consider the product of F^T and the column vector $\mathbf{Y}$:

(17)
$$F^T\mathbf{Y} = \begin{pmatrix} f_1(x_1) & f_1(x_2) & f_1(x_3) & \cdots & f_1(x_N) \\ f_2(x_1) & f_2(x_2) & f_2(x_3) & \cdots & f_2(x_N) \\ \vdots & \vdots & & & \vdots \\ f_M(x_1) & f_M(x_2) & f_M(x_3) & \cdots & f_M(x_N) \end{pmatrix} \begin{pmatrix} y_1 \\ y_2 \\ \vdots \\ y_N \end{pmatrix}.$$

The element in the ith row of the product $F^T\mathbf{Y}$ in (17) is the same as the ith element in the column vector in equation (16), that is,

(18)
$$\sum_{k=1}^{N} f_i(x_k) y_k = \left(\text{row}_i \ F^T(y_1, y_2, \ldots, y_N)^T\right).$$

Now consider the product F^TF, which is an M by M matrix:

$$F^TF = \begin{pmatrix} f_1(x_1) & f_1(x_2) & f_1(x_3) & \cdots & f_1(x_N) \\ f_2(x_1) & f_2(x_2) & f_2(x_3) & \cdots & f_2(x_N) \\ \vdots & \vdots & \vdots & & \vdots \\ f_M(x_1) & f_M(x_2) & f_M(x_3) & \cdots & f_M(x_N) \end{pmatrix} \begin{pmatrix} f_1(x_1) & f_2(x_1) & \cdots & f_M(x_1) \\ f_1(x_2) & f_2(x_2) & \cdots & f_M(x_2) \\ f_1(x_3) & f_2(x_3) & \cdots & f_M(x_3) \\ \vdots & \vdots & & \vdots \\ f_1(x_N) & f_2(x_N) & \cdots & f_M(x_N) \end{pmatrix}.$$

The element in the ith row and jth column of F^TF is the coefficient of c_j in the ith row in equation (16), that is,

(19)
$$\sum_{k=1}^{N} f_i(x_k) f_j(x_k) = f_i(x_1)f_j(x_1) + f_i(x_2)f_j(x_2) + \ldots + f_i(x_N)f_j(x_N).$$

When M is small, a computationally efficient way to calculate the linear least-squares coefficients for (13) is to store the matrix F and compute F^TF and $F^T\mathbf{Y}$ and then solve the linear system

(20)
$$F^TFC = F^T\mathbf{Y} \qquad \text{for the coefficient vector } \mathbf{C}.$$

Polynomial Fitting

When the foregoing method is adapted using the functions $\{f_j(x) = x^j\}$ and the index of summation ranges from $j = 0$ to $j = M$, the function $f(x)$ will be a polynomial of degree M:

(21)
$$f(x) = c_1 + c_2x + c_3x^2 + \ldots + c_{M+1}x^M.$$

We now show how to find the *least-squares parabola* and the extension to a polynomial of higher degree is easily made. Given N points $(x_1, y_1), \ldots, (x_N, y_N)$, we wish to find the parabola

(22)
$$y = f(x) = Ax^2 + Bx + C,$$

where the coefficients A, B, and C will minimize the quantity

(23)
$$E(A, B, C) = \sum_{k=1}^{N} (Ax_k^2 + Bx_k + C - y_k)^2.$$

The partial derivatives of E with respect to $A, B,$ and C are

$$\frac{\partial E}{\partial A} = 2 \sum_{k=1}^{N} (Ax_k^2 + Bx_k + C - y_k)(x_k^2),$$

(24)
$$\frac{\partial E}{\partial B} = 2 \sum_{k=1}^{N} (Ax_k^2 + Bx_k + C - y_k)(x_k),$$

$$\frac{\partial E}{\partial C} = 2 \sum_{k=1}^{N} (Ax_k^2 + Bx_k + C - y_k).$$

Using the distributive property of addition, we can move the values A, B, C outside the summations to obtain the normal equations:

(25)
$$\left(\sum_{k=1}^{N} x_k^4\right)A + \left(\sum_{k=1}^{N} x_k^3\right)B + \left(\sum_{k=1}^{N} x_k^2\right)C = \sum_{k=1}^{N} y_k x_k^2,$$

$$\left(\sum_{k=1}^{N} x_k^3\right)A + \left(\sum_{k=1}^{N} x_k^2\right)B + \left(\sum_{k=1}^{N} x_k\right)C = \sum_{k=1}^{N} y_k x_k,$$

$$\left(\sum_{k=1}^{N} x_k^2\right)A + \left(\sum_{k=1}^{N} x_k\right)B + NC = \sum_{k=1}^{N} y_k.$$

Example 5.5 Find the least-squares parabola for the four points $(-3, 3), (0, 1), (2, 1),$ and $(4, 3)$.

Solution. The entries in Table 5.7 are used to compute the summations required in the linear system (25).

TABLE 5.7 Obtaining the coefficients for the least-squares parabola of Example 5.5

x_k	y_k	x_k^2	x_k^3	x_k^4	$x_k y_k$	$x_k^2 y_k$
-3	3	9	-27	81	-9	27
0	1	0	0	0	0	0
2	1	4	8	16	2	4
4	3	16	64	256	12	48
3	8	29	45	353	5	79

The linear system (25) for finding A, B, and C becomes

$$353A + 45B + 29C = 79$$

$$45A + 29B + 3C = 5$$

$$29A + 3B + 4C = 8.$$

The solution to the linear system is $A = 585/3{,}278$, $B = -631/3{,}278$, and $C = 1{,}394/1{,}639$, and the desired parabola is (see Figure 5.3)

$$y = \frac{585}{3{,}278}x^2 - \frac{631}{3{,}278}x + \frac{1{,}394}{1{,}639} = 0.178462x^2 - 0.192495x + 0.850519.$$

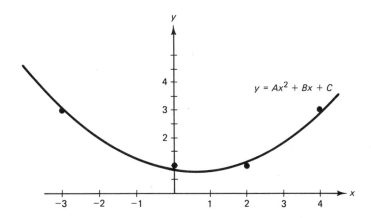

Figure 5.3. The "least-squares parabola" for Example 5.5.

Polynomial Wiggle

It is tempting to use least-squares polynomials to fit data that is nonlinear. But beware, a polynomial of degree N can have $N - 1$ local extrema. If the data points do not actually lie on a polynomial curve, then the least-squares polynomials may exhibit large oscillations. This phenomenon, called *polynomial wiggle*, becomes more pronounced with higher-degree polynomials. Hence we should seldom use a polynomial of degree 6 or above unless it is known that the true function we are working with is a polynomial.

For example, the function $f(x) = x^{-2} + (|x - 1|)^{1/2}$ was used to generate the six data points in Figure 5.4. The least-squares polynomials $P_2(x)$, $P_3(x)$, $P_4(x)$, and $P_5(x)$ are also shown in the figure. Notice that $P_4(x)$ and $P_5(x)$ exhibit a large wiggle in the interval $[3, 5]$. Indeed, $P_5(x)$ goes through the six data points and it is the worst approximation. If we had to use a polynomial, then $P_2(x)$ or $P_3(x)$ should be our choice.

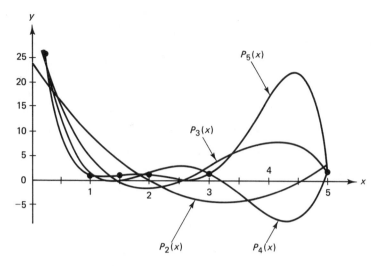

Figure 5.4. The phenomenon of "polynomial wiggle."

Algorithm 5.2 [*Least-Squares Polynomial*] Suppose that N points (x_1, y_1), ..., (x_N, y_N) are given. Find the least-squares polynomial of degree M that fits the data points:

$$P_M(x) = c_1 + c_2 x + c_3 x^2 + c_4 x^3 + \ldots + c_M x^{M-1} + c_{M+1} x^M.$$

Remark. If much work is to be done with polynomial curve fitting, then the topic of orthogonal polynomials should be researched.

```
READ N                                              {Number of points}
READ M                                              {Degree of polynomial}
DIMENSION  X[1 .. N],Y[1 .. N],B[1 .. M+1],C[1 .. M+1],P[0 .. 2M]
DIMENSION  A[1 .. M+1, 1 .. M+1]


FOR    K = 1  TO  N  DO                             {Get the N points}
 └───     READ X(K), Y(K)


FOR    R = 1 TO M+1 DO: B(R) := 0                   {Zero the array}
FOR    K =  TO  N  DO                               {Compute
 │        Y := Y(K), X := X(K), P := 1                  the
 │        FOR    R = 1  TO  M+1  DO                     column
 │         │       B(R) := B(R) + Y*P                   vector}
 └───      └───  P := P*X
```

```
FOR    J = 1 TO 2*M DO: P(J) := 0                    {Zero the array}
       P(0) := N
FOR    K = 1  TO  N  DO                              {Compute
       X := X(K), P := X(K)                          the sum of
       FOR    J = 1  TO  2*M  DO                      powers
              P(J) := P(J) + P                        of xₖ}
              P := P*X


FOR    R = 1  TO  M+1  DO                            {Determine
       FOR    C = 1  TO  M+1  DO                      the matrix
              A(R, C) = P(R+C−2)                      entries}
```

USE a linear systems algorithm to solve the M+1 equations:
$$A*C = B \text{ for coefficient vector } C = (c_1, c_2, \ldots, c_M, c_{M+1}).$$

PRINT "Least squares polynomial of degree M"   {Output}
PRINT "The coefficients are"
PRINT C(1), C(2), ..., C(M+1)

Algorithm 5.3 [*Nonlinear Curve Fitting*] Suppose that N points $(x_1, y_1), \ldots,$ (x_N, y_N) are given. Find selected nonlinear fits.

READ N   {Number of points}
FOR J = 1 TO N DO: READ X(J), Y(J)   {Get the N points}

CASES {Select the nonlinear form you wish to fit.}
   (i) $y = A/X + B$ (ii) $y = D/[x + C]$
   (iii) $y = 1/[Ax + B]$ (iv) $y = x/[A + Bx]$
   (v) $y = A \ln(x) + B$ (vi) $y = C*\exp(Ax)$
   (vii) $y = C*x^A$ (viii) $y = [Ax + B]^{-2}$
END

CASE {Make transformation(s) to linearize the data $j = 1, 2, \ldots, N.$}
   (i) $X_j = 1/x_j, \ Y_j = y_j$ (ii) $X_j = x_j y_j, \ Y_j = y_j$
   (iii) $X_j = x_j, \ Y_j = 1/y_j$ (iv) $X_j = 1/x_j, \ Y_j = 1/y_j$
   (v) $X_j = \ln(x_j), \ Y_j = y_j$ (vi) $X_j = x_j, \ Y_j = \ln(y_j)$
   (vii) $X_j = \ln(x_j), \ Y_j = \ln(y_j)$ (viii) $X_j = x_j, \ Y_j = y_j^{-1/2}$
END

USE Algorithm 3.1 to find the least squares line $Y = A*X + B$
     for the transformed data points $\{X_j, \ Y_j\}$.

CASES {Make coefficient transformation(s) if necessary.}

| (ii) $C = -1/A$, $D = -B/A$
| (vi) $C = \exp(B)$
| (vii) $C = \exp(B)$
END

PRINT "The least-squares curve for the data is:"   {Output}
CASES {Print appropriate equation and coefficients}

| (i) $y = A/X + B$ | (ii) $y = D/[x + C]$
| (iii) $y = 1/[Ax + B]$ | (iv) $y = x/[A + Bx]$
| (v) $y = A \ln(x) + B$ | (vi) $y = C*\exp(Ax)$
| (vii) $y = C*x^A$ | (viii) $y = [Ax + B]^{-2}$
END

Exercises for Curve Fitting

1. Find the least-squares parabolic fit $y = Ax^2 + Bx + C$.

(a)

x_k	y_k
-3	15
-1	5
1	1
3	5

(b)

x_k	y_k
-3	-1
-1	25
1	25
3	1

2. Find the least-squares parabolic fit $y = Ax^2 + Bx + C$.

(a)

x_k	y_k
-2	-5.8
-1	1.1
0	3.8
1	3.3
2	-1.5

(b)

x_k	y_k
-2	2.8
-1	2.1
0	3.25
1	6.0
2	11.5

(c)

x_k	y_k
-2	10
-1	1
0	0
1	2
2	9

3. Use $E_1(f)$ to determine which curve fits best.

(a)

x_k	y_k	$y_k \approx 0.5102x^2$	$y_k \approx 0.4304 \times 2^x$
1	0.7	0.51	0.86
2	2.0	.	.
3	4.2	_	_
4	8.0	.	.
5	13.0	$\overline{12.76}$	$\overline{13.77}$

(b) Change the last entry to $(5, 15.0)$ and determine the best fit $y \approx 0.5613x^2$ or $y \approx 0.4773 \times 2^x$.

4. (a) Find the curve fit $y = C \exp(Ax)$ by using the change of variables $X = x$, $Y = \ln(y)$, $C = \exp(B)$ to linearize the data points.
   (b) Find the curve fit $y = Cx^A$ by using the change of variables $X = \ln(x)$, $Y = \ln(y)$, and $C = \exp(B)$ to linearize the data points.
   (c) Use $E_1(f)$ and determine which curve fit in part (a) or (b) is best.

x_k	y_k	$\ln(x_k)$	$\ln(y_k)$
1	0.6	0.0000	−0.5108
2	1.9	0.6931	0.6419
3	4.3	1.0986	1.4586
4	7.6	1.3863	2.0281
5	12.6	1.6094	2.5337

5. Follow the instructions for Exercise 4.

x_k	y_k	$\ln(x_k)$	$\ln(y_k)$
1	0.7	0.0000	−0.3567
2	1.7	0.6931	0.5306
3	3.4	1.0986	1.2238
4	6.7	1.3863	1.9021
5	12.7	1.6094	2.5416

6. (a) Find the curve fit $y = C \exp(Ax)$ by using the change of variables $X = x$, $Y = \ln(y)$, $C = \exp(B)$ to linearize the data points.
   (b) Find the curve fit $y = 1/(Ax + B)$ by using the change of variables $X = x$, $Y = 1/y$ to linearize the data points.
   (c) Use $E_1(f)$ and determine which curve fit in part (a) or (b) is best.

x_k	y_k	$\ln(y_k)$	$1/y_k$
−1	6.62	1.8901	0.1511
0	3.94	1.3712	0.2538
1	2.17	0.7747	0.4608
2	1.35	0.3001	0.7407
3	0.89	−0.1165	1.1236

7. Follow the instructions for Exercise 6.

x_k	y_k	$\ln(y_k)$	$1/y_k$
-1	6.62	1.8901	0.1511
0	2.78	1.0225	0.3597
1	1.51	0.4121	0.6623
2	1.23	0.2070	0.8130
3	0.89	-0.1165	1.1236

8. (a) Find the curve fit $y = C \exp(Ax)$ by using the transformations $X = x$, $Y = \ln(y)$, $C = \exp(B)$.
   (b) Find the curve fit $y = (Ax + B)^{-2}$ by using the transformations $X = x$, $Y = y^{-1/2}$.
   (c) Use $E_1(f)$ and determine which curve fit in part (a) or (b) is best.

(i)

x_k	y_k
-1	13.45
0	3.01
1	0.67
2	0.15

(ii)

x_k	y_k
-1	13.65
0	1.38
1	0.49
3	0.15

9. *Logistic Population Growth.* When the population $P(t)$ is bounded by the limiting value L, it follows a logistic curve and has the form $P(t) = L/[1 + C \exp(Ax)]$. Find A and C for the following data, where L is a known value.
   (a) $(0, 200)$, $(1, 400)$, $(2, 650)$, $(3, 850)$, $(4, 950)$, and $L = 1,000$
   (b) $(0, 500)$, $(1, 1000)$, $(2, 1800)$, $(3, 2800)$, $(4, 3700)$, and $L = 5,000$

10. Use the data for the U.S. population and find the logistic curve $P(t)$. Estimate the population in the year 2000.
    (a) Assume that $L = 800$ (million).   (b) Assume that $L = 800$ (million).

Year	t_k	P_k
1800	-10	5.3
1850	-5	23.2
1900	0	76.1
1950	5	152.3

Year	t_k	P_k
1900	0	76.1
1920	2	106.5
1940	4	132.6
1960	6	180.7
1980	8	226.5

11. Write a report on orthogonal polynomials and how they are used to construct polynomials used in curve fitting.

12. The least-squares plane $z = Ax + By + C$ for the N points $(x_1, y_1, z_1), \dots,$ (x_N, y_N, z_N) is obtained by minimizing

$$E(A, B, C) = \sum_{k=1}^{N} (Ax_k + By_k + C - z_k)^2.$$

Derive the normal equations:

$$\left(\sum_{k=1}^{N} x_k^2\right)A + \left(\sum_{k=1}^{N} x_k y_k\right)B + \left(\sum_{k=1}^{N} x_k\right)C = \sum_{k=1}^{N} z_k x_k$$

$$\left(\sum_{k=1}^{N} x_k y_k\right)A + \left(\sum_{k=1}^{N} y_k^2\right)B + \left(\sum_{k=1}^{N} y_k\right)C = \sum_{k=1}^{N} z_k y_k$$

$$\left(\sum_{k=1}^{N} x_k\right)A + \left(\sum_{k=1}^{N} y_k\right)B + NC = \sum_{k=1}^{N} z_k.$$

13. Find the least-squares plane for the following data.
(a) $(1, 1, 7)$, $(1, 2, 9)$, $(2, 1, 10)$, $(2, 2, 11)$, $(2, 3, 12)$
(b) $(1, 2, 6)$, $(2, 3, 7)$, $(1, 1, 8)$, $(2, 2, 8)$, $(2, 1, 9)$
(c) $(3, 1, -3)$, $(2, 1, -1)$, $(2, 2, 0)$, $(1, 1, 1)$, $(1, 2, 3)$

14. Consider the following table of data.

x_k	y_k
1.0	2.0
2.0	5.0
3.0	10.0
4.0	17.0
5.0	26.0

When the change of variables $X = xy$ and $Y = y$ are used with the function $y = D/(x + C)$, the transformed least squares fit is:

$$y = \frac{-17.719403}{x - 5.476617}.$$

When the change of variables $X = x$ and $Y = 1/y$ are used with the function $y = 1/(Ax + B)$, the transformed least squares fit is:

$$y = \frac{1}{-0.1064253x + 0.4987330}.$$

Determine which fit is best and why one of the solutions is completely absurd.

In Exercises 15–22, show how to linearize the given formula.

15. $y = \dfrac{A}{x} + B$

16. $y = \dfrac{D}{x + C}$

17. $y = \dfrac{1}{Ax + B}$

18. $y = \dfrac{x}{A + Bx}$

19. $y = A \ln(x) + B$

20. $Y = Cx^A$

21. $y = (Ax + B)^{-2}$

22. $y = Cx \exp(-Dx)$

5.3 Interpolation by Spline Functions

Polynomial interpolation for a set of $N + 1$ points $\{(x_k, y_k)\}$ is frequently unsatisfactory. As discussed in Section 5.2, a polynomial of degree N can have $N - 1$ relative maxima and minima and the graph can wiggle in order to pass through the points. Another method is to "piece together" the graphs of lower-degree polynomials $S_k(x)$ and interpolate between the successive nodes (x_k, y_k) and (x_{k+1}, y_{k+1}) (see Figure 5.5). The two adjacent portions of the curve $y = S_k(x)$ and $y = S_{k+1}(x)$, which lie above $[x_k, x_{k+1}]$ and $[x_{k+1}, x_{k+2}]$, respectively, pass through the common *knot* (x_{k+1}, y_{k+1}). The two portions of the graph are "tied together" at the knot (x_{k+1}, y_{k+1}) and the set of functions $\{S_k(x)\}$ form a piecewise polynomial curve which is denoted by $S(x)$.

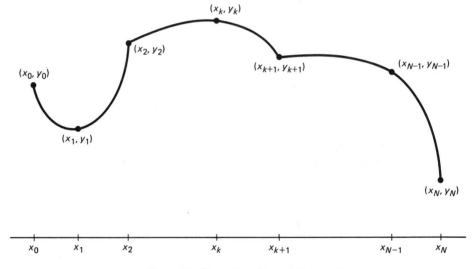

Figure 5.5. Piecewise polynomial interpolation.

Piecewise Linear Interpolation

The simplest polynomial to use, a polynomial of degree 1, produces a polygonal path that consists of line segments that pass through the points. The

Lagrange polynomial from Section 4.2 is used to represent this piecewise linear curve:

(1) $$S_k(x) = y_k \frac{x - x_{k+1}}{x_k - x_{k+1}} + y_{k+1} \frac{x - x_k}{x_{k+1} - x_k} \qquad \text{for } x_k \leq x \leq x_{k+1}.$$

The resulting curve looks like a "broken line" (see Figure 5.6).

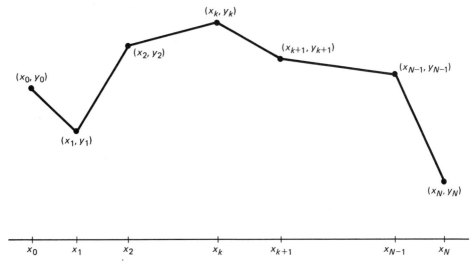

Figure 5.6. Piecewise linear interpolation (a linear spline).

An equivalent expression can be obtained if we use the point-slope formula for a line segment

$$S_k(x) = y_k + d_k(x - x_k),$$

where $d_k = (y_{k+1} - y_k)/(x_{k+1} - x_k)$. The resulting linear spline function can be written in the form

(2) $$S(x) = \begin{cases} y_0 + d_0(x - x_0) & \text{for } x \text{ in } [x_0, x_1], \\ y_1 + d_1(x - x_1) & \text{for } x \text{ in } [x_1, x_2], \\ \quad \vdots & \quad \vdots \\ y_k + d_k(x - x_k) & \text{for } x \text{ in } [x_k, x_{k+1}], \\ \quad \vdots & \quad \vdots \\ y_{N-1} + d_{N-1}(x - x_{N-1}) & \text{for } x \text{ in } [x_{N-1}, x_N]. \end{cases}$$

The form of equation (2) is better than equation (1) for the explicit calculation of $S(x)$. It is assumed that the abscissas are ordered $x_0 < x_1 < \cdots < x_{N-1} < x_N$. For a fixed value of x, the interval $[x_k, x_{k+1}]$ containing x can be found by successively computing the differences $x - x_1, \ldots, x - x_k,$ $x - x_{k+1}$ until $k + 1$ is the smallest integer such that $x - x_{k+1} < 0$. Hence we have found k so that $x_k \leq x \leq x_{k+1}$, and the value of the spline function $S(x)$ is

(3) $$S(x) = S_k(x) = y_k + d_k(x - x_k) \qquad \text{for } x_k \leq x \leq x_{k+1}.$$

These techniques can be extended to higher-degree polynomials. For example, if an odd number of nodes $x_0, x_1, \ldots, x_{2M}$ are given, then a piecewise quadratic polynomial can be constructed on each subinterval $[x_{2k}, x_{2k+2}]$, for $k = 0, 1, \ldots, M - 1$. A shortcoming of the resulting quadratic spline is that the curvature at the even nodes x_{2k} changes abruptly and this can cause an undesired bend or distortion in the graph. The second derivative of a quadratic spline is discontinuous at the even nodes. If we use piecewise cubic polynomials, then both the first and second derivatives can be made continuous.

Piecewise Cubic Splines

The fitting of a polynomial curve to a set of data points has applications in the areas of drafting and computer graphics. The drafter wants to draw a "smooth curve" through data points that are not subject to error. It is common to use a french curve and subjectively draw a curve that looks smooth when viewed by the eye. Mathematically, it is possible to construct a cubic functions $S_k(x)$ on each interval $[x_k, x_{k+1}]$ so that the resulting piecewise curve $y = S(x)$ and its first and second derivatives are all continuous on the larger interval $[x_0, x_N]$. The continuity of $S'(x)$ means that the graph $Y = S(x)$ will not have sharp corners. The continuity of $S''(x)$ means that the "radius of curvature" is defined at each point.

Definition 5.2 Consider the $N + 1$ points $\{(x_k, y_k)\}$ where the abscissas are ordered $x_0 < x_1 < \ldots < x_N$. The function $S(x)$ is called a *cubic spline* if there exist N cubic polynomials $S_k(x)$ such that:

   I. $S(x) = S_k(x) = s_{k,0} + s_{k,1}(x - x_k) + s_{k,2}(x - x_k)^2 + s_{k,3}(x - x_k)^3$ for x in $[x_k, x_{k+1}]$ and $k = 0, 1, \ldots, N - 1$.
  II. $S(x_k) = y_k$ for $k = 0, 1, \ldots, N$.
      The spline passes through each data point.
 III. $S_k(x_{k+1}) = S_{k+1}(x_{k+1})$ for $k = 0, 1, \ldots, N - 2$.
      The spline forms a continuous function.
  IV. $S'_k(x_{k+1}) = S'_{k+1}(x_{k+1})$ for $k = 0, 1, \ldots, N - 2$.
      The spline forms a smooth function.
   V. $S''_k(x_{k+1}) = S''_{k+1}(x_{k+1})$ for $k = 0, 1, \ldots, N - 2$.
      The second derivative is continuous.

Existence of Cubic Splines

Let us try to ascertain if it is possible to construct a cubic spline that satisfies properties I-V. Each cubic polynomial $S_k(x)$ has four unknown constants, hence there are $4N$ coefficients to be determined. Loosely speaking, we have $4N$ degrees of freedom or conditions that must be specified. The data points supply $N + 1$ conditions, and properties III, IV, and V each supply $N - 1$ conditions. Hence, $N + 1 + 3(N - 1) = 4N - 2$ conditions are specified. This leaves us two additional degrees of freedom. We will call them *endpoint constraints*; they will involve either $S'(x)$ or $S''(x)$ at x_0 and x_N and will be discussed later. We now proceed with the construction.

Since $S(x)$ is piecewise cubic, its second derivative $S''(x)$ is piecewise linear on $[x_0, x_N]$. The linear Lagrange interpolation formula gives the following representation for $S''(x) = S_k''(x)$:

(4)
$$S_k''(x) = S''(x_k)\frac{x - x_{k+1}}{x_k - x_{k+1}} + S''(x_{k+1})\frac{x - x_k}{x_{k+1} - x_k}.$$

Use $m_k = S''(x_k)$, $m_{k+1} = S''(x_{k+1})$, and $h_k = x_{k+1} - x_k$ in (4) to get

(5)
$$S_k''(x) = \frac{m_k}{h_k}(x_{k+1} - x) + \frac{m_{k+1}}{h_k}(x - x_k)$$

for $x_k \leq x \leq x_{k+1}$ and $k = 0, 1, \ldots, N - 1$. Integrating (5) twice will introduce two constants of integration, and the result can be manipulated so that it has the form

(6)
$$S_k(x) = \frac{m_k}{6h_k}(x_{k+1} - x)^3 + \frac{m_{k+1}}{6h_k}(x - x_k)^3 + p_k(x_{k+1} - x) + q_k(x - x_k).$$

Substituting x_k and x_{k+1} into equation (6) and using the values $y_k = S_k(x_k)$ and $y_{k+1} = S_k(x_{k+1})$ yields the following equations that involve p_k and q_k, respectively:

(7)
$$y_k = \frac{m_k}{6}h_k^2 + p_k h_k \quad \text{and} \quad y_{k+1} = \frac{m_{k+1}}{6}h_k^2 + q_k h_k.$$

These two equations are easily solved for p_k and q_k, and when these values are substituted into equation (6), the result is the following expression for the cubic function $S_k(x)$:

(8)
$$S_k(x) = \frac{m_k}{6h_k}(x_{k+1} - x)^3 + \frac{m_{k+1}}{6h_k}(x - x_k)^3 + \left(\frac{y_k}{h_k} - \frac{m_k h_k}{6}\right)(x_{k+1} - x)$$
$$+ \left(\frac{y_{k+1}}{h_k} - \frac{m_{k+1} h_k}{6}\right)(x - x_k).$$

Notice that the representation (8) has been reduced to a form that involves only the unknown coefficients $\{m_k\}$. To find these values we must use the derivative of (8), which is

(9)
$$S'_k(x) = -\frac{m_k}{2h_k}(x_{k+1} - x)^2 + \frac{m_{k+1}}{2h_k}(x - x_k)^2 - \left(\frac{y_k}{h_k} - \frac{m_k h_k}{6}\right)$$

$$+ \frac{y_{k+1}}{h_k} - \frac{m_{k+1} h_k}{6}.$$

Evaluating (9) at x_k and simplifying the result yields

(10)
$$S'_k(x_k) = -\frac{m_k}{3}h_k - \frac{m_{k+1}}{6}h_k + d_k, \qquad \text{where } d_k = \frac{y_{k+1} - y_k}{h_k}.$$

Similarly, we can replace k by $k - 1$ in (9) to get the expression for $S'_{k-1}(x)$ and evaluate it at x_k to obtain

(11)
$$S'_{k-1}(x_k) = \frac{m_k}{3}h_{k-1} + \frac{m_{k-1}}{6}h_{k-1} + d_{k-1}.$$

Now use property IV and equations (10) and (11) to obtain an important relation involving m_{k-1}, m_k, and m_{k+1}:

(12) $h_{k-1}m_{k-1} + 2(h_{k-1} + h_k)m_k + h_k m_{k+1} = u_k$, \qquad where $u_k = 6(d_k - d_{k-1})$

$$\text{for } k = 1, 2, \ldots, N - 1.$$

Construction of Cubic Splines

Observe that the unknowns in (12) are the desired values $\{m_k\}$ and the other terms are constants obtained by performing simple arithmetic with the data points $\{(x_k, y_k)\}$. Therefore, in reality system (12) is an underdetermined system of $N - 1$ linear equations involving $N + 1$ unknowns. Hence two additions equations must be supplied. They are used to eliminate m_0 from equation 1 and m_N from equation $N - 1$, in system (12). The standard strategies for the endpoints constraints are summarized in Table 5.8.

Consider strategy (v) in Table 5.8. If m_0 is given, then $h_0 m_0$ can be computed, and the first equation (when $k = 1$) of (12) is

(13)
$$2[h_0 + h_1]m_1 + h_1 m_2 = u_1 - h_0 m_0.$$

Similarly, if m_N is given, then $h_{N-1}m_N$ can be computed, and the last equation (when $k = N - 1$) of (12) is

(14)
$$h_{N-2}m_{N-2} + 2(h_{N-2} + h_{N-1})m_{N-1} = u_{N-1} - h_{N-1}m_N.$$

Equations (13) and (14) with (12) used for $k = 2, 3, \ldots, N - 2$ form $N - 1$ linear equations involving the coefficients $m_1, m_2, \ldots, m_{N-1}$.

TABLE 5.8 Endpoint constraints for a cubic spline

Description of the Strategy	Equations Involving m_0 and m_N
(i) "Clamped cubic spline": specify $S'(x_0)$, $S'(x_N)$ (the "best choice" if the derivatives are known)	$m_0 = \dfrac{3}{h_0}[d_0 - S'(x_0)] - \dfrac{m_1}{2}$, $m_N = \dfrac{3}{h_{N-1}}[S'(x_N) - d_{N-1}] - \dfrac{m_{N-1}}{2}$
(ii) "Natural cubic spline" (a "relaxed curve")	$m_0 = 0,\ m_N = 0$
(iii) Extrapolate $S''(x)$ to the endpoints.	$m_0 = m_1 - \dfrac{h_0(m_2 - m_1)}{h_1}$, $m_N = m_{N-1} + \dfrac{h_{N-1}(m_{N-1} - m_{N-2})}{h_{n-2}}$
(iv) $S''(x)$ is constant near the endpoints.	$m_0 = m_1,\ m_N = m_{N-1}$
(v) Specify $S''(x)$ at each endpoint.	$m_0 = S''(x_0),\ m_N = S''(x_N)$

Regardless of the particular strategy chosen in Table 5.8, we can rewrite equations 1 and $N - 1$ in (12) and obtain a tridiagonal linear system of the form $HM = V$, which involves $m_1, m_2, \ldots, m_{N-1}$:

(15)
$$
\begin{pmatrix}
b_1 & c_1 & & & & \\
a_1 & b_2 & c_2 & & & \\
& & \ddots & & & \\
& & & a_{N-3} & b_{N-2} & c_{N-2} \\
& & & & a_{N-2} & b_{N-1}
\end{pmatrix}
\begin{pmatrix}
m_1 \\ m_2 \\ \vdots \\ m_{N-2} \\ m_{N-1}
\end{pmatrix}
=
\begin{pmatrix}
v_1 \\ v_2 \\ \vdots \\ v_{N-2} \\ v_{N-1}
\end{pmatrix}.
$$

The linear system in (15) is diagonally dominant and has a unique solution (see Chapter 8 for details). After the coefficients $\{m_k\}$ are determined, the spline coefficients $\{s_{k,j}\}$ for $S_k(x)$ are computed using the formulas

(16)
$$
s_{k,0} = y_k, \qquad s_{k,1} = d_k - \frac{h_k(2m_k + m_{k+1})}{6},
$$

$$
s_{k,2} = \frac{m_k}{2}, \qquad s_{k,3} = \frac{m_{k+1} - m_k}{6h_k}.
$$

Each cubic polynomial $S_k(x)$ can be written in nested multiplication form for efficient computation:

(17)
$$S_k(x) = [(s_{k,3}w + s_{k,2})w + s_{k,1}]w + y_k, \qquad \text{where } w = x - x_k$$

and $S_k(x)$ is used on the interval $x_k \leq x \leq x_{k+1}$.

Theorem 5.1 [*Description of Cubic Splines*] Let $(x_0, y_0), \ldots, (x_N, y_N)$ be $N + 1$ points that form a cubic spline. Equations (12) together with an endpoint strategy from Table 5.8 can be used to obtain a linear system of the form (15). The tridiagonal system is solved for the coefficients $m_1, \ldots, m_{N-1}$ and the formulas used in Table 5.8 determine m_0 and m_N. Then equations (16) are used to find the spline coefficients. The following linear systems are used for the cases in Table 5.8.

Case (i) Clamped Cubic Spline. Supply $S'(x_0)$ and $S'(x_N)$.
$$\left(\tfrac{3}{2}h_0 + 2h_1\right)m_1 + h_1 m_2 = u_1 - 3[d_0 - S'(x_0)],$$
$$h_{k-1}m_{k-1} + 2(h_{k-1} + h_k)m_k + h_k m_{k+1} = u_k \qquad \text{for } k = 2, 3, \ldots, N - 2,$$
$$h_{N-2}m_{N-2} + \left(2h_{N-2} + \tfrac{3}{2}h_{N-1}\right)m_{N-1} = u_{N-1} - 3[S'(x_N) - d_{N-1}].$$

Case (ii) Natural Spline. Set $S''(x_0) = 0$ and $S''(x_N) = 0$.
$$2(h_0 + h_1)m_1 + h_1 m_2 = u_1,$$
$$h_{k-1}m_{k-1} + 2(h_{k-1} + h_k)m_k + h_k m_{k+1} = u_k \qquad \text{for } k = 2, 3, \ldots, N - 2,$$
$$h_{N-2}m_{N-2} + 2(h_{N-2} + h_{N-1})m_{N-1} = u_{N-1}.$$

Case (iii) Extrapolate $S''(x)$ to the endpoints.
$$\left(3h_0 + 2h_1 + \frac{h_0^2}{h_1}\right)m_1 + \left(h_1 - \frac{h_0^2}{h_1}\right)m_2 = u_1,$$
$$h_{k-1}m_{k-1} + 2(h_{k-1} + h_k)m_k + h_k m_{k+1} = u_k$$
$$\text{for } k = 2, 3, \ldots, N - 2,$$
$$\left(h_{N-2} - \frac{h_{N-1}^2}{h_{N-2}}\right)m_{N-2} + \left(2h_{N-2} + 3h_{N-1} + \frac{h_{N-1}^2}{h_{N-2}}\right)m_{N-1} = u_{N-1}.$$

Case (iv) Assume that $S''(x)$ is constant near the endpoints.
$$(3h_0 + 2h_1)m_1 + h_1 m_2 = u_1,$$
$$h_{k-1}m_{k-1} + 2(h_{k-1} + h_k)m_k + h_k m_{k+1} = u_k \qquad \text{for } k = 2, 3, \ldots, N - 2,$$
$$h_{N-2}m_{N-2} + (2h_{N-2} + 3h_{N-1})m_{N-1} = u_{N-1}.$$

Case (v) Specify $S''(x_0)$ and $S''(x_N)$.
$$2(h_0 + h_1)m_1 + h_1 m_2 = u_1 - h_0 S''(x_0),$$
$$h_{k-1}m_{k-1} + 2(h_{k-1} + h_k)m_k + h_k m_{k+1} = u_k \qquad \text{for } k = 2, 3, \ldots, N - 2,$$
$$h_{N-2}m_{N-2} + 2(h_{N-2} + h_{N-1})m_{N-1} = u_{N-1} - h_{N-1}S''(x_N).$$

Strategy (i) involves the slope. This spline can be visualized as the curve obtained by a "flexible elastic bar" that is forced to pass through the points, and the bar is clamped at each end (i.e., the slope at each end is given). Strategy (ii) is the curve obtained by forcing a flexible elastic bar through the points but letting the slope at the ends be free to equilibrate to a position that minimizes the oscillatory behavior of the spline.

Strategy (iii) is equivalent to assuming that the end cubic is an extension of the adjacent cubic, that is, the spline forms a single cubic curve over the interval $[x_0, x_2]$ and another single cubic over the interval $[x_{N-2}, x_N]$. Strategy (iv) is equivalent to assuming that the end cubics degenerate to quadratics. Strategy (v) imposes the desired second derivative (hence curvature) at each end.

Example 5.6 Find the five different cubic splines for the four points $(0, 0)$, $(1, 0.5)$, $(2, 2)$, and $(3, 1.5)$. For the clamped spline use $S'(0) = 0.2$ and $S'(3) = -1$ and for case (v) use the values $S''(0) = -0.3$ and $S''(3) = 3.3$. See Figure 5.7 on page 246 for the graphs.

Solution

Case (i) The Clamped Spline. Using $h_0 = 1$, $h_1 = 1$, $h_2 = 1$, $d_0 = 0.5$, $d_1 = 1.5$, $d_2 = -0.5$, $u_1 = 6(d_1 - d_0)$, and $u_2 = 6(d_2 - d_1)$, we obtain the two equations

$$(\tfrac{3}{2} + 2)m_1 + m_2 = 6(1.5 - 0.5) - 3(0.5 - 0.2),$$

$$m_1 + (2 + \tfrac{3}{2})m_2 = 6(-0.5 - 1.5) - 3(-1.0 + 0.5).$$

When these equations are simplified the result is

$$\begin{pmatrix} 3.5 & 1.0 \\ 1.0 & 3.5 \end{pmatrix} \begin{pmatrix} m_1 \\ m_2 \end{pmatrix} = \begin{pmatrix} 5.1 \\ -10.5 \end{pmatrix}.$$

It is easy to compute the solution $m_1 = 2.52$ and $m_2 = -3.72$. Now the equations (i) in Table 5.8 are used to compute m_0 and m_3.

$$m_0 = 3(0.5 - 0.2) - \frac{2.52}{2} = -0.36,$$

$$m_3 = 3(-1.0 + 0.5) + \frac{3.72}{2} = 0.36.$$

Next, the values $m_0 = -0.36$, $m_1 = 2.52$, $m_2 = -3.72$, and $m_3 = 0.36$ are used in equations (16) to find the spline coefficients, and we get

$$S_0(x) = 0.48x^3 - 0.18x^2 + 0.2x \qquad \text{for } 0 \le x \le 1,$$

(18) $\quad S_1(x) = -1.04(x - 1)^3 + 1.26(x - 1)^2 + 1.28(x - 1) + 0.5 \qquad \text{for } 1 \le x \le 2,$

$$S_2(x) = 0.68(x - 2)^3 - 1.86(x - 2)^2 + 0.68(x - 2) + 2 \qquad \text{for } 2 \le x \le 3.$$

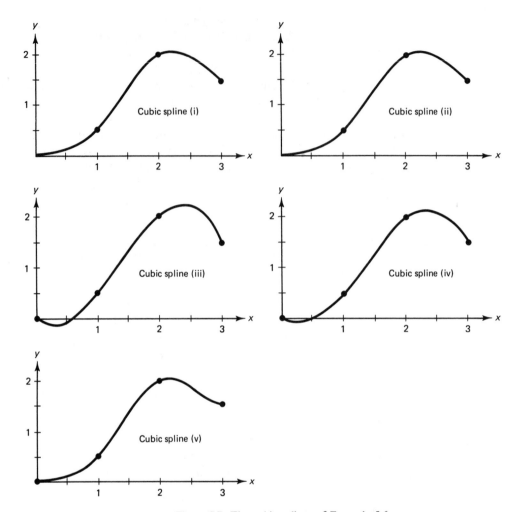

Figure 5.7. The cubic splines of Example 5.6.
  (i) "Clamped" $S'(0) = 0.2$ and $S'(3) = -1.0$.
 (ii) "Natural" $S''(0) = 0$ and $S''(3) = 0$.
(iii) Extrapolate $S''(x)$ near the endpoints.
(iv) Parabolic near the endpoints.
 (v) Specify $S''(0) = -0.3$ and $S''(3) = 3.3$.

Case (ii) Natural Spline. Using the same values $\{h_k\}$, $\{d_k\}$, and $\{u_k\}$ in case (i) we obtain the equations

$$2(1 + 1)m_1 + m_2 = 6(1.5 - 0.5)$$

$$m_1 + 2(1 + 1)m_2 = 6(-0.5 - 1.5).$$

When these equations are simplified the result is

$$\begin{pmatrix} 4.0 & 1.0 \\ 1.0 & 4.0 \end{pmatrix} \begin{pmatrix} m_1 \\ m_2 \end{pmatrix} = \begin{pmatrix} 6.0 \\ -12.0 \end{pmatrix}.$$

It is easy to compute the solution $m_1 = 2.4$ and $m_2 = -3.6$. Since $m_0 = S''(x_0) = 0$ and $m_3 = S''(x_3) = 0$, when equations (16) are used to find the spline coefficients the result is

$$\begin{array}{lll} & S_0(x) = 0.4x^3 + 0.1x & \text{for } 0 \le x \le 1, \\ (19) & S_1(x) = -(x-1)^3 + 1.2(x-1)^2 + 1.3(x-1) + 0.5 & \text{for } 1 \le x \le 2, \\ & S_2(x) = 0.6(x-2)^3 - 1.8(x-2)^2 + 0.7(x-2) + 2.0 & \text{for } 2 \le x \le 3. \end{array}$$

Case (iii) Extrapolation for m_0 and m_3. The linear system is

$$\begin{pmatrix} 6.0 & 0.0 \\ 0.0 & 6.0 \end{pmatrix} \begin{pmatrix} m_1 \\ m_2 \end{pmatrix} = \begin{pmatrix} 6.0 \\ -12.0 \end{pmatrix}.$$

The solution $m_1 = 1.0$ and $m_2 = -2.0$. Now equations (iii) in Table 5.8 are used to compute m_0 and m_3:

$$m_0 = 1.0 - (-2.0 - 1.0) = 4.0,$$
$$m_3 = -2.0 + (-2.0 - 1.0) = -5.0.$$

The spline coefficients are computed and we have

$$\begin{array}{lll} & S_0(x) = -0.5x^3 + 2.0x^2 - x & \text{for } 0 \le x \le 1, \\ (20) & S_1(x) = -0.5(x-1)^3 + 0.5(x-1)^2 + 1.5(x-1) + 0.5 & \text{for } 1 \le x \le 2, \\ & S_2(x) = -0.5(x-2)^3 - (x-2)^2 + (x-2) + 2.0 & \text{for } 2 \le x \le 3. \end{array}$$

Case (iv) $S''(x)$ is constant near the ends. The linear system is

$$\begin{pmatrix} 5.0 & 1.0 \\ 1.0 & 5.0 \end{pmatrix} \begin{pmatrix} m_1 \\ m_2 \end{pmatrix} = \begin{pmatrix} 6.0 \\ -12.0 \end{pmatrix}.$$

The solution is $m_1 = 1.75$ and $m_2 = -2.75$. Since $S''(x)$ is constant near the ends, we choose $m_0 = 1.75$ and $m_3 = -2.75$ and the spline is

$$\begin{array}{lll} & S_0(x) = 0.875x^2 - 0.375x & \text{for } 0 \le x \le 1, \\ (21) & S_1(x) = -0.75(x-1)^3 + 0.875(x-1)^2 + 1.375(x-1) + 0.5 & \text{for } 1 \le x \le 2, \\ & S_2(x) = -1.375(x-2)^2 + 0.875(x-2) + 2.0 & \text{for } 2 \le x \le 3. \end{array}$$

Case (v) $m_0 = S''(x_0)$ and $m_3 = S''(x_3)$. The linear system is

$$\begin{pmatrix} 4.0 & 1.0 \\ 1.0 & 1.0 \end{pmatrix} \begin{pmatrix} m_1 \\ m_2 \end{pmatrix} = \begin{pmatrix} 6.3 \\ -15.3 \end{pmatrix}.$$

The solution is $m_1 = 2.7$ and $m_2 = -4.5$. Since $m_0 = S''(x_0) = -0.3$ and $m_3 = S''(x_3) = 3.3$, when the spline coefficients are computed we have

$$
\begin{aligned}
S_0(x) &= 0.5x^3 - 0.15x^2 + 0.15x & \text{for } 0 \leq x \leq 1, \\
(22) \quad S_1(x) &= -1.2(x-1)^3 + 1.35(x-1)^2 + 1.35(x-1) + 0.5 & \text{for } 1 \leq x \leq 2, \\
S_2(x) &= 1.3(x-2)^3 - 2.25(x-2)^2 + 0.45(x-2) + 2 & \text{for } 2 \leq x \leq 3.
\end{aligned}
$$

Algorithm 5.4 [*Cubic Spline Construction*] Suppose that $N+1$ points $(x_0, y_0), \ldots, (x_N, y_N)$ are given and an endpoint constraint is chosen from Table 5.8. The coefficients $\{s_{k,j}\}$ are found so that formula (17) can be used to evaluate the piecewise cubic spline.

```
        H(0) := X(1) − X(0)                              {Difference in abscissa}
        D(0) := [Y(1) − Y(0)]/H(0)                       {Difference quotient}


FOR     K = 1  TO  N−1  DO
        H(K) := X(K+1) − X(K)                            {Differences in abscissa}
        D(K) := [Y(K+1)−Y(K)]/H(K)                       {Difference quotients}
        A(K) := H(K)                                     {Subdiagonal elements}
        B(K) := 2*[H(K−1)+H(K)]                          {Diagonal elements}
        C(K) := H(K)                                     {Superdiagonal elements}


        FOR    K = 1  TO  N−1  DO                         {Determine the
               V(K) := 6*[D(K)−D(K−1)]                     column vector}


CASES   {Modify the matrix and/or column vector}

   (i)  Set  B(1) := B(1) − H(0)/2
             V(1) := V(1) − 3*[D(0) − S'(x₀)]            {Input S'(x₀)}
             B(N−1) := B(N−1) − H(N−1)/2
             V(N−1) := V(N−1) − 3*[S'(xₙ) − D(N−1)]      {Input S'(xₙ)}

  (ii)  Set  M(0) := 0 and M(N) := 0

 (iii)  Set  B(1) := B(1) + H(0) + H(0)*H(0)/H(1)
             C(1) := C(1) − H(0)*H(0)/H(1)
             B(N−1) := B(N−1) + H(N−1) + H(N−1)*H(N−1)/H(N−2)
             A(N−2) := A(N−2) − H(N−1)*H(N−1)/H(N−2)

  (iv)  Set  B(1) := B(1) + H(0) and B(N−1) := B(N−1) + H(N−1)

   (v)  Set  V(1) := V(1) − H(0)*S''(x₀)                 {Input S''(x₀)}
             V(N−1) := V(N−1) − H(N−1)*S''(xₙ)           {Input S''(xₙ)}
END
```

```
FOR    K = 2  TO  N−1   DO                    {Gaussian elimination is used
  │    T := A(K−1)/B(K−1)                       to produce an upper-triangular
  │    B(K) := B(K)−T*C(K−1)                     system with "two diagonals"}
  └    V(K) := V(K) − T*V(K−1)


       M(N−1) := V(N−1)/B(N−1)                 {Back substitution
FOR    K = N−2  DOWNTO  1  DO                    is used to find m_k}
       M(K) := [V(K) − C(K)*M(K+1)]/B(K)
```

CASES {Determine the values M(0) and M(N)}

   (i) Set $M(0) := 3*[D(0) − S'(x_0)]/H(0) − M(1)/2$
               $M(N) := 3*[S'(x_N) − D(N−1)]/H(N−1) − M(N−1)/2$

  (ii) Set $M(0) := 0$ and $M(N) := 0$

 (iii) Set $M(0) := M(1) − H(0)*[M(2)−M(1)]/H(1)$
               $M(N) := M(N−1) + H(N−1)*[M(N−1)−M(N−2)]/H(N−2)$

 (iv) Set $M(0) := M(1)$ and $M(N) := M(N−1)$

  (v) Set $M(0) := S''(x_0)$ and $M(N) := S''(x_N)$

END

```
FOR    K = 0  TO  N−1   DO
  │    S(K, 0) := Y(K)                         {Compute and store
  │    S(K, 1) := D(K) − H(K)*[2*M(K)+M(K+1)]/6    the coefficients
  │    S(K, 2) := M(K)/2                           for each cubic
  └    S(K, 3) := [M(K+1)−M(K)]/[6*H(K)]         polynomial S_k(x)}
```

{Procedure for evaluating the cubic spline above on $[x_0, x_N]$}

```
INPUT  X                                       {Input the independent variable}
FOR    J = 1  TO  N  DO                         {Find the interval
  │    IF  X ≤ X(J)  THEN                          so that x lies
  └    └ Set  K := J−1  and  J := N              in [x_k, x_{k+1}]}
       IF  X = X(0)  THEN  Set K := 0
       W := X − X(K)                            {Evaluate the spline}
       Z := [[S(K,3)*W + S(K,2)]*W + S(K,1)]*W + S(K,0)
PRINT  "The value of the spline S(x) is"  Z    {Output}
```

Exercises for Interpolation by Spline Functions

1. Consider the points $(0, 4)$, $(1, 1)$, and $(2, 2)$ and the natural spline
$$S_0(x) = 4 + b_0 x + x^3 \quad \text{and} \quad S_1(x) = 1 + b_1(x − 1) + 3(x − 1)^2 − (x − 1)^3.$$
(a) Verify that $S_0''(0) = 0$ and $S_1''(2) = 0$.
(b) Use $S_0(1) = 1$ and solve for b_0 in the first equation.

(c) Use $S_1(2) = 2$ and solve for b_1 in the second equation.

(d) Use the values obtained for b_0 and b_1 and verify that

$$S_0'(1) = S_1'(1) \quad \text{and} \quad S_0''(1) = S_1''(1).$$

2. Consider the points $(0, 3)$, $(1, 4)$, and $(2, 9)$ and the natural spline

$$S_0(x) = 3 + b_0 x + x^3 \quad \text{and} \quad S_1(x) = 4 + b_1(x - 1) + 3(x - 1)^2 - (x - 1)^3.$$

(a) Verify that $S_0''(0) = 0$ and $S_1''(2) = 0$.

(b) Use $S_0(1) = 4$ and solve for b_0 in the first equation.

(c) Use $S_1(2) = 9$ and solve for b_1 in the second equation.

(d) Use the values obtained for b_0 and b_1 and verify that

$$S_0'(1) = S_1'(1) \quad \text{and} \quad S_0''(1) = S_1''(1).$$

In Exercises 3–5, consider five different splines. Show that the linear system in (i)–(v) arises when cases (i)–(v) of Theorem 5.1 are applied, respectively. After you have found $\{m_k\}$, use formula (16) to find the coefficients $\{s_{k,j}\}$ of $\{S_k(x)\}$. For part (i) use $S'(0)$ and $S'(3)$ and for (v) use $S''(0)$ and $S''(3)$.

3. Use the points $(0, 7)$, $(1, 2)$, $(2, 0)$, and $(3, -5)$, and use

$$S'(0) = -5, \quad S'(3) = -5 \quad \text{and} \quad S''(0) = 3, \quad S''(3) = -6.$$

(i) $3.5m_1 + m_2 = 18$, $m_1 + 3.5m_2 = -18$

(ii) $4m_1 + m_2 = 18$, $m_1 + 4m_2 = -18$

(iii) $6m_1 = 18$, $ 6m_2 = -18$

(iv) $5m_1 + m_2 = 18$, $m_1 + 5m_2 = -18$

(v) $4m_1 + m_2 = 15$, $m_1 + 4m_2 = -12$

4. Use the points $(0, 1)$, $(1, 4)$, $(2, 0)$, and $(3, -2)$, and use

$$S'(0) = 2, \quad S'(3) = 2 \quad \text{and} \quad S''(0) = -1.5, \quad S''(3) = 3.$$

(i) $3.5m_1 + m_2 = -45$, $m_1 + 3.5m_2 = 0$

(ii) $4m_1 + m_2 = -42$, $m_1 + 4m_2 = 12$

(iii) $6m_1 = -42$, $ 6m_2 = 12$

(iv) $5m_1 + m_2 = -42$, $m_1 + 5m_2 = 12$

(v) $4m_1 + m_2 = -40.5$, $m_1 + 4m_2 = 9$

5. Use the function $f(x) = x + 2/x$ for $x_0 = 1$, $x_1 = 2$, $x_2 = 3$, and $x_3 = 4$, and

$$S'(0) = -1, \quad S'(3) = 0.9 \quad \text{and} \quad S''(0) = 1.6, \quad S''(3) = 0.1.$$

(i) $3.5m_1 + m_2 = 1$, $m_1 + 3.5m_2 = 0.8$

(ii) $4m_1 + m_2 = 4$, $m_1 + 4m_2 = 1$

(iii) $6m_1 = 4$, $ 6m_2 = 1$

(iv) $5m_1 + m_2 = 4$, $m_1 + 5m_2 = 1$

(v) $4m_1 + m_2 = 2.4$, $m_1 + 4m_2 = 0.9$

6. Use the substitutions

$$x_{k+1} - x = h_k + (x_k - x)$$

and

$$(x_{k+1} - x)^3 = h_k^3 + 3h_k^2(x_k - x) + 3h_k(x_k - x)^2 + (x_k - x)^3$$

to show that when equation (8) is expanded into powers of $(x - x_k)$, the coefficients are those given in equations (16).

7. Consider each cubic function $S_k(x)$ over $[x_k, x_{k+1}]$.

   (a) Give a formula for $\int_{x_k}^{x_{k+1}} S_k(x) \, dx$.

   Then evaluate $\int_0^3 S(x) \, dx$ for case (ii), the natural spline, in part (ii) of

   (b) Exercise 3 (c) Exercise 4 (d) Exercise 5

   In Exercises 8–10, consider five different splines. Show that the linear system in (i)–(v) arises when cases (i)–(v) of Theorem 5.1 are applied, respectively. Find $\{m_k\}$ and $\{S_k(x)\}$. For (i) use $S'(x_0)$, $S'(x_4)$ and for (v) use $S''(x_0)$, $S''(x_4)$.

8. Use the points $(0, 5)$, $(1, 2)$, $(2, 1)$, $(3, 3)$, and $(4, 1)$, and use

   $$S'(0) = -2, \quad S'(4) = -1 \quad \text{and} \quad S''(0) = 0.5, \quad S''(4) = -1.9.$$

   (i) $3.5m_1 + m_2 = 15,$ $m_1 + 4m_2 + m_3 = 18,$ $m_2 + 3.5m_3 = -27$
   (ii) $\quad 4m_1 + m_2 = 12,$ $m_1 + 4m_2 + m_3 = 18,$ $m_2 + \quad 4m_3 = -24$
   (iii) $\quad 6m_1 \quad\quad = 12,$ $m_1 + 4m_2 + m_3 = 18,$ $\quad\quad 6m_3 = -24$
   (iv) $\quad 5m_1 + m_2 = 12,$ $m_1 + 4m_2 + m_3 = 18,$ $m_2 + \quad 5m_3 = -24$
   (v) $\quad 4m_1 + m_2 = 11.5,$ $m_1 + 4m_2 + m_3 = 18,$ $m_2 + \quad 4m_3 = -22.1$

9. Use the points $(0, 2)$, $(1, 3)$, $(2, 2)$, $(3, 2)$, and $(4, 1)$, and use

   $$S'(0) = 1.6, \quad S'(4) = -1 \quad \text{and} \quad S''(0) = -1.4, \quad S''(4) = -1.4.$$

   (i) $3.5m_1 + m_2 = -10.2,$ $m_1 + 4m_2 + m_3 = 6,$ $m_2 + 3.5m_3 = -6$
   (ii) $\quad 4m_1 + m_2 = -12,$ $m_1 + 4m_2 + m_3 = 6,$ $m_2 + \quad 4m_3 = -6$
   (iii) $\quad 6m_1 \quad\quad = -12,$ $m_1 + 4m_2 + m_3 = 6,$ $\quad\quad 6m_3 = -6$
   (iv) $\quad 5m_1 + m_2 = -12,$ $m_1 + 4m_2 + m_3 = 6,$ $m_2 + \quad 5m_3 = -6$
   (v) $\quad 4m_1 + m_2 = -10.6,$ $m_1 + 4m_2 + m_3 = 6,$ $m_2 + \quad 4m_3 = -4.6$

10. Use the function $f(x) = \sin(\pi x)$ with $x_0 = 0$, $x_1 = \frac{1}{6}$, $x_2 = \frac{1}{2}$, $x_3 = \frac{5}{6}$, and $x_4 = 1$, and use

    $$S'(0) = 3, \quad S'(1) = -3 \quad \text{and} \quad S''(0) = -1, \quad S''(1) = -1.$$

    (i) $\frac{11}{12}m_1 + \frac{1}{3}m_2 = -9,$ $\frac{1}{3}m_1 + \frac{4}{3}m_2 + \frac{1}{3}m_3 = -18,$ $\frac{1}{3}m_2 + \frac{11}{12}m_3 = -9$
    (ii) $\quad m_1 + \frac{1}{3}m_2 = -9,$ $\frac{1}{3}m_1 + \frac{4}{3}m_2 + \frac{1}{3}m_3 = -18,$ $\frac{1}{3}m_2 + \quad m_3 = -9$
    (iii) $\frac{5}{4}m_1 + \frac{1}{4}m_2 = -9,$ $\frac{1}{3}m_1 + \frac{4}{3}m_2 + \frac{1}{3}m_3 = -18,$ $\frac{1}{4}m_2 + \frac{5}{4}m_3 = -9$
    (iv) $\frac{7}{6}m_1 + \frac{1}{3}m_2 = -9,$ $\frac{1}{3}m_1 + \frac{4}{3}m_2 + \frac{1}{3}m_3 = -18,$ $\frac{1}{3}m_2 + \frac{7}{6}m_3 = -9$
    (v) $\quad m_1 + \frac{1}{3}m_2 = \frac{-265}{30},$ $\frac{1}{3}m_1 + \frac{4}{3}m_2 + \frac{1}{3}m_3 = -18,$ $\frac{1}{3}m_2 + \quad m_3 = \frac{-265}{30}$

11. Show how strategy (i) in Table 5.8 and equations (12) are combined to obtain the equations in case (i) of Theorem 5.1.

12. Show how strategy (iii) in Table 5.8 and equations (12) are combined to obtain the equations in case (iii) of Theorem 5.1.

13. The distance d_k that a car has traveled at time t_k is given in the table below. Use the values $S'(0) = 0$ and $S'(8) = 98$, and find the clamped spline for the points.

Time, t_k	0	2	4	6	8
Distance, d_k	0	40	160	300	480

In Exercises 14–16, use a computer to find the five different splines in cases (i)-(v) for the given points.

14. Use the points $(0, 1)$, $(1, 0)$, $(2, 0)$, $(3, 1)$, $(4, 2)$, $(5, 2)$, and $(6, 1)$ and

$$S'(0) = -0.6, \quad S'(6) = -1.8 \quad \text{and} \quad S''(0) = 1, \quad S''(6) = -1.$$

15. Use the points $(0, 0)$, $(1, 4)$, $(2, 8)$, $(3, 9)$, $(4, 9)$, $(5, 8)$, and $(6, 6)$ and

$$S'(0) = 1, \quad S'(6) = -2 \quad \text{and} \quad S''(0) = 1, \quad S''(6) = -1.$$

16. Use the points $(0, 0)$, $(1, 2)$, $(2, 3)$, $(3, 2)$, $(4, 2)$, $(5, 1)$, and $(6, 0)$ and

$$S'(0) = 1.5, \quad S'(6) = -0.3 \quad \text{and} \quad S''(0) = -1, \quad S''(6) = 1.$$

17. Write a report on quadratic splines.

18. Write a report on basic splines (sometimes called B-splines).

19. Linear search is not efficient. Write a binary search procedure that will find the integer k so that x lies in $[x_k, x_{k+1}]$. This is similar to the bisection method, except that we use the subscript $k = INT[(i + j)/2]$ to decide whether x lies in $[x_i, x_k]$ or in $[x_k, x_j]$.

5.4 Fourier Series and Trigonometric Polynomials

Scientists and engineers often study physical phenomena, such as light and sound, which have a periodic character. They are described by functions $f(x)$ which are periodic, that is,

(1)
$$g(x + P) = g(x) \qquad \text{for all } x.$$

The number P is called a *period* of the function.

It will suffice to consider functions that have period 2π. If $g(x)$ has period P, then $f(x) = g(Px/2\pi)$ will be periodic with period 2π. This is verified by the observation

(2)
$$f(x + 2\pi) = g\left(\frac{Px}{2\pi} + P\right) = g\left(\frac{Px}{2\pi}\right) = f(x).$$

Henceforth in this section we shall assume that $f(x)$ is a function that is periodic with period 2π, that is,

(3)
$$f(x + 2\pi) = f(x) \qquad \text{for all } x.$$

The graph $y = f(x)$ is obtained by repeating the portion of the graph in any interval of length 2π, as shown in Figure 5.8.

Examples of functions with period 2π are $\sin(jx)$ and $\cos(jx)$, where j is an integer. This raises the question: Can a periodic function can be respresented by the sum of terms involving $a_j \cos(jx)$ and $b_j \sin(jx)$? We will soon see that the answer is yes.

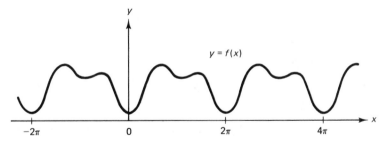

Figure 5.8. A function $f(x)$ with period 2π.

Definition 5.3   [Piecewise continuous] The function $f(x)$ is said to be *piecewise continuous* on $[a, b]$ if there exists values $t_0, t_1, \ldots, t_K$ with $a = t_0 < t_1 < \ldots < t_{K-1} < t_K = b$ such that $f(x)$ is continuous on each open interval $t_{i-1} < x < t_i$ for $i = 1, 2, \ldots, K$, and $f(x)$ has left- and right-hand limits at each of the points t_i. The situation is illustrated in Figure 5.9.

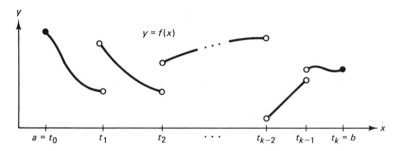

Figure 5.9. A piecewise continuous function over $[a, b]$.

Theorem 5.2   [Description of Fourier series] If $f(x)$ is periodic with period 2π and if $f(x)$ and $f'(x)$ are piecewise continuous, then the Fourier series for $f(x)$ is

(4)
$$\frac{a_0}{2} + \sum_{j=1}^{\infty} [a_j \cos(jx) + b_j \sin(jx)],$$

where the coefficients $\{a_j\}$ and $\{b_j\}$ are given by the *Euler formulas*,

(5)
$$a_j = \frac{1}{\pi} \int_{-\pi}^{\pi} f(x) \cos(jx)\, dx \qquad \text{for } j = 0, 1, 2, \ldots$$

and

(6)
$$b_j = \frac{1}{\pi} \int_{-\pi}^{\pi} f(x) \sin(jx)\, dx \qquad \text{for } j = 1, 2, 3, \ldots.$$

The factor $\frac{1}{2}$ in the constant term $a_0/2$ in the Fourier series (4) has been introduced for convenience so that a_0 could be obtained from the general formula (5) by setting $j = 0$. The Fourier series (4) converges for all values of x. The series converges to the sum $f(x)$ at points where $f(x)$ is continuous, and its sum is $[f(x^-) + f(x^+)]/2$ at points of discontinuity of $f(x)$. With this understanding, we obtain the Fourier series representation

(7)
$$f(x) = \frac{a_0}{2} + \sum_{j=1}^{\infty} [a_j \cos (jx) + b_j \sin (jx)].$$

The proof of formula (4) is given at the end of the section.

Example 5.7 Show that the function $f(x) = x/2$ for $-\pi < x < \pi$, extended periodically by the equation $f(x + 2\pi) = f(x)$, has the Fourier series representation

$$f(x) = \sum_{j=1}^{\infty} \frac{(-1)^{j+1}}{j} \sin (jx) = \sin (x) - \frac{\sin (2x)}{2} + \frac{\sin (3x)}{3} - \cdots .$$

Solution. Using Euler's formulas and integration by parts, we get

$$a_j = \frac{1}{\pi} \int_{-\pi}^{\pi} \frac{x}{2} \cos (jx) \, dx = \frac{x \sin (jx)}{2\pi j} + \frac{\cos (jx)}{2\pi j^2} \Big|_{-\pi}^{\pi} = 0 \qquad \text{for } j = 1, 2, 3, \ldots$$

and

$$b_j = \frac{1}{\pi} \int_{-\pi}^{\pi} \frac{x}{2} \sin (jx) \, dx = \frac{-x \cos (jx)}{2\pi j} + \frac{\sin (jx)}{2\pi j^2} \Big|_{-\pi}^{\pi} = \frac{(-1)^{j+1}}{j} \qquad \text{for } j = 1, 2, 3, \ldots .$$

The coefficient a_0 is obtained by a separate calculation:

$$a_0 = \frac{1}{\pi} \int_{-\pi}^{\pi} \frac{x}{2} \, dx = \frac{x^2}{4\pi} \Big|_{-\pi}^{\pi} = 0.$$

The calculations above show that all the coefficients of the cosine functions are zero. The graphs of $f(x)$ and the partial sums

$$S_2(x) = \sin (x) - \frac{\sin (2x)}{2},$$

$$S_3(x) = \sin (x) - \frac{\sin (2x)}{2} + \frac{\sin (3x)}{3}$$

and

$$S_4(x) = \sin (x) - \frac{\sin (2x)}{2} + \frac{\sin (3x)}{3} - \frac{\sin (4x)}{4}$$

are shown in Figure 5.10.

We now state some general properties of Fourier series. The proofs are left as exercises.

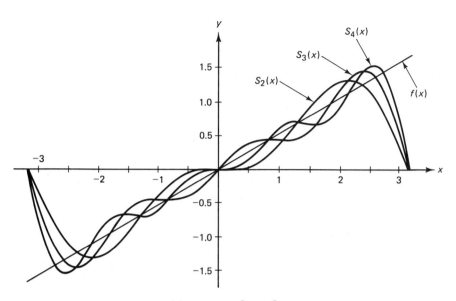

Figure 5.10. The function $f(x) = x/2$ over $[-\pi, \pi]$ and its trigonometric approximations $S_2(x)$, $S_3(x)$, and $S_4(x)$.

Theorem 5.3 [Addition of Fourier series] If $f(x)$ and $g(x)$ have Fourier series representations, then the coefficients of $h(x) = f(x) + g(x)$ can be obtained by adding the coefficients for $f(x)$ and $g(x)$.

Theorem 5.4 [Cosine series] Let $f(x)$ be an even function, that is, $f(-x) = f(x)$ holds for all x. If $f(x)$ has period 2π and if $f(x)$ and $f'(x)$ are piecewise continuous, then the Fourier series for $f(x)$ involves only cosine terms:

(8)
$$f(x) = \frac{a_0}{2} + \sum_{j=1}^{\infty} a_j \cos(jx),$$

where

(9)
$$a_j = \frac{2}{\pi} \int_0^{\pi} f(x) \cos(jx) \, dx \qquad \text{for } j = 0, 1, 2, \ldots.$$

Theorem 5.5 [Sine series] Let $f(x)$ be an odd function, that is, $f(-x) = -f(x)$ holds for all x. If $f(x)$ has period 2π and if $f(x)$ and $f'(x)$ are piecewise continuous, then the Fourier series for $f(x)$ involves only sine terms:

(10)
$$f(x) = \sum_{j=1}^{\infty} b_j \sin(jx),$$

where

(11)
$$b_j = \frac{2}{\pi} \int_0^\pi f(x) \sin (jx) \, dx \qquad \text{for } j = 1, 2, 3, \ldots.$$

Example 5.8 Show that the function $f(x) = |x|$ for $-\pi < x < \pi$, extended periodically by the equation $f(x + 2\pi) = f(x)$, has the Fourier cosine representation

(12)
$$f(x) = \frac{\pi}{2} - \frac{4}{\pi} \sum_{j=1}^\infty \frac{\cos ((2j - 1)x)}{(2j - 1)^2}$$

$$= \frac{\pi}{2} - \frac{4}{\pi} \left[\cos (x) + \frac{\cos (3x)}{3^2} + \frac{\cos (5x)}{5^2} + \cdots \right].$$

Solution. The function $f(x)$ is an even function, so we can use Theorem 5.4 and need only to compute the coefficients $\{a_j\}$:

$$a_j = \frac{2}{\pi} \int_0^\pi x \cos (jx) \, dx = \frac{2x \sin (jx)}{\pi j} + \frac{2 \cos (jx)}{\pi j^2} \bigg|_0^\pi$$

$$= \frac{2 \cos (j\pi) - 2}{\pi j^2} = \frac{2[(-1)^j - 1]}{\pi j^2} \qquad \text{for } j = 1, 2, 3, \ldots.$$

Since $[(-1)^j - 1] = 0$ when j is even, the cosine series will involve only the odd terms. The odd coefficients have the pattern

$$a_1 = \frac{-4}{\pi}, \quad a_3 = \frac{-4}{\pi 3^2}, \quad a_5 = \frac{-4}{\pi 5^2}, \quad \ldots.$$

The coefficient a_0 is obtained by the separate calculation

$$a_0 = \frac{2}{\pi} \int_0^\pi x \, dx = \frac{x^2}{\pi} \bigg|_0^\pi = \pi.$$

Therefore, we have found the desired coefficients in (12).

Proof of Euler's Formulas for Theorem 5.2. The following heuristic argument assumes the existence and convergence of the Fourier series representation. To determine a_0, we can integrate both sides of (7) and get

(13)
$$\int_{-\pi}^\pi f(x) \, dx = \int_{-\pi}^\pi \left\{ \frac{a_0}{2} + \sum_{j=1}^\infty [a_j \cos (jx) + b_j \sin (jx)] \right\} dx$$

$$= \int_{-\pi}^\pi \frac{a_0}{2} \, dx + \sum_{j=1}^\infty a_j \int_{-\pi}^\pi \cos (jx) \, dx + \sum_{j=1}^\infty b_j \int_{-\pi}^\pi \sin (jx) \, dx$$

$$= a_0 + 0 + 0.$$

Justification for switching the order of integration and summation requires a detailed treatment of uniform convergence and can be found in advanced texts. Hence we have shown that

(14)
$$a_0 = \frac{1}{\pi} \int_{-\pi}^{\pi} f(x) \, dx.$$

To determine a_m, we let $m > 0$ be a fixed integer and multiply both sides of (7) by $\cos(mx)$ and integrate both sides to obtain

(15)
$$\int_{-\pi}^{\pi} f(x) \cos(mx) \, dx = \frac{a_0}{2} \int_{-\pi}^{\pi} \cos(mx) \, dx + \sum_{j=1}^{\infty} a_j \int_{-\pi}^{\pi} \cos(jx) \cos(mx) \, dx$$
$$+ \sum_{j=1}^{\infty} b_j \int_{-\pi}^{\pi} \sin(jx) \cos(mx) \, dx.$$

Equation (15) can be simplified by using the orthogonal properties of the trigonometric functions, which are now stated. The value of the first term on the right-hand side of (15) is

(16)
$$\frac{a_0}{2} \int_{-\pi}^{\pi} \cos(mx) \, dx = \frac{a_0 \sin(mt)}{2m} \bigg|_{-\pi}^{\pi} = 0.$$

The value of the term involving $\cos(jx) \cos(mx)$ is found by using the trigonometric identity

(17)
$$\cos(jx) \cos(mx) = \tfrac{1}{2} \cos((j+m)x) + \tfrac{1}{2} \cos((j-m)x).$$

When $j \neq m$, then (17) is used to get

(18)
$$a_j \int_{-\pi}^{\pi} \cos(jx) \cos(mx) \, dx = \tfrac{1}{2} a_j \int_{-\pi}^{\pi} \cos((j+m)x) \, dx$$
$$+ \tfrac{1}{2} a_j \int_{-\pi}^{\pi} \cos((j-m)x) \, dx = 0 + 0 = 0.$$

When $j = m$, the value of the integral is:

(19)
$$a_m \int_{-\pi}^{\pi} \cos(jx) \cos(mx) \, dx = a_m \pi.$$

The value of the term on the right side of (15) involving $\sin(jx) \cos(mx)$ is found by using the trigonometric identity

(20)
$$\sin(jx) \cos(mx) = \tfrac{1}{2} \sin((j+m)x) + \tfrac{1}{2} \sin((j-m)x).$$

For all values of j and m in (20) we obtain

(21)
$$b_j \int_{-\pi}^{\pi} \sin(jx) \cos(mx) \, dx = \tfrac{1}{2} b_j \int_{-\pi}^{\pi} \sin((j+m)x) \, dx$$
$$+ \tfrac{1}{2} b_j \int_{-\pi}^{\pi} \sin((j-m)x) \, dx = 0 + 0 = 0.$$

Therefore, using the results of (16), (18), (19), and (21) in equation (15), we conclude that

(22)
$$\pi a_m = \int_{-\pi}^{\pi} f(x) \cos (mx) \, dx, \qquad \text{for } m = 1, 2, \ldots .$$

Therefore, Euler's formula (5) is established.
Euler's formula (6) is proven similarly.

Trigonometric Polynomial Approximation

Let $[-\pi, \pi]$ be subdivided by the equally spaced values

(23)
$$-\pi = x_0 < x_1 < \ldots < x_N = \pi, \qquad \text{where } x_k = -\pi + 2\pi k/N.$$

The data points $\{(x_k, y_k): k = 0, 1, \ldots, N\}$, where $y_0 = y_N$, are to be fit by a periodic function. If $2M + 1 \leq N$, then there are enough points to use a trigonometric polynomial of the form

(24)
$$P(x) = \frac{a_0}{2} + \sum_{j=1}^{M} [a_j \cos (jx) + b_j \sin (jx)].$$

The least-squares procedure can be used to find the coefficients $\{a_j\}$ and $\{b_j\}$ that minimize the quantity

(25)
$$E(P(x)) = \sum_{k=1}^{N} [P(x_k) - y_k]^2.$$

Because the points in (23) are equally spaced, the coefficients in (24) are easy to compute, as shown in the next theorem.

Theorem 5.6 [Trigonometric Polynomial Approximation] Suppose that the $N + 1$ equally spaced points in (23) are the abscissas for the $N + 1$ data points $\{(x_k, y_k)\}$, where $y_0 = y_N$. If $2M + 1 \leq N$, then the coefficients for the least-squares trigonometric polynomial (24) that minimizes the expression in (25) are

(26)
$$a_j = \frac{2}{N} \sum_{k=1}^{N} f(x_k) \cos (jx_k) \qquad \text{for } j = 0, 1, 2, \ldots, M$$

and

(27)
$$b_j = \frac{2}{N} \sum_{k=1}^{N} f(x_k) \sin (jx_k) \qquad \text{for } j = 1, 2, 3, \ldots, M.$$

Although formulas (26) and (27) are defined with the least-squares procedure, they can also be viewed as numerical approximations to the integrals in Euler's formulas (5) and (6). Euler's formulas give the coefficients for the Fourier series of a continuous function, whereas formulas (26) and (27) give the trigonometric polynomial coefficients for curve fitting to a data points. The next example uses data points generated by the function $f(x) = x/2$ at discrete points. When more points are used, the trigonometric polynomial coefficients get closer to the Fourier series coefficients.

Example 5.9 Use the 12 equally spaced points $x_k = -\pi + k\pi/6$ for $k = 1, 2, \ldots, 12$ and find the trigonometric polynomial approximation for $M = 5$ to the 12 data points $\{(x_k, f(x_k))\}$, where $f(x) = x/2$. Also, compare the results when 60 and 360 points are used, and with the first five terms of the Fourier series expansion for $f(x)$ that is given in Example 5.7.

Solution. Since the periodic extension is assumed, at a point of discontinuity, the function value $f(\pi)$ must be computed using the formula

(28)
$$f(\pi) = \frac{f(\pi^-) + f(\pi^+)}{2} = \frac{\pi/2 - \pi/2}{2} = 0.$$

The function $f(x)$ is an odd function, hence the coefficients for the sine terms are all zero (i.e., $a_j = 0$ for all j). The trigonometric polynomial of degree $M = 5$ involves only the cosine terms and when formula (26) is used with (28), we get

(29)
$$P(x) = 0.9770486 \cos(x) - 0.4534498 \cos(2x) + 0.26179938 \cos(3x)$$
$$- 0.1511499 \cos(4x) + 0.0701489 \cos(5x).$$

The graph of $P(x)$ is shown in Figure 5.11.

The coefficients of the fifth-degree trigonometric polynomial change slightly when the number of interpolation points increases to 60 and 360. As the number of points increases, they get closer to the coefficients of the Fourier series expansion of $f(x)$. The results are compared in Table 5.9.

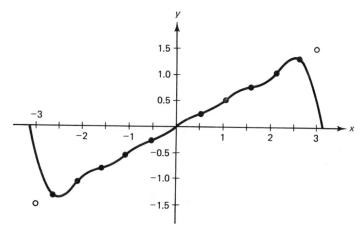

Figure 5.11. The trigonometric polynomial $P_5(x)$ of degree $M = 5$, based on 12 data points that lie on the line $y = x/2$.

TABLE 5.9   Comparison of trigonometric polynomial coefficients
for approximations to $f(x) = x/2$ over $[-\pi, \pi]$

	Trigometric Polynomial Coefficients			Fourier Series Coefficients
	12 Points	60 Points	360 Points	
b_1	0.97704862	0.99908598	0.99997462	1.0
b_2	−0.45344984	−0.49817096	−0.49994923	−0.5
b_3	0.26179939	0.33058726	0.33325718	0.33333333
b_4	−0.15114995	−0.24633386	−0.24989845	−0.25
b_5	0.07014893	0.19540972	0.19987306	0.2

Algorithm 5.5 [*Trigonometric Polynomial Construction*] Let $[-\pi, \pi]$ be subdivided by the equally spaced values $x_k = -\pi + 2\pi k/N$. The data points $\{(x_k, y_k)\}$, where $y_0 = y_N$ are fit with a least-squares trigonometric polynomial. If $2M + 1 \leq N$, then we can compute the $2M + 1$ coefficients $a_0, a_1, \ldots, a_M$ and $b_1, b_2, \ldots, b_M$ for

$$P(x) = \frac{a_0}{2} + \sum_{k=1}^{M} [a_j \cos (jx) + b_j \sin (jx)].$$

```
INPUT N                                          {Number of points}
INPUT  X(1), . . . , X(N),  Y(1), Y(2), . . . , Y(N)          {Data points}
INPUT M                          {Degree of trigonometric polynomial}
Max := INT((N − 1)/2)                     {Maximum degree possible}
IF   M > Max   THEN   M := Max


FOR    K = 1  TO  N  DO                              {Calculate sums
       A(0) := A(0) + Y(K)                            needed for
       FOR    J = 1  TO  M  DO                         coefficients}
              T := J*X(K)
              A(J) := A(J) + Y(K)*COS(T)
              B(J) := B(J) + Y(K)*SIN(T)


       FOR    J = 0  TO  M  DO                            {Multiple of
              A(J) := 2*A(J)/N                             sum forms
              B(J) := 2*B(J)/N                            coefficient}
```

{Procedure for evaluating the trigonometric polynomial above}

```
INPUT   X                                          {Input independent variable}

P := A(0)/2                                         {Sum terms of
FOR     J = 1  TO  M  DO                             trigonometric
└─── P := P + A(J)*COS(J*X) + B(J)*SIN(J*X)          polynomial}

PRINT "The value of the trig. poly. P(x) is" P      {Output}
```

Exercises for Fourier Series

1. Find the Fourier sine series for the function

$$f(x) = \begin{cases} 1 & \text{for } 0 < x < \pi, \\ -1 & \text{for } -\pi < x < 0. \end{cases}$$

2. Find the Fourier cosine series for the function

$$f(x) = \begin{cases} \dfrac{\pi}{2} - x & \text{for } 0 \leq x < \pi, \\[2mm] \dfrac{\pi}{2} + x & \text{for } -\pi \leq x < 0. \end{cases}$$

3. In Exercise 1 set $x = \pi/2$ and conclude that

$$\frac{\pi}{4} = 1 - \frac{1}{3} + \frac{1}{5} - \frac{1}{7} + \cdots.$$

4. In Exercise 2 set $x = 0$ and conclude that

$$\frac{\pi^2}{8} = 1 + \frac{1}{3^2} + \frac{1}{5^2} + \frac{1}{7^2} + \cdots.$$

5. Find the Fourier series for the function

$$f(x) = \begin{cases} x & \text{for } 0 \leq x < \pi, \\ 0 & \text{for } -\pi < x < 0. \end{cases}$$

6. Find the Fourier cosine series for the function

$$f(x) = \begin{cases} -1 & \text{for } \dfrac{\pi}{2} < x < \pi, \\[2mm] 1 & \text{for } \dfrac{-\pi}{2} < x < \dfrac{\pi}{2}, \\[2mm] -1 & \text{for } -\pi < x < \dfrac{-\pi}{2}. \end{cases}$$

7. Find the Fourier sine series for the function

$$f(x) = \begin{cases} \pi - x & \text{for } \dfrac{\pi}{2} \leqq x < \pi, \\[2ex] x & \text{for } \dfrac{-\pi}{2} \leqq x < \dfrac{\pi}{2}, \\[2ex] -\pi - x & \text{for } -\pi \leqq x < \dfrac{-\pi}{2}. \end{cases}$$

8. Find the Fourier cosine series for the function $f(x) = x^2/4$ extended periodically by the equation $f(x + 2\pi) = f(x)$.

9. Prove Theorem 5.3.

10. Prove Theorem 5.4.

11. Prove Theorem 5.5.

12. Write a report on half-range expansions of Fourier series.

13. Write a report on the Gibbs phenomenon.

14. Write a report on the fast Fourier transform.

15. Modify Algorithm 5.5 so that it will find the trigonometric polynomial of period $P = B - A$ when the data points are equally spaced over the interval $[A, B]$.

16. Write a report on how to adapt the linear least-squares procedure to find the trigonometric polynomial when the data points are not equally spaced in the interval $[-\pi, \pi]$.

17. Use a numerical integration program and find approximations for the coefficients $a_0, a_1, a_2, a_3, a_4,$ and a_5 in Exercise 8. At least four significant digits of accuracy are required.

18. Use a numerical integration program and find approximations for the coefficients $a_0, a_1, \ldots, a_5$ and $b_1, b_2, \ldots, b_5$ in Exercise 5. At least four significant digits of accuracy are required.

19. Use Algorithm 5.5 with $N = 12$ points follow Example 5.9 to find the trigonometric polynomial of degree $M = 5$ for the equally spaced points $\{(x_k, f(x_k)): k = 1, \ldots, 12\}$, where $f(x)$ is the function in (a) Exercise 1 (b) Exercise 2 (c) Exercise 5 (d) Exercise 6 (e) Exercise 7.

20. The temperature cycle in a suburb of Los Angeles on November 8 is given in the table below. There are 24 data points.
    (a) Find the trigonometric polynomial for the temperature that involves terms up to $\cos(2\pi x/24)$ and $\sin(2\pi x/24)$.
    (b) Compare the values of the trigonometric polynomial in part (a) with the values in the table.

Time p.m.	Degrees Fahrenheit	Time a.m.	Degrees Fahrenheit
1	66	1	58
2	66	2	58
3	65	3	58
4	64	4	58
5	63	5	57
6	63	6	57
7	62	7	57
8	61	8	58
9	60	9	60
10	60	10	64
11	59	11	67
Midnight	58	Noon	68

(c) Repeat (a) and (b) using temperatures from your locale.

21. The yearly temperature cycle for Fairbanks, Alaska, is given in the table below. There are 13 equally spaced data points, which corresponds to a measurement every 28 days.

(a) Find the trigonometric polynomial for the temperature that involves terms up to $\cos(2\pi x/13)$ and $\sin(2\pi x/13)$.

(b) Compare the values of the trigonometric polynomial in part (a) with the values in the table.

Calendar Date	Average Degrees Fahrenheit
Jan. 1	−14
Jan. 29	−9
Feb. 26	2
Mar. 26	15
Apr. 23	35
May 21	52
June 18	62
July 16	63
Aug. 13	58
Sept. 10	50
Oct. 8	34
Nov. 5	12
Dec. 3	−5

6

Numerical Differentiation
and Optimization

6.1 Approximating the Derivative

The Limit of the Difference Quotient

We now turn our attention to the numerical process for approximating the derivative of $f(x)$:

(1)
$$f'(x) = \lim_{h \to 0} \frac{f(x+h) - f(x)}{h}.$$

The method seems straightforward; choose a sequence $\{h_k\}$ so that $h_k \to 0$ and compute the limit of the sequence:

(2)
$$D_k = \frac{f(x + h_k) - f(x)}{h_k} \qquad \text{for } k = 1, 2, \ldots, n, \ldots.$$

The reader may notice that we will only compute a finite number of terms $D_1, D_2, \ldots, D_N$ in the sequence (2), and it appears that we should use D_N for our answer. The following question is often posed: Why compute $D_1, D_2, \ldots, D_{N-1}$? Equivalently, we could ask: What value h_N should be chosen so that D_N is a good approximation to the derivative $f'(x)$? To answer this question, we must look at an example to see why there is no simple solution.

Example 6.1 Let $f(x) = \exp(x)$ and $x = 1$. Compute the difference quotients D_k using the step sizes $h_k = 10^{-k}$ for $k = 1, 2, \ldots, 10$. Carry nine decimal places in all the calculations.

Solution. The table of values $f(1 + h_k)$ and $[f(1 + h_k) - f(1)]/h_k$ that are used in the computation of D_k are shown in Table 6.1.

Table 6.1 Finding the difference quotients $D_k = [\exp(1 + h_k) - e]/h_k$

h_k	$f_k = f(1 + h_k)$	$f_k - e$	$D_k = [f_k - e]/h_k$
$h_1 = 0.1$	3.004166024	0.285884196	2.858841960
$h_2 = 0.01$	2.745601015	0.027319187	2.731918700
$h_3 = 0.001$	2.721001470	0.002719642	2.719642000
$h_4 = 0.0001$	2.718553670	0.000271842	2.718420000
$h_5 = 0.00001$	2.718309011	0.000027183	2.718300000
$h_6 = 10^{-6}$	2.718284547	0.000002719	2.719000000
$h_7 = 10^{-7}$	2.718282100	0.000000272	2.720000000
$h_8 = 10^{-8}$	2.718281856	0.000000028	2.800000000
$h_9 = 10^{-9}$	2.718281831	0.000000003	3.000000000
$h_{10} = 10^{-10}$	2.718281828	0.000000000	0.000000000

The largest value $h_1 = 0.1$ does not produce a good approximation $D_1 \approx f'(1)$, because the step size h_1 is too large and the difference quotient is the slope of the secant line through two points that are not close enough to each other. When formula (2) is used with a fixed precision of nine decimal places, h_9 produced the approximation $D_9 = 3$, and h_{10} produced $D_{10} = 0$. If h_k is too small, then the computed function values $f(x + h_k)$ and $f(x)$ are very close together. The difference $f(x + h_k) - f(x)$ can exhibit the problem of loss of significance due to the subtraction of quantities that are nearly equal. The value $h_{10} = 10^{-10}$ is so small that the stored values of $f(x + h_{10})$ and $f(x)$ are the same, and hence the computed difference quotient is zero. In Example 6.1 the mathematical value for the limit is $f'(1) \approx 2.718281828$. Observe that the value $h_5 = 10^{-5}$ gives the best approximation, $D_5 = 2.7183$.

Example 6.1 shows that it is not easy to find numerically the limit in equation (2). The sequence starts to converge to e, and D_5 is the closest; then the terms move away from e. In Algorithm 6.1 it is suggested that terms in the sequence $\{D_k\}$ should be computed until $|D_{N+1} - D_N| \geq |D_N - D_{N-1}|$. This is an attempt to determine the best approximation before the terms start to move away from the limit. When this criterion is applied to Example 6.1, we have $0.0007 = |D_6 - D_5| > |D_5 - D_4| = 0.00012$; hence D_5 is the answer we choose. We want to develop formulas that give a reasonable amount of accuracy for larger values of h.

The Central-Difference Formulas

If the function $f(x)$ can be evaluated at values that lie to the left and right of x, then the best two-point formula will involve abscissas that are chosen symmetrically on both sides of x. First we derive that central-difference formula:

(3)
$$f'(x) = \frac{f(x + h) - f(x - h)}{2h} + E_{\text{trunc}}(f, h).$$

The truncation error term $E_{\text{trunc}}(f, h)$ in (3) has the form

(4)
$$E_{\text{trunc}}(f, h) = \frac{-h^2 f^{(3)}(c)}{6} = O(h^2),$$

where c is a value that lies in the interval $[x - h, x + h]$.

Start with the Taylor series expansions for $f(x + h)$ and $f(x - h)$:

(5)
$$f(x + h) = f(x) + f'(x)h + \frac{f^{(2)}(x)h^2}{2} + \frac{f^{(3)}(x)h^3}{6} + \cdots$$

and

(6)
$$f(x - h) = f(x) - f'(x)h + \frac{f^{(2)}(x)h^2}{2} - \frac{f^{(3)}(x)h^3}{6} + \cdots .$$

After (6) is subtracted from (5), we obtain

(7)
$$f(x + h) - f(x - h) = 2f'(x)h + \frac{2f^{(3)}(x)h^3}{6} + \frac{2f^{(5)}(x)h^5}{120} + \cdots .$$

If the series in (7) is truncated at the third derivative, Taylor's theorem states that there exists a value c satisfying $|c - x| < h$ so that

(8)
$$f(x + h) - f(x - h) = 2f'(x)h + \frac{2f^{(3)}(c)h^3}{6}.$$

When equation (8) is solved for $f'(x)$, the result is

(9)
$$f'(x) = \frac{f(x + h) - f(x - h)}{2h} - \frac{f^{(3)}(c)h^2}{6}.$$

The first term on the right side of (9) is the central-difference formula given in (3) and the second term is the truncation error. Notice that the truncation error $E_{\text{trunc}}(f, h) = -f^{(3)}(c)h^2/6$ goes to zero as h goes to zero. But we shall see that when this formula is used computationally and h is small, the error incurred in using formula (3) is dominated by round-off error in calculating the function and not truncation error. For this reason it is important to consider also the higher-order central-difference formula:

(10)
$$f'(x) = \frac{-f(x + 2h) + 8f(x + h) - 8f(x - h) + f(x - 2h)}{12h} + E_{\text{trunc}}(f, h).$$

The truncation error term $E_{trunc}(f, h)$ in (10) has the form

(11) $$E_{trunc}(f, h) = \frac{h^4 f^{(5)}(c)}{30} = O(h^4),$$

where c is a value that lies in the interval $[x - 2h, x + 2h]$.

One way to derive formula (10) is as follows. Rewrite (7) using step size $2h$ instead of h, and use Taylor's theorem to truncate the series at the fifth derivative:

(12) $$f(x + 2h) - f(x - 2h) = 4f'(x)h + \frac{16f^{(3)}(x)h^3}{6} + \frac{64f^{(5)}(c_2)h^5}{120}.$$

Then multiply formula (7) by 8, and use Taylor's theorem to truncate the series at the fifth derivative:

(13) $$8f(x + h) - 8f(x - h) = 16f'(x)h + \frac{16f^{(3)}(x)h^3}{6} + \frac{16f^{(5)}(c_1)h^5}{120}.$$

When (12) is subtracted from (13), the terms involving the third derivative will be eliminated, and we get

$$-f(x + 2h) + 8f(x + h) - 8f(x - h) + f(x - 2h)$$

(14) $$= 12f'(x)h + \frac{[16f^{(5)}(c_1) - 64f^{(5)}(c_2)]h^5}{120}.$$

If $f^{(5)}(x)$ has one sign and if its magnitude does not change rapidly, we can find a value c that lies in $[x - 2h, x + 2h]$ so that

(15) $$16f^{(5)}(c_1) - 64f^{(5)}(c_2) = -48f^{(5)}(c).$$

After (15) is substituted into (14) and the result is solved for $f'(x)$, we obtain

(16) $$f'(x) = \frac{-f(x + 2h) + 8f(x + h) - 8f(x - h) + f(x - 2h)}{12h} + \frac{h^4 f^{(5)}(c)}{30}.$$

The first term on the right side of (16) is the central-difference formula in (10) and the second term is the error term (11). If $f(x)$ has five continuous derivatives and if the magnitude of $|f^{(3)}(x)|$ and $|f^{(3)}(x)|$ is about the same, then the truncation error for the fourth-order scheme (10) is $O(h^4)$ and will go to zero faster than the truncation error $O(h^2)$ for the second order scheme (3). This will often permit the use of a larger step size to achieve a desired accuracy and will reduce the effect of round-off error, which was the major difficulty in Example 6.1.

Example 6.2 Let $f(x) = \cos(x)$.
(a) Use formulas (3) and (10) with step sizes $h = 0.1, 0.01, 0.001$, and 0.0001 and calculate approximations for $f'(0.8)$. Carry nine decimal places in all the calculations.
(b) Compare with the true value $f'(0.8) = -\sin(0.8)$.

Solution. (a) Using formula (3) with $h = 0.01$, we get

$$f'(0.8) \approx \frac{f(0.81) - f(0.79)}{0.02} \approx \frac{0.689498433 - 0.703845316}{0.02} \approx -0.717344150.$$

Using formula (10) with $h = 0.01$, we get

$$f'(0.8) \approx \frac{-f(0.82) + 8f(0.81) - 8f(0.79) + f(0.78)}{0.12}$$

$$\approx \frac{-0.682221207 + 8 \times 0.689498433 - 8 \times 0.703845316 + 0.710913538}{0.12}$$

$$\approx -0.717356108.$$

(b) The error in approximation for formulas (3) and (10) is -0.000011941 and 0.000000017, respectively. In this example formula (10) gives a better approximation to $f'(0.8)$ than formula (3) when $h = 0.01$. The error analysis will illuminate this example and show why this happened. The other calculations are summarized in Table 6.2.

Table 6.2   Numerical differentiation using formulas (3) and (10)

Step Size	Approximation by Formula (3)	Error Using Formula (3)	Approximation by Formula (10)	Error Using Formula (10)
0.1	−0.716161095	−0.001194996	−0.717353703	−0.000002389
0.01	−0.717344150	−0.000011941	−0.717356108	0.000000017
0.001	−0.717356000	−0.000000091	−0.717356167	0.000000076
0.0001	−0.717360000	−0.000003909	−0.717360833	0.000004742

Error Analysis

An important topic in the study of numerical differentiation is the effect of round-off error. Let us examine the formulas more closely. Suppose that in computing $f(x + h)$ and $f(x - h)$ we use the numerical values y_1 and y_{-1} which have round-off errors e_1 and e_{-1}, respectively, that is, we have the relationships

$$f(x + h) = y_1 + e_1 \quad \text{and} \quad f(x - h) = y_{-1} + e_{-1}.$$

Then difference quotient in formula (3) can be written

$$\frac{f(x + h) - f(x - h)}{2h} = \frac{y_1 + e_1 - y_{-1} - e_{-1}}{2h} = \frac{y_1 - y_{-1}}{2h} + \frac{e_1 - e_{-1}}{2h}.$$

The round-off error in calculating this difference quotient is defined to be

$$E_{\text{round}}(f, h) = \frac{e_1 - e_{-1}}{2h}.$$

The computer's evaluation for the difference quotient is the quantity $(y_1 - y_{-1})/2h$, and formula (3) must be rewritten

(17)
$$f'(x) = \frac{y_1 - y_{-1}}{2h} + E_{round}(f, h) + E_{trunc}(f, h).$$

The total error term $E(f, h)$ for (17) will have a part due to round-off error plus a part due to truncation error:

(18)
$$E(f, h) = \frac{e_1 - e_{-1}}{2h} - \frac{h^2 f^{(3)}(c)}{6}.$$

When h is small, the portion of (18) involving $(e_1 - e_{-1})/2h$ can be relatively large. In Example 6.2, when $h = 0.0001$, this difficulty was encountered. The round-off errors are

$$f(0.8001) = 0.696634970 + e_1, \quad \text{where } e_1 \approx -0.0000000003,$$
$$f(0.7999) = 0.696778442 + e_2, \quad \text{where } e_2 \approx 0.0000000005.$$

The truncation error term is

$$\frac{-h^2 f^{(3)}(c)}{6} \approx -(0.0001)^2 \frac{\sin(0.8)}{6} \approx 0.000000001.$$

The error term $E(f, h)$ in (18) can now be estimated:

$$E(f, h) \approx \frac{-0.0000000003 - 0.0000000005}{0.0002} - 0.000000001 = -0.000004001.$$

Indeed, the computed numerical approximation for the derivative using $h = 0.0001$ is found by the calculation

$$f'(0.8) \approx \frac{f(0.8001) - f(0.7999)}{0.0002} = \frac{0.696634970 - 0.696778442}{0.0002} = -0.717360000$$

and a loss of about four significant digits is evident. The error is -0.000003909 and this is close to the predicted error, -0.000004001.

Optimal Step Size

If the errors e_1 and e_{-1} are of the magnitude e and of opposite sign, and if $|f^{(3)}(c)| \leq M$, we obtain the following error bound:

(19)
$$|E(f, h)| \leq \frac{e}{h} + \frac{Mh^2}{6}.$$

When h is small, then the contribution due to round-off is e/h and could be large. When h is large, the contribution $Mh^2/6$ due to truncation could be large. The optimal step size can be approximated by minimizing the quantity

(20)
$$g(h) = \frac{e}{h} + \frac{Mh^2}{6}.$$

Setting $g'(h) = 0$ results in $-e/h^2 + Mh/3 = 0$, which yields the equation $h^3 = 3e/M$, from which we obtain the optimal value:

(21) $$h = \left(\frac{3e}{M}\right)^{1/3} \qquad \text{[optimum } h \text{ for formula (3)].}$$

When formula (21) is applied to Example 6.2, we can use the bound $|f^{(3)}(x)| \le |\sin(x)| \le 1 = M$ and the value $e = 0.5 \times 10^{-9}$ for the magnitude of the round-off error. The optimal value for h is easily calculated: $h = (1.5 \times 10^{-9}/1)^{1/3} = 0.001144714$. The step size $h = 0.001$ was closest to the optimal value 0.001144714 and it gave the best approximation to $f'(0.8)$ among the four choices involving formula (3) (see Table 6.2).

An error analysis for formula (10) is similar. Suppose that the round-off error in computing the function values satisfies $f(x + hk) = y_k + e_k$, where $|e_k| \le e$; then portion of (10) due to round-off is defined to be

$$E_{\text{round}}(f, h) = \frac{-e_2 + 8e_1 - 8e_{-1} + e_{-2}}{12h}.$$

Formula (10) must be rewritten

(22) $$f'(x) = \frac{-y_2 + 8y_1 - 8y_{-1} + y_{-2}}{12h} + E_{\text{round}}(f, h) + E_{\text{trunc}}(f, h).$$

The total error term $E(f, h)$ for (22) will have a part due to round-off error plus a part due to truncation error:

(23) $$E(f, h) = \frac{-e_2 + 8e_1 - 8e_{-1} + e_{-2}}{12h} + \frac{h^4 f^{(5)}(c)}{30}.$$

If $|f^{(5)}(c)| \le M$, then we obtain the error bound:

(24) $$|E(f, h)| = \frac{3e}{2h} + \frac{Mh^4}{30}.$$

The reader can verify that the optimal value of h is

(25) $$h = \left(\frac{45e}{4M}\right)^{1/5} \qquad \text{[optimum } h \text{ for formula (10)].}$$

When formula (25) is applied to Example 6.2, we can use the bound $|f^{(5)}(x)| \le |\sin(x)| \le 1 = M$ and the value $e = 0.5 \times 10^{-9}$. The optimal step size is $h = (22.5 \times 10^{-9}/4)^{1/5} = 0.022388475$. The step size $h = 0.01$ was closest to the optimal value, 0.022388475, and it gave the best approximation to $f'(0.8)$ among the four choices involving formula (10) (see Table 6.2).

Richardson's Extrapolation

In this section we emphasize the relationship between formulas (3) and (10). Let $f_k = f(x_k)$ and use the notation $D(h, 0)$ and $D(2h, 0)$ to denote the approximations to $f'(x)$ that are obtained from (3) with step sizes h and $2h$, respectively:

(26)
$$f'(x) \approx D(h, 0) = \frac{f_1 - f_{-1}}{2h}$$

and

(27)
$$f'(x) \approx D(2h, 0) = \frac{f_2 - f_{-2}}{4h}.$$

If we multiply relation (26) by 4 and subtract relation (27) from this product, then the result is

(28)
$$3f'(x) \approx 4D(h, 0) - D(2h, 0)$$
$$= \frac{4(f_1 - f_{-1})}{2h} - \frac{f_2 - f_{-2}}{4h}.$$

When we solve for $f'(x)$ in (28) the result is

(29)
$$f'(x) \approx \frac{4D(h, 0) - D(2h, 0)}{3}$$
$$= \frac{-f_2 + 8f_1 - 8f_{-1} + f_{-2}}{12h}.$$

The last expression in (29) is the central-difference formula (10) for approximating $f'(x)$.

Example 6.3 Let $f(x) = \cos(x)$. Use (26) and (27) with $h = 0.01$ and show how the linear combination $[4D(h, 0) - D(2h, 0)]/3$ in (29) can be used to obtain the approximation to $f'(0.8)$ given in (10). Carry nine decimal places in all the calculations.

Solution. Use (26) and (27) with $h = 0.01$ to get

$$D(h, 0) \approx \frac{f(0.81) - f(0.79)}{0.02} \approx \frac{0.689498433 - 0.703845316}{0.02} \approx -0.717344150$$

and

$$D(2h, 0) \approx \frac{f(0.82) - f(0.78)}{0.04} \approx \frac{0.682221207 - 0.710913538}{0.04} \approx -0.717308275.$$

Now the linear combination in (29) is computed:

$$f'(0.8) \approx \frac{4D(h, 0) - D(2h, 0)}{3} \approx \frac{4(-0.717344150) - (-0.717308275)}{3}$$
$$\approx -0.717356108.$$

This is exactly the same as the solution in Example 6.2 that used (10) directly to approximate to $f'(0.8)$.

The method of obtaining a formula for $f'(x)$ of higher order from a formula of lower order is called *extrapolation*. A detailed proof starts with the error term for (3) expanded in a series containing only even powers of h. Step sizes h and $2h$ are used to remove the terms involving h^2, then those involving h^4 are removed, and so on. To see how h^4 is removed, let $D(h, 1)$ and $D(2h, 1)$ denote the approximations to $f'(x)$ of order $O(h^4)$ obtained from (16), using step sizes h and $2h$, respectively. Then

$$(30) \qquad f'(x) = D(h, 1) + \frac{h^4 f^{(5)}(c_1)}{30}$$

and

$$(31) \qquad f'(x) = D(2h, 1) + \frac{16h^4 f^{(5)}(c_2)}{30}.$$

If $f^{(5)}(x)$ has one sign and does not change too rapidly, then the assumption that $f^{(5)}(c_1) \approx f^{(5)}(c_2)$ can be used to eliminate the terms involving h^4 in (30) and (31) and the result is

$$(32) \qquad f'(x) \approx \frac{16D(h, 1) - D(2h, 1)}{15}.$$

Richardson's extrapolation scheme for derivatives relies on the theoretical base that the truncation error in formula (3) can be expanded in a series involving only even powers of h, that is,

$$(33) \qquad f'(x) = \frac{f(x + h) - f(x - h)}{2h} + E_{\text{trunc}}(f, h),$$

where

$$(34) \qquad E_{\text{trunc}}(f, h) = a_1 h^2 + a_2 h^4 + a_3 h^6 + \ldots .$$

With this assumption we see that the truncation error for the improvement formula (29) is

$$\frac{4E_{\text{trunc}}(f, h) - E_{\text{trunc}}(f, 2h)}{3}$$

$$= \frac{4a_1 h^2 + 4a_2 h^4 + 4a_3 h^6 + \ldots - 4a_1 h^2 - 16a_2 h^4 - 64a_3 h^6 - \ldots}{3}$$

$$= -4a_2 h^4 - 20a_3 h^6 - \ldots = c_1 h^4 + c_2 h^6 + \ldots .$$

A general pattern for improving the calculated value for the derivative is stated in Lemma 6.1. It gives the formula that is used in Algorithm 6.2.

Lemma 6.1 [Richardson's improvement for derivatives] Given two approximations $D(h, k - 1)$ and $D(2h, k - 1)$ for the quantity D that satisfy

$$(35) \qquad D = D(h, k - 1) + c_1 h^{2k} + c_2 h^{2k+2} + \ldots ,$$

$$(36) \qquad D = D(2h, k - 1) + c_1 4^k h^{2k} + c_2 4^{k+1} h^{2k+2} + \ldots ,$$

an improved approximation has the form

(37)
$$D = \frac{4^k D(h, k-1) - D(2h, k-1)}{4^k - 1} + O(h^{2k+2}).$$

Algorithm 6.1 [*Differentiation Using Limits*] Assume that $f(x)$ is differentiable on $[x-h, x+h]$. Use $h_k = h/2^k$ for $k = 0, 1, \ldots, n, \ldots$ in formula (3) and compute the difference quotients:

$$D_n = \frac{f(x+h_n) - f(x-h_n)}{2h_n}; \quad \text{then } f'(x) = \lim_{n \to \infty} D_n.$$

Terms D_n are generated until $|D_{n+1} - D_n| \geq |D_n - D_{n-1}|$ or $|D_n - D_{n-1}| < \text{Tol}$. This is an attempt to determine the best value D_n for the approximation $f'(x) \approx D_n$.

Tol := 10^{-5}		{Convergence criterion}
H := 1		{Initial step size}
Max := 20		{Maximum number of terms}

INPUT X   {Abscissa for f'(x)}
D(0) := .5 * [f(X+H) − f(X−H)]/H   {Compute the quotient}

FOR   N =1 TO 2 DO   {Compute D_1 and D_2}
         H := H/2   {Reduce step size}
         D(N) := .5 * [f(X+H) − f(X−H)]/H   {Compute the quotient}
         E(N) := |D(N) − D(N−1)| {Error estimate}
         R(N) := 2 * E(N)/[|D(N)|+|D(N−1)|+Tol]   {Relative error}
         N := 1   {Initialize the counter}
WHILE   [E(N)>E(N+1) OR R(N)≥Tol] AND N<Max DO
         H := H/2   {Reduce the step size}
         D(N+2) := .5 * [f(X+H) − f(X−H)]/H   {Compute the quotient}
         E(N+2) := |D(N+2) − D(N+1)| {Error estimate}
         R(N+2) := 2 * E(N+2)/[|D(N+2)|+|D(N+1)|+Tol]
         N := N+1   {Increment the counter}

PRINT "The approximation for f'(x) is" D(N−1)   {Output}
PRINT "The accuracy is" +−E(N−1)

Algorithm 6.2 [*Numerical Differentiation Using Extrapolation*] Assume that $f(x)$ is differentiable on $[x-h, x+h]$. An approximation for the derivative, $f'(x) \approx D(N, N)$, is computed as follows:

Successive	D(0,0)			
approximations	D(1,0)	D(1,1)		
are stored	D(2,0)	D(2,1)	D(2,2)	
in the array	⋮	⋮	⋮	
D(J,K) K≤J	D(N,0)	D(N,1)	D(N,2) ... D(N,N)	

The element $D(J, 0)$ in column 0 is computed using formula (3) and the abscissas $x + h * 2^{-J}$ and $x - h * 2^{-J}$:

$$D(J,0)=[f(x+h*2^{-J}) \ - \ f(x-h*2^{-J})]/[h*2^{-J+1}].$$

The elements in column K are computed using the extrapolation rule

$$D(J,K) \ = \ D(J,K-1) \ + \ [D(J,K-1) \ - \ D(J-1,K-1)/[4^K-1].$$

Tol := 10^{-5}, Delta := 10^{-7}	{Termination criterion}
Error :=1, RelErr := 1	{Initialize the variables}
H := 1	{Initial step size}
J := 1	{Initialize the counter}
INPUT X	{Abscissa for f'(x)}
D(0,0) := .5*[f(X+H) $-$ f(X$-$H)]/H	{Compute the quotient}

WHILE   RelErr>Tol AND Error>Delta AND J≤12   DO
```
  H := H/2                                        {Reduce the step size}
  D(J,0) := .5*[f(X+H) − f(X−H)]/H                {Compute the quotient}
  FOR  K = 1  TO  J  DO
    D(J,K) := D(J,K−1) + [D(J,K−1)−D(J−1,K−1)]/(4^K − 1)
  Error := |D(J,J)−D(J−1,J−1)|                    {Error estimate}
  RelErr := 2*Error/[|D(J,J)|+|D(J−1,J−1)|+Tol]
  J := J + 1, N := J
```

PRINT "The approximate value of f'(x) using   {Output}
        Richardson's extrapolation is" D(N−1,N−1)
PRINT "The accuracy is" +−Error

Exercises for Approximating the Derivative

1. Let $f(x) = \sin(x)$, where x is measured in radians.
   (a) Calculate approximations to $f'(0.8)$ using formula (3) with $h = 0.1, h = 0.01$, $h = 0.001$. Carry eight or nine decimal places.
   (b) Compare with the value $f'(0.8) = \cos(0.8)$.
   (c) Compute bounds for the truncation error (4). Use

   $$|f^{(3)}(c)| \leqq \cos(0.7) \approx 0.764842187$$

   for all the cases.

2. Let $f(x) = \exp(x)$.
   (a) Calculate approximations to $f'(2.3)$ using formula (3) with $h = 0.1, h = 0.01$, $h = 0.001$. Carry eight or nine decimal places.
   (b) Compare with the value $f'(2.3) = \exp(2.3)$.

(c) Compute bounds for the truncation error (4). Use

$$|f^{(3)}(c)| \leq \exp(2.4) \approx 11.02317638$$

for all the cases.

3. Let $f(x) = \sin(x)$, where x is measured in radians.
   (a) Calculate approximations to $f'(0.8)$ using formula (10) with $h = 0.1$ and $h = 0.01$, and compare with $f'(0.8) = \cos(0.8)$.
   (b) Use the extrapolation formula in (29) to compute the approximations to $f'(0.8)$ in part (a).
   (c) Compute bounds for the truncation error (11). Use

$$|f^{(5)}(c)| \leq \cos(0.6) \approx 0.825335615$$

for both cases.

4. Let $f(x) = \exp(x)$.
   (a) Calculate approximations to $f'(2.3)$ using (10) with $h = 0.1$ and $h = 0.01$, and compare with $f'(2.3) = \exp(2.3)$.
   (b) Use the extrapolation formula in (29) to compute the approximations to $f'(2.3)$ in part (a).
   (c) Compute bounds for the truncation error (11). Use

$$|f^{(5)}(c)| \leq \exp(2.5) \approx 12.18249396$$

for both cases.

5. Compare the numerical differentiation formulas (3) and (10). Let $f(x) = x^3$ and find approximations for $f'(2)$.
   (a) Use formula (3) with $h = 0.05$.
   (b) Use formula (10) with $h = 0.05$.
   (c) Compute bounds for the truncation errors (4) and (11).

6. (a) Use Taylor's theorem to show that

$$f(x + h) = f(x) + hf'(x) + \frac{h^2 f''(c)}{2}, \qquad \text{where } |c - x| < h.$$

   (b) Use part (a) to show that the difference quotient in equation (2) has error of order $O(h) = -hf''(c)/2$.
   (c) Why is formula (3) better to use than formula (2)?

7. *Partial Differentiation Formulas.* The partial derivative $f_x(x, y)$ of $f(x, y)$ with respect to x is obtained holding y fixed and differentiating with respect to x. Similarly, $f_y(x, y)$ is found by holding x fixed and differentiating with respect to y. The functions $f_x(x, y)$ and $f_y(x, y)$ are defined by the limits

$$f_x(x, y) = \lim_{h \to 0} \frac{f(x + h, y) - f(x, y)}{h},$$

$$f_y(x, y) = \lim_{h \to 0} \frac{f(x, y + h) - f(x, y)}{h}.$$

Formula (3) can be adapted to partial derivatives:

$$f_x(x, y) = \frac{f(x + h, y) - f(x - h, y)}{2h} + O(h^2),$$

(38)

$$f_y(x, y) = \frac{f(x, y + h) - f(x, y - h)}{2h} + O(h^2).$$

(a) Let $f(x, y) = xy/(x + y)$. Calculate approximations to $f_x(2, 3)$ and $f_y(2, 3)$ using the formulas in (38) with $h = 0.1, 0.01$, and 0.001. Compare with the values obtained by differentiating $f(x, y)$ partially.

(b) Let $z = f(x, y) = \arctan(y/x)$; z is in radians. Calculate approximations to $f_x(3, 4)$ and $f_y(3, 4)$ using the formulas in (38) with $h = 0.1, 0.01$, and 0.001. Compare with the values obtained by differentiating $f(x, y)$ partially.

8. Complete the details that show how (32) is obtained from equations (30) and (31).

9. The voltage $E = E(t)$ in an electrical circuit obeys the equation $E(t) = L(dI/dt) + RI(t)$, where R is resistance and L is inductance. Use $L = 0.05$ and $R = 2$ and values for $I(t)$ in the table below.

t	$I(t)$
1.0	8.2277
1.1	7.2428
1.2	5.9908
1.3	4.5260
1.4	2.9122

(a) Find $I'(1.2)$ by numerical differentiation, and use it to compute $E(1.2)$.

(b) Compare your answer with $I(t) = 10 \exp(-t/10) \sin(2t)$.

10. The distance $D = D(t)$ traveled by an object is given in the table below.

t	$D(t)$
8.0	17.453
9.0	21.460
10.0	25.752
11.0	30.301
12.0	35.084

(a) Find the velocity $V(10)$ by numerical differentiation.

(b) Compare your answer with $D(t) = -70 + 7t + 70 \exp(-t/10)$.

11. Let $f(x)$ be given by the table below. The inherent round-off error has the bound $|e_k| \leq 5 \times 10^{-6}$. Use the rounded values in your calculations.

x	$f(x) = \cos(x)$
1.100	0.45360
1.190	0.37166
1.199	0.36329
1.200	0.36236
1.201	0.36143
1.210	0.35302
1.300	0.26750

(a) Find approximations for $f'(1.2)$ using formula (17) with $h = 0.1$, $h = 0.01$, and $h = 0.001$.

(b) Compare with $f'(1.2) = -\sin(1.2) \approx -0.93204$.

(c) Find the total error bound (19) for the three cases in part (a).

12. Let $f(x)$ be given by the table below. The inherent round-off error has the bound $|e_k| \leq 5 \times 10^{-6}$. Use the rounded values in your calculations.

x	$f(x) = \ln(x)$
2.900	1.06471
2.990	1.09527
2.999	1.09828
3.000	1.09861
3.001	1.09895
3.010	1.10194
3.100	1.13140

(a) Find approximations for $f'(3.0)$ using formula (17) with $h = 0.1$, $h = 0.01$, and $h = 0.001$.

(b) Compare with $f'(3.0) = \frac{1}{3} \approx 0.33333$.

(c) Find the total error bound (19) for the three cases in part (a).

13. Let $f(x)$ be given by the table below. The inherent round-off error has the bound $|e_k| \leq 5 \times 10^{-6}$. Use the rounded values in your calculations.

x	$f(x) = x^{1/2}$
0.400	0.63246
0.490	0.70000
0.499	0.70640
0.500	0.70711
0.501	0.70781
0.510	0.71414
0.600	0.77460

(a) Find approximations for $f'(0.5)$ using formula (17) with $h = 0.1$, $h = 0.01$, and $h = 0.001$.

(b) Compare with $f'(0.5) = 0.70711$.

(c) Find the total error bound (19) for the three cases in part (a).

14. Suppose that a table of the function $f(x_k)$ is computed where the values are rounded off to three decimal places, and the inherent round-off error is 5×10^{-4}. Also, assume that $|f^{(3)}(c)| \leq 1.5$ and $|f^{(5)}(c)| \leq 1.5$.

    (a) Find the best step size h for formula (17).

    (b) Find the best step size h for formula (22).

15. Let $f(x)$ be given by the table below. The inherent round-off error has the bound $|e_k| \leq 5 \times 10^{-6}$. Use the rounded values in your calculations.

x	$f(x) = \cos(x)$
1.000	0.54030
1.100	0.45360
1.198	0.36422
1.199	0.36329
1.200	0.36236
1.201	0.36143
1.202	0.36049
1.300	0.26750
1.400	0.16997

    (a) Approximate $f'(1.2)$ using (22) with $h = 0.1$ and $h = 0.001$.

    (b) Find the total error bound (24) for the two cases in part (a).

16. Let $f(x)$ be given by the table below. The inherent round-off error has the bound $|e_k| \leq 5 \times 10^{-6}$. Use the rounded values in your calculations.

x	$f(x) = \ln(x)$
2.800	1.02962
2.900	1.06471
2.998	1.09795
2.999	1.09828
3.000	1.09861
3.001	1.09895
3.002	1.09928
3.100	1.13140
3.200	1.16315

    (a) Approximate $f'(3.0)$ using (22) with $h = 0.1$ and $h = 0.001$.

    (b) Find the total error bound (24) for the two cases in part (a).

17. Let $f(x)$ be given by the table below. The inherent round-off error has the bound $|e_k| \leq 5 \times 10^{-6}$. Use the rounded values in your calculations.

x	$f(x) = x^{1/2}$
0.300	0.54772
0.400	0.63246
0.498	0.70569
0.499	0.70640
0.500	0.70711
0.501	0.70781
0.502	0.70852
0.600	0.77460
0.700	0.83666

(a) Approximate $f'(0.5)$ using (22) with $h = 0.1$, $h = 0.001$.
(b) Find the total error bound (24) for the two cases in part (a).

6.2 Numerical Differentiation Formulas

More Central-Difference Formulas

The formulas for $f'(x)$ in the preceding section required that the function can be computed at abscissas that lie on both sides of x, and were referred to as central-difference formulas. Taylor series can be used to obtain central-difference formulas for the higher derivatives. The popular choices are those of order $O(h^2)$ and $O(h^4)$ and are given in Tables 6.3 and 6.4. In these tables we use the convention that $f_k = f(x + hk)$ for $k = -3, -2, 1, 0, 1, 2, 3$.

Table 6.3 Central-difference formulas of order $O(h^2)$.

$$f'(x) \approx \frac{f_1 - f_{-1}}{2h}$$

$$f''(x) \approx \frac{f_1 - 2f_0 + f_{-1}}{h^2}$$

$$f^{(3)}(x) \approx \frac{f_2 - 2f_1 + 2f_{-1} - f_{-2}}{2h^3}$$

$$f^{(4)}(x) \approx \frac{f_2 - 4f_1 + 6f_0 - 4f_{-1} + f_{-2}}{h^4}$$

Table 6.4   Central-difference formulas of order $O(h^4)$

$$f'(x) \approx \frac{-f_2 + 8f_1 - 8f_{-1} + f_{-2}}{12h}$$

$$f''(x) \approx \frac{-f_2 + 16f_1 - 30f_0 + 16f_{-1} - f_{-2}}{12h^2}$$

$$f^{(3)}(x) \approx \frac{-f_3 + 8f_2 - 13f_1 + 13f_{-1} - 8f_{-2} + f_{-3}}{8h^3}$$

$$f^{(4)}(x) \approx \frac{-f_3 + 12f_2 - 39f_1 + 56f_0 - 39f_{-1} + 12f_{-2} - f_{-3}}{6h^4}$$

For illustration, we will derive the formula for $f''(x)$ of order $O(h^2)$ in Table 6.3. Start with the Taylor expansions

(1) $$f(x + h) = f(x) + hf'(x) + \frac{h^2 f''(x)}{2} + \frac{h^3 f^{(3)}(x)}{6} + \frac{h^4 f^{(4)}(x)}{24} + \dots$$

and

(2) $$f(x - h) = f(x) - hf'(x) + \frac{h^2 f''(x)}{2} - \frac{h^3 f^{(3)}(x)}{6} + \frac{h^4 f^{(4)}(x)}{24} - \dots.$$

Adding equations (1) and (2) will eliminate the terms involving the odd derivatives $f'(x), f^{(3)}(x), f^{(5)}(x), \dots$:

(3) $$f(x + h) + f(x - h) = 2f(x) + \frac{2h^2 f''(x)}{2} + \frac{2h^4 f^{(4)}(x)}{24} + \dots.$$

Solving equation (3) for $f''(x)$ yields

(4)
$$f''(x) = \frac{f(x + h) - 2f(x) + f(x - h)}{h^2} - \frac{2h^2 f^{(4)}(x)}{4!}$$
$$- \frac{2h^4 f^{(6)}(x)}{6!} - \dots - \frac{2h^{2k-2} f^{(2k)}(x)}{(2k)!} - \dots -.$$

If the series in (4) is truncated at the fourth derivative, there exists a value c that lies in $[x - h, x + h]$ so that

(5) $$f''(x) = \frac{f_1 - 2f_0 + f_{-1}}{h^2} - \frac{h^2 f^{(4)}(c)}{12}.$$

This gives us the desired formula for approximating $f''(x)$:

(6) $$f''(x) \approx \frac{f_1 - 2f_0 + f_{-1}}{h^2}.$$

Example 6.4 Let $f(x) = \cos(x)$.

(a) Use formula (6) with $h = 0.1, 0.01$, and 0.001 and find approximations to $f''(0.8)$. Carry nine decimal places in all the calculations.

(b) Compare with the true value $f''(0.8) = -\cos(0.8)$.

Solution. (a) The calculation for $h = 0.01$ is

$$f''(0.8) \approx \frac{f(0.81) - 2f(0.80) + f(0.79)}{0.0001}$$

$$\approx \frac{0.689498433 - 2(0.696706709) + 0.703845316}{0.0001} \approx -0.696690000.$$

(b) The error in this approximation is -0.000016709. The other calculations are summarized in Table 6.5. The error analysis will illuminate this example and show why $h = 0.01$ was best.

Table 6.5 Numerical approximations to $f''(x)$
for Example 6.4

Step Size	Approximation by Formula (6)	Error Using Formula (6)
$h = 0.1$	-0.696126300	-0.000580409
$h = 0.01$	-0.696690000	-0.000016709
$h = 0.001$	-0.696000000	-0.000706709

Error Analysis

Let $f_k = y_k + e_k$, where e_k is the error in computing $f(x_k)$, including noise in measurement and round-off error. Then formula (6) can be written

(7)
$$f''(x) = \frac{y_1 - 2y_0 + y_{-1}}{h^2} + E(f, h).$$

The error term $E(f, h)$ for the numerical derivative (7) will have a part due to round-off and a part due to truncation:

(8)
$$E(f, h) = \frac{e_1 - 2e_0 + e_{-1}}{h^2} - \frac{h^2 f^{(4)}(c)}{12}.$$

If it is assumed that each error e_k is of the magnitude e with signs which accumulate errors, and the $|f^{(4)}(x)| \leq M$, then we get the following error bound:

(9)
$$|E(f, h)| \leq \frac{4e}{h^2} + \frac{Mh^2}{12}.$$

If h is small, then the contribution $4e/h^2$, due to round-off, is large. When h is large, the contribution $Mh^2/12$ is large. The optimum step size will minimize the quantity

$$(10) \qquad\qquad g(h) = \frac{4e}{h^2} + \frac{Mh^2}{12}.$$

Setting $g'(h) = 0$ results in $-8e/h^3 + Mh/6 = 0$, which yields the equation $h^4 = 48e/M$, from which we obtain the optimal value:

$$(11) \qquad\qquad h = \left(\frac{48e}{M}\right)^{1/4}.$$

When formula (11) is applied to Example 6.4, we can use the bound $|f^{(4)}(x)| \le |\cos(x)| \le 1 = M$ and the value $e = 0.5 \times 10^{-9}$. The optimal step size is $h = (24 \times 10^{-9}/1)^{1/4} = 0.01244666$, and we see that $h = 0.01$ was closest to the optimal value.

Since the portion of the error due to round-off is inversely proportional to the square of h, this term grows when h gets small. This is some-times referred to as the *step-size dilemma*. One partial solution to this problem is to use a formula of higher order so that a larger value of h will produce the desired accuracy. The formula for $f''(x)$ of order $O(h^4)$ in Table 6.4 is

$$(12) \qquad\qquad f''(x) = \frac{-f_2 + 16f_1 - 30f_0 + 16f_{-1} - f_{-2}}{12h^2} + E(f, h).$$

The error term for (12) has the form

$$(13) \qquad\qquad E(f, h) = \frac{16e}{3h^2} + \frac{h^4 f^{(6)}(c)}{90},$$

where c lies in the interval $[x - 2h, x + 2h]$. A bound for $|E(f, h)|$ is

$$(14) \qquad\qquad |E(f, h)| \le \frac{16e}{3h^2} + \frac{Mh^4}{90},$$

where $|f^{(6)}(x)| \le M$. The optimal value for h given by the formula

$$(15) \qquad\qquad h = \left(\frac{240e}{M}\right)^{1/6}$$

Example 6.5 Let $f(x) = \cos(x)$.
(a) Use formula (12) with $h = 1.0, 0.1$, and 0.01 and find approximations to $f''(0.8)$. Carry nine decimal places in all the calculations.
(b) Compare with the true value $f''(0.8) = -\cos(0.8)$.
(c) Determine the optimal step size.

Solution. (a) The calculation for $h = 0.1$ is

$f''(0.8)$

$$\approx \frac{-f(1.0) + 16f(0.9) - 30f(0.8) + 16f(0.7) - f(0.6)}{0.12}$$

$$\approx \frac{-0.540302306 + 9.945759488 - 20.90120127 + 12.23747499 - 0.825335615}{0.12}$$

$$\approx -0.696705958.$$

(b) The error in this approximation is -0.000000751. The other calculations are summarized in Table 6.6.

Table 6.6 Numerical approximations to $f''(x)$ for Example 6.5

Step Size	Approximation by Formula (12)	Error Using Formula (12)
$h = 1.0$	-0.689625413	-0.007081296
$h = 0.1$	-0.696705958	-0.000000751
$h = 0.01$	-0.696690000	-0.000016709

(c) When formula (15) is applied, we can use the bound $|f^{(6)}(x)| \leq |\cos(x)| \leq 1 = M$ and the value $e = 0.5 \times 10^{-9}$. These values give the optimal step size $h = (120 \times 10^{-9}/1)^{1/6} = 0.070231219$.

Generally speaking, if numerical differentiation is performed, only about half the accuracy of which the computer is capable is obtained. This severe loss of significant digits will almost always occur unless we are fortunate to find a step size that is optimal. Hence we must always proceed with caution when numerical differentiation is performed. The difficulties are more pronounced when working with experimental data, where the function values have been rounded to only a few digits. If a numerical derivatives must be obtained from data, then we should consider curve fitting, by using least-squares techniques, and differentiate the formula for the curve.

Other Formulas

If the function cannot be evaluated at abscissas that lie on both sides of x, then the central-difference formula cannot be used to approximate the derivatives. When the function can be evaluated at equally spaced abscissas that lie to the right {left} of x, then the forward (backward)-difference formula can be used. The formulas can be derived using different methods, and the proofs can rely on Taylor series, Lagrange interpolation polynomials, or Newton interpolation polynomials. Some of the common forward/backward-difference formulas are give in Table 6.7.

Table 6.7 Forward/backward-differences formulas of order $O(h^2)$

$$f'(x) \approx \frac{-3f_0 + 4f_1 - f_2}{2h} \qquad \binom{\text{forward}}{\text{difference}}$$

$$f'(x) \approx \frac{3f_0 - 4f_{-1} + f_{-2}}{2h} \qquad \binom{\text{backward}}{\text{difference}}$$

$$f''(x) \approx \frac{2f_0 - 5f_1 + 4f_2 - f_3}{h^2} \qquad \binom{\text{forward}}{\text{difference}}$$

$$f''(x) \approx \frac{2f_0 - 5f_{-1} + 4f_{-2} - f_{-3}}{h^2} \qquad \binom{\text{backward}}{\text{difference}}$$

$$f^{(3)}(x) \approx \frac{-5f_0 + 18f_1 - 24f_2 + 14f_3 - 3f_4}{2h^3}$$

$$f^{(3)}(x) \approx \frac{5f_0 - 18f_{-1} + 24f_{-2} - 14f_{-3} + 3f_{-4}}{2h^3}$$

$$f^{(4)}(x) \approx \frac{3f_0 - 14f_1 + 26f_2 - 24f_3 + 11f_4 - 2f_5}{h^4}$$

$$f^{(4)}(x) \approx \frac{3f_0 - 14f_{-1} + 26f_{-2} - 24f_{-3} + 11f_{-4} - 2f_{-5}}{h^4}$$

Differentiation of the Newton Polynomial

In this section we show the relationship between the three formulas of order $O(h^2)$ for approximating $f'(x)$, and a general algorithm is given for computing the numerical derivative. In Section 4.3 we saw that the Newton polynomial $P(t)$ of degree $N = 2$ that approximates $f(t)$ using the nodes t_0, t_1, and t_2 is

(16)
$$P(t) = a_0 + a_1(t - t_0) + a_2(t - t_0)(t - t_1),$$

where $a_0 = f(t_0)$, $a_1 = [f(t_1) - f(t_0)]/(t_1 - t_0)$, and

$$a_2 = \left[\frac{f(t_2) - f(t_1)}{t_2 - t_1} - \frac{f(t_1) - f(t_0)}{t_1 - t_0} \right] / (t_2 - t_0).$$

The derivative of $P(t)$ is

(17)
$$P'(t) = a_1 + a_2[(t - t_0) + (t - t_1)].$$

and when it is evaluated at $t = t_0$ the result is

(18)
$$P'(t_0) = a_1 + a_2(t_0 - t_1) \approx f'(t_0)$$

Observe that the nodes $\{t_k\}$ do not need to be equally spaced for formulas (16)–(18) to hold. Choosing the abscissas in different orders will produce different formulas for approximating $f'(x)$.

Case (i) If $t_0 = x$, $t_1 = x + h$, and $t_2 = x + 2h$, then

$$a_1 = \frac{f(x+h) - f(x)}{h},$$

$$a_2 = \frac{f(x) - 2f(x+h) + f(x+2h)}{2h^2}.$$

When these values are substituted into (18) we get

$$P'(x) = \frac{f(x+h) - f(x)}{h} + \frac{-f(x) + 2f(x+h) - f(x+2h)}{2h}.$$

This is simplified to obtain

(19)
$$P'(x) = \frac{-3f(x) + 4f(x+h) - f(x+2h)}{2h} \approx f'(x)$$

which is the second-order forward-difference formula for $f'(x)$.

Case (ii) If $t_0 = x$, $t_1 = x + h$, and $t_2 = x - h$, then

$$a_1 = \frac{f(x+h) - f(x)}{h},$$

$$a_2 = \frac{f(x+h) - 2f(x) + f(x-h)}{2h^2}.$$

When these values are substituted into (18) we get

$$P'(x) = \frac{f(x+h) - f(x)}{h} + \frac{-f(x+h) + 2f(x) - f(x-h)}{2h}.$$

This is simplified to obtain

(20)
$$P'(x) = \frac{f(x+h) - f(x-h)}{2h} \approx f'(x)$$

which is the second-order central-difference formula for $f'(x)$.

Case (iii) If $t_0 = x$, $t_1 = x - h$, and $t_2 = x - 2h$, then

$$a_1 = \frac{f(x) - f(x-h)}{h},$$

$$a_2 = \frac{f(x) - 2f(x-h) + f(x-2h)}{2h^2}.$$

These values are substituted into (18) and simplified to get

$$(21) \qquad P'(x) = \frac{3f(x) - 4f(x-h) + f(x-2h)}{2h} \approx f'(x)$$

which is the second-order backward-difference formula for $f'(x)$.

The Newton polynomial $P(t)$ of degree N that approximates $f(t)$ using the nodes $t_0, t_1, \ldots, t_N$ is

$$(22) \qquad P(t) = a_0 + a_1(t - t_0) + a_2(t - t_0)(t - t_1)$$
$$+ a_3(t - t_0)(t - t_1)(t - t_2) + \ldots + a_N(t - t_0)(t - t_1)(t - t_{N-1}).$$

The derivative of $P(t)$ is

$$(23) \qquad P'(t) = a_1 + a_2[(t - t_0) + (t - t_1)]$$
$$+ a_3[(t - t_0)(t - t_1) + (t - t_0)(t - t_2) + (t - t_1)(t - t_2)] + \ldots$$
$$+ a_N \sum_{\substack{k=0}}^{N-1} \prod_{\substack{j=0 \\ j \neq k}}^{N-1} (t - t_j).$$

When $P'(t)$ is evaluated at $t = t_0$ several of the terms in the summation are zero, and $P'(t_0)$ has the simpler form

$$(24) \qquad P'(t_0) = a_1 + a_2(t_0 - t_1) + a_3(t_0 - t_1)(t_0 - t_2) + \ldots$$
$$+ a_N(t_0 - t_1)(t_0 - t_2)(t_0 - t_3) \ldots (t_0 - t_{N-1}).$$

The kth partial sum on the right side of equation (24) is the derivative of the Newton polynomial of degree k based on the first k nodes. If

$$|t_0 - t_1| \leq |t_0 - t_2| \leq \ldots \leq |t_0 - t_N|, \quad \text{and if} \quad \{(t_j, 0)\}_{j=0}^{j=N}$$

forms a set of $N + 1$ equally spaced points on the real axis, the kth partial sum is an approximation to $f'(t_0)$ of order $O(h^{k-1})$.

Suppose that $N = 5$. If the five nodes are $t_k = x + hk$, then (24) is an equivalent way to compute the forward-difference formula for $f'(x)$ of order $O(h^4)$. If the five nodes $\{t_k\}$ are chosen to be $t_0 = x$, $t_1 = x + h$, $t_2 = x - h$, $t_3 = x + 2h$, and $t_4 = x - 2h$, then (24) is the central-difference formula for $f'(x)$ of order $O(h^4)$. When the five nodes are $t_k = x - hk$, then (24) is the backward-difference formula for $f'(x)$ of order $O(h^4)$.

The following algorithm is an extension of Algorithm 4.4 and can be used to implement formula (24). It uses the space-saving modifications mentioned in Exercise 13 of Section 4.3. Note that the nodes do not need to be equally spaced. Also, it computes the derivative at only one point, $f'(x_0)$.

Algorithm 6.3 [*Numerical Differentiation Based on $N + 1$ Nodes*] Let $P(x)$ be the Newton polynomial approximation to $f(x)$ of degree N based on the $N + 1$ nodes $x_0, x_1, \ldots, x_N$:

$$P(x) = a_0 + a_1(x - x_0) + a_2(x - x_0)(x - x_1) + a_3(x - x_0)(x - x_1)(x - x_2)$$
$$+ \ldots + a_N(x - x_0)(x - x_1) \ldots (x - x_{N-1}).$$

The algorithm computes $P'(x_0)$, which is the approximation for the numerical derivative, that is, $f'(x_0) \approx P'(x_0)$.

```
INPUT N                              {Degree of P(x)}
INPUT X(0), X(1), ..., X(N)          {Interpolation of nodes}
INPUT Y(0), Y(1), ..., Y(N)          {Ordinates of the points}
FOR  K = 0  TO  N  DO                {Initialize the coefficient array}
└─── A(K) := Y(K)
FOR  J = 1  TO  N  DO                {Compute the divided
    FOR  K = N  DOWNTO  J  DO                      differences}
    └──── └─── A(K) := [A(K) − A(K−1)]/[X(K) − X(K−J)]


X := X(0)                            {Compute the derivative at x₀}
Df := A(1), Prod := 1                {Initialize the variables}
FOR  K = 2  TO  N  DO
    |   Prod := Prod * [X − X(K−1)]
    └── Df := Df + Prod * A(K)


PRINT "The numerical derivative f'(x₀) is" Df          {Output}
```

Exercises for Numerical Differentiation Formulas

1. Let $f(x) = \ln(x)$ and carry eight or nine decimal places.
   (a) Use formula (6) with $h = 0.05$ to approximate $f''(5)$.
   (b) Use formula (6) with $h = 0.01$ to approximate $f''(5)$.
   (c) Use formula (12) with $h = 0.1$ to approximate $f''(5)$.
   (d) Which answer—(a), (b), or (c)—is more accurate?

2. Let $f(x) = \cos(x)$ and carry eight or nine decimal places.
   (a) Use formula (6) with $h = 0.05$ to approximate $f''(1)$.
   (b) Use formula (6) with $h = 0.01$ to approximate $f''(1)$.
   (c) Use formula (12) with $h = 0.1$ to approximate $f''(1)$.
   (d) Which answer—(a), (b), or (c)—is more accurate?

3. Consider the table for $f(x) = \ln(x)$ rounded to four decimal places.

x	$f(x) = \ln(x)$
4.90	1.5892
4.95	1.5994
5.00	1.6094
5.05	1.6194
5.10	1.6292

(a) Use formula (6) and $h = 0.05$ to approximate $f''(5)$.
(b) Use formula (6) and $h = 0.10$ to approximate $f''(5)$.
(c) Use formula (12) and $h = 0.05$ to approximate $f''(5)$.
(d) Which answer—(a), (b), or (c)—is more accurate?

4. Consider the table for $f(x) = \cos(x)$ rounded to four decimal places.

x	$f(x) = \cos(x)$
0.90	0.6216
0.95	0.5817
1.00	0.5403
1.05	0.4976
1.10	0.4536

(a) Use formula (6) and $h = 0.05$ to approximate $f''(1)$.
(b) Use formula (6) and $h = 0.10$ to approximate $f''(1)$.
(c) Use formula (12) and $h = 0.05$ to approximate $f''(1)$.
(d) Which answer—(a), (b), or (c)—is more accurate?

5. Use the numerical differentiation formula (6) and $h = 0.01$ to approximate $f''(1)$ for the functions
(a) $f(x) = x^2$ (b) $f(x) = x^4$

6. Use the numerical differentiation formula (12) and $h = 0.1$ to approximate $f''(1)$ for the functions
(a) $f(x) = x^4$ (b) $f(x) = x^6$

7. Use the Taylor expansions for $f(x + h), f(x - h), f(x + 2h)$, and $f(x - 2h)$ and derive the central-difference formula:

$$f^{(3)}(x) \approx \frac{f(x + 2h) - 2f(x + h) + 2f(x - h) - f(x - 2h)}{2h^3}.$$

8. Use the Taylor expansions for $f(x + h), f(x - h), f(x + 2h)$, and $f(x - 2h)$ and derive the central-difference formula:

$$f^{(4)}(x) \approx \frac{f(x + 2h) - 4f(x + h) + 6f(x) - 4f(x - h) + f(x - 2h)}{h^4}.$$

9. Find approximations to $f'(x_k)$ of order $O(h^2)$ at each of the four points in the tables.

(a)

x	$f(x)$
0.0	0.989992
0.1	0.999135
0.2	0.998295
0.3	0.987480

(b)

x	$f(x)$
0.0	0.141120
0.1	0.041581
0.2	-0.058374
0.3	-0.157746

10. Use the approximations

$$f'\left(\frac{x+h}{2}\right) \approx \frac{f_1 - f_0}{h} \quad \text{and} \quad f'\left(\frac{x-h}{2}\right) \approx \frac{f_0 - f_{-1}}{h}$$

and derive the approximation

$$f''(x) \approx \frac{f_1 - 2f_0 + f_{-1}}{h^2}.$$

11. Use formulas (16)–(18) and derive a formula for $f'(x)$ based on the abscissas $t_0 = x$, $t_1 = x + h$, and $t_2 = x + 3h$.

12. Use formulas (16)–(18) and derive a formula for $f'(x)$ based on the abscissas $t_0 = x$, $t_1 = x - h$, and $t_2 = x + 2h$.

13. The numerical solution of a certain differential equation requires an approximation to $f''(x) + f'(x)$ of order $O(h^2)$.
    (a) Find the central-difference formula for $f''(x) + f'(x)$ by adding the formulas for $f'(x)$ and $f''(x)$ of order $O(h^2)$.
    (b) Find the forward-difference formula for $f''(x) + f'(x)$ by adding the formulas for $f'(x)$ and $f''(x)$ of order $O(h^2)$.
    (c) What would happen if a formula for $f'(x)$ of order $O(h^4)$ were added to a formula for $f''(x)$ of order $O(h^2)$?

14. Critique the following argument. Taylor's formula can be used to get the representations

$$f(x+h) = f(x) + hf'(x) + \frac{h^2 f''(x)}{2} + \frac{h^3 f'''(c)}{6}$$

and

$$f(x-h) = f(x) - hf'(x) + \frac{h^2 f''(x)}{2} - \frac{h^3 f'''(c)}{6}.$$

Adding these quantities results in

$$f(x+h) + f(x-h) = 2f(x) + h^2 f''(x).$$

which can be solved to obtain an exact formula for $f''(x)$:

$$f''(x) = \frac{f(x+h) - 2f(x) + f(x-h)}{h^2}.$$

15. Write a report on the forward-difference operator for numerical differentiation.

16. Write a report on the backward-difference operator for numerical differentiation.

17. Write a report on lozenge diagrams for numerical differentiation.

18. Modify Algorithm 6.3 so that it will calculate $P'(x_M)$ for $M = 1, 2, \ldots, N + 1$.

6.3 Minimization of a Function

An important topic in calculus is finding the local extrema of a function. The technique employed in calculus is to set the derivative equal to zero and solve the resulting equation. We can then adapt a root-finding technique such as the secant method to locate the zeros of the derivative. We shall first review some definitions and theorems.

Functions of One Variable

Definition 6.1 [Local extremum] The function $f(x)$ is said to have a *local minimum* at p if there exists an open interval I containing p such that $f(p) \leq f(x)$ for all x in I. Similarly, $f(x)$ is said to have a *local maximum* at p if $f(p) \geq f(x)$ for all x in I. If $f(x)$ has either a local minimum or local maximum at p, then $f(x)$ is said to have a *local extremum* at p.

Definition 6.2 [Increasing and decreasing] A function $f(x)$ defined on an interval I is said to be *increasing* on I if and only if $f(x_1) < f(x_2)$ whenever $x_1 < x_2$ and x_1, x_2 are any numbers in I. Similarly, $f(x)$ is said to be *decreasing* on I if and only if $f(x_1) > f(x_2)$ whenever $x_1 < x_2$ and x_1, x_2 are any numbers in I.

Theorem 6.1 [Increasing and decreasing] Suppose that $f(x)$ is differentiable on $I = (a, b)$. If $f'(x) > 0$ for all x in I, then $f(x)$ is increasing on I. Similarly, if $f'(x) < 0$ for all x in I, then $f(x)$ is decreasing on I.

Theorem 6.2 [Horizontal tangent at an extremum] Suppose that $f(x)$ is differentiable on (a, b). If $f(x)$ has a local extremum at the point p in the interval (a, b), then $f'(p) = 0$.

Theorem 6.3 [Local extrema] Let $f(x)$ be continuous on (a, b). If there exists a value p in the interval and:

Case (i) If $f(x)$ is decreasing on (a, p) and increasing on (p, b), then $f(x)$ has a local minimum at p.

Case (ii) If $f(x)$ is increasing on (a, p) and decreasing on (p, b), then $f(x)$ has a local maximum at p.

Corollary 6.1 Suppose that $f'(x)$ is continuous on (a, b). If there exists a value p in the interval and:

Case (i) If $f'(x) < 0$ on (a, p) and $f'(x) > 0$ on (p, b), then $f(x)$ has a local minimum at p.

Case (ii) If $f'(x) > 0$ on (a, p) and $f'(x) < 0$ on (p, b), then $f(x)$ has a local maximum at p.

Theorem 6.4 [Second derivative test] Assume that $f(x), f'(x), f''(x)$ are defined on the interval (a, b). Suppose that p can be found in the interval and $f'(p) = 0$.

Case (i)   If $f'(p) = 0$ and $f''(p) > 0$, then $f(x)$ has a local minimum at $x = p$.
Case (ii) If $f'(p) = 0$ and $f''(p) < 0$, then $f(x)$ has a local maximum at $x = p$.
Case (iii) If $f'(p) = 0$ and $f''(p) = 0$, then the test is inconclusive.

Example 6.6 Use the second derivative test to classify the local extrema of $f(x) = x^3 + x^2 - x + 1$ on the interval $[-2, 2]$.

Solution. The derivative is $f'(x) = 3x^2 + 2x - 1 = (3x - 1)(x + 1)$, and the second derivative is $f''(x) = 6x + 2$. There are two points where $f'(x) = 0$ (i.e. $x = \frac{1}{3}$, -1).

Case (i) At $x = \frac{1}{3}$ we find that $f'(\frac{1}{3}) = 0$ and $f''(\frac{1}{3}) = 4 > 0$, so that $f(x)$ has a local minimum at $x = \frac{1}{3}$.

Case (ii) At $x = -1$ we find that $f'(-1) = 0$ and $f''(-1) = -4 < 0$, so that $f(x)$ has a local maximum at $x = -1$.

Search Method

Another method for finding the minimum of $f(x)$ is to evaluate the function many times and search for a local minimum. In order to reduce the number of function evaluations, it is important to have a good strategy for determining where $f(x)$ is evaluated. One of the most efficient methods is called the *golden ratio search*, which is named for the ratio's involvement in selecting the points.

The Golden Ratio

Let the initial interval be $[0, 1]$. If $0.5 < r < 1$, then $0 < 1 - r < 0.5$ and the interval is divided into three subintervals $[0, 1 - r], [1 - r, r]$ and $[r, 1]$. A decision process is used to either squeeze from the right and get the new interval $[0, r]$ or squeeze from the left and get $[1 - r, 1]$. Then this new subinterval is divided into three subintervals in the same ratio as was $[0, 1]$.

We want to choose r so that one of the old points will be in the correct position with respect to the new interval as shown in Figure 6.1. This implies that the ratio $(1 - r):r$ be the same as $r:1$. Hence r satisfies the equation $1 - r = r^2$, which can be expressed as a quadratic equation $r^2 + r - 1 = 0$. The solution r satisfying $0.5 < r < 1$ is found to be $r = (5^{1/2} - 1)/2$.

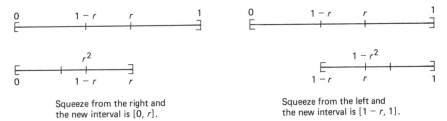

Squeeze from the right and   Squeeze from the left and
the new interval is $[0, r]$.   the new interval is $[1 - r, 1]$.

Figure 6.1. The intervals involved in the golden ratio search.

To use the golden search for finding the minimum of $f(x)$, a special condition must be met to ensure that there is a proper minimum in the interval.

Definition 6.3 [Unimodal function] The function $f(x)$ is *unimodal* on $[a, b]$ if there exists a unique number p in $[a, b]$ such that

(1) $$f(x) \text{ is decreasing on } [a, p]$$

and

(2) $$f(x) \text{ is increasing on } [p, b].$$

If $f(x)$ is known to be unimodal on $[a, b]$, it is possible to replace the interval with a subinterval on which $f(x)$ takes on its minimum value. The golden search requires that two interior points $c = a + (1 - r)(b - a)$ and $d = a + r(b - a)$ be used, where r is the golden ratio mentioned above. This results in $a < c < d < b$. The condition that $f(x)$ is unimodal guarantees that the function values $f(c)$ and $f(d)$ are less than max $\{f(a), f(b)\}$. We have two cases to consider (see Figure 6.2).

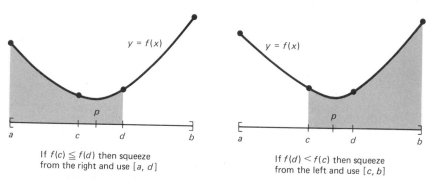

Figure 6.2. The decision process for the golden search.

If $f(c) \leq f(d)$, then the minimum must occur in the subinterval $[a, d]$ and we replace b with d and continue the search in the new subinterval. If $f(d) < f(c)$, then the minimum must occur in $[c, b]$ and we replace a with c and continue the search. The next example compares the root-finding method with the golden search method.

Example 6.7 Find the minimum of the unimodal function $f(x) = x^2 - \sin(x)$ in $[0, 1]$.

Solution by solving $f'(x) = 0$. A root-finding method can be used to determine where the derivative $f'(x) = 2x - \cos(x)$ is zero. Since $f'(0) = -1$ and $f'(1) = 1.4596977$, a root of $f'(x)$ lies in the interval $[0, 1]$. Starting with $p_0 = 0$ and $p_1 = 1$, Table 6.8 shows the iterations.

Table 6.8   Secant method for solving
$f'(x) = 2x - \cos(x) = 0$

k	p_k	$2p_k - \cos(p_k)$
0	0.0000000	−1.00000000
1	1.0000000	1.45969769
2	0.4065540	−0.10538092
3	0.4465123	−0.00893398
4	0.4502137	0.00007329
5	0.4501836	−0.00000005

The conclusion from applying the secant method is that $f'(0.4501836) = 0$. The second derivative is $f''(x) = x + \sin(x)$ and we compute $f''(0.4501863) = 0.8853196$ > 0. Hence the minimum value is $f(0.4501863) = -0.2324656$.

Solution using the golden search. At each step, the function values $f(c)$ and $f(d)$ are compared and a decision is made as to whether to continue the search in $[a, d]$ or $[c, b]$. Some of the computations are shown in Table 6.9.

Table 6.9 Golden search for the minimum of $f(x) = x^2 - \sin(x)$

k	a_k	c_k	d_k	b_k	$f(c_k)$	$f(d_k)$
0	0.0000000	0.3819660	0.6180340	1	−0.22684748	−0.19746793
1	0.0000000	0.2360680	0.3819660	0.6180340	−0.17815339	−0.22684748
2	0.2360680	0.3819660	0.4721360	0.6180340	−0.22684748	−0.23187724
3	0.3819660	0.4721360	0.5278640	0.6180340	−0.23187724	−0.22504882
4	0.3819660	0.4376941	0.4721360	0.5278640	−0.23227594	−0.23187724
5	0.3819660	0.4164079	0.4376941	0.4721360	−0.23108238	−0.23227594
6	0.4164079	0.4376941	0.4508497	0.4721360	−0.23227594	−0.23246503
⋮	⋮	⋮	⋮	⋮	⋮	⋮
21	0.4501574	0.4501730	0.4501827	0.4501983	−0.23246558	−0.23246558
22	0.4501730	0.4501827	0.4501886	0.4501983	−0.23246558	−0.23246558
23	0.4501827	0.4501886	0.4501923	0.4501983	−0.23246558	−0.23246558

At the twenty-third iteration the interval has been narrowed down to $[a_{23}, b_{23}] = [0.4501827, 0.4501983]$. This interval has width 0.0000156. However, the computed function values at the endpoints agree to eight decimal places [i.e., $f(a_{23}) \approx -0.23246558 \approx f(b_{23})$], hence the algorithm is terminated. A problem in using search methods is that the function may be flat near the minimum and this limits the accuracy that can be obtained. The secant method was able to find the more accurate answer $p_5 = 0.4501836$.

Although the golden search method is slower in this example, it has the desirable feature that it can apply in cases where $f(x)$ is not differentiable.

Algorithm 6.4 [*Golden Search for a Minimum*] Given that $f(x)$ is unimodal on the interval $[a, b]$, the (unique) minimum of $f(x)$ can be approximated by a golden search.

Delta := 10^{-5}	{Tolerance for interval width}
Epsilon := 10^{-7}	{Tolerance for $\|f(b) - f(a)\|$}
Rone := $[5^{1/2} - 1]/2$	{Determine constants
Rtwo := Rone∗Rone	for golden search}
INPUT A, B	{Input endpoints of interval}
YA := F(A), YB := F(B), H := B−A	{Compute function values}
C := A+Rtwo∗H, YC := F(C)	{Compute two
D := A+Rone∗H, YD := F(D)	interior points}

```
WHILE  |YB−YA|>Epsilon  OR  H>Delta   DO
         IF  YC < YD   THEN
               B := D, YB := YD                          {Squeeze from the right}
               D := C, YD := YC, H := B−A
               C := A + Rtwo*H, YC := F(C)
         ELSE
               A := C, YA := YC                          {Squeeze from the left}
               C := D, YC := YD, H := B−A
               D := A + Rone*H, YD := F(D)
         ENDIF

         P := A, YP := YA
         IF  YB < YA   THEN   P := B, YP := YB

         PRINT 'The minimum of f(x) occurs at' P          {Output}
         PRINT 'The accuracy is +−' H
         PRINT 'The minimum of the function f(P) is' YP
```

Finding Extreme Values of f(x, y)

Definition 6.1 is easily extended to functions of several variables. Suppose that $f(x, y)$ is defined in the region

(3)
$$R = \{(x, y) : (x - p)^2 + (y - q)^2 < r^2\}.$$

The function has a local minimum at (p, q) provided that

(4)
$$f(p, q) \leq f(x, y) \qquad \text{for each point } (x, y) \text{ in } R.$$

The function has a local maximum at (p, q) provided that

(5)
$$f(p, q) \geq f(x, y) \qquad \text{for each point } (x, y) \text{ in } R.$$

The second derivative test for an extreme value is an extension of Theorem 6.4.

Theorem 6.5 [Second derivative test] Let $f(x, y)$ and its first- and second-order partial derivatives be continuous on a region R. Suppose that (p, q) is a point in R where both $f_x(p, q) = 0$ and $f_y(p, q) = 0$. The higher-order partial derivatives are used to determine which case we have.

Case (i)   If $f_{xx}(p, q)f_{yy}(p, q) - f_{xy}^2(p, q) > 0$ and $f_{xx}(p, q) > 0$, then $f(x, y)$ has a local minimum at (p, q).

Case (ii)   If $f_{xx}(p, q)f_{yy}(p, q) - f_{xy}^2(p, q) > 0$ and $f_{xx}(p, q) < 0$, then $f(x, y)$ has a local maximum at (p, q).

Case (iii)   If $f_{xx}(p, q)f_{yy}(p, q) - f_{xy}^2(p, q) < 0$, then $f(x, y)$ does not have a local extremum at (p, q).

Case (iv)   If $f_{xx}(p, q)f_{yy}(p, q) - f_{xy}^2(p, q) = 0$, then the test is inconclusive.

Example 6.8 Find the minimum of $f(x, y) = x^2 - 4x + y^2 - y - xy$.

Solution. The first-order partial derivatives are

(6)
$$f_x(x, y) = 2x - 4 - y \quad \text{and} \quad f_y(x, y) = 2y - 1 - x.$$

Setting these derivatives equal to zero yields the linear system

(7)
$$2x - y = 4 \qquad -x + 2y = 1.$$

The solution point to (7) is found to be $x = 3, y = 2$. The second-order partial derivatives of $f(x, y)$ are

$$f_{xx}(x, y) = 2 \qquad f_{yy}(x, y) = 2 \qquad f_{xy}(x, y) = -1.$$

It is easy to see that we have case (i) of Theorem 6.5, that is,

$$f_{xx}(3, 2) f_{yy}(3, 2) - f_{xy}^2(3, 2) = 3 > 0 \quad \text{and} \quad f_{xx}(3, 2) = 2 > 0.$$

Hence $f(x, y)$ has a local minimum $f(3, 2) = -7$ at the point $(3, 2)$.

The Nelder–Mead Method

A simplex method for finding a local minimum of a function of several variables has been devised by Nelder and Mead. For two variables, a simplex is a triangle, and the method is a pattern search that compares function values at the three vertices of a triangle. The worst vertex, where $f(x, y)$ is largest, is rejected and replaced with a new vertex. A new triangle is formed and the search is continued. The process generates a sequence of triangles (which might have different shapes), for which the function values at the vertices get smaller and smaller. The size of the triangles is reduced and the coordinates of the minimum point are found.

The algorithm is stated using the term simplex (a generalized triangle in N dimensions) and will find the minimum of a function of N variables. It is effective and computationally compact.

The Initial Triangle BGW

Let $f(x, y)$ be the function that is to be minimized. To start we are given three vertices of a triangle; $\mathbf{V}_k = (x_k, y_k), k = 1, 2, 3$. The function $f(x, y)$ is then evaluated at each of the three points $z_k = f(x_k, y_k)$ for $k = 1, 2, 3$. The subscripts are then reordered so that $z_1 \leq z_2 \leq z_3$. We use the notation

(8)
$$\mathbf{B} = (x_1, y_1) \qquad \mathbf{G} = (x_2, y_2) \qquad \mathbf{W} = (x_3, y_3)$$

to help remember that $\mathbf{B}$ is the best vertex, $\mathbf{G}$ is good (next to best), and $\mathbf{W}$ is the worst vertex.

Midpoint of the Good Side

The construction process uses the midpoint of the line segment joining $\mathbf{B}$ and $\mathbf{G}$. It is found by averaging the coordinates;

(9)
$$\mathbf{M} = \frac{\mathbf{B} + \mathbf{G}}{2} = \left(\frac{x_1 + x_2}{2}, \frac{y_1 + y_2}{2} \right).$$

Reflection Using the Point **R**

The function decreases as we move along the side of the triangle from **W** to **B**, and it decreases as we move along the side from **W** to **G**. Hence it is feasible that $f(x, y)$ takes on smaller values at points that lie away from **W** on the opposite side of the line between **B** and **G**. We choose a test point **R** that is obtained by "reflecting" the triangle through the side $\widehat{BG}$. To determine **R**, we first find the midpoint **M** of the side $\widehat{BG}$. Then draw the line segment from **W** to **M** and call its length d. This last segment is extended a distance d through **M** to locate the point **R** (see Figure 6.3). The vector formula for **R** is

(10)
$$\mathbf{R} = \mathbf{M} + (\mathbf{M} - \mathbf{W}) = 2\mathbf{M} - \mathbf{W}.$$

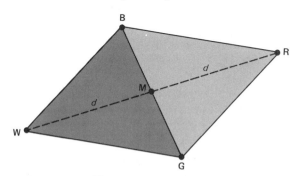

Figure 6.3. The triangle $\widehat{BGW}$ and midpoint **M** and reflected point **R** for the Nelder-Mead method.

Expansion Using the Point **E**

If the function value at **R** is smaller than the function value at **W** then we have moved in the correct direction toward the minimum. Perhaps the minimum is just a bit farther than the point **R**. So we extend the line segment through **M** and **R** to the point **E**. This forms an expanded triangle BGE. The point **E** is found by moving an additional distance d along the line joining **M** and **R** (see Figure 6.4). If the function value at **E** is less than the function

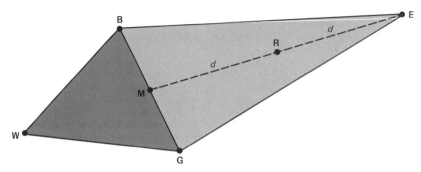

Figure 6.4. The triangle $\widehat{BGW}$ and point **R** and extended point **E**.

value at **R**, then we have found a better vertex than **R**. The vector formula for **E** is

(11)
$$\mathbf{E} = \mathbf{R} + (\mathbf{R} - \mathbf{M}) = 2\mathbf{R} - \mathbf{M}.$$

Contraction Using the Point **C**

If the function values at **R** and **W** are the same, another point must be tested. Perhaps the function is smaller at **M**, but we cannot replace **W** with **M** because we must have a triangle. Consider the two midpoints $\mathbf{C}_1$ and $\mathbf{C}_2$ of the line segments $\widehat{WM}$ and $\widehat{MR}$, respectively (see Figure 6.5). The point with the smaller function value is called **C**, and the new triangle is BGC. *Note:* The choice between $\mathbf{C}_1$ and $\mathbf{C}_2$ might seem inappropriate for the two-dimensional case, but it is important in higher dimensions.

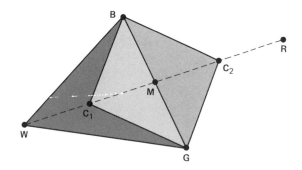

Figure 6.5. The contraction point $\mathbf{C}_1$ or $\mathbf{C}_2$ for the Nelder-Mead method.

Shrink toward **B**

If the function value at **C** is not less than the value at **W**, the points **G** and **W** must be shrunk toward **B** (see Fig. 6.6). The point **G** is replaced with **M**, and **W** is replaced with **S**, which is the midpoint of the line segment joining **B** with **W**.

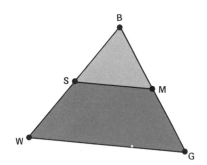

Figure 6.6. Shrinking the triangle toward **B**.

Logical Decisions for Each Step

A computationally efficient algorithm should perform function evaluations only if needed. In each step, a new vertex is found which replaces **W**. As soon as it is found, further investigation is not needed, and the iteration step is completed. In the two-dimensional cases, the logical details are explained in Table 6.10.

Table 6.10 Logical decisions for the Nelder–Mead algorithm

IF f(R) < f(G), THEN Perform Case (i) {either reflect or extend}	
ELSE Perform Case (ii) {either contract or shrink}	
BEGIN {Case (i).}	BEGIN {Case (ii).}
IF f(B) < f(R) THEN	IF f(R) < f(W) THEN
replace W with R	└─ replace W with R
ELSE	Compute C = [W+M]/2 and f(C)
Compute E and f(E)	IF f(C) < f(W) THEN
IF f(E) < f(B) THEN	replace W with C
replace W with E	ELSE
ELSE	Compute S and f(S)
replace W with R	replace W with S
ENDIF	replace G with M
ENDIF	ENDIF
END {Case (i).}	END {Case (ii).}

Example 6.9 Use the Nelder–Mead algorithm to find the minimum of $f(x, y) = x^2 - 4x + y^2 - y - xy$. Start with the three vertices

$$\mathbf{V}_1 = (0, 0) \qquad \mathbf{V}_2 = (1.2, 0.0) \qquad \mathbf{V}_3 = (0.0, 0.8).$$

Solution. The function $f(x, y)$ takes on the values

$$f(0, 0) = 0.0 \qquad f(1.2, 0.0) = -3.36 \qquad f(0.0, 0.8) = -0.16.$$

The function values must be compared to determine **B**, **G**, and **W**;

$$\mathbf{B} = (1.2, 0.0) \qquad \mathbf{G} = (0.0, 0.8) \qquad \mathbf{W} = (0, 0).$$

The vertex $\mathbf{W} = (0, 0)$ will be replaced. The points **M** and **R** are

$$\mathbf{M} = \frac{\mathbf{B} + \mathbf{G}}{2} = (0.6, 0.4) \quad \text{and} \quad \mathbf{R} = 2\mathbf{M} - \mathbf{W} = (1.2, 0.8).$$

The function value $f(\mathbf{R}) = f(1.2, 0.8) = -4.48$ is less than $f(\mathbf{G})$, so the situation is case (i). Since $f(\mathbf{R}) \leq f(\mathbf{B})$ we have moved in the right direction and the vertex **E** must be constructed;

$$\mathbf{E} = 2\mathbf{R} - \mathbf{M} = 2(1.2, 0.8) - (0.6, 0.4) = (1.8, 1.2).$$

The function value $f(\mathbf{E}) = f(1.8, 1.2) = -5.88$ is less than $f(\mathbf{B})$, and the new triangle has vertices

$$\mathbf{V}_1 = (1.8, 1.2) \qquad \mathbf{V}_2 = (1.2, 0.0) \qquad \mathbf{V}_3 = (0.0, 0.8).$$

The process continues and generates a sequence of triangles that converge down on solution point $(3, 2)$ (see Figure 6.7). Table 6.11 gives the function values at vertices of the triangle for several steps in the iteration. A computer implementation of the algorithm continued until the thirty-third step, where the best vertex was $\mathbf{B} = (2.99996456, 1.99983839)$ and $f(\mathbf{B}) = -6.99999998$. These values are approximations to $f(3, 2) = -7$ found in Example 6.8. The reason that the iteration quit before $(3, 2)$ was obtained is that the function is flat near the minimum. The function values $f(\mathbf{B})$, $f(\mathbf{G})$, and $f(\mathbf{W})$ were checked and found to be the same (this is an example of round-off error), and the algorithm was terminated.

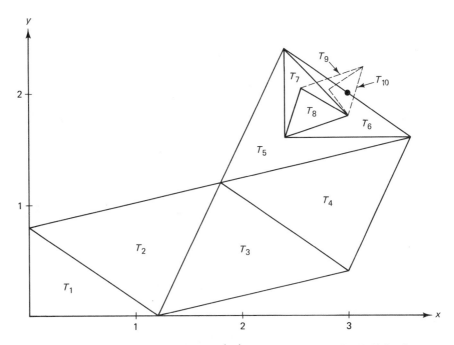

Figure 6.7. The sequence of triangles $\{T_k\}$ converging to the point $(3, 2)$ for the Nelder-Mead method.

Algorithm 6.5 [*Nelder–Mead Method for N Dimensions*] Let $f(x_1, x_2, \ldots, x_N)$ be a continuous function of N variables. A local minimum is found, given the $N + 1$ initial points

$$V_k = (v_{k,1}, \ldots, v_{k,N}) \qquad \text{for } k = 0, 1, \ldots, N.$$

Remark. The function values at the point V_k is stored in $Y(k)$.

Table 6.11   Function values at various triangles for Example 6.9

k	Best Point	Good Point	Worst Point
1	$f(1.2, 0.0) = -3.36$	$f(0.0, 0.8) = -0.16$	$f(0.0, 0.0) = 0.00$
2	$f(1.8, 1.2) = -5.88$	$f(1.2, 0.0) = -3.36$	$f(0.0, 0.8) = -0.16$
3	$f(1.8, 1.2) = -5.88$	$f(3.0, 0.4) = -4.44$	$f(1.2, 0.0) = -3.36$
4	$f(3.6, 1.6) = -6.24$	$f(1.8, 1.2) = -5.88$	$f(3.0, 0.4) = -4.44$
5	$f(3.6, 1.6) = -6.24$	$f(2.4, 2.4) = -6.24$	$f(1.8, 1.2) = -5.88$
6	$f(2.4, 1.6) = -6.72$	$f(3.6, 1.6) = -6.24$	$f(2.4, 2.4) = -6.24$
7	$f(3.0, 1.8) = -6.96$	$f(2.4, 1.6) = -6.72$	$f(2.4, 2.4) = -6.24$
8	$f(3.0, 1.8) = -6.96$	$f(2.55, 2.05) = -6.7725$	$f(2.4, 1.6) = -6.72$
9	$f(3.0, 1.8) = -6.96$	$f(3.15, 2.25) = -6.9525$	$f(2.55, 2.05) = -6.7725$
10	$f(3.0, 1.8) = -6.96$	$f(2.8125, 2.0375) = -6.95640625$	$f(3.15, 2.25) = -6.9525$

```
SUBROUTINE   Size(V,Lo,Norm)                                    {Size of simplex}
      Norm := 0
FOR   J = 0   TO   N   DO
      │  S := 0
      │  FOR   K = 1   TO   N   DO   S := S+[V(Lo,K)−V(J,K)]²
      └─ IF      S > Norm   THEN   Norm := S
         Norm := [Norm]¹ᐟ²                                   {End of procedure Size}


SUBROUTINE   Order(Y,Lo,Li,Ho,Hi)                              {Find indices:
      Lo := 0,  Hi := 0                                       {Lo, best vertex;
FOR   J = 1   TO   N   DO                                      Hi, worst vertex}
      │  IF   Y(J)<Y(Lo)   THEN   Lo := J
      └─ IF   Y(J)>Y(Hi)   THEN   Hi := J


      Li := Hi,  Ho := Lo                                     {Li, next to best;
FOR   J = 0   TO   N   DO                                      Ho, next to worst}
      │  IF   J≠Lo   AND   Y(J)<Y(Li)   THEN   Li := J
      └─ IF   J≠Hi   AND   Y(J)>Y(Ho)   THEN   Ho := J  {End of procedure Order}


SUBROUTINE   Newpoints(V,Hi,M,R,YR)                            {Compute M and R}
FOR   K = 1   TO   N   DO
      │  S := 0
      │  FOR   J=0   TO   N   DO   S := S+V(J,K)               {Construct median M}
      └─ M(K) := [S − V(Hi,K)]/N
         FOR   K = 1   TO   N   DO   R(K) := 2∗M(K)−V(Hi,K)       {Construct R}
         YR := F(R(1), . . . , R(N))                         {End of procedure Newpoints}
```

```
SUBROUTINE   Shrink(V, Y,Lo)                          {Shrink simplex}
FOR  J = 0  TO  N  DO
     IF  J≠Lo  THEN
          FOR  K=1  TO  N  DO  V(J,K) := [V(J,K)+V(Lo,K)]/2
          Y(J) := F(V(J,1), . . . , V(J,N))           {End of procedure Shrink}


SUBROUTINE   Replace(V,R,YR,Hi)                       {Replace W with R}
     FOR  K=1  TO  N  DO  V(Hi,K) := R(K)
     Y(Hi) := YR                                      {End of procedure Replace}


SUBROUTINE   Improve(V, Y,M,R,YR,Lo,Li,Ho,Hi)        {Improve worst vertex}
IF  YR < Y(Ho)  THEN
    IF  Y(Li) < YR  THEN                              {N−M: use Y(Lo), not Y(Li)}
         CALL Procedure Replace(V,R,YR,Hi)           {Replace W with R}
       ELSE
         FOR  K=1  TO  N  DO  E(K) := 2*R(K)−M(K)     {Construct E}
         YE := F(E(1), . . . , E(N))
       IF  YE < Y(Li)  THEN                           {N−M: use Y(Lo), not Y(Li)}
            FOR  K=1  TO  N  DO  V(Hi,K) := E(K)      {Replace W with E}
            Y(Hi) := YE
          ELSE
            CALL Procedure Replace(V,R,YR,Hi)         {Replace W with R}
       ENDIF
    ENDIF
  ELSE
    IF  YR < Y(Hi)  THEN
        CALL Procedure Replace(V,R,YR,Hi)            {Replace W with R}
        FOR  K=1  TO  N  DO  C(K) := [V(Hi,K)+M(K)]/2  {Construct C}
        YC := F((C1), . . . , C(N))
    IF  YC < Y(Hi)  THEN
         FOR  K=1  TO  N  DO  V(Hi,K) := C(K)         {Replace W with C}
         Y(Hi) := YC
       ELSE
         CALL Procedure Shrink(V, Y,Lo)              {Shrink simplex}
    ENDIF
ENDIF                                                 {End of the procedure Improve}
```

{The main program starts here.}

```
     Tol := 10⁻⁶                                 {Convergence criterion}
     Min := 10                             {Minimum number of iterations}
     Max := 200                            {Maximum number of iterations}
     M := 0                                       {Initialize the counter}
     INPUT  N                                       {Number of variables}
```

```
FOR   J = 0  TO  N   DO                                    {Get N+1 initial points}
  └── GET  V(J,1), . . . , V(J,N)                          {Get N coordinates of V_J}


FOR   J = 0  TO  N   DO                                    {Compute function
  └── Y(J) := F(V(J,1), . . . , V(J,N))                        value at V_J}


CALL Procedure Order(Y,Lo,Li,Ho,Hi)


WHILE   [Y(Hi)>Y(Lo)+Tol  AND   M<Max]  OR   [M<Min]  DO
  │  CALL Procedure Newpoints(V,Hi,M,R,YR)
  │  CALL Procedure Improve(V, Y,M,R,YR,Lo,Li,Ho,Hi)
  │  CALL Procedure Order(Y,Lo,Li,Ho,Hi)
  └── M := M+1                                             {Increment the counter}


CALL Procedure Size(V,Lo,Norm)


PRINT "Coordinates of local minimum are"                   {Output}
        V(Lo,1), V(Lo,2), . . . , V(Lo,N)
PRINT "The function value at this point is" Y(Lo)
PRINT "The size of the final simplex is" Norm
```

Minimization Using Derivatives

Suppose that $f(x)$ is unimodal over $[a, b]$ and has a unique minimum at $x = p$. Also, assume that $f'(x)$ is defined at all points in (a, b). Let the starting value p_0 lie in (a, b). If $f'(p_0) < 0$, the minimum point p lies to the right of p_0. If $f'(p_0) > 0$, then p lies to the left of p_0 (see Figure 6.8).

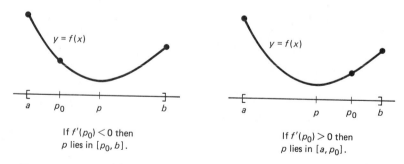

If $f'(p_0) < 0$ then
p lies in $[p_0, b]$.

If $f'(p_0) > 0$ then
p lies in $[a, p_0]$.

Figure 6.8. Using $f'(x)$ to find the minimum value of the unimodal function $f(x)$ on the interval $[a, b]$.

Bracketing the Minimum

Our first task is to obtain three test values:

(12)
$$p_0, \quad p_1 = p_0 + h \quad \text{and} \quad p_2 = p_0 + 2h$$

so that

(13)
$$f(p_0) > f(p_1) \quad \text{and} \quad f(p_1) < f(p_2).$$

Suppose that $f'(p_0) < 0$; then $p_0 < p$ and the step size h should be chosen positive. It is an easy task to find a value for h so that the three points in (12) satisfy (13). Start with $h = 1$ in formula (12).

Case (i)   If (13) is satisfied, then we are done.

Case (ii) If $f(p_0) > f(p_1)$ and $f(p_1) > f(p_2)$, then $p_2 < p$. We need to check points that lie farther to the right. Double the step size and repeat the process.

Case (iii) If $f(p_0) \leq f(p_1)$, then we have jumped over p and h is too large. We need to check values closer to p_0. Reduce the step size by a factor of $\frac{1}{2}$ and repeat the process.

When $f'(p_0) > 0$ the step size h should be chosen negative and then cases similar to (i)–(iii) can be used.

Quadratic Approximation to Find p

Finally, we have three points (12) that satisfy (13). We will use quadratic interpolation to find $p_{\min}$, which is an approximation to p. The Lagrange polynomial based on the nodes in (12) is

(14)
$$Q(x) = \frac{y_0(x - p_1)(x - p_2)}{2h^2} - \frac{y_1(x - p_0)(x - p_2)}{h^2} + \frac{y_2(x - p_0)(x - p_1)}{2h^2}.$$

The derivative of $Q(x)$ is

(15)
$$Q'(x) = \frac{y_0(2x - p_1 - p_2)}{2h^2} - \frac{y_1(2x - p_0 - p_2)}{h^2} + \frac{y_2(2x - p_0 - p_1)}{2h^2}.$$

Solving $Q'(x) = 0$ in the form $Q'(p_0 + h_{\min}) = 0$ yields

$$0 = \frac{y_0[2(p_0 + h_{\min}) - p_1 - p_2]}{2h^2} - \frac{y_1[4(p_0 + h_{\min}) - 2p_0 - 2p_2]}{2h^2}$$

(16)
$$+ \frac{y_2[2(p_0 + h_{\min}) - p_0 - p_1]}{2h^2}.$$

Multiply each term in (16) $2h^2$ and collect terms involving h_{him}:

$$-h_{\min}(2y_0 - 4y_1 + 2y_2) = y_0(2p_0 - p_1 - p_2) - y_1(4p_0 - 2p_0 - 2p_2)$$
$$+ y_2(2p_0 - p_0 - p_1)$$
$$= y_0(-3h) - y_1(-4h) + y_2(-h).$$

This last quantity is easily solved for $h_{\min}$:

(17)
$$h_{\min} = \frac{h(4y_1 - 3y_0 - y_2)}{4y_1 - 2y_0 - 2y_2}.$$

The value $p_{\min} = p_0 + h_{\min}$ is a better approximation to p than p_0. Hence we can replace p_0 with $p_{\min}$ and repeat the two processes outlined above to determine a new h and a new $h_{\min}$. Continue the iteration until the desired accuracy is achieved. The details are outlined in Algorithm 6.6.

Steepest Descent or Gradient Method

Now let us turn to the minimization of a function $f(\mathbf{X})$ of N variables, where $\mathbf{X} = (x_1, x_2, \ldots, x_N)$. The gradient of $f(\mathbf{X})$ is a vector function defined as follows:

(18)
$$\operatorname{grad} f(\mathbf{X}) = (f_1, f_2, \ldots, f_N),$$

where the partial derivatives $f_k = \partial f / \partial x_k$ are evaluated at $\mathbf{X}$.

Recall that the gradient vector (18) points locally in the direction of the greatest rate of increase of $f(\mathbf{X})$. Hence $-\operatorname{grad} f(\mathbf{X})$ points locally in the direction of the greatest decrease. Start at the point $\mathbf{P}_0$ and search along the line through $\mathbf{P}_0$ in the direction $\mathbf{S}_0 = -\mathbf{G}/\|\mathbf{G}\|$, where $\mathbf{G} = \operatorname{grad}(\mathbf{P}_0)$. You will arrive at a point $\mathbf{P}_1$ which is a local minimum when the point $\mathbf{X}$ is constrained to lie on the line $\mathbf{X} = \mathbf{P}_0 + t\mathbf{S}_0$.

Next, we can compute $\mathbf{G} = \operatorname{grad}(\mathbf{P}_1)$ and move in the search direction $\mathbf{S}_1 = -\mathbf{G}/\|\mathbf{G}\|$. You will come to $\mathbf{P}_2$, which is a local minimum when $\mathbf{X}$ is constrained to lie on the line $\mathbf{X} = \mathbf{P}_1 + t\mathbf{S}_1$. Iteration will produce a sequence $\{\mathbf{P}_k\}$ of points with the property $f(\mathbf{P}_0) > f(\mathbf{P}_1) > \ldots > f(\mathbf{P}_k) > \ldots$. If $\lim_{k \to \infty} \mathbf{P}_k = \mathbf{P}$, then $f(\mathbf{P})$ will be a local minimum for $f(\mathbf{X})$.

Outline of the Gradient Method

Suppose that $\mathbf{P}_k$ has been obtained.

Step 1. Evaluate the gradient vector $\mathbf{G} = \operatorname{grad} f(\mathbf{P}_k)$.

Step 2. Compute the search direction $\mathbf{S} = -\mathbf{G}/\|\mathbf{G}\|$.

Step 3. Perform a single parameter minimization of $\phi(t) = f(\mathbf{P}_k + t\mathbf{S})$ on the interval $[0, b]$, where b is large. This will produce a value $t = h_{\min}$ that is a local minimum for $\phi(t)$. The relation $\phi(h_{\min}) = f(\mathbf{P}_k + h_{\min}\mathbf{S})$ shows that this is a minimum for $f(\mathbf{X})$ along the search line $\mathbf{X} = \mathbf{P}_k + t\mathbf{S}$.

Step 4. Construct the next point $\mathbf{P}_{k+1} = \mathbf{P}_k + h_{\min}\mathbf{S}$.

Step 5. Perform the termination test for minimization; that is, are the function values $f(\mathbf{P}_k)$ and $f(\mathbf{P}_{k+1})$ sufficiently close and the distance $\|\mathbf{P}_{k+1} - \mathbf{P}_k\|$ small enough?

Repeat the process.

Algorithm 6.6 [*Search for a Local Minimum Using Quadratic Interpolation*]
Assume that $f(x)$ and $f'(x)$ are defined and that $f(x)$ has a local minimum on
either the interval $[a, p_0]$ or $[p_0, b]$. A local minimum p is found on one of the
intervals above.

```
Delta := 10⁻⁵                                    {Convergence tolerance for the abscissas}
Epsilon := 10⁻⁷                                  {Convergence tolerance for the ordinates}
Jmax := 20, Kmax := 50                           {Maximum number of iterations}


INPUT  P0                                        {Get the starting value}


Err := 1                                         {Initialize the variable}
H := 1                                           {Initialize the step size}
K := 0                                           {Initialize the counter}


WHILE  (K<Kmax) AND (Err>Epsilon)  DO            {Loop to find Ymin}
   IF  f'(P0)>0  THEN  H := −|H|                 {Slope points downhill}
   P1 := P0+H, P2 := P0+2*H                      {Two more test points}
   Y0 := F(P0), Y1 := F(P1), Y2 := F(P2)         {Compute the function values}
   Cond := 0, J := 0                             {Initialize the variables}
   WHILE  (J<Jmax) AND (H>Delta) AND (Cond=0)  DO    {Determine H so
      IF  Y0 ≤ Y1  THEN                                that Y1<Y0 & Y1<Y2}
         P2 := P1, Y2 := Y1, H := H/2                 {Make H smaller}
         P1 := P0+H, Y1 := F(P1)
      ELSE
         IF  Y2 < Y1  THEN
            P1 := P2, Y1 := Y2, H := 2*H             {Make H larger}
            P2 := P0+2*H, Y2 := F(P2)
         ELSE
            Cond := −1                               {Loop termination}
      ENDIF
      J := J+1                                       {Increment the counter}
   ENDWHILE                                          {End loop to determine proper H}
   D := 4*Y1−2*Y0−2*Y2                               {Use quadratic
   IF  D<0  THEN                                      interpolation to
      Hmin := H*(4*Y1−3*Y0−Y2)/D                      compute the abscissa
   ELSE                                               at the minimum}
      Hmin := H/3                                    {Check division by zero}
      Cond := 4
```

```
      Pmin := P0+Hmin                                          {Coordinates of
      Ymin := F(Pmin)                                           the minimum}
      H := |H|                                             {How determine the
      IF  |Hmin|<H   THEN   H := |Hmin|                        magnitude of
      IF  |Hmin−H|<H   THEN   H := |Hmin−H|                     the next H}
      IF  |Hmin−2*H|<H   THEN   H := |Hmin−2*H|
      IF  H < Delta   THEN   Cond := 1             {Convergence of abscissa}
      Err := |Y0−Ymin|
      IF  |Y1−Ymin|<Err   THEN   Err := |Y1−Ymin|
      IF  |Y2−Ymin|<Err   THEN   Err := |Y2−Ymin|
      IF  Err<Epsilon   THEN   Cond := 2           {Convergence of ordinate}
      K := K+1
      P0 := Pmin                                        {Update the value P0}
      IF  (Cond=2) AND (H<Delta)   THEN   Cond := 3
ENDWHILE                                              {End loop to find Ymin}
      PRINT 'An approximation for the minimum is', Pmin             {Output}
      PRINT 'An estimate for its accuracy is', |H|
      PRINT 'The function value f(Pmin) is', Ymin


IF  Cond = 0   THEN
      PRINT 'Convergence is doubtful because the maximum'
            'number of iterations was exceeded.'
IF  Cond = 1   THEN
      PRINT 'Convergence of the abscissas has been achieved.'
IF  Cond = 2   THEN
      PRINT 'Convergence of the ordinates has been achieved.'
IF  Cond = 3   THEN
      PRINT 'Convergence of both coordinates has been achieved.'
IF  Cond = 4   THEN
      PRINT 'Convergence is doubtful because'
            'division by zero was encountered.'
```

Algorithm 6.7 [*Steepest Descent Gradient Method*] Assume that $f(\mathbf{X})$ is a function of N variables where $\mathbf{X} = (x_1, x_2, \ldots, x_N)$. Suppose that the first-order partial derivatives of $f(\mathbf{X})$ are continuous. The gradient method is used to locate a relative minimum of $f(\mathbf{X})$.

Remark. The steepest descent method has a slow rate of convergence (linear). The user is encouraged to read about the quasi-Newton methods which converge faster (quadratic).

Delta := 10^{-5} {Convergence tolerance for points}
Epsilon := 10^{-7} {Tolerance for the function values}
Jmax := 20, Max := 50   {Maximum number of iterations}

PROCEDURE GRADIENT   {$\vec{S} = (s_1, s_2, \ldots, s_N)$ is a
$\quad$ $\vec{G}$:= grad f($\vec{P}$)   unit vector pointing
$\quad$ $\vec{S}$:= $-\vec{G}/\|\vec{G}\|$ toward the minimum}

PROCEDURE QUADMIN
$\quad$ Cond := 0, J := 0
$\quad$ $\vec{P}_1$:= $\vec{P}_0$ + H*$\vec{S}$, $\vec{P}_2$:= $\vec{P}_0$ + 2*H*$\vec{S}$
$\quad$ Y1 := f($\vec{P}_1$), Y2 := f($\vec{P}_2$)
$\quad$ WHILE (J<Jmax) AND (Cond=0) DO   {Determine H so that
$\quad\quad$ IF Y0 $\leq$ Y1 THEN   Y1<Y0 and Y1<Y2}
$\quad\quad\quad$ $\vec{P}_2$:= $\vec{P}_1$, Y2 := Y1, H := H/2   {Make H smaller}
$\quad\quad\quad$ $\vec{P}_1$:= $\vec{P}_0$ + H*$\vec{S}$, Y1 := f($\vec{P}_1$)
$\quad\quad$ ELSE
$\quad\quad\quad$ IF Y2 < Y1 THEN
$\quad\quad\quad\quad$ $\vec{P}_1$:= $\vec{P}_2$, Y1 := Y2, H := 2*H   {Make H larger}
$\quad\quad\quad\quad$ $\vec{P}_2$:= $\vec{P}_0$ + 2*H*$\vec{S}$, Y2 := f($\vec{P}_2$)
$\quad\quad\quad$ ELSE
$\quad\quad\quad\quad$ Cond := -1
$\quad$ IF H<Delta THEN Cond := 1
$\quad$ D := 4*Y1$-$2*Y0$-$2*Y2
$\quad$ IF D<0 THEN   {Quadratic interpolation
$\quad\quad$ Hmin := H*(4*Y1$-$3*Y0$-$Y2)/D   to find Hmin}
$\quad$ ELSE
$\quad\quad$ Cond := 4, Hmin := H/3   {Check division by zero}
$\quad$ $\vec{P}_{min}$:= $\vec{P}_0$ + Hmin*$\vec{S}$, Ymin := f($\vec{P}_{min}$)
$\quad$ H0 := $|$Hmin$|$, H1 := $|$Hmin$-$H$|$, H2 := $|$Hmin$-$2*H$|$ {Convergence
$\quad$ IF H0<H THEN H := H0
$\quad$ IF H1<H THEN H := H1   test for
$\quad$ IF H2<H THEN H := H2
$\quad$ IF H<Delta THEN Cond := 1   the points}
$\quad$ E0 := $|$Y0$-$Ymin$|$, E1 := $|$Y1$-$Ymin$|$, E2 := $|$Y2$-$Ymin$|$ {Convergence
$\quad$ IF E0 < Err THEN Err := E0
$\quad$ IF E1 < Err THEN Err := E1   test for the
$\quad$ IF E2 < Err THEN Err := E2
$\quad$ IF (E0=0) AND (E1=0) AND (E2=0) THEN Err := 0   function values}
$\quad$ IF Err<Epsilon THEN Cond := 2
$\quad$ IF (Cond=2) AND (H<Delta) THEN Cond := 3
$\quad$ J := J+1
END   {End of Procedure Qmin}

{The main Gradient Search program starts here.}
INPUT $\vec{P}_0$ {Get the starting point}
Count := 0, H := 1, Err := 1   {Initialize variables}
WHILE   (Count<Max) AND ((H>Delta) OR (Err>Epsilon))   DO
    | CALL PROCEDURE GRADIENT
    | CALL PROCEDURE QUADMIN
    | $\vec{P}_0$:= $\vec{P}_{\text{min}}$, Y0 := Ymin, Count := Count+1

PRINT 'The local minimum is', $\vec{P}_{\text{min}}$ {Output}
PRINT 'The minimum value of the function is', Ymin

See Algorithm 6.6 for the output remarks when Cond $= 0, 1, 2, 3, 4$.

Exercises for Minimization of a Function

1. Use Definition 6.2 to show that the following functions are increasing on the interval $[0, 4]$.
   (a) $f(x) = x^2$ (b) $f(x) = x^{1/2}$

2. Use Theorem 6.1 to show that the following functions are decreasing on the interval $[1, 3]$.
   (a) $f(x) = x^{-1}$ (b) $f(x) = \cos(x)$

3. Use Definition 6.3 to show that the following functions are unimodal on the interval $[0, 4]$.
   (a) $f(x) = x^2 - 2x + 1$ (b) $f(x) = \cos(x)$

   In Exercises 4–11:
   (a) Use Theorem 6.4 to find the local minimum.
   (b) Use Algorithm 6.4 to find the local minimum.
   (c) Use Algorithm 6.6 to find the local minimum, starting with the midpoint of the given interval.

4. $f(x) = 3x^2 - 2x + 5$ on the interval $[0, 1]$.

5. $f(x) = 2x^3 - 3x^2 - 12x + 1$ on the interval $[0, 3]$.

6. $f(x) = 4x^3 - 8x^2 - 11x + 5$ on the interval $[0, 2]$.

7. $f(x) = x + 3/x^2$ on the interval $[0.5, 3]$.

8. $f(x) = (x + 2.5)/(4 - x^2)$ on the interval $[-1.9, 1.9]$.

9. $f(x) = \exp(x)/x^2$ on the interval $[0.5, 3]$.

10. $f(x) = -\sin(x) - \sin(3x)/3$ on the interval $[0, 2]$.

11. $f(x) = -2\sin(x) + \sin(2x) - 2\sin(3x)/3$ on $[1, 3]$.

12. Find the point on the parabola $y = x^2$ that is closest to the point $(3, 1)$.

13. Find the point on the curve $y = \sin(x)$ that closest to the point $(2, 1)$.

   In Exercises 14–21:
   (a) Use Theorem 6.5 to find the local minimum.

(b) Use Algorithm 6.5 to find the local minimum.

(c) Use Algorithm 6.7 to find the local minimum, starting with the first vertex given.

14. (a) $f(x, y) = x^3 + y^3 - 3x - 3y + 5$

(b) Use the starting vertices $(1, 2), (2, 0)$, and $(2, 2)$.

15. (a) $f(x, y) = x^2 + y^2 + x - 2y - xy + 1$

(b) Use the starting vertices $(0, 0), (2, 0)$, and $(2, 1)$.

16. (a) $f(x, y) = x^2 y + xy^2 - 3xy$

(b) Use the starting vertices $(0, 0), (2, 0)$, and $(2, 1)$.

17. (a) $f(x, y) = x^4 - 8xy + 2y^2$

(b) i. Use the starting vertices $(2, 2), (3, 2)$, and $(3, 3)$.

ii. Use the starting vertices $(-1, -2), (-1, -3)$, and $(-2, -2)$.

18. (a) $f(x, y) = (x - y)/(2 + x^2 + y^2)$

(b) Use the starting vertices $(0, 0), (0, 1)$, and $(1, 1)$.

19. (a) $f(x, y) = x^4 + y^4 - (x + y)^2$

(b) i. Use the starting vertices $(0, 0), (0, 1)$, and $(2, 0)$.

ii. Use the starting vertices $(0, 0), (-2, -2)$, and $(-2, 0)$.

20. (a) $f(x, y) = (x - y)^4 + (x + y - 2)^2$

(b) Use the starting vertices $(0, 0), (0, 2)$, and $(1, 0)$.

21. [Rosenbrock's parabolic valley, circa 1960]

(a) $f(x, y) = 100(y - x^2)^2 + (1 - x)^2$

(b) Use the starting vertices $(0, 0), (1, 0)$, and $(0, 2)$.

22. Let $\mathbf{B} = (2, -3), \mathbf{G} = (1, 1)$, and $\mathbf{W} = (5, 2)$. Find the points $\mathbf{M}, \mathbf{R}$, and $\mathbf{E}$ and sketch the triangles that are involved.

23. Let $\mathbf{B} = (-1, 2), \mathbf{G} = (-2, -5)$, and $\mathbf{W} = (3, 1)$. Find the points $\mathbf{M}, \mathbf{R}$, and $\mathbf{E}$ and sketch the triangles that are involved.

24. Give a vector proof that $\mathbf{M} = (\mathbf{B} + \mathbf{G})/2$ is the midpoint of the line segment joining the points $\mathbf{B}$ and $\mathbf{G}$.

25. Give a vector proof of equation (10).

26. Give a vector proof of equation (11).

27. Give a vector proof that the medians of any triangle intersect at a point which is two-thirds of the distance from each vertex to the midpoint of the opposite side.

28. Let $\mathbf{B} = (0, 0, 0), \mathbf{G} = (1, 1, 0), \mathbf{P} = (0, 0, 1)$, and $\mathbf{W} = (1, 0, 0)$.

(a) Sketch the tetrahedron *BGPW*.

(b) Find $\mathbf{M} = (\mathbf{B} + \mathbf{G} + \mathbf{P})/3$.

(c) Find $\mathbf{R} = 2\mathbf{M} - \mathbf{W}$ and sketch the tetrahedron *BGPR*.

(d) Find $\mathbf{E} = 2\mathbf{R} - \mathbf{M}$ and sketch the tetrahedron *BGPE*.

29. Let $\mathbf{B} = (0, 0, 0), \mathbf{G} = (0, 2, 0), \mathbf{P} = (0, 1, 1)$, and $\mathbf{W} = (2, 1, 0)$. Follow the instructions in Exercise 28.

30. Write a report on the Fibonacci search algorithm.

31. Write a report on the method of steepest descent.

In Exercises 32–37.
   (a) Set all partial derivatives equal to zero and solve for the local minimum.
   (b) Use Algorithm 6.5 to find the local minimum.
   (c) Use Algorithm 6.7 to find the local minimum, starting with the first vertex given.

32. (a) $f(x, y, z) = 2x^2 + 2y^2 + z^2 - 2xy + yz - 7y - 4z$
    (b) Start with $(1, 1, 1)$, $(0, 1, 0)$, $(1, 0, 1)$, and $(0, 0, 1)$.

33. (a) $f(x, y, z) = 2x^2 + 2y^2 + z^2 + xy - xz - 2x - 5y - 5z$
    (b) Start with $(1, 1, 1)$, $(0, 1, 0)$, $(1, 0, 1)$, and $(0, 0, 1)$.

34. (a) $f(x, y, z) = x^4 + y^4 + z^4 - 4xyz$
    (b) Use $(0.5, 0.5, 0.5)$, $(1, 0.5, 0.5)$, $(0.5, 2, 0.5)$, and $(0.5, 0.5, 3)$.

35. (a) $f(x, y, z, u) = 2(x^2 + y^2 + z^2 + u^2)$
$$- x(y + z - u) + yz - 3x - 8y - 5z - 9u$$
    (b) Start the search near $(1, 1, 1, 1)$.

36. (a) $f(x, y, z, u) = 2(x^2 + y^2 + z^2 + u^2)$
$$+ x(y + z - u) + yz - 5x - 6y - 6z - 3u$$
    (b) Start the search near $(0, 0, 0, 0)$.

37. (a) $f(x, y, z, u) = xyzu + 1/x + 1/y + 1/z + 1/u$
    (b) Use $(0.7, 0.7, 0.7, 0.7)$, $(1.2, 0.7, 0.7, 0.7)$, $(0.7, 1.2, 0.7, 0.7)$, $(0.7, 0.7, 1.2, 0.7)$, and $(0.7, 0.7, 0.7, 1.2)$.

38. Write a report on minimization of multivariable functions. Include a discussion of the quasi-Newton methods.

7

Numerical Integration

7.1 Introduction

We now approach the subject of numerical integration. The goal is to approximate the definite integral of $f(x)$ over the interval $a \leq x \leq b$ using $N + 1$ sample points $(x_0, f_0), (x_1, f_1), \ldots, (x_N, f_N)$, where $f_k = f(x_k)$. The approximating formula has the form

$$\int_a^b f(x)\, dx \approx w_0 f_0 + w_1 f_1 + \ldots + w_N f_N.$$

The values $w_0, w_1, \ldots, w_N$ are constants and are called *weights*. Depending on the desired application, the *nodes* x_k are chosen in various ways. For the trapezoidal rule, Simpson's rule, and Boole's rule, the nodes $x_k = a + hk$ are chosen to be equally spaced. For Gauss-Legendre integration the nodes are chosen to be zeros of certain Legendre polynomials. When the integration formula is used to derive an explicit algorithm for solving differential equations, the nodes are all chosen less than b.

Some common formulas based on polynomial interpolation are referred to as Newton–Cotes integration formulas. Suppose that $f(x)$ is approximated by the Lagrange interpolating polynomial $P_N(x)$ of degree at most N, based on the $N + 1$ nodes $x_0, x_1, \ldots, x_N$; then

$$f(x) = P_N(x) + R_N(x),$$

where

$$P_N(x) = f_0 L_0(x) + f_1 L_1(x) + \ldots + f_N L_N(x)$$

and

$$R_N(x) = (x - x_0)(x - x_1) \ldots (x - x_N) \frac{f^{(N+1)}(c)}{(N+1)!}.$$

The error $E_N(f)$ is defined to be the difference between the integrals of $f(x)$ and $P_N(x)$:

$$E_N(f) = \int_a^b f(x) \, dx - \int_a^b P_N(x) \, dx.$$

A more useful expression for $E_N(f)$ involves a higher derivative of $f(x)$. For illustration purposes, it will be developed in the case of the trapezoidal rule.

The Newton–Cotes closed integration formulas that arise from approximating $y = f(x)$ with polynomials of degree $N = 1, 2, 3, 4$ on $[x_0, x_N]$, where $x_N = x_0 + hN$ are summarized in Table 7.1 and are illustrated in Figure 7.1.

Table 7.1 Newton–Cotes integration formulas with error term

Trapezoidal rule:	$\displaystyle\int_{x_0}^{x_1} f(x) \, dx = \frac{h}{2}(f_0 + f_1)$	$-\dfrac{h^3}{12} f^{(2)}(c)$
Simpson's rule:	$\displaystyle\int_{x_0}^{x_2} f(x) \, dx = \frac{h}{3}(f_0 + 4f_1 + f_2)$	$-\dfrac{h^5}{90} f^{(4)}(c)$
Simpson's 3/8 rule:	$\displaystyle\int_{x_0}^{x_3} f(x) \, dx = \frac{3h}{8}(f_0 + 3f_1 + 3f_2 + f_3)$	$-\dfrac{3h^5}{80} f^{(4)}(c)$
Boole's rule:	$\displaystyle\int_{x_0}^{x_4} f(x) \, dx = \frac{2h}{45}(7f_0 + 32f_1 + 12f_2 + 32f_3 + 7f_4)$	$-\dfrac{8h^7}{945} f^{(6)}(c)$

7.2 Composite Trapezoidal Rule

An intuitive method of finding the area under a curve is by approximating that area with a series of trapezoids that lie above the intervals $[x_{k-1}, x_k]$. When several trapezoids are used, we call it the composite trapezoidal rule (see Figure 7.2).

Theorem 7.1 [*Description of the Composite Trapezoidal Rule*] Let $f(x), f'(x)$ and $f''(x)$ be continuous on the interval $[a, b]$ and let M be a positive integer. If we subdivide the interval $[a, b]$ into M subintervals of equal width $h = (b - a)/M$

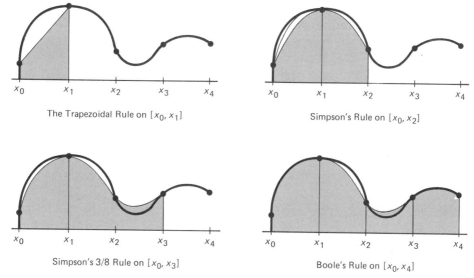

Figure 7.1. The four elementary integration rules.

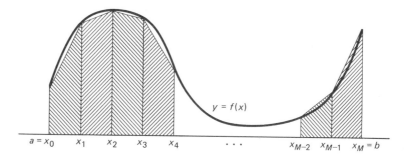

Figure 7.2. The composite trapezoidal rule on the interval $[a, b]$.

and use the nodes $a = x_0 < x_1 < \ldots < x_M = b$, where $x_k = a + hk$, the *trapezoidal rule* for $f(x)$ with step size h is

(1)
$$T(f, h) = \frac{h}{2}[f(a) + f(b)] + h \sum_{k=1}^{M-1} f(x_k).$$

This is an approximation to the integral of $f(x)$ and we write

(2)
$$\int_a^b f(x) \, dx = T(f, h) + E(f, h).$$

Furthermore, there exists a value c with $a < c < b$ so that the error term $E(f, h)$ has the form

(3)
$$E(f, h) = -\frac{(b - a)h^2}{12} f^{(2)}(c) = \mathbf{O}(h^2).$$

Remark. Observe that we only need to know the function values of $f(x_k)$ at equally spaced abscissas x_k to compute $T(f, h)$. Hence (1) can be used to integrate a function known only as a table of values.

Example 7.1 Suppose that the velocity of a submarine that is under the polar ice cap is given in Table 7.2. The values for the approximate distance traveled in the table are all obtained by using the trapezoidal rule. Verify that the approximation for the total distance traveled in the time interval $[0, 2]$ is 16.5 km.

Table 7.2

Time, t (hr)	Velocity, $v(t)$ (km/hr)	Approximate Distance Traveled in the Interval $[0, t]$ (km)
0.00	6.0	0.0000
0.25	7.5	1.6875
0.50	8.0	3.6250
0.75	9.0	5.7500
1.00	8.5	7.9375
1.25	10.5	10.3125
1.50	9.5	12.8125
1.75	7.0	14.8750
2.00	6.0	16.5000

Solution. We show how to obtain the last entry. Since

$$\text{distance} = \int_0^2 v(t)\, dt,$$

we use the trapezoidal rule (1) with $M = 8$, $h = 0.25$ and compute

$$T(v, h) = \frac{0.25}{2}(6 + 6) + 0.25(7.5 + 8 + 9 + 8.5 + 10.5 + 9.5 + 7).$$

Therefore, the approximate distance traveled in the interval $[0, 2]$ is

$$\text{distance} \approx T(v, h) = 16.5 \text{ km.}$$

When we consider the order of the error term, $E(f, h) = O(h^2)$, in (3), we see that the error is proportional to h^2 and expect that the error in approximating the integral should be reduced by a factor of $\frac{1}{4}$ if the step size is decreased by a factor of $\frac{1}{2}$.

Example 7.2 Consider the definite integrals

$$\int_1^3 \frac{x\, dx}{1 + x^2} = \frac{\ln(5)}{2} \quad \text{and} \quad \int_1^3 \frac{dx}{7 - 2x} = \frac{\ln(5)}{2}.$$

Tables 7.3 and 7.4 show the approximations obtained using the composite trapezoidal rule with $M = 1, 2, 4, 8, 16, 32$ subintervals and the error in the approximation. It is instructive to note that as M is made larger h is successively divided by 2 and the successive errors become about $\frac{1}{4}$ of the previous error. Notice that the error for the first integral seems to be going to zero faster. One might be led to think that the trapezoidal rule is giving a better answer for the first integer. But look closely and see that the errors in both cases are diminishing at the same rate, and the trapezoidal rule is working equally well in both cases!

Table 7.3 Trapezoidal rule approximation

$$\text{for } \int_1^3 \frac{x \, dx}{1 + x^2}$$

M	h	$T(f, h)$	$E(f, h)$
1	2.0	0.800000	0.004719
2	1.0	0.800000	0.004719
4	0.5	0.803183	0.001536
8	0.25	0.804311	0.000408
16	0.125	0.804615	0.000104
32	0.0625	0.804693	0.000026

Table 7.4 Trapezoidal rule approximation

$$\text{for } \int_1^3 \frac{dx}{7 - 2x}$$

M	h	$T(f, h)$	$E(f, h)$
1	2.0	1.200000	-0.395281
2	1.0	0.933333	-0.128614
4	0.5	0.841667	-0.036948
8	0.25	0.814484	-0.009765
16	0.125	0.807203	-0.002484
32	0.0625	0.805343	-0.000624

The error term

$$E(f, h) = \frac{-(b - a)h^2 f^{(2)}(c)}{12}$$

can be used to predict the number of subintervals M that are needed to compute an approximation to the integral with a specified accuracy.

Example 7.3 Find the number of subintervals M and the step size h so that the absolute value of the error for the trapezoidal rules is less than 5×10^{-9} when it is used to approximate the integral

$$\int_2^7 \frac{dx}{x} = \ln(7) - \ln(2) \approx 1.252762968.$$

Solution. Using calculus we find that $f'(x) = -1/x^2, f''(x) = 2/x^3$. The maximum value of $|f''(c)|$ on $[2, 7]$ is seen to occur at $c = 2$ and thus $|f''(c)| \leq 1/4$ for $2 \leq c \leq 7$, and we obtain

$$|E(f, h)| \leq \left| \frac{-(b-a)h^2 f^{(2)}(c)}{12} \right| \leq \frac{5h^2}{48}.$$

We seek the value h so that $5h^2/48 \leq 5 \times 10^{-9}$ or $h^2 \leq 48 \times 10^{-9}$. But we really want to find the integer M, so we use $h = 5/M$ and solve $M^2 \geq (25/48) \times 10^9$ and obtain $M > 22,821.77$. Since M must be an integer, we choose $M = 22,822$ and $h \approx 0.000219087$. Indeed, when the computation is carried out, it is

$$T\left(f, \frac{5}{22,822}\right) = 1.252762969,$$

and the accuracy achieved is actually 1×10^{-9}. The actual error is smaller than the estimate because a bound for the derivative was used. However, it would still take about 10,000 subintervals to achieve the stated accuracy of 5×10^{-9} and calculation reveals that

$$T\left(f, \frac{5}{10,000}\right) = 1.252762973.$$

The trapezoidal rule usually requires a large number of function evaluations to achieve an accurate answer. In Section 7.3 we will see that Simpson's rule will require significantly fewer calculations.

Algorithm 7.1 [*Composite Trapezoidal Rule*] Approximate the integral

$$\int_A^B f(x)\, dx \approx \frac{h}{2}[f(A) + f(B)] + h \sum_{k=1}^{M-1} f(x_k)$$

where the interval $[A, B]$ is divided into M subintervals of equal width $h = (B-A)/M$ and $x_k = A + hk$ for $k = 0, 1, 2, \ldots, M$ and $A = x_0 < x_1 < \cdots < x_{M-1} < x_M = B$. Notice that $x_M = B$.

```
    USES the function F(X)
    INPUT   A, B, M
    H := [B−A]/M
    SUM := 0

FOR   K = 1   TO   M−1   DO
  ┌   X := A + H∗K
  └   SUM := SUM + F(X)
    SUM := H∗[F(A) + F(B) + 2∗SUM]/2

PRINT 'The approximate value of the integral of f(X)' 'on the interval' A, B
      'using' M 'subintervals' 'computed using the trapezoidal rule is' SUM
```

Proof of Theorem 7.1. In Figure 7.2 we see that the trapezoid that lies above the interval $[x_{k-1}, x_k]$ has the altitude $f(x_{k-1})$ on the left and its altitude is $f(x_k)$ on the right. The width of the base is h, and the area of this trapezoid is given by the formula $(h/2)[f(x_{k-1}) + f(x_k)]$. When the areas of all the trapezoids are summed up for $k = 1, 2, \ldots, M$, the result is

$$(4) \qquad \frac{h}{2} \sum_{k=1}^{M} [f(x_{k-1}) + f(x_k)] = \frac{h}{2} [f(x_0) + f(x_M)] + h \sum_{k=1}^{M-1} f(x_k).$$

This quantity appears on the right side of equation (1). The error term $E(f, h)$ on the right side of (2) requires more justification.

Proof of the error expression (3). First we establish the rule for $F(t)$ on the interval $[0, 1]$. The linear Lagrange interpolating polynomial with remainder term gives us the representation

$$(5) \qquad F(t) = F(0)\frac{t-1}{0-1} + F(1)\frac{t-0}{1-0} + \frac{(t-0)(t-1)}{2} F^{(2)}(s(t)),$$

where $s(t)$ is a continuous function of t [see Figure 7.3(a)]. Integrating both sides of (5), we obtain

$$(6) \qquad \int_0^1 F(t)\, dt = F(0) \int_0^1 (1-t)\, dt + F(1) \int_0^1 t\, dt + \tfrac{1}{2} \int_0^1 (t^2 - t) F^{(2)}(s(t))\, dt.$$

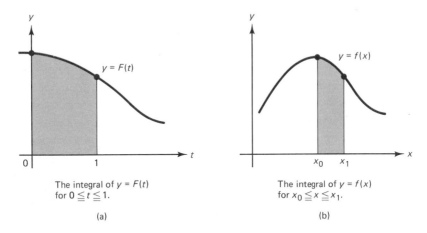

The integral of $y = F(t)$ for $0 \leq t \leq 1$.

(a)

The integral of $y = f(x)$ for $x_0 \leq x \leq x_1$.

(b)

Figure 7.3. The two regions involved in the proof of the trapezoidal rule.

The right side of (6) can be simplified by using

$$(7) \qquad \int_0^1 (1-t)\, dt = \tfrac{1}{2} \quad \text{and} \quad \int_0^1 t\, dt = \tfrac{1}{2}.$$

The value of the integral involving the derivative $F^{(2)}(s(t))$ is more difficult to simplify. If we may be permitted to use the second mean value theorem from calculus, then a simplification can be made. Since the first term $(t^2 - t)$ does not change sign on the interval $[0, 1]$ and $F^{(2)}(s(t))$ is continuous, there exists a value t_1 so that

(8) $$\int_0^1 (t^2 - t) F^{(2)}(s(t))\, dt = F^{(2)}(s(t_1)) \int_0^1 (t^2 - t)\, dt = \frac{-1}{6} F^{(2)}(s(t_1)).$$

Substituting (7) and (8) into equation (6), we obtain the trapezoidal rule for the interval $[0, 1]$ where $M = 1$ and $h = 1$:

(9) $$\int_0^1 F(t)\, dt = \tfrac{1}{2}[F(0) + F(1)] - \tfrac{1}{12} F^{(2)}(s(t_1)).$$

Next we derive the rule on $[x_0, x_1]$ [See Figure 7.3(b)]. Factors of h and h^3 will be needed in the terms on the right side of equation (9). To show how to get them we use the change of variables

(10) $$t = \frac{1}{h}(x - x_0) \qquad F(t) = f(x_0 + ht) \qquad c_1 = x_0 + hs(t_1).$$

Therefore

(11) $$dt = \frac{1}{h}\, dx \qquad F(0) = f(x_0) \qquad F(1) = f(x_1)$$

and by the chain rule for differentiation of $f(x_0 + ht)$,

(12) $$F^{(2)}(t) = h^2 f^{(2)}(x_0 + ht).$$

Using (10), the integral of $f(x)$ taken over $[x_0, x_1]$ is transformed:

(13) $$\int_{x_0}^{x_1} f(x)\, dx = \int_0^1 f(x_0 + ht) h\, dt = h \int_0^1 F(t)\, dt$$

Using (9)–(12) to integrate $F(t)$ in (13) results in the trapezoidal rule for $[x_0, x_1]$ where $M = 1$:

(14) $$\int_{x_0}^{x_1} f(x)\, dx = \frac{h}{2}[f(x_0) + f(x_1)] - \frac{h^3}{12} f^{(2)}(c_1).$$

Finally, we are ready to derive the composite trapezoidal rule. The total integral over $[a, b]$ is the sum of M integrals taken over the subintervals $[x_{k-1}, x_k]$ for $k = 1, 2, \dots, M$:

$$\int_a^b f(x)\, dx = \sum_{k=1}^M \int_{x_{k-1}}^{x_k} f(x)\, dx.$$

For $k = 2, \dots, M$ we apply (14) shifted over to the interval $[x_{k-1}, x_k]$ and add the contributions in each interval, to obtain

(15) $$\int_a^b f(x)\, dx = \frac{h}{2} \sum_{k=1}^M [f(x_{k-1}) + f(x_k)] - \frac{h^3}{12} \sum_{k=1}^M f^{(2)}(c_k).$$

It is plain to see that the trapezoidal rule is the first summation on the right side of equation (15). We need only verify that $E(f, h)$ is the other sum on the right side of (15). Using the Intermediate Value Theorem from calculus, there exists a number c so that the average value of the M terms $f^{(2)}(c_k)$ is equal to $f^{(2)}(c)$, that is,

(16)
$$\frac{1}{M} \sum_{k=1}^{M} f^{(2)}(c_k) = f^{(2)}(c).$$

Using the fact that $h = (b - a)/M$ and substituting (16) into (15), we obtain the desired representation for $E(f, h)$:

$$E(f, h) = \frac{-h^3}{12} \sum_{k=1}^{M} f^{(2)}(c_k) = \frac{-(b-a)h^2}{12} \frac{1}{M} \sum_{k=1}^{M} f^{(2)}(c_k) = \frac{-(b-a)h^2}{12} f^{(2)}(c)$$

The proof of the composite trapezoidal rule is now complete.

Exercises for the Composite Trapezoidal Rule

In Exercises 1–6:
   (a) Approximate the integral using the composite trapezoidal rule with one subinterval, $M = 1$.
   (b) Approximate the integral using the composite trapezoidal rule with two subintervals, $M = 2$.
   (c) Approximate the integral using the composite trapezoidal rule with four subintervals, $M = 4$.
   (d) Use the Fundamental Theorem of Integral Calculus to find the exact answer, that is, find an antiderivative $F(x)$ of the integrand $f(x)$ and compute $F(b) - F(a)$.
   (e) Use a computer to find a trapezoidal rule approximation to the integral that has eight accurate digits.

1. $\int_1^3 3x^2 \, dx$

2. $\int_1^3 4x^3 \, dx$

2. $\int_1^3 x^{-2} \, dx$

4. $\int_1^3 x^{-1} \, dx$

5. $\int_1^3 64(x - 1)4^{-x} \, dx$. The antiderivative is

$$F(x) = 64 \left[\frac{1 - x}{\ln (4)} - (\ln (4))^{-2} \right] 4^{-x}.$$

6. $\int_0^1 4/(1 + x^2) \, dx = \pi$. The antiderivative is

$$F(x) = 4 \arctan (x).$$

7. Use the trapezoidal rule to approximate the length of the curve $y = \cos (x)$ on the interval $0 \le x \le 0.8$.

Remark. cos (x) is to be evaluated with x in radians.
(a) Use $M = 4$ and $h = 0.2$.
(b) Use $M = 8$ and $h = 0.1$.
(c) Use a computer to obtain an answer with eight accurate digits.
Hint. A result from calculus states that the length of the curve $y = f(x)$ for $a \leqq x \leqq b$ is given by the definite integral

$$\text{length} = \int_a^b (1 + [f'(x)]^2)^{1/2} \, dx$$

provided that $f'(x)$ is continuous.

8. Find the approximate length of the parabola $y = x^2/2$ on the interval $-1.6 \leqq x \leqq 1.6$ by using the trapezoidal rule on the interval $[0, 1.6]$ and doubling the answer.
   (a) Use $M = 4$ and $h = 0.4$.
   (b) Use $M = 8$ and $h = 0.2$.
   (c) Use a computer to obtain an answer with eight accurate digits.

9. The velocity of a submarine that is under the polar ice cap is given by the following table. Use the trapezoidal rule to find the distance traveled during the time interval $[0, t]$.

Time, t (hr)	Velocity, v (km/hr)	Approximate Distance Traveled in the Interval $[0, t]$ (km)
0.0	6.0	0.0
0.5	8.8	--.--
1.0	9.2	--.--
1.5	10.8	--.--
2.0	12.4	--.--
2.5	10.4	--.--
3.0	8.4	--.--
3.5	7.2	--.--
4.0	6.4	--.--

10. The trapezoidal rule can be used to integrate a function known only as a table of values even when the values are unequally spaced. Approximate the integral of the function over the interval $[0, 10]$ that passes through the points $(0, 10)$, $(1, 6)$, $(2, 4)$, $(6, 2)$, and $(10, 1)$. *Hint.* Use $h = 1$ on $[0, 2]$ and $h = 4$ on $[2, 10]$.

11. (a) Verify that the trapezoidal rule with $M = 1$ is exact for polynomial functions of degree $\leqq 1$, $f(x) = c_1 x + c_0$, on $[0, 1]$.
    (b) Verify that the trapezoidal rule is not exact for $f(x) = x^2$ on $[0, 1]$ and that the formula for the error term

$$E(f, h) = \frac{-(b - a)h^2 f^{(2)}(c)}{12}$$

is indeed true.

12. Derive the trapezoidal rule when $M = 1$ using the following steps:
    (a) Find w_0 and w_1 so that

    $$\int_0^1 g(t)\, dt = w_0 g(0) + w_1 g(1)$$

    is exact for the functions $g(t) = 1$ and $g(t) = t$. *Hint.* You must first get a linear system involving w_0 and w_1.
    (b) Use the relation $f(x_0 + ht) = g(t)$ and the change of variables $x = x_0 + ht$ and $dx = h\, dt$ to derive the trapezoidal rule with $M = 1$ for f on the interval $x_0 \leq x \leq x_1$, where $x_1 = x_0 + h$.

In Exercises 13–15:
    (a) Determine the number M and the interval width h so that the trapezoidal rule for M intervals can be used to compute the given integral with an accuracy of 5×10^{-9}.
    (b) Use a computer to obtain an answer with eight accurate digits.

13. $\int_{-\pi/6}^{\pi/6} \cos(x)\, dx$. *Hint.* Show that $|f^{(2)}(c)| \leq 1$ on the interval $[-\pi/6, \pi/6]$.

14. $\int_2^3 (5 - x)^{-1}\, dx$. *Hint.* Show that $|f^{(2)}(c)| \leq \frac{1}{4}$ on the interval $[2, 3]$.

15. $\int_0^2 x \exp(-x)\, dx$. *Hint.* $f^{(2)}(x) = (x - 2)\exp(-x)$.

In Exercises 16–18, complete the table to verify that the error term for the composite trapezoidal rule obeys the relation $E(f, h) \approx Kh^2$. What happens to $E(f, h)$ when h is reduced to $h/2$?

16. $E(f, h) = \int_{-0.1}^{0.1} \cos(x)\, dx - T(f, h) \approx 0.1996668 - T(f, h)$

h	$T(f, h)$	$E(f, h)$	$Kh^2 = 0.01664h^2$
0.2000	0.1990008	0.0006660	0.0006656
0.1000	0.1995004	0.0001664	
0.0500	0.1996252	_____	0.0000416
0.0250	0.1996564	0.0000104	
0.0125	0.1996642	_____	_____

17. $E(f, h) = \int_0^{0.46} \exp(-x)\, dx - T(f, h) \approx 0.3687164 - T(f, h)$

h	$T(f, h)$	$E(f, h)$	$Kh^2 = -0.03072h^2$
0.46000	0.3751952	−0.0064788	−0.0065004
0.23000	0.3703403	−0.0016239	
0.11500	0.3691226	_____	−0.0004063
0.05750	0.3688179	−0.0001015	
0.02875	0.3687418	_____	_____

18. $E(f, h) = \int_{3.0}^{3.6} \ln(x)\, dx - T(f, h) \approx 0.7155250 - T(f, h)$

h	$T(f, h)$	$E(f, h)$	$Kh^2 = 0.00462h^2$
0.6000	0.7138638	0.0016612	0.0016632
0.3000	0.7151087	0.0004163	
0.1500	0.7154208	_____	0.0001040
0.0750	0.7154989	0.0000261	_____
0.0375	0.7155185	_____	_____

19. Write a report on the corrected trapezoidal rule.

20. *Midpoint Rule.* The midpoint rule on $[x_0, x_1]$ is

$$\int_{x_0}^{x_1} f(x)\, dx = hf\left(x_0 + \frac{h}{2}\right) + \frac{h^3 f''(c_1)}{24},$$

where $h = (x_1 - x_0)/2$.

(a) Expand $F(x)$, the antiderivative of $f(x)$, in a Taylor series about $x_0 + h/2$ and establish the midpoint rule on $[x_0, x_1]$.

(b) Use part (a) and show that the composite midpoint rule for approximating the integral of $f(x)$ over $[a, b]$ is

$$\int_a^b f(x)\, dx = M(f, h) + E(f, h), \qquad \text{where } h = \frac{b - a}{N} \quad \text{and}$$

$$M(f, h) = h \sum_{k=1}^{N} f(z_k) \quad \text{and} \quad z_k = a + (k - \tfrac{1}{2})h \qquad \text{for } k = 1, 2, \ldots, N.$$

(c) Show that the error term $E(f, h)$ for part (b) is

$$E(f, h) = \sum_{k=1}^{N} \frac{f''(c_k)h^3}{24} = \frac{b - a}{24} f''(c)h^2 = O(h^2).$$

21. Modify Algorithm 7.1 so that it uses the composite midpoint rule to approximate the integral of $f(x)$ over $[a, b]$.

22. Use the midpoint rule with $N = 4$ subintervals to approximate the integrals in (a) Exercise 1, (b) Exercise 2, (c) Exercise 3, (d) Exercise 4, (e) Exercise 5, (f) Exercise 6.

7.3 Composite Simpson's Rule

Assume that the curve $y = f(x)$ is approximated by a parabola that passes through three points (x_0, y_0), (x_1, y_1), (x_2, y_2) and an interpolating polynomial $P(x)$ of degree ≤ 2 is obtained. If we integrate $P(x)$ over $[x_0, x_2]$ and use this value to approximate the integral of $f(x)$, then we are using *Simpson's rule* for numerical integration. Simpson's rule is often significantly more accurate than the trapezoidal rule, because a parabola bends and takes on the shape of the curve. When this process is repeated on subintervals of $[a, b]$, we call it a *composite rule* (see Figure 7.4).

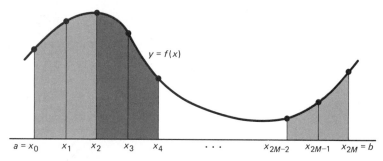

Figure 7.4. The composite Simpson's rule on the interval $[a, b]$.

Theorem 7.2 [*Description of the Composite Simpson's Rule*] Let $f, f', f'', f^{(3)}$, and $f^{(4)}$ be continuous on the interval $[a, b]$. If we subdivide the interval into $2M$ subintervals of equal width $h = (b - a)/2M$ and use the nodes $a = x_0 < x_1 < \ldots < x_{2M} = b$, where $x_k = a + hk$ for $k = 0, 1, \ldots, 2M$, Simpson's rule for $f(x)$ with $2M$ intervals of width h is

(1)
$$S(f, h) = \frac{h}{3} \sum_{k=1}^{M} \left[f(x_{2k-2}) + 4f(x_{2k-1}) + f(x_{2k}) \right]$$

or

$$S(f, h) = \frac{h}{3}\left[f_0 + 4f_1 + 2f_2 + 4f_3 + \ldots + 2f_{2M-2} + 4f_{2M-1} + f_{2M} \right].$$

This is an approximation to the integral of $f(x)$ and we write

(2)
$$\int_a^b f(x)\, dx = S(f, h) + E(f, h).$$

Furthermore, there exists a value c with $a < c < b$ so that the error term $E(f, h)$ has the form

(3)
$$E(f, h) = -\frac{(b - a)h^4}{180} f^{(4)}(c) = O(h^4).$$

Remarks. Notice in (1) that the weights w_k are the product of $h/3$ and a term in the sequence $1, 4, 2, 4, \ldots, 4, 1$. For this reason the method is sometimes called *Simpson's 1/3 rule*. Also, the interval $[a, b]$ must be divided into an even number of subintervals. When Simpson's rule is applied to integrate a function known only at a table of values, there should be an odd number of equally spaced points.

Example 7.4 Use Simpson's rule to find the area under the curve $y = f(x)$ that passes through the three points $(0, 2), (1, 3)$, and $(2, 2)$.

Solution. Since $M = 1$ and $h = 1, f(0) = 2, f(1) = 3$, and $f(2) = 2$,

$$\text{area} \approx \frac{h}{3}[f(0) + 4f(1) + f(2)] = \frac{1}{3}[2 + 12 + 2] = \frac{16}{3}.$$

Observation. The parabola passing through the three points is

$$P(x) = 3 - (x - 1)^2 = -x^2 + 2x + 2 \qquad \text{(see Figure 7.5).}$$

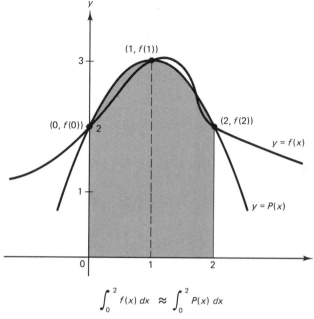

$$\int_0^2 f(x)\, dx \approx \int_0^2 P(x)\, dx$$

Figure 7.5. Simpson's rule for integrating $y = f(x)$ over $[0, 2]$.

Since Simpson's rule is based on the quadratic approximation $f(x) \approx P(x) = -x^2 + 2x + 2$, we can integrate $P(x)$ and obtain

$$\int_0^2 f(x)\, dx \approx \int_0^2 (-x^2 + 2x + 2)\, dx = \left(\frac{-x^3}{3} + x^2 + 2x\right)\Bigg|_0^2 = \frac{16}{3}.$$

Engineers sometimes use numerical integration to approximate the volume of a solid object. The volume of the solid of revolution is

$$\text{volume} = \pi \int_a^b [R(x)]^2\, dx,$$

where the solid is obtained by rotating the region under the curve $y = R(x)$, $a \leqq x \leqq b$, about the x-axis.

Example 7.5 Use Simpson's rule with $M = 3$ and $h = 1$ to approximate the volume of the solid of revolution, where the radius $R(x)$ of at the position x along the x-axis is given in Table 7.5 (see Figure 7.6).

Table 7.5

x	$R(x)$ (cm)	$[R(x)]^2$ (cm)
0	6.2	38.44
1	5.8	33.64
2	4.0	16.00
3	4.6	21.16
4	5.0	25.00
5	7.6	57.76
6	8.2	67.24

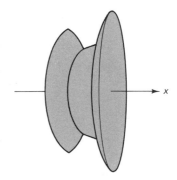

Figure 7.6. The solid of revolution for Example 7.5.

Solution. Using Simpson's rule, the approximate value of the integral is given by the calculation

$$\text{volume} \approx \frac{\pi}{3}(38.44 + 4 \times 33.64 + 2 \times 16.00$$

$$+ 4 \times 21.16 + 2 \times 25.00 + 4 \times 57.76 + 67.24) \approx 668.03.$$

When we consider the order of the error term, $E(f, h) = O(h^4)$, in (3), then we see that the error is proportional to h^4 and expect that the error in approximating the integral should be reduced by a factor of $\frac{1}{16}$ if the step size is decreased by a factor of $\frac{1}{2}$.

Example 7.6 Consider the definite integrals

$$\int_1^3 \frac{x\,dx}{1 + x^2} = \frac{\ln(5)}{2} \quad \text{and} \quad \int_1^3 \frac{dx}{7 - 2x} = \frac{\ln(5)}{2}.$$

Tables 7.6 and 7.7 show the approximations obtained using the composite Simpson's rule with $M = 1, 2, 4, 8, 16$ and the error in the approximation. It is worthwhile to note that as M is made larger h is successively divided by 2 and in the limit the successive errors become about $\frac{1}{16}$ of the previous error. Notice that the error for the first integral seems to be going to zero faster. But on closer inspection we see that the errors are diminishing at the same rate, and Simpson's rule is working equally well in both cases!

Table 7.6 Simpson's rule approximation

for $\displaystyle\int_1^3 \frac{x\,dx}{1+x^2}$

M	h	S(f, h)	E(f, h)
1	1.	0.80000000	0.00471896
2	0.5	0.80424403	0.00047493
4	0.25	0.80468635	0.00003261
8	0.125	0.80471690	0.00000206
16	0.0625	0.80471883	0.00000013

Table 7.7 Simpson's rule approximation

for $\displaystyle\int_1^3 \frac{dx}{7-2x}$

M	h	S(f, h)	E(f, h)
1	1.	0.84444444	−0.03972548
2	0.5	0.81111111	−0.00639215
4	0.25	0.80542328	−0.00070432
8	0.125	0.80477617	−0.00005721
16	0.0625	0.80472288	−0.00000392

The error term $E(f, h) = -(b-a)h^4 f^{(4)}(c)/180$ can be used to predict the number of subintervals $2M$ that are needed to compute an approximation to the integral with a specified accuracy.

Example 7.7 Find the number M and the step size h so that the absolute value of the error for Simpson's rule is less than 5×10^{-9} when it is used to approximate the integral

$$\int_2^7 \frac{dx}{x} = \ln(7) - \ln(2) \approx 1.252762968.$$

Solution. Compute the fourth derivative of $f(x) = 1/x, f^{(4)}(x) = 24/x^5$. The maximum value of $f^{(4)}(c)$ on $[2, 7]$ is seen to occur at $c = 2$ and we can use the bound $|f^{(4)}(c)| \leq \frac{3}{4}$ for $2 \leq x \leq 7$ and obtain

$$|E(f, h)| \leq \left| \frac{-(b-a)h^4 f^{(4)}(c)}{180} \right| \leq \frac{h^4}{48}$$

We seek the value h so that $h^4/48 \leq 5 \times 10^{-9}$ or $h^4 \leq 24 \times 10^{-8}$. But we really want to find the integer M, so we use $h = 5/(2M)$ and solve $M^4 \geq (625/384)10^8$ and obtain $M > 112.95$. Since M must be an integer, we choose $M = 113, 2M = 226$, and $h \approx 0.022123894$. When the computation is done, it is $S(f, 5/226) = 1.252762969$, and the accuracy is 1×10^{-9}. Since a bound was used in obtaining the estimate $M = 113$, it is possible that a smaller value of M might give an accuracy of 5×10^{-9}. Experimentation will lead to $M = 64$ and the approximation $S(f, 5/128) = 1.252762973$.

Compare Example 7.7 with Example 7.3 to see that Simpson's rule requires 227 evaluations of $f(x)$ and the trapezoidal rule requires 22823 evaluations of $f(x)$, for both methods to achieve an accuracy of 1×10^{-9}. For these examples, Simpson's rule required less than $1/100$ of the number of function evaluations to achieve the same accuracy!

We now turn to the derivation of Simpson's rule. As mentioned before, we will derive the integration formula. A proof that the error term $E(f, h)$ has the form given in equation (3) can be found in advanced books (see Reference 66).

Proof of the composite Simpson's rule. First we derive the rule for $F(t)$ on the interval $[0, 2]$. The quadratic Lagrange interpolating polynomial gives us the following approximation for $F(t)$:

(4) $$F(t) \approx F(0) \frac{(t-1)(t-2)}{(0-1)(0-2)} + F(1) \frac{(t-0)(t-2)}{(1-0)(1-2)} + F(2) \frac{(t-0)(t-1)}{(2-0)(2-1)}.$$

The approximation (4) can be simplified and written as

(5) $$F(t) \approx \frac{F(0)(t^2 - 3t + 2)}{2} - F(1)(t^2 - 2t) + \frac{F(2)(t^2 - t)}{2}.$$

Integrating both sides of (5), we obtain

(6) $$\int_0^2 F(t)\, dt \approx \frac{F(0)}{2} \int_0^2 (t^2 - 3t + 2)\, dt - F(1) \int_0^2 (t^2 - 2t)\, dt + \frac{F(2)}{2} \int_0^2 (t^2 - t)\, dt.$$

Also, as the reader may verify,

(7) $$\int_0^2 (t^2 - 3t + 2)\, dt = \tfrac{2}{3} \quad \int_0^2 (t^2 - 2t)\, dt = -\tfrac{4}{3} \quad \int_0^2 (t^2 - t)\, dt = \tfrac{2}{3}.$$

Now substitute the results of (7) into approximation (6) and obtain Simpson's rule for the interval $[0, 2]$ where $M = 1$ and $h = 1$:

(8) $$\int_0^2 F(t)\, dt \approx \tfrac{1}{3}[F(0) + 4F(1) + F(2)].$$

Next we derive the rule on the interval $[x_0, x_2]$ of width $2h$. Starting with the change of variables

(9) $$t = \frac{1}{h}(x - x_0) \quad \text{and} \quad F(t) = f(x_0 + ht),$$

we obtain

(10) $$dt = \frac{1}{h}\, dx \quad \text{and} \quad F(0) = f(x_0), \quad F(1) = f(x_1), \quad F(2) = f(x_2).$$

The integral of $f(x)$ taken over $[x_0, x_2]$ is transformed as follows:

(11)
$$\int_{x_0}^{x_2} f(x)\, dx = \int_0^2 f(x_0 + ht)h\, dt = h \int_0^2 F(t)\, dt.$$

Using (8) and (10) to integrate $F(t)$ in (11) results in Simpson's rule for $[x_0, x_2]$ where $M = 1$:

(12)
$$\int_{x_0}^{x_2} f(x)\, dx \approx \frac{h}{3}[f(x_0) + 4f(x_1) + f(x_2)].$$

Finally, we are ready to derive the composite Simpson's rule. The total integral over $[a, b]$ is the sum of M integrals taken over the subintervals $[x_{2k-1}, x_{2k}]$, $k = 1, 2, \ldots, M$. Thus

$$\int_a^b f(x)\, dx = \sum_{k=1}^M \int_{x_{2k-1}}^{x_{2k}} f(x)\, dx.$$

For $k = 2, \ldots, M$ we apply (12) shifted over to the interval $[x_{2k-2}, x_{2k}]$ and add the contributions in each interval, to obtain

(13)
$$\int_a^b f(x)\, dx \approx \frac{h}{3} \sum_{k=1}^M [f(x_{2k-2}) + 4f(x_{2k-1}) + f(x_{2k})].$$

The right side of (13) is $S(f, h)$ given in equation (1) and the proof of the composite Simpson's rule is complete.

Algorithm 7.2 [*Composite Simpson's Rule*] Approximate the integral

$$\int_A^B f(x)\, dx \approx \frac{h}{3} \sum_{k=1}^M [f(x_{2k-2}) + 4f(x_{2k-1}) + f(x_{2k})]$$

or

$$\int_A^B f(x)\, dx \approx \frac{h}{3}[f(A) + f(B)] + \frac{2h}{3} \sum_{k=1}^{M-1} f(x_{2k}) + \frac{4h}{3} \sum_{k=1}^M f(x_{2k-1}),$$

where the interval $[A, B]$ is divided into $2M$ subintervals of equal width $h = (B - A)/(2M)$ and $x_k = A + hk$ for $k = 0, 1, \ldots, 2M$ and $A = x_0 < x_1 < \ldots < x_{2M-1} < x_{2M} = B$. Notice that $x_{2M} = B$.

```
        USES the function F(X)
        INPUT  A, B, M
        H := [B−A]/[2*M]
        SumEven := 0
FOR   K = 1  TO  M−1  DO                    {Loop for even subscripts}
  │     X := A + H*2*K
  └──   SumEven := SumEven + F(X)
        SumOdd := 0
```

```
FOR  K = 1  TO  M  DO                              {Loop for odd subscripts}
      X := A + H*(2*K−1)
      SumOdd := SumOdd + F(X)
     SUM := H*[F(A) + F(B) + 2*SumEven + 4*SumOdd]/3
```

PRINT 'The approximate value of the integral of f(x)' 'on the interval' A, B
      'using' 2*M 'subintervals' 'computed using the Simpson's rule is' SUM

Exercises for the Composite Simpson's Rule

In Exercises 1–6:
- (a) Approximate the integral using the composite Simpson's rule with two subintervals, $M = 1$.
- (b) Approximate the integral using the composite Simpson's rule with four subintervals, $M = 2$.
- (c) Use the Fundamental Theorem of Integral Calculus to find the exact answer, that is, find an antiderivative $F(x)$ of the integrand $f(x)$ and compute $F(b) - F(a)$.
- (d) Use a computer to find a Simpson's rule approximation to the integral that has eight accurate digits.

1. $\int_1^3 4x^3 \, dx$ **2.** $\int_1^3 5x^4 \, dx$

3. $\int_1^3 x^{-2} \, dx$ **4.** $\int_1^3 x^{-1} \, dx$

5. $\int_1^3 64(x - 1)4^{-x} \, dx$. The antiderivative is

$$F(x) = 64\left[\frac{1 - x}{\ln(4)} - (\ln(4))^{-2}\right]4^{-x}.$$

6. $\int_0^1 \frac{4}{1 + x^2} \, dx = \pi$. The antiderivatives is

$$F(x) = 4\arctan(x).$$

7. Use Simpson's rule with $M = 3$ and $h = 1$ to approximate the volume of the solid of revolution, where the radius $R(x)$ of at the position x along the x-axis is given in the table.

x	$R(x)$ (cm)	$[R(x)]^2$ (cm²)
0	6.0	.
1	5.5	.
2	4.0	.
3	4.0	.
4	5.0	.
5	7.5	.
6	9.0	.

8. Use Simpson's rule to approximate the surface area of the solid obtained by rotating $y = \sin(x), 0 \le x \le 0.8$ about the x-axis.

   Remark. $\sin(x)$ is to be evaluated with x in radians.
   (a) Use $M = 1$ and $h = 0.4$.
   (b) Use $M = 2$ and $h = 0.2$.
   (c) Use a computer to obtain an answer with eight accurate digits.

   Hint. The surface area of a solid of revolution is given by

   $$\text{surface area} = 2\pi \int_a^b R(x)(1 + [R'(x)]^2)^{1/2} \, dx$$

   when the solid is obtained by rotating the region under the curve $y = R(x)$, $a \le x \le b$, about the x-axis.

9. Use Simpson's rule to approximate the surface area of the solid obtained by rotating $y = \exp(-x), 0 \le x \le 1$, about the x-axis.
   (a) Use $M = 1$ and $h = 0.5$.
   (b) Use $M = 2$ and $h = 0.25$.
   (c) Use a computer to obtain an answer with eight accurate digits.

10. Simpson's rule can be used to integrate a function known as a table of values provided that the abscissa occur in groups which form an even number of equal subintervals. Approximate the integral of the function over the interval $[0, 10]$ that passes through the points $(0, 10)$, $(1, 6)$, $(2, 4)$, $(6, 2)$, and $(10, 1)$. *Hint.* Use $h = 1$ on $[0, 2]$ and $h = 4$ on $[2, 10]$.

11. (a) Verify that Simpson's rule with $M = 1$ is exact for the functions $f(x) = 1$, $f(x) = x, f(x) = x^2$, and $f(x) = x^3$ on $[0, 2]$.
    (b) Verify that Simpson's rule is not exact for $f(x) = x^4$ on $[0, 2]$ and that the formula for remainder term $-(b - a)h^4 f^{(4)}(c)/180$ is indeed true.

12. Derive Simpson's rule when $M = 1$ using the following steps:
    (a) Find w_0, w_1, and w_2 so that

    $$\int_0^2 g(t) \, dt = w_0 g(0) + w_1 g(1) + w_2 g(2)$$

    for the functions $g(t) = 1, g(t) = t$, and $g(t) = t^2$. *Hint.* You must first get a linear system involving w_0, w_1, and w_2.
    (b) Use the relation $f(x_0 + ht) = g(t)$ and the change of variables $x = x_0 + ht$ and $dx = h \, dt$ to derive Simpson's rule with $M = 1$ for $f(x)$ on the interval $x_0 \le x \le x_2$, where $x_k = x_0 + ht$.

    In Exercises 13–15:
    (a) Determine the number M and the interval width h so that Simpson's rule for $2M$ intervals can be used to compute the given integral with an accuracy of 5×10^{-9}.
    (b) Use a computer to obtain an answer with eight accurate digits.

13. $\int_{-\pi/6}^{\pi/6} \cos(x) \, dx$. *Hint.* Show that $|f^{(4)}(c)| \le 1$ on the interval $[-\pi/6, \pi/6]$.

14. $\int_2^3 (5 - x)^{-1} \, dx$. *Hint.* Show that $|f^{(4)}(c)| \le \frac{3}{4}$ on the interval $[2, 3]$.

15. $\int_0^2 x \exp(-x) \, dx$. *Hint.* $f^{(4)}(x) = (x - 4) \exp(-x)$.

In Exercises 16–18, complete the table to verify that the error term for the composite Simpson's rule obeys the relation $E(f, h) \approx Kh^4$. What happens to $E(f, h)$ when h is reduced to $h/2$?

16. $E(f, h) = \int_{-0.75}^{0.75} \cos(x) \, dx - S(f, h) \approx 1.3632775 - S(f, h)$

h	$S(f, h)$	$E(f, h)$	$Kh^4 = -0.00770h^4$
0.75000	1.3658444	−0.0025669	
0.37500	1.3634298		−0.0001523
0.18750	1.3632869	−0.0000094	
0.09375	1.3632781		

17. $E(f, h) = \int_{0}^{1.75} \exp(-x) \, dx - S(f, h) \approx 0.8262261 - S(f, h)$

h	$S(f, h)$	$E(f, h)$	$Kh^4 = -0.00445h^4$
0.875000	0.8286898	−0.0024637	
0.437500	0.8263905		−0.0001630
0.218750	0.8262365	−0.0000104	
0.109375	0.8262267		

18. $E(f, h) = \int_{2.0}^{3.4} \ln(x) \, dx - S(f, h) \approx 1.37454211 - S(f, h)$

h	$S(f, h)$	$E(f, h)$	$Kh^4 = 0.00107h^4$
0.7000	1.37431693	0.00022518	
0.3500	1.37452628		0.00001606
0.1750	1.37454108	0.00000103	
0.0875	1.37454204		

19. Write a report on Newton–Cotes integration formulas.

20. Write a report on the corrected Simpson method.

7.4 Sequential Integration Rules

In this section we show how to compute Simpson's approximation with a special linear combination of trapezoidal rules. The approximation will have greater accuracy if a larger number of subintervals are used. How many should we choose? The sequential process helps answer this question by trying two subintervals, four subintervals, and so on, until the desired accuracy is achieved. A sequence $\{T(J)\}$ of trapezoidal rule approximations are generated. When the number of subintervals is doubled, the number of function values is roughly doubled, because the function must be evaluated at all the previous points and at the midpoints of the previous subintervals, as shown in Figure 7.7. Theorem 7.3 explains how to eliminate redundant function evaluations and additions by using $T(J-1)$ in the computation of $T(J)$.

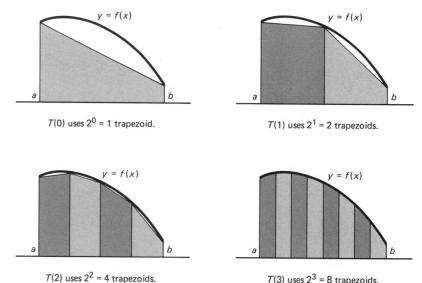

$T(0)$ uses $2^0 = 1$ trapezoid. $T(1)$ uses $2^1 = 2$ trapezoids.

$T(2)$ uses $2^2 = 4$ trapezoids. $T(3)$ uses $2^3 = 8$ trapezoids.

Figure 7.7. The trapezoidal rules $T(0)$, $T(1)$, $T(2)$ and $T(3)$.

Theorem 7.3 [Description of the sequential trapezoidal rule] For any integer $J \geq 1$, let the interval $[a, b]$ be divided into $2^J = 2M$ subintervals of equal width $h = (b-a)/2^J$. The trapezoidal rules $T(f, h)$ and $T(f, 2h)$ for step sizes h and $2h$, respectively, obey the relationship

$$(1) \qquad T(f, h) = \frac{T(f, 2h)}{2} + h \sum_{k=1}^{M} f(x_{2k-1}).$$

We define $T(0) = h[f(a) + f(b)]/2$. Then for any positive integer J we define $T(J) = T(f, h)$ and $T(J-1) = T(f, 2h)$, where $h = (b-a)/2^J$. This permits us to write (1) in a form that is more useful for computational purposes.

$$(2) \qquad T(J) = \frac{T(J-1)}{2} + h \sum_{k=1}^{M} f(x_{2k-1}) \qquad (\text{for } J = 1, 2, \ldots).$$

Proof. For the even nodes $x_0 < x_2 < \ldots < x_{2M-2} < x_{2M}$ we have the trapezoidal rule with step size $2h$:

$$(3) \qquad T(J-1) = \frac{2h}{2}(f_0 + 2f_2 + 2f_4 + \ldots + 2f_{2M-4} + 2f_{2M-2} + f_{2M}).$$

Using all the nodes $x_0 < x_1 < \ldots < x_{2M-2} < x_{2M-1} < x_{2M}$, we have the trapezoidal rule with step size h:

$$(4) \qquad T(J) = \frac{h}{2}(f_0 + 2f_1 + 2f_2 + \ldots + 2f_{2M-2} + 2f_{2M-1} + f_{2M}).$$

Collecting the even and odd subscripts in (4) yields

$$(5) \qquad T(J) = \frac{h}{2}(f_0 + 2f_2 + \ldots + 2f_{2M-2} + f_{2M}) + h\sum_{k=1}^{M} f_{2k-1}.$$

Substituting (3) in (5) results in $T(J) = T(J-1)/2 + h\sum_{k=1}^{M} f_{2k-1}$ and the proof of the theorem is complete.

Example 7.8 Use the sequential trapezoidal rule to find the approximations $T(0)$, $T(1)$, $T(2)$, and $T(3)$ for the integral

$$\int_1^5 \frac{dx}{x} = \ln(5) = 1.609438.$$

Solution. Table 7.8 shows the nine values required to compute $T(3)$ and the midpoints required to compute $T(1)$, $T(2)$, and $T(3)$.

Table 7.8

x	$f(x) = \dfrac{1}{x}$	Values at the Endpoints	Midpoint Values for $T(1)$	Midpoint Values for $T(2)$	Midpoint Values for $T(3)$
1.0	1.000000	1.000000			
1.5	0.666667				0.666667
2.0	0.500000			0.500000	
2.5	0.400000				0.400000
3.0	0.333333		0.333333		
3.5	0.285714				0.285714
4.0	0.250000			0.250000	
4.5	0.222222				0.222222
5.0	0.200000	0.200000			

When $h = 4$: $\quad T(0) = \dfrac{4}{2}(1.000000 + 0.200000) = 2.400000.$

When $h = 2$: $\quad T(1) = \dfrac{T(0)}{2} + 2(0.333333)$

$$= 1.200000 + 0.666666 = 1.866666.$$

When $h = 1$: $\quad T(2) = \dfrac{T(1)}{2} + 1(0.500000 + 0.250000)$

$$= 0.933333 + 0.750000 = 1.683333.$$

When $h = 0.5$: $T(3) = \dfrac{T(2)}{2} + \dfrac{1}{2}(0.666667 + 0.400000 + 0.285714 + 0.222222)$

$$= 0.841667 + 0.787302 = 1.628968.$$

Our next result shows an important relationship between the trapezoidal rule and Simpson's rule. When the trapezoidal rule is computed using step sizes $2h$ and h the result is $T(f, 2h)$ and $T(f, h)$. These values can be combined to obtain Simpson's rule:

$$S(f, h) = \frac{4T(f, h) - T(f, 2h)}{3}$$

Theorem 7.4 [Sequential Simpson's rule] For any integer $J \geq 1$ let the interval $[a, b]$ be divided into $2^J = 2M$ subintervals of equal width $h = (b - a)/2^J$. Assume that the sequential trapezoidal rule is used to obtain $T(0), T(1), T(2), \ldots, T(J)$; then Simpson's rule $S(J) = S(f, h)$ for 2^J subintervals is obtained from $T(J)$ and $T(J - 1)$ by the formula

(6) $$S(J) = \frac{4T(J) - T(J - 1)}{3} \qquad \text{(for } J \geq 1\text{)}.$$

Proof. The trapezoidal rule $T(J)$ with step size h is

(7) $$\int_a^b f(x)\, dx \approx \frac{h}{2}(f_0 + 2f_1 + 2f_2 + \ldots + 2f_{2M-2} + 2f_{2M-1} + f_{2M}) = T(J).$$

The trapezoidal rule $T(J - 1)$ with step size $2h$ is

(8) $$\int_a^b f(x)\, dx \approx \frac{2h}{2}(f_0 + 2f_2 + 2f_4 + \ldots + 2f_{2M-2} + f_{2M}) = T(J - 1).$$

Multiplying relation (7) by 4 results in

(9) $$4\int_a^b f(x)\, dx \approx \frac{4h}{2}(f_0 + 2f_1 + 2f_2 + \ldots + 2f_{2M-2} + 2f_{2M-1} + f_{2M}) = 4T(J).$$

Now subtract (8) from (9) and the result is

(10) $$3\int_a^b f(x)\, dx \approx h(f_0 + 4f_1 + 2f_2 + 4f_3 + \ldots + 2f_{2M-2} + 4f_{2M-1} + f_{2M}).$$

This can be expressed in the $T(J)$ and $T(J - 1)$ notation as

(11) $$3\int_a^b f(x)\, dx \approx 4T(J) - T(J - 1).$$

When relation (10) is divided by 3, the result on the right side is easily seen to be Simpson's rule. Therefore, the right side of (11) is equivalent to $3S(J)$ and the theorem is proven.

Example 7.9 Use the sequential Simpson's rule to compute the approximations $S(1), S(2)$, and $S(3)$ for the integral of Example 7.8.

Solution. Using formula (6) with $J = 1, 2, 3$, we compute

$$S(1) = \frac{4 \times 1.866666 - 2.400000}{3} = 1.688888,$$

$$S(2) = \frac{4 \times 1.683333 - 1.866666}{3} = 1.622222,$$

$$S(3) = \frac{4 \times 1.628968 - 1.683333}{3} = 1.610846.$$

The Lagrange interpolating polynomial of degree 4 can be used to obtain a polynomial approximation $P(x)$ to $f(x)$ based on the nodes x_0, x_1, x_2, x_3, and x_4. The polynomial $P(x)$ can be integrated term by term as was done in the proof of Simpson's rule and the result is called *Boole's rule* for the interval $[x_0, x_4]$:

(12)
$$\int_{x_0}^{x_4} f(x)\, dx \approx \frac{2h}{45}(7f_0 + 32f_1 + 12f_2 + 32f_3 + 7f_4).$$

The details of this derivation are long and involved and will not be given. An alternative method for establishing the formula in (12) is mentioned in Exercise 9. When this process is repeated over $4M$ subintervals of $[a, b]$ we call it the *composite Boole's rule*:

$$B(J) = \frac{2h}{45} \sum_{k=1}^{M} (7f_{4k-4} + 32f_{4k-3} + 12f_{4k-2} + 32f_{4k-1} + 7f_{4k}).$$

Theorem 7.5 [Sequential Boole's rule] For any integer $J \geq 2$ let the interval $[a, b]$ be divided into 2^J subintervals of equal width $h = (b - a)/2^J$. Assume that the sequential Simpson's rule is used to obtain $S(1), S(2), \ldots, S(J)$. Then Boole's rule $B(J)$ for 2^J subintervals is obtained from $S(J)$ and $S(J - 1)$ by the formula

(13)
$$B(J) = \frac{16S(J) - S(J - 1)}{15} \qquad \text{(for } J \geq 2\text{)}.$$

The proof is left for the reader (see Exercise 10).

Example 7.10 Use the sequential Boole's rule to compute the approximations $B(2)$ and $B(3)$ for the integral in Example 7.9.

Solution. Using formula (12) with $J = 2, 3$, we compute

$$B(2) = \frac{16 \times 1.622222 - 1.688888}{15} = 1.617778,$$

$$B(3) = \frac{16 \times 1.610846 - 1.622222}{15} = 1.610088.$$

The reader may wonder what we are leading up to. In Section 7.5 we will see that formulas (6) and (12) are special cases of the process of Romberg integration. Let us announce that the next level of approximation for the integral in Example 7.10 is

$$\frac{64B(3) - B(2)}{63} = \frac{64 \times 1.610088 - 1.617778}{63} = 1.609490,$$

and this answer gives an accuracy of five decimal places.

Algorithm 7.3 [*Sequential Trapezoidal Rule*] The composite trapezoidal rule is used repeatedly with an increasing number of subintervals to approximate the integral

$$\int_A^B f(x)\, dx \approx \frac{h}{2} \sum_{k=1}^{2^J} \left[f(x_{k-1}) + f(x_k) \right],$$

where the Jth iteration subdivides the interval $[A, B]$ into 2^J subintervals of equal width $h = (B - A)/2^J$ and $x_k = A + hk$ and $A = x_0 < x_1 < \ldots < x_{2^J-1} < x_{2^J} = B$. Notice that $x_{2^J} = B$.

```
    INPUT   A, B                                      {Endpoints of the interval}
    INPUT   N                     {Number of times subintervals are bisected}
    VARiable declaration REAL   T[0 .. N]
    M := 1
    H := B − A
    T(0) := H*[F(A) + F(B)]/2                        {Compute one trapezoid}

PRINT 'The approximate value of the integral using'   1
      'subinterval and the trapezoidal rule is'   T(0)

FOR   J = 1  TO  N   DO                          {Do sequential calculations}
 │       M := 2*M                               {Use twice as many subintervals}
 │       H := H/2                               {Reduce the step size one-half}
 │       SUM := 0
 │       FOR  K = 1  TO  M/2  DO                     {Loop for odd subscripts}
 │        └── SUM := SUM + F(A + H*(2*K−1))
 │       T(J) := T(J−1)/2  + H*SUM
 │       PRINT 'The approximate value of the integral using'   M
 └──           'subintervals and the trapezoidal rule is'   T(J)
```

Exercises for Sequential Integration Rules

In Exercises 1–6:
  (a) Use the values $T(0), T(1), \ldots, T(J)$ from the sequential trapezoidal rule and compute the values for the sequential Simpson's rule and sequential Boole's rule.
  (b) Use a computer to find the approximations in part (a).

1. $\int_1^3 6x^5 \, dx = 728$

J	$T(J)$ Trapezoid Rule	$S(J)$ Simpson's Rule	$B(J)$ Boole's Rule
0	1,464.000000		
1	924.000000		
2	777.750000	_____	
3	740.484375	_____	_____

2. $\int_1^3 7x^6 \, dx = 2{,}186$

J	$T(J)$ Trapezoid Rule	$S(J)$ Simpson's Rule	$B(J)$ Boole's Rule
0	5,110.0000		
1	3,003.0000		
2	2,395.8594	_____	
3	2,238.8191	_____	_____
4	2,199.2270	_____	_____

3. $\int_1^3 x^{-2} \, dx = \frac{2}{3} \approx 0.666667$

J	$T(J)$ Trapezoid Rule	$S(J)$ Simpson's Rule	$B(J)$ Boole's Rule
0	1.111111		
1	0.805556		
2	0.705000	_____	
3	0.676573	_____	_____
4	0.669166	_____	_____

4. $\int_1^3 x^{-1} \, dx = \ln(3) \approx 1.098612$

J	$T(J)$ Trapezoid Rule	$S(J)$ Simpson's Rule	$B(J)$ Boole's Rule
0	1.333333		
1	1.166667		
2	1.116667	_____	
3	1.103211	_____	_____
4	1.099768	_____	_____

5. $\int_1^3 64(x-1)4^{-x}\,dx \approx 6.362439$

J	$T(J)$ Trapezoid Rule	$S(J)$ Simpson's Rule	$B(J)$ Boole's Rule
0	2.000000		
1	5.000000		
2	6.000000	_____	
3	6.270369	_____	_____
4	6.339328	_____	_____

6. $\int_0^1 \dfrac{4}{1+x^2}\,dx = \pi \approx 3.14159265$

J	$T(J)$ Trapezoid Rule	$S(J)$ Simpson's Rule	$B(J)$ Boole's Rule
0	3.0000000		
1	3.1000000		
2	3.1311765	_____	
3	3.1389885	_____	_____
4	3.1409416	_____	_____

7. Assume that the sequential trapezoidal rule converges to L, that is, suppose that $\lim\limits_{J \to \infty} T(J) = L$.

(a) Show that $\lim\limits_{J \to \infty} S(J) = L$.

(b) Show that $\lim\limits_{J \to \infty} B(J) = L$.

Hint. For part (a) use $\lim\limits_{J \to \infty} \dfrac{4T(J) - T(J-1)}{3}$.

Fact. Boole's rule and error term for the interval $[x_0, x_4]$ is

$$\int_{x_0}^{x_4} f(x)\,dx = \frac{2h}{45}(7f_0 + 32f_1 + 12f_2 + 32f_3 + 7f_4) - \frac{8h^7}{945}f^{(6)}(c).$$

8. (a) Verify that Boole's rule is exact for the functions $f(x) = 1, x, x^2, x^3, x^4, x^5$ on the interval $[0, 4]$.

(b) Verify that Boole's rule is not exact for $f(x) = x^6$ on $[0, 4]$ and that the formula for the remainder term $-8h^7 f^{(6)}(c)/945$ is indeed true.

9. Boole's rule can be derived by finding the weights w_0, w_1, w_2, w_3, and w_4 so that the integration formula

$$\int_0^4 g(t)\, dt = w_0 g(0) + w_1 g(1) + w_2 g(2) + w_3 g(3) + w_4 g(4)$$

is exact for the functions $g(t) = 1, t, t^2, t^3, t^4$.

(a) Verify that the linear system involving the weights is

$$w_0 + w_1 + \quad w_2 + \quad w_3 + \quad w_4 = 4$$

$$w_1 + \quad 2w_2 + \quad 3w_3 + \quad 4w_4 = 8$$

$$w_1 + \quad 4w_2 + \quad 9w_3 + \quad 16w_4 = \frac{64}{3}$$

$$w_1 + \quad 8w_2 + 27w_3 + \quad 64w_4 = 64$$

$$w_1 + 16w_2 + 81w_3 + 256w_4 = \frac{1{,}024}{5}.$$

(b) Verify that solution to the linear system in part (a) is

$$w_0 = \tfrac{14}{45} \qquad w_1 = \tfrac{64}{45} \qquad w_2 = \tfrac{24}{45} \qquad w_3 = \tfrac{64}{45} \qquad w_4 = \tfrac{14}{45}.$$

10. Establish the relation (13): $B(J) = [16S(J) - S(J-1)]/15$ for the case $J = 2$. Use the following information:

$$S(1) = \frac{2h}{3}(f_0 + 4f_2 + f_4),$$

$$S(2) = \frac{h}{3}(f_0 + 4f_1 + 2f_2 + 4f_3 + f_4).$$

Obtain the following expression for $B(2)$:

$$B(2) = \frac{2h}{45}(7f_0 + 32f_1 + 12f_2 + 32f_3 + 7f_4).$$

11. Modify Algorithm 7.3 so that it will stop when consecutive values $T(K-1)$ and $T(K)$ for sequential trapezoidal rule differ by less than 5×10^{-6}.

12. Extend Algorithm 7.3 so that it will compute the values for the sequential Simpson's rule and sequential Boole's rule.

13. *Simpson's 3/8 Rule.* Consider the following trapezoidal rules applied to the interval $[x_0, x_3]$, where $x_k = x_0 + hk$:

$$T(f, 3h) = \frac{3h}{2}(f_0 + f_3) \text{ with step size } 3h,$$

$$T(f, h) = \frac{h}{2}(f_0 + 2f_1 + 2f_2 + f_3) \text{ with step size } h.$$

Show that the linear combination

$$\frac{9T(f, h) - T(f, 3h)}{8}$$

yields Simpson's 3/8 rule

$$\frac{3h}{8}(f_0 + 3f_1 + 3f_2 + f_3).$$

7.5 Romberg Integration

In Section 7.4 we saw how to construct successively more accurate approximations from the values of the sequential trapezoidal rule. Let $T(f, h)$, $S(f, h)$, and $B(f, h)$ denote the trapezoidal rule, Simpson's rule and Boole's rule, respectively, with step size h. The order of their error terms is

(1)
$$\int_a^b f(x)\, dx = T(f, h) + O(h^2),$$

$$\int_a^b f(x)\, dx = S(f, h) + O(h^4),$$

$$\int_a^b f(x)\, dx = B(f, h) + O(h^6),$$

where the error terms are

$$O(h^2) = \frac{-(b-a)h^2 f^{(2)}(c)}{12},$$

$$O(h^4) = \frac{-(b-a)h^4 f^{(4)}(c)}{180},$$

and

$$O(h^6) = \frac{-2(b-a)h^6 f^{(6)}(c)}{945}.$$

The pattern in (1) can be extended in the following sense. If an approximation rule is used with step sizes h and $2h$, then an algebraic manipulation of the two answers can be used to produce an improved answer. Each successive level of improvement increases the order of the error term from $O(h^{2N})$ to $O(h^{2N+2})$.

This process is called *Romberg integration* and has its strengths and weaknesses. Once we are past the level involving Boole's rule, the successive rules do not correspond to Newton–Cotes integration formulas, that is, they do not continue to be integration formulas based on polynomial interpolation formulas. Since twice as many function evaluations are needed to increase the order of the error term by a power of 2, the method is seen to require many calculations.

On the other hand, the Romberg method has the advantage that all of the weights w_k are positive and the abcissas x_k are equally spaced. If higher-order Newton–Cotes integration formulas are used, then starting with the polynomial approximation of degree 8, some of the weights are negative, and their magnitude is large. This causes loss of significance errors due to round-off.

The development of Romberg integration relies on the theoretical assumption that $f(x)$ is smooth enough so that the error in the trapezoidal rule can be expanded in a series involving only even powers of h, that is,

$$(2) \qquad \int_a^b f(x)\,dx = T(f, h) + E(f, h),$$

where

$$(3) \qquad E(f, h) = a_1 h^2 + a_2 h^4 + a_3 h^6 + \ldots .$$

A derivation of (3) is given in Reference [109].

Since only even powers of h occur in (3), the Richardson improvement process can be used to successively eliminate the coefficients $a_1, a_2, \ldots$ and obtain numerical integration formulas whose error terms have the even orders $O(h^4), O(h^6), \ldots$. To show that the first improvement is Simpson's rule for $2M$ intervals, start with $T(f, h)$ and $T(f, 2h)$ and the equations

$$(4) \qquad \int_a^b f(x)\,dx = T(f, h) + a_1 h^2 + a_2 h^4 + a_3 h^6 + \ldots ,$$

$$(5) \qquad \int_a^b f(x)\,dx = T(f, 2h) + 4a_1 h^2 + 16a_2 h^4 + 64a_3 h^6 + \ldots .$$

Multiply equation (4) by 4 and obtain

$$(6) \qquad 4\int_a^b f(x)\,dx = 4T(f, h) + 4a_1 h^2 + 4a_2 h^4 + 4a_3 h^6 + \ldots .$$

Now eliminate a_1 by subtracting (5) from (6). The result is

$$(7) \qquad 3\int_a^b f(x)\,dx = 4T(f, h) - T(f, 2h) - 12a_2 h^4 - 60a_3 h^6 + \ldots .$$

Now divide equation (7) by 3 and rename the coefficients in the series

$$(8) \qquad \int_a^b f(x)\,dx = \frac{4T(f, h) - T(f, 2h)}{3} + b_1 h^4 + b_2 h^6 + \ldots .$$

The first quantity on the right side of (8) is Simpson's rule:

$$(9) \qquad \int_a^b f(x)\,dx = S(f, h) + b_1 h^4 + b_2 h^6 + \ldots .$$

To show that the second improvement is Boole's rule for $4M$ intervals, start with $S(f, h)$ in (9) and $S(f, 2h)$ given in

$$(10) \qquad \int_a^b f(x)\, dx = S(f, 2h) + 16 b_1 h^4 + 64 b_2 h^6 + \dots.$$

When b_1 is eliminated from (9) and (10), the result is Boole's rule:

$$(11) \qquad \int_a^b f(x)\, dx = \frac{16 S(f, h) - S(f, 2h)}{15} + c_1 h^6 + c_2 h^8 + \dots.$$

The general pattern for Romberg integration relies on Lemma 7.1.

Lemma 7.1 [Richardson improvement for Romberg integration] Given two approximations $R(2h, K - 1)$ and $R(h, K - 1)$ for the quantity Q that satisfy

$$(12) \qquad Q = R(h, K - 1) + c_1 h^{2K} + c_2 h^{2K+2} + \dots,$$

$$(13) \qquad Q = R(2h, K - 1) + c_1 4^K h^{2K} + c_2 4^{K+1} h^{2K+2} + \dots$$

an improved approximation has the form

$$(14) \qquad Q = \frac{4^K \times R(h, K - 1) - R(2h, K - 1)}{4^K - 1} + O(h^{2K+2}).$$

The proof is straightforward and is left as an exercise.

For convenience we use the notation

$$(15) \qquad \begin{aligned} R(J, 0) &= T(J) &&\text{for the sequential Trapezoidal rule (for } J \geq 0\text{),} \\ R(J, 1) &= S(J) &&\text{for the sequential Simpson's rule (for } J \geq 1\text{),} \\ R(J, 2) &= B(J) &&\text{for the sequential Boole's rule (for } J \geq 2\text{),} \end{aligned}$$

where J corresponds to the given rule applied with 2^J subintervals of equal width $h = (b - a)/2^J$. Here our thoughts are that $R(J, 1)$ is the first improvement, $R(J, 2)$ is the second improvement, and so on. So far we have already seen the patterns

$$(16) \qquad \begin{aligned} R(J, 1) &= \frac{4 R(J, 0) - R(J - 1, 0)}{3} &&\text{(for } J \geq 1\text{),} \\[2mm] R(J, 2) &= \frac{16 R(J, 1) - R(J - 1, 1)}{15} &&\text{(for } J \geq 2\text{).} \end{aligned}$$

which are the rules (8) and (11) stated using the notation in (15). Using Lemma 7.1 the improvement can be continued inductively and we define the *Romberg integration rule* as follows:

$$(17) \qquad R(J, K) = \frac{4^K R(J, K - 1) - R(J - 1, K - 1)}{4^K - 1} \qquad \text{(for } J \geq K\text{).}$$

For computational purposes, the values $R(J, K)$ are arranged in the Romberg integration table given in Table 7.9.

Table 7.9   Romberg integration table

J	$R(J, 0)$ Trapezoidal Rule	$R(J, 1)$ Simpson's Rule	$R(J, 2)$ Boole's Rule	$R(J, 3)$ Third Improvement	$R(J, 4)$ Fourth Improvement
0	$R(0, 0)$				
1	$R(1, 0)$	$R(1, 1)$			
2	$R(2, 0)$	$R(2, 1)$	$R(2, 2)$		
3	$R(3, 0)$	$R(3, 1)$	$R(3, 2)$	$R(3, 3)$	
4	$R(4, 0)$	$R(4, 1)$	$R(4, 2)$	$R(4, 3)$	$R(4, 4)$

Example 7.11   Use Romberg integration to find approximations for

$$\int_0^4 \sin (x) \, dx = -\cos (4) + \cos (0) \approx 1.65364362.$$

Solution. The Romberg integration table is

J	$R(J, 0)$ Trapezoidal Rule	$R(J, 1)$ Simpson's Rule	$R(J, 2)$ Boole's Rule	$R(J, 3)$ Third Improvement	$R(J, 4)$ Fourth Improvement
0	-1.51360499				
1	1.06179236	1.92025814			
2	1.51348717	1.66405211	1.64697171		
3	1.61904831	1.65423535	1.65358090	1.65368581	
4	1.64502191	1.65367978	1.65364274	1.65364372	1.65364355

The table of numbers in Example 7.11 is not easy to interpret without knowing the behavior of the error terms. When the Romberg method is used to integrate f over the interval $[a, b]$ where $H = b - a$, the error terms have the ideal form given in Table 7.10. In each column K the constants $C_{J, K}$ are the same magnitude and as one moves down a row of column K the error diminishes by a factor of about 4^{K+1}.

Table 7.10   Romberg error table

Step Size, h	Error in $R(J, 0)$, $O(h^2)$	Error in $R(J, 1)$, $O(h^4)$	Error in $R(J, 2)$, $O(h^6)$	Error in $R(J, 3)$, $O(h^8)$	Error in $R(J, 4)$, $O(h^{10})$
H	$C_{0,0}H^2$				
$\dfrac{H}{2}$	$\dfrac{C_{1,0}H^2}{4}$	$\dfrac{C_{1,1}H^4}{4^2}$			
$\dfrac{H}{2^2}$	$\dfrac{C_{2,0}H^2}{4^2}$	$\dfrac{C_{2,1}H^4}{4^4}$	$\dfrac{C_{2,2}H^6}{4^6}$		
$\dfrac{H}{2^3}$	$\dfrac{C_{3,0}H^2}{4^3}$	$\dfrac{C_{3,1}H^4}{4^6}$	$\dfrac{C_{3,2}H^6}{4^9}$	$\dfrac{C_{3,3}H^8}{4^{12}}$	
$\dfrac{H}{2^4}$	$\dfrac{C_{4,0}H^2}{4^4}$	$\dfrac{C_{4,1}H^4}{4^8}$	$\dfrac{C_{4,2}H^6}{4^{12}}$	$\dfrac{C_{4,3}H^8}{4^{16}}$	$\dfrac{C_{4,4}H^{10}}{4^{20}}$

Example 7.12   The Romberg error table for Example 7.11 is

Step Size, h	Error in $R(J, 0)$, $O(h^2)$	Error in $R(J, 1)$, $O(h^4)$	Error in $R(J, 2)$, $O(h^6)$	Error in $R(J, 3)$, $O(h^8)$	Error in $R(J, 4)$, $O(h^{10})$
0	3.16724861				
2	0.59185126	−0.26661452			
1	0.14015645	−0.01040849	0.00667191		
0.5	0.03459531	−0.00059173	0.00006272	−0.00004219	
0.25	0.00862171	−0.00003616	0.00000088	−0.00000010	0.00000007

It is known that the error term for Romberg integration satisfies

$$(18) \qquad \int_a^b f(x)\, dx = R(J, K) + b_K h^{2K+2} f^{(2K+2)}(c_{J,K}),$$

where $h = (b - a)/2^J$, b_k is a constant that depends on K, and $c_{J,K}$ is a value satisfying $a \leqq c_{J,K} \leqq b$ (see Ref. [109], p.126). An important implication of (18) is the following theorem.

Theorem 7.6   [Precision of Romberg integration] If $f(x)$ is a polynomial of degree M and N is an integer satisfying $M \leqq 2N + 1$, then

$$(19) \qquad \int_a^b f(x)\, dx = R(N, N).$$

Example 7.13   Apply Theorem 7.6 and show that

$$\int_0^2 10x^9\, dx = 1{,}024 = R(4, 4).$$

Solution. Here the integrand is $f(x) = 10x^9$ and $f^{(10)}(x) = 0$. The degree is $M = 9$ and $9 \le 2N + 1$ implies that $4 \le N$ so that (19) holds true for $N = 4$.

Algorithm 7.4 [*Romberg Integration*] Approximate the integral

$$\int_A^B f(x)\, dx \approx R(J, J).$$

Successive	R(0,0)			
approximations	R(1,0)	R(1,1)		
are stored	R(2,0)	R(2,1)	R(2,2)	
in the array	$\vdots$	$\vdots$	$\vdots$	
R(J,K) K$\le$J	R(J,0)	R(J,1)	R(J,2) ... R(J,J)	

The elements R(J,0) of column 0 are computed using the sequential Trapezoidal Rule based on 2^J subintervals of [A,B]. The elements of column K are computed using the Romberg integration rule,

$$R(J,K) = R(J,K-1) + [R(J,K-1) - R(J-1,K-1)] / [4^K - 1].$$

At least four rows of the table are computed. The algorithm is terminated in the Jth row when the approximations R(J−1,J−1) and R(J,J) differ by less than the preassigned value Tol.

```
INPUT A, B                              {Endpoints of the interval}
INPUT N                         {Maximum number of rows, e.g., 14}
INPUT Tol                       {Termination criterion, e.g., 10⁻⁷}

VARiable declaration REAL  R[0..N,0..N]   {Array passed to the subroutine}

SUBROUTINE  TrRule (A,H,J,M,R)              {Sequential trapezoidal rule}
    H := H/2                      {Reduce step size for Jth refinement}
    SUM := 0
    FOR  P = 1  TO  M  DO                    {Loop for odd subscripts}
        SUM := SUM + F(A + H*(2*P−1))
    R(J,0) := R(J−1,0)/2 + H*SUM
    H := 2*M                       {Update number of subintervals}
    END                             {End of the procedure TrRule}

{The main program starts here.}

    M := 1                    {Initialize the number of subintervals}
    H := B − A                        {Initialize the step size}
    Close := 1                        {Initialize the variable}
    J := 0                            {Initialize the counter}

    R(0,0) := H*[F(A) + F(B)]/2          {Compute one trapezoid}
```

```
WHILE  [Close>Tol and J<N] or [J<4]  DO              {Do sequential
     J := J + 1                                       calculations}
     CALL Procedure  TrRule (A,H,J,M,R)               {Compute new R(J,0)}

     FOR  K = 1  TO  J  DO              {Richardson's improvements}
         R(J,K) := R(J,K−1) + [R(J,K−1)−R(J−1,K−1)]/[4^K − 1]

     PRINT 'The entries in row' J 'for Romberg integration are'
           R(J,0), R(J,1), R(J,2), ..., R(J,J)
     Close := |R(J−1,J−1) − R(J,J)|

     PRINT 'The best approximation for the value of the integral
           using Romberg integration is'; R(J,J)
```

Exercises for Romberg Integration

In Exercises 1–4:
   (a) Complete the Romberg integration table for approximating the given integrals.
   (b) Use a computer to find the approximations in part (a).

1. $\int_{-2}^{2} 7x^6 \, dx = 256$

J	$R(J,0)$ Trapezoidal Rule	$R(J,1)$ Simpson's Rule	$R(J,2)$ Boole's Rule	$R(J,3)$ Third Improvement
0	1792.0000			
1	896.00000	_____		
2	462.00000	_____	_____	
3	310.84375	_____	_____	_____

2. The normal probability density function is $\exp\left(-x^2/2\right)/(2\pi)^{1/2}$. The probability that the random variable X lies between 0 and 1 is

$$\Pr(0 \le X \le 1) = (2\pi)^{-1/2} \int_{0}^{1} \exp\left(\frac{-x^2}{2}\right) dx \approx 0.3413447.$$

J	$R(J,0)$ Trapezoidal Rule	$R(J,1)$ Simpson's Rule	$R(J,2)$ Boole's Rule	$R(J,3)$ Third Improvement
0	0.3204565			
1	0.3362609	_____		
2	0.3400818	_____	_____	
3	0.3410295	_____	_____	_____

3. Romberg integration can be used to evaluate the following improper integral if the value of the integrand f at 0 is defined to be $f(0) = 1$.

$$\int_0^1 \frac{\sin(x)}{x}\, dx \approx 0.9460831.$$

J	R(J,0) Trapezoidal Rule	R(J,1) Simpson's Rule	R(J,2) Boole's Rule	R(J,3) Third Improvement
0	0.9207355			
1	0.9397933	_____		
2	0.9445135	_____	_____	
3	0.9456909	_____	_____	_____

4. Sometimes integrals are used to define special mathematical functions. The Bessel function of order zero $J_0(X)$ satisfies

$$J_0(X) = \pi^{-1} \int_0^\pi \cos\left[X \sin(t)\right] dt.$$

The following Romberg table is used to find $J_0(0.6) = 0.9120049$.

J	R(J,0) Trapezoidal Rule	R(J,1) Simpson's Rule	R(J,2) Boole's Rule	R(J,3) Third Improvement
0	1.0000000			
1	0.9126678	_____		
2	0.9120049	_____	_____	
3	0.9120049	_____	_____	_____

5. Use equations (9) and (10) to establish equation (11).

6. Use equations (12) and (13) to establish equation (14).

7. Determine the smallest integer N for which

   (a) $\int_0^2 8x^7\, dx = 256 = R(N,N)$ (b) $\int_0^2 11x^{10}\, dx = 2{,}048 = R(N,N)$

8. Modify Algorithm 7.4 so that it uses a relative error stopping criterion instead of an absolute error criterion.

9. Romberg integration can be modified to evaluate improper integrals $\int_a^b f(x)\, dx$, where $f(x)$ has a removable discontinuity at $x = a$; i.e., $f(a)$ is not defined, but

$$\lim_{x \to a^+} f(x) = L.$$

The crucial step is to use the formula

$$R(0,0) = \frac{H[L + F(B)]}{2}.$$

Modify Algorithm 7.4 by including a procedure for finding L.

10. We know that

$$I = \int_{-0.3}^{0.3} \cos(x) \, dx = 2 \sin(0.3) \approx 0.591040413.$$

The second row from Romberg integration with $h = 0.15$ is $R(2,0) = 0.589931797$, $R(2,1) = 0.591042080$, and $R(2,2) = 0.591040399$. Use the approximate value for the derivatives of f,

$$K = -0.986 = f^{(2)}(c_0) = -f^{(4)}(c_1) = f^{(6)}(c_2),$$

to verify that the error terms are described by the relations

$$I - R(2,0) \approx -0.6\frac{h^2 K}{12},$$

$$I - R(2,1) \approx \ \ 0.6\frac{h^4 K}{180},$$

$$I - R(2,2) \approx -1.2\frac{h^6 K}{945}.$$

11. We know that

$$I = \int_{-0.3}^{0.3} \exp(x) \, dx = \exp(0.3) - \exp(-0.3) \approx 0.609040587.$$

The second row from Romberg integration with $h = 0.15$ is $R(2,0) = 0.610182110$, $R(2,1) = 0.609042295$, and $R(2,2) = 0.609040602$. Use the approximate value for the derivatives of f,

$$K = 1.013 = f^{(2)}(c_0) = f^{(4)}(c_1) = f^{(6)}(c_2),$$

to verify that the error terms are described by the relations

$$I - R(2,0) \approx -0.6\frac{h^2 K}{12},$$

$$I - R(2,1) \approx -0.6\frac{h^4 K}{180},$$

$$I - R(2,2) \approx -1.2\frac{h^6 K}{945}.$$

12. (a) Use the change of variables $x = t^2$ and $dx = 2t \, dt$ and show that the definite integrals have the same value.

(b) Romberg integration is used to evaluate each integral. Speculate as to why convergence is slower for the integral on the left.

(c) Use a computer to find the approximations in part (b).

$$\int_0^1 x^{1/2}\,dx \quad = \quad \int_0^1 2t^2\,dt$$

Integral on the Left	Integral on the Right
$R(0,0) = 0.5000000$	$R(0,0) = 1.0000000$
$R(1,1) = 0.6380712$	$R(1,1) = 0.6666667$
$R(2,2) = 0.6577566$	$R(2,2) = 0.6666667$
$R(3,3) = 0.6636076$	$R(3,3) = 0.6666667$
$R(4,4) = 0.6655929$	$R(4,4) = 0.6666667$

13. Verify that the following modification of Algorithm 7.4 will sequentially compute the rows of the Romberg integration table in the one dimensional array R[0 .. N], hence it saves space. The algorithm will stop when the relative error in the consecutive approximations is less than the preassigned value Tol (e.g., 10^{-7}).

```
SUBROUTINE  TrRule (A,H,M,R)                    {Sequential trapezoidal rule}
    H := H/2                                     {Reduce the step size for refinement}
    SUM := 0
    FOR  P = 1  TO  M  DO                        {Loop for odd subscripts}
        SUM := SUM + F(A + H*(2*P−1))
    M := 2*M                                     {Update the number of subintervals}
    R(0) := R(0)/2 + H*SUM                       {End of the procedure TrRule}


{The main program starts here.}

    M := 1                                       {Initialize the number of subintervals}
    H := B − A                                   {Initialize the step size}
    RelErr := 1                                  {Initialize the variable}
    J := 0                                       {Initialize the counter}
    R(0) := H*[F(A + F(B)]/2                      {Compute one trapezoid}

WHILE  [RelErr>Tol and J<N] or [J<4]  DO
    OldR := R(J)                                 {Old Romberg approximation}
    J := J + 1
    Temp := R(0)
    CALL Procedure  TrRule (A,H,M,R)             {Compute new R(0)}
    FOR  K = 1  TO  J  DO                        {Richardson's improvements}
        Last := R(K)
        R(K) := R(K−1) + [R(K−1)−Temp]/[4^K − 1]
        Temp := Last
    PRINT 'The entries in row' J 'for Romberg integration are'
        R(0), R(1), R(2), ..., R(J)
    RelErr := 2*|OldR − R(J)|/[|OldR| + |R(J)|]
PRINT 'The Romberg approximation for the integral is'  R(J)
```

14. *Romberg Integration Based on the Midpoint Rule.* The composite midpoint rule is competitive with the composite trapezoidal rule with respect to efficiency and error term. Use the midpoint rule:

$$\int_a^b f(x)\, dx = M(f, h) + E(f, h), \qquad \text{where } h = \frac{b - a}{N}$$

and

$$M(f, h) = h \sum_{k=1}^{N} f(z_k) \quad \text{and} \quad z_k = a + (k - \tfrac{1}{2})h, \qquad \text{for } k = 1, 2, \ldots, N.$$

$$E(f, h) = \frac{b - a}{24} f''(c)h^2 = a_1 h^2 + a_2 h^4 + a_3 h^6 + \cdots.$$

(a) Develop the sequential midpoint rule

$$M(0) = h f\left(a + \frac{h}{2}\right), \qquad \text{where } h = b - a$$

$$M(J) = h \sum_{k=1}^{2^J} f(z_k), \qquad \text{where } h = \frac{b - a}{2^J} \quad \text{and} \quad z_k = a + (k - \tfrac{1}{2})h$$

$$\text{for } k = 1, 2, \ldots, 2^J.$$

(b) Show how the sequential midpoint rule can be used in place of the sequential trapezoidal rule in Romberg integration.

15. Modify Algorithm 7.4 so that it uses the sequential midpoint rule to perform Romberg integration (see Exercise 14).

16. Write a report on adaptive integration, which is sometimes called "adaptive quadrature."

17. Write a report on singular integrals.

18. Write a report on multiple integrals.

7.6 Gauss–Legendre Integration (Optional)

We wish to find the area under the curve

$$y = f(x), \qquad -1 \leqq x \leqq 1.$$

What method gives the best answer if only two function evaluations are to be made? We have already seen that the trapezoidal rule is a method for finding the area under the curve and that it uses two function evaluations at the end-points $(-1, f(-1))$, $(1, f(1))$. But if the graph $y = f(x)$ is concave down, the error in approximation is the entire region that lies between the curve and the line segment joining the points [see Figure 7.8(a)].

If we can use nodes x_1, x_2 that lie inside the interval, the line through the two points $(x_1, f(x_1))$, $(x_2, f(x_2))$ crosses the curve and the area under the line

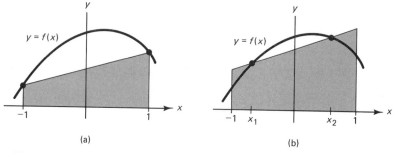

Figure 7.8. (a) Trapezoidal approximation using the abscissas -1 and 1. (b) Trapezoidal approximation using the abscissas x_1 and x_2.

more closely approximates the area under the curve [see Figure 7.8(b)]. The equation of this line is

(1)
$$y = f(x_1) + \frac{(x - x_1)[f(x_2) - f(x_1)]}{x_2 - x_1}$$

and the area of the trapezoid under this line is

(2)
$$\text{area} = \frac{2x_2}{x_2 - x_1} f(x_1) - \frac{2x_1}{x_2 - x_1} f(x_2).$$

Notice that the trapezoidal rule is a special case of (2). When we choose $x_1 = -1$, $x_2 = 1$, and $h = 2$, then

$$T(f, h) = \frac{2}{2} f(x_1) - \frac{-2}{2} f(x_2) = f(x_1) + f(x_2).$$

We shall use the method of undetermined coefficients to find the abscissas x_1, x_2 and weights w_1, w_2 so that the formula

(3)
$$\int_{-1}^{1} f(x) \, dx \approx w_1 f(x_1) + w_2 f(x_2)$$

is exact for cubic polynomials [i.e., $f(x) = a_3 x^3 + a_2 x^2 + a_1 x + a_0$]. Since four coefficients w_1, w_2, x_1, and x_2 need to be determined in equation (3), we can select four conditions to be satisfied. Using the fact that integration is additive, it will suffice to require that (3) is exact for the four functions $f(x) = 1, x, x^2, x^3$. The four integral conditions are:

(4)
$$f(x) = 1: \quad \int_{-1}^{1} 1 \, dx = 2 = w_1 + w_2.$$

$$f(x) = x: \quad \int_{-1}^{1} x \, dx = 0 = w_1 x_1 + w_2 x_2.$$

$$f(x) = x^2: \quad \int_{-1}^{1} x^2 \, dx = \tfrac{2}{3} = w_1 x_1^2 + w_2 x_2^2.$$

$$f(x) = x^3: \quad \int_{-1}^{1} x^3 \, dx = 0 = w_1 x_1^3 + w_2 x_2^3.$$

Now solve the system of nonlinear equations

(5)
$$w_1 + w_2 = 2$$

(6)
$$w_1 x_1 = -w_2 x_2$$

(7)
$$w_1 x_1^2 + w_2 x_2^2 = \tfrac{2}{3}$$

(8)
$$w_1 x_1^3 = -w_2 x_2^3.$$

We can divide (8) by (6), and the result is

(9)
$$x_1^2 = x_2^2 \quad \text{or} \quad x_1 = -x_2.$$

Use (9) and divide (6) by x_1 on the left and $-x_2$ on the right to get

(10)
$$w_1 = w_2.$$

Substituting (10) into (5) results in $w_1 + w_1 = 2$. Hence

(11)
$$w_1 = w_2 = 1.$$

Now using (11) and (9) in (7), we write

(12)
$$w_1 x_1^2 + w_2 x_2^2 = x_2^2 + x_2^2 = \tfrac{2}{3} \quad \text{or} \quad x_2^2 = \tfrac{1}{3}.$$

Finally, from (12) and (9), we see that the nodes are

(13)
$$-x_1 = x_2 = 1/3^{1/2} \approx 0.5773502692.$$

We have found the nodes and the weights that make up the two point Gauss–Legendre rule. Since the formula is exact for cubic equations, the error term will involve the fourth derivative. A discussion of the error term can be found in Reference 29.

Theorem 7.7 [Description of the two-point Gauss–Legendre rule] If $f(x)$ and its first four derivatives are continuous on $[-1, 1]$, then the Gauss–Legendre rule based on the two nodes $x_1 = -1/3^{1/2}$ and $x_2 = 1/3^{1/2}$ is

(14)
$$G(f, 2) = f(-1/3^{1/2}) + f(1/3^{1/2}).$$

If $f(x)$ has four continuous derivatives, then this is an approximation to the integral of $f(x)$ and we write

(15)
$$\int_{-1}^{1} f(x)\, dx = G(f, 2) + E(f, 2)$$

where the truncation error term $E(f, 2)$ has the form

$$E(f, 2) = \frac{f^{(4)}(c)}{135}.$$

Example 7.14 Use the two-point Gauss–Legendre rule to approximate

$$\int_{-1}^{1} \frac{dx}{x + 2} = \ln(3) \approx 1.09861$$

and compare the result with the trapezoidal rule $T(f, h)$ with $h = 2$, and Simpson's rule $S(f, h)$ with $h = 1$.

    Solution. Let $G(f, 2)$ denote the two-point Gauss–Legendre rule; then

$$G(f, 2) = f(-0.57735) + f(0.57735) = 0.70291 + 0.38800 = 1.09091,$$

$$T(f, 2) = f(-1.0000) + f(1.0000) = 1.00000 + 0.33333 = 1.33333,$$

$$S(f, 1) = \frac{f(-1) + 4f(0) + f(1)}{3} = \frac{1 + 2 + \frac{1}{3}}{3} = 1.11111.$$

The errors are 0.00770, -0.23472, and -0.01250, respectively, so that the Gauss–Legendre rule is seen to be best. Notice that the Gauss–Legendre rule required only two function evaluations and Simpson's rule required three. In this example the size of the error for $G(f, 2)$ is about 61% of the size of the error for $S(f, 1)$.

    The general N-point Gauss–Legendre rule is exact for polynomials of degree $\leq 2N - 1$ and the numerical integration formula is

(16) $$G(f, N) = w_{N,1}f(x_{N,1}) + w_{N,2}f(x_{N,2}) + \ldots + w_{N,N}f(x_{N,N}).$$

    The abscissas $x_{N,k}$ and weights $w_{N,k}$ to be used have been tabulated and are easily available; Table 7.11 gives the values up to eight points. Also included in the table is the form of the error term $E(f, N)$ that corresponds to $G(f, N)$, and it can be used to determine the accuracy of the Gauss–Legendre integration formula.

    The values in Table 7.11 in general have no easy representation. This fact makes the method less attractive for human beings to use when hand calculations are required. But once the values are stored in a computer it is easy to call them up when needed. The nodes are actually roots of the Legendre polynomials and the corresponding weights must be obtained by solving a system of equations. For the three-point Gauss–Legendre rule the nodes are $-(0.6)^{1/2}$, 0, and $(0.6)^{1/2}$ and the corresponding weights are $5/9, 8/9$, and $5/9$.

Theorem 7.8 [Description of the three-point Gauss–Legendre rule] If $f(x)$ and its first six derivatives are continuous on $[-1, 1]$, then the Gauss–Legendre rule based on the three nodes $x_1 = -(0.6)^{1/2}$, $x_2 = 0$, and $x_3 = (0.6)^{1/2}$ is

(17) $$G(f, 3) = \frac{5f(-(0.6)^{1/2}) + 8f(0) + 5f((0.6)^{1/2})}{9}.$$

If $f(x)$ has six continuous derivatives, then this is an approximation to the integral of $f(x)$ and we write

(18) $$\int_{-1}^{1} f(x)\, dx = G(f, 3) + E(f, 3),$$

where the truncation error term $E(f, 3)$ has the form

(19) $$E(f, 3) = \frac{f^{(6)}(c)}{15{,}750}.$$

Table 7.11 Gauss–Legendre abscissas and weights

$$\int_{-1}^{1} f(x)\, dx = \sum_{k=1}^{N} w_{N,\,k} f(x_{N,\,k}) + E(f, N)$$

N	Abscissas, $x_{N,\,k}$	Weights, $w_{N,\,k}$	Truncation Error, $E(f, N)$
2	-0.5773502692 0.5773502692	1.0000000000 1.0000000000	$\dfrac{f^{(4)}(c)}{135}$
3	±0.7745966692 0.0000000000	0.5555555556 0.8888888888	$\dfrac{f^{(6)}(c)}{15{,}750}$
4	±0.8611363116 ±0.3399810436	0.3478548451 0.6521451549	$\dfrac{f^{(8)}(c)}{3{,}472{,}875}$
5	±0.9061798459 ±0.5384693101 0.0000000000	0.2369268851 0.4786286705 0.5688888888	$\dfrac{f^{(10)}(c)}{1{,}237{,}732{,}650}$
6	±0.9324695142 ±0.6612093865 ±0.2386191861	0.1713244924 0.3607615730 0.4679139346	$\dfrac{f^{(12)}(c)2^{13}[6!]^4}{[12!]^3 13!}$
7	±0.9491079123 ±0.7415311856 ±0.4058451514 0.0000000000	0.1294849662 0.2797053915 0.3818300505 0.4179591837	$\dfrac{f^{(14)}(c)2^{15}[7!]^4}{[14!]^3 15!}$
8	±0.9602898565 ±0.7966664774 ±0.5255324099 ±0.1834346425	0.1012285363 0.2223810345 0.3137066459 0.3626837834	$\dfrac{f^{(16)}(c)2^{17}[8!]^4}{[16!]^3 17!}$

Example 7.15 Show that the three-point Gauss–Legendre rule is exact for

$$\int_{-1}^{1} 5x^4\, dx = 2 = G(f, 3).$$

Solution. Since the integrand is $f(x) = 5x^4$ and $f^{(6)}(x) = 0$, we can use (19) to see that $E(f, 3) = 0$. But it is instructive use (17) and do the calculation in this case.

$$G(f, 3) = \frac{5(5)(0.6)^2 + 0 + 5(5)(0.6)^2}{9} = \frac{18}{9} = 2.$$

The next result shows how to change the variable of integration so that the Gauss–Legendre rules can be used on the interval $[a, b]$.

Theorem 7.9 [Interval transformation for Gauss–Legendre rules] Let $x_{N,k}$ and $w_{N,k}$ be the abscissa and weights for the N-point Gauss–Legendre rule on the interval $-1 \leq x \leq 1$. To apply the rule to integrate $f(t)$ on $a \leq t \leq b$, use the change of variables

$$(20) \qquad t = \frac{a+b}{2} + x\frac{b-a}{2} \quad \text{and} \quad dt = \frac{b-a}{2} dx.$$

Then the integral relationship is

$$(21) \qquad \int_a^b f(t)\, dt = \int_{-1}^1 f\left(\frac{a+b}{2} + x\frac{b-a}{2}\right)\frac{b-a}{2} dx,$$

and the Gauss–Legendre rule for $[a, b]$ is

$$(22) \qquad \int_a^b f(t)\, dt = \frac{b-a}{2} \sum_{k=1}^N w_{N,k} f\left(\frac{a+b}{2} + x_{N,k}\frac{b-a}{2}\right),$$

Example 7.16 Use the three-point Gauss–Legendre rule to approximate

$$\int_1^5 \frac{dt}{t} = \ln(5) \approx 1.609438$$

and compare the result with Boole's rule $B(2)$ with $h = 1$.

Solution. Here $a = 1$ and $b = 5$, so that the rule in (22) yields

$$G(f, 3) = (2)\frac{5f(3 - 2(0.6)^{1/2}) + 8f(3 + 0) + 5f(3 + 2(0.6)^{1/2})}{9}$$

$$= (2)\frac{3.446359 + 2.666667 + 1.099096}{9} = 1.602694.$$

In Example 7.10 we saw that Boole's rule gave $B(2) = 1.617778$. The errors are 0.006744 and -0.008340, respecitvely, so that the Gauss–Legendre rule is slightly better in this case. Notice that the Gauss–Legendre rule requires three function evaluations and Boole's rule requires five. In this example the size of the two errors is about the same.

Gauss–Legendre integration formulas are extremely accurate; and they should be considered seriously when many integrals of a similar nature are to be evaluated. In this case proceed as follows. Pick a few representative integrals, including some with the worst behavior that is likely to occur. Determine the number of sample points N that are needed to obtain accuracy. Then fix the value N, and use the Gauss–Legendre rule with N sample points for all the integrals.

Algorithm 7.5 [*Gauss-Legendre Rule*] Approximate the integral

$$\int_A^B f(t)\, dt \approx \frac{B - A}{2}\left[w_{N,1}f(t_{N,1}) + w_{N,2}f(t_{N,2}) + \ldots + w_{N,N}f(t_{N,N})\right]$$

where N sample points $t_{N,1}, t_{N,2}, \ldots, t_{N,N}$ are computed using equation (20) and the abscissas in Table 7.11. The corresponding weights $w_{N,1}, w_{N,2}, \ldots, w_{N,N}$ are given in Table 7.11.

```
USES the function   F(X)
USES the Table of Abscissa   X(N,K)
USES the Table of Weights   W(N,K)
INPUT  A,  B,  N
SUM := 0


FOR  K = 1  TO  N  DO
     T := [A+B]/2 + X(N,K)*[B−A]/2
     SUM := SUM + W(N,K)*F(T)


     SUM := SUM*[B−A]/2


PRINT 'The approximate value of the integral of f(t)'
      'on the interval' A, B 'using' N 'sample points'
      'computed using the' N 'point Gauss−Legendre rule is'   SUM
```

Exercises for Gauss–Legendre Integration

In Exercises 1–4.
  (a) Show that the two integrals are equivalent.
  (b) Find $G(f, 2)$. (c) Find $G(f, 3)$. (d) Find $G(f, 4)$.
  (e) Use a computer to find the values in parts (b)-(d).

1. $\displaystyle\int_0^2 6t^5\, dt = \int_{-1}^1 6(x + 1)^5\, dx = 64$

2. $\displaystyle\int_0^2 \sin(t)\, dt = \int_{-1}^1 \sin(x + 1)\, dx = -\cos(2) + \cos(0) \approx 1.416147$

3. $\displaystyle\int_0^1 \frac{\sin(t)}{t}\, dt = \int_{-1}^1 \frac{\sin((x + 1)/2)}{x + 1}\, dx \approx 0.9460831$

4. $\displaystyle(2\pi)^{-1/2}\int_0^1 \exp\left(\frac{-t^2}{2}\right) dt = (2\pi)^{-1/2}\int_{-1}^1 \frac{\exp(-(x + 1)^2/8)}{2}\, dx$

5. $\pi^{-1} \int_0^\pi \cos(0.6 \sin(t)) \, dt = 0.5 \int_{-1}^1 \cos\left(0.6 \sin\left((x+1)\dfrac{\pi}{2}\right)\right) dx$

(a) Show that the two integrals are equivalent.
(b) Find $G(f, 4)$. (c) Find $G(f, 6)$.
(d) Use a computer to find the values in parts (b)-(c).

6. Use $E(f, N)$ in Table 7.11 and the change of variable $dt = dx$ in Theorem 7.9 to find the smallest integer N so that $E(f, N) = 0$ for

(a) $\displaystyle\int_0^2 8x^7 \, dx = 256 = G(f, N)$ (b) $\displaystyle\int_0^2 11x^{10} \, dx = 2{,}048 = G(f, N)$

7. We know that

$$I = \int_{-1}^1 \cos(x) \, dx = 2 \sin(1) \approx 1.6829420$$

(a) Show that Gauss-Legendre quadrature for $N = 2, 3, 4$ yields $G(f, 2) = 1.6758237$, $G(f, 3) = 1.6830036$, and $G(f, 4) = 1.6829417$.
(b) Use the approximate value for the derivatives of f,

$$K = 0.968 = f^{(4)}(c_2) = -f^{(6)}(c_3) = f^{(8)}(c_4),$$

to verify that the error terms are described by the relations

$$I - G(f, 2) \approx \frac{K}{135},$$

$$I - G(f, 3) \approx -\frac{K}{15{,}750},$$

$$I - G(f, 4) \approx \frac{K}{3{,}472{,}875}.$$

8. We know that $I = \displaystyle\int_{-1}^1 \exp(x) \, dx = \exp(1) - \exp(-1) \approx 2.3504024$.

(a) Show that Gauss-Legendre quadrature for $N = 2, 3, 4$ yields $G(f, 2) = 2.3426961$, $G(f, 3) = 2.3503369$, and $G(f, 4) = 2.3504021$.
(b) Use the approximate value for the derivatives of f,

$$K = 1.031 = f^{(4)}(c_2) = f^{(6)}(c_3) = f^{(8)}(c_4),$$

to verify that the error terms are described by the relations

$$I - G(f, 2) \approx \frac{K}{135},$$

$$I - G(f, 3) \approx \frac{K}{15{,}750},$$

$$I - G(f, 4) \approx \frac{K}{3{,}472{,}875}.$$

9. Find the roots of the following Legendre polynomials and compare them with the abscissa in Table 7.11.
   (a) $P_2(x) = (3x^2 - 1)/2$
   (b) $P_3(x) = (5x^3 - 3x)/2$
   (c) $P_4(x) = (35x^4 - 30x^2 + 3)/8$

10. The truncation error term for two-point Gauss–Legendre rule on the interval $[-1, 1]$ is $f^{(4)}(c_1)/135$. The truncation error term for Simpson's rule on $[a, b]$ is $-h^5 f^{(4)}(c_2)/90$. Compare the truncation error terms when $[a, b] = [-1, 1]$. Which method do you think is best. Why?

11. The three-point Gauss–Legendre rule is

$$\int_{-1}^{1} f(x) \, dx \approx \frac{5f(-(0.6)^{1/2}) + 8f(0) + 5f((0.6)^{1/2})}{9}.$$

   Show that the formula is exact for $f(x) = 1, x, x^2, x^3, x^4, x^5$. *Hint.* If f is an odd function [i.e., $f(-x) = -f(x)$], the integral of f over $[-1, 1]$ is zero.

12. The truncation error term for the three-point Gauss–Legendre rule on the interval $[-1, 1]$ is $f^{(6)}(c_1)/15{,}750$. The truncation error term for Boole's rule on $[a, b]$ is $-8h^7 f^{(6)}(c_2)/945$. Compare the error terms when $[a, b] = [-1, 1]$. Which method do you think is better? Why?

13. Derive the three-point Gauss–Legendre rule using the following steps. Use the fact that the abscissas are the roots of the Legendre polynomial of degree 3.

$$x_1 = -(0.6)^{1/2} \qquad x_2 = 0 \qquad x_3 = (0.6)^{1/2}$$

   Find the weights w_1, w_2, and w_3 so that the relation

$$\int_{-1}^{1} f(x) \, dx \approx w_1 f(-(0.6)^{1/2}) + w_2 f(0) + w_3 f((0.6)^{1/2})$$

   is exact for the functions $f(x) = 1, x, x^2$. *Hint.* First obtain, then solve, the linear system of equations

$$w_1 + w_2 + \qquad w_3 = 2$$
$$-(0.6)^{1/2} w_1 \qquad + (0.6)^{1/2} w_3 = 0$$
$$0.6 w_1 \qquad + \qquad 0.6 w_3 = \tfrac{2}{3}.$$

14. Modify Algorithm 7.5 so that it will compute $G(f, 1), G(f, 2), \ldots$, and stop when the relative error in the approximations $G(f, N - 1)$ and $G(f, N)$ is less than the preassigned value Tol, that is,

$$\frac{2 |G(f, N - 1) - G(f, N)|}{|G(f, N - 1)| + |G(f, N)|} < \text{Tol.}$$

15. In practice, if many integrals of a similar type are evaluated, a preliminary analysis is made to determine the number of function evaluations required to obtain the desired accuracy. Suppose that 17 function evaluations are to be made. Compare the Romberg answer $R(4, 4)$ with the Gauss–Legendre answer $G(f, 17)$.

8

Iterative Methods for
Systems of Equations

8.1 Iterative Methods for Linear Systems

The goal of this chapter is to extend some of the iterative methods introduced in Chapters 1 and 2 to higher dimensions. We first consider an extension of fixed-point iteration that applies to systems of linear equations.

Jacobi Iteration

Example 8.1 Consider the system of equations

(1)
$$\begin{aligned} 4x - y + z &= 7 \\ 4x - 8y + z &= -21 \\ -2x + y + 5z &= 15. \end{aligned}$$

These equations can be written in the form

(2)
$$\begin{aligned} x &= \frac{7 + y - z}{4} \\ y &= \frac{21 + 4x + z}{8} \\ z &= \frac{15 + 2x - y}{5}. \end{aligned}$$

This suggests the following Jacobi iterative process:

$$x_{k+1} = \frac{7 + y_k - z_k}{4}$$

(3)

$$y_{k+1} = \frac{21 + 4x_k + z_k}{8}$$

$$z_{k+1} = \frac{15 + 2x_k - y_k}{5}.$$

See that if we start with $(x_0, y_0, z_0) = (1, 2, 2)$, then the iteration in (3) appears to converge to the solution $(2, 4, 3)$.

Solution. Substitute $x_0 = 1, y_0 = 2, z_0 = 2$ into the right-hand side of each equation in (3) to obtain the new values

$$x_1 = \frac{7 + 2 - 2}{4} = 1.75,$$

$$y_1 = \frac{21 + 4 + 2}{8} = 3.375,$$

$$z_1 = \frac{15 + 2 - 2}{5} = 3.00.$$

The new point $P_1 = (1.75, 3.375, 3.00)$ is closer to $(2, 4, 3)$ than P_0. Iteration using (3) generates a sequence of points $\{P_k\}$ which converges to the solution $(2, 4, 3)$ (see Table 8.1).

TABLE 8.1   Convergent Jacobi iteration
for the linear system (1)

k	x_k	y_k	z_k
0	1.0	2.0	2.0
1	1.75	3.375	3.0
2	1.84375	3.875	3.025
3	1.9625	3.925	2.9625
4	1.99062500	3.97656250	3.00000000
5	1.99414063	3.99531250	3.00093750
⋮	⋮	⋮	⋮
15	1.99999993	3.99999985	2.99999993
⋮	⋮	⋮	⋮
19	2.00000000	4.00000000	3.00000000

This process is called *Jacobi iteration* and can be used to solve certain types of linear systems. After 19 steps, the iteration has converged to the 9 digit machine approximation

$$(2.00000000, 4.00000000, 3.00000000).$$

This requires more computational effort than Gaussian elimination.

Linear systems with as many as 100,000 variables often arise in the solution of partial differential equations. The coefficient matrix for these systems are sparse, that is, the non-zero entries form a pattern. An iterative process provides an efficient method for solving these large systems.

Sometimes, the Jacobi method does not work. Let us experiment and see that a rearrangement of the original linear system can result in a system of iteration equations that will produce a divergent sequence of points.

Example 8.2 Let the linear system (1) be rearranged as follows:

(4)
$$-2x + y + 5z = 15$$
$$4x - 8y + z = -21$$
$$4x - y + z = 7.$$

These equations can be written in the form

(5)
$$x = \frac{-15 + y + 5z}{2},$$
$$y = \frac{21 + 4x + z}{8},$$
$$z = 7 - 4x + y.$$

This suggests the following Jacobi iterative process:

(6)
$$x_{k+1} = \frac{-15 + y_k + 5z_k}{2},$$
$$y_{k+1} = \frac{21 + 4x_k + z_k}{8},$$
$$z_{k+1} = 7 - 4x_k + y_k.$$

See that if we start with $(x_0, y_0, z_0) = (1, 2, 2)$, then iteration using (6) will diverge away from the solution $(2, 4, 3)$.

Solution. Substitute $x_0 = 1$, $y_0 = 2$, and $z_0 = 2$ into the right-hand side of each equation in (6) to obtain the new values x_1, y_1, and z_1:

$$x_1 = \frac{-15 + 2 + 10}{2} = -1.5,$$

$$y_1 = \frac{21 + 4 + 2}{8} = 3.375,$$

$$z_1 = 7 - 4 + 2 = 5.00.$$

The new point $P_1 = (-1.5, 3.375, 5.00)$ is farther away from the solution $(2, 4, 3)$ than P_0. Iteration using equations (6) produces a divergent sequence (see Table 8.2).

TABLE 8.2   Divergent Jacobi iteration
for the linear system (4)

k	x_k	y_k	z_k
0	1.0	2.0	2.0
1	−1.5	3.375	5.0
2	6.6875	2.5	16.375
3	34.6875	8.015625	−17.25
4	−46.617188	17.8125	−123.73438
5	−307.929688	−36.150391	211.28125
6	502.62793	−124.929688	1,202.56836
⋮	⋮	⋮	⋮

Gauss–Seidel Iteration

Sometimes the convergence can be speeded up. Observe that $\{x_k\}$, $\{y_k\}$, and $\{z_k\}$ converge to 2, 4, and 3, respectively. It seems reasonable that x_{k+1} could be used in place of x_k in the computation of y_{k+1}. Similarly, the values x_{k+1} and y_{k+1} might be used in the computation of z_{k+1}. The next example shows what happens when this is applied to the equations in Example 8.1.

Example 8.3 Consider the system of equations given in (1), and the Gauss-Seidel iterative process suggested by (2):

(7)
$$x_{k+1} = \frac{7 + y_k - z_k}{4},$$

$$y_{k+1} = \frac{21 + 4x_{k+1} + z_k}{8},$$

$$z_{k+1} = \frac{15 + 2x_{k+1} - y_{k+1}}{5}.$$

See that if we start with $P_0 = (x_0, y_0, z_0) = (1, 2, 2)$, then iteration using (7) will converge to the solution $(2, 4, 3)$.

Solution. Substitute $y_0 = 2$, $z_0 = 2$ into the first equation and obtain

$$x_1 = \frac{7 + 2 - 2}{4} = 1.75.$$

Then substitute $x_1 = 1.75$, $z_0 = 2$ into the second equation and get

$$y_1 = \frac{21 + 4 \times 1.75 + 2}{8} = 3.75.$$

Finally, substitute $x_1 = 1.75$, $y_1 = 3.75$ into the third equation and get

$$z_1 = \frac{15 + 2 \times 1.75 - 3.75}{5} = 2.95.$$

The new point $P_1 = (1.75, 3.75, 2.95)$ is closer to $(2, 4, 3)$ than P_0 and is better than the value given in Example 8.1. Iteration using (7) generates a sequence $\{P_k\}$ converges to $(2, 4, 3)$ (see Table 8.3).

TABLE 8.3 Convergent Gauss–Seidel iteration for the system (1)

k	x_k	y_k	z_k
0	1.0	2.0	2.0
1	1.75	3.75	2.95
2	1.95	3.96875	2.98625
3	1.995625	3.99609375	2.99903125
⋮	⋮	⋮	⋮
8	1.99999983	3.99999988	2.99999996
9	1.99999998	3.99999999	3.00000000
10	2.00000000	4.00000000	3.00000000

In view of Examples 8.1 and 8.2 it is necessary to have some criterion to determine whether the Jacobi iteration will converge. Hence we make the following definition.

Definition 8.1 [Diagonally dominant] A matrix A of dimension N by N is said to be *diagonally dominant* provided that

$$(8) \quad |a_{k,k}| > |a_{k,1}| + \ldots + |a_{k,k-1}| + |a_{k,k+1}| + \ldots + |a_{k,N}|$$

$$\text{for } k = 1, 2, \ldots, N.$$

This means that in each row of the matrix, the magnitude of the diagonal coefficient must exceed the sum of the magnitudes of the other coefficients in the row. The matrix in Example 8.1 is diagonally dominant because

$$\text{In row 1:} \quad |4| > |-1| + |1|.$$
$$\text{In row 2:} \quad |-8| > |4| + |1|.$$
$$\text{In row 3:} \quad |5| > |-2| + |1|.$$

All the rows satisfy relation (8) in Definition 8.1; therefore, the matrix A for the linear system (1) is diagonally dominant.

The matrix in Example 8.2 is not diagonally dominant because

$$\text{In row 1:} \quad |-2| < |1| + |5|.$$
$$\text{In row 2:} \quad |-8| > |4| + |1|.$$
$$\text{In row 3:} \quad |1| < |4| + |-1|.$$

Rows 1 and 3 do not satisfy relation (8) in Definition 8.1; therefore, the matrix A for the linear system (4) is not diagonally dominant.

Suppose that the given linear system is

$$
\begin{aligned}
a_{1,1}x_1 + a_{1,2}x_2 + \ldots + a_{1,j}x_j + \ldots + a_{1,N}x_N &= b_1 \\
a_{2,1}x_1 + a_{2,2}x_2 + \ldots + a_{2,j}x_j + \ldots + a_{2,N}x_N &= b_2 \\
&\vdots \\
a_{j,1}x_1 + a_{j,2}x_2 + \ldots + a_{j,j}x_j + \ldots + a_{j,N}x_N &= b_j \\
&\vdots \\
a_{N,1}x_1 + a_{N,2}x_2 + \ldots + a_{N,j}x_j + \ldots + a_{N,N}x_N &= b_N.
\end{aligned}
$$

(9)

Let the kth point be $\mathbf{P}_k = \left(x_1^{(k)}, x_2^{(k)}, \ldots, x_j^{(k)}, \ldots, x_N^{(k)}\right)$; then the next point is $\mathbf{P}_{k+1} = \left(x_1^{(k+1)}, x_2^{(k+1)}, \ldots, x_j^{(k+1)}, \ldots, x_N^{(k+1)}\right)$. The superscript (k) on the coordinates of $\mathbf{P}_k$ enables us to identify the coordinates that belong to this point. The iteration formulas use row j to solve for $x_j^{(k+1)}$ in terms of a linear combination of the previous values $x_1^{(k)}, x_2^{(k)}, \ldots, x_{j-1}^{(k)}, x_{j+1}^{(k)}, \ldots, x_N^{(k)}$:

(10)   Jacobi iteration:

$$
x_j^{(k+1)} = \frac{b_j - a_{j,1}x_1^{(k)} - \ldots - a_{j,j-1}x_{j-1}^{(k)} - a_{j,j+1}x_{j+1}^{(k)} - \ldots - a_{j,N}x_N^{(k)}}{a_{j,j}},
$$

for $j = 1, 2, \ldots, N$.

Jacobi iteration uses all old coordinates to generate all new coordinates, whereas Gauss–Seidel iteration uses the new coordinates as they become available:

(11)   Gauss-Seidel iteration:

$$
x_j^{(k+1)} = \frac{b_j - a_{j,1}x_1^{(k+1)} - \ldots - a_{j,j-1}x_{j-1}^{(k+1)} - a_{j,j+1}x_{j+1}^{(k)} - \ldots - a_{j,N}x_N^{(k)}}{a_{j,j}},
$$

for $j = 1, 2, \ldots, N$.

Theorem 8.1   [Jacobi iteration] Let A be a diagonally dominant matrix. Then the system of linear equations $A\mathbf{X} = \mathbf{B}$ has a unique solution $\mathbf{P}$. If the iteration formula (10) is used, then the Jacobi iteration will converge to $\mathbf{P}$ for any choice of the starting point $\mathbf{P}_0$.

Proof. See Reference [29].

It can be proven that the Gauss–Seidel method will converge when the matrix A is diagonally dominant. In many cases the Gauss–Seidel method will converge faster than the Jacobi method; hence it is usually preferred (compare Examples 8.1 and 8.2). It is important to understand the slight modification of formula (10) that has been made to obtain formula (11). In some cases the Jacobi method will converge even though the Gauss–Seidel method will not.

Convergence

A measure of the closeness between points is needed so that we can determine if $\{P_k\}$ is converging to P. The Euclidean distance between

$$P = (x_1, x_2, \ldots, x_N) \quad \text{and} \quad Q = (y_1, y_2, \ldots, y_N)$$

is

(12) $$D(P, Q) = [(x_1 - y_1)^2 + (x_2 - y_2)^2 + \ldots + (x_N - y_N)^2]^{1/2}.$$

Its disadvantage is that it requires considerable computing effort. Hence we introduce a different norm $\|X\|_1$ that will be used to define the separation between points.

Definition 8.2 Let X be an N-dimensional vector; then we define the function $\|X\|_1$ as follows:

(13) $$\|X\|_1 = \sum_{j=1}^{N} |x_j|.$$

The following result ensures that $\|X\|_1$ has the mathematical structure of a metric, hence is suitable to use as a generalized "distance fomula."

Theorem 8.2 Let X and Y be N-dimensional vectors and c be a real number. Then the function $\|X\|_1$ has the following properties:

(14) $$\|X\|_1 \geq 0.$$

(15) $$\|X\|_1 = 0 \text{ if and only if } X = 0.$$

(16) $$\|cX\|_1 = |c| \|X\|_1.$$

(17) $$\|X + Y\|_1 \leq \|X\|_1 + \|Y\|_1.$$

Proof. We prove (17) and leave the others as exercises. For each j, the triangle inequality for real numbers states that $|x_j + y_j| \leq |x_j| + |y_j|$. Summing these yields inequality (17):

$$\|X + Y\|_1 = \sum_{j=1}^{N} |x_j + y_j| \leq \sum_{j=1}^{N} |x_j| + \sum_{j=1}^{N} |y_j| = \|X\|_1 + \|Y\|_1.$$

The norm given by (13) can be used to define the "separation between" points.

Definition 8.3 Let X and Y be points in N-dimensional space. The *separation* between X and Y is defined as follows:

(18) $$\|Y - X\|_1 = \sum_{j=1}^{N} |y_j - x_j|.$$

Example 8.4 Determine the Euclidean distance and separation between the points $\mathbf{P} = (2, 4, 3)$ and $\mathbf{Q} = (1.75, 3.75, 2.95)$.

Solution. The Euclidean distance is

$$D(\mathbf{P}, \mathbf{Q}) = [(2 - 1.75)^2 + (4 - 3.75)^2 + (3 - 2.95)^2]^{1/2} = 0.3570714214.$$

The separation is

$$\|\mathbf{P} - \mathbf{Q}\|_1 = |2 - 1.75| + |4 - 3.75| + |3 - 2.95| = 0.55.$$

The separation is easier to compute and use for determining convergence in N-dimensional space.

Algorithm 8.1 [*Jacobi Iteration*] Assume that A is diagonally dominant. Solve the linear system $AX = B$ by iteration. Start with $\mathbf{P}_0 = \mathbf{0}$, and generate a sequence $\{\mathbf{P}_k\}$ that converges to the solution $\mathbf{P}$.

```
       Tol := 10⁻⁶                                              {Tolerance}
       Sep := 1, K := 1, Max := 99                              {Initialize}
       Cond := 1                                       {Condition of matrix}


       INPUT  N, A[1 .. N, 1 .. N], B[1 .. N]                     {Input}


FOR  R = 1 TO N DO
   │   Row := 0                                                  {Check for
   │   FOR  C = 1 TO N DO   Row := Row+|A(R, C)|                  diagonal
   └── IF  Row ≥ 2*|A(R, R)|  THEN  Cond := 0                   dominance}


IF    Cond = 0   THEN
   │       PRINT "The matrix is not diagonally dominant."
   │       TERMINATE  ALGORITHM
ENDIF


       FOR  J = 0 TO N DO  P(J) := 0, Pnew(J) := 0             {Initialize}
WHILE  K < Max AND Sep > Tol   DO
   │   FOR  R = 1 TO N DO                                        {Perform
   │      │   Sum := B(R)                                         Jacobi
   │      │   FOR  C = 1 TO N DO                                 iteration}
   │      └── IF C ≠ R THEN   Sum := Sum−A(R, C)*P(C)
   │      └── Pnew(R) := Sum/A(R, R)
   │      Sep := 0                                            {Convergence
   │   FOR  J = 1 TO N DO   Sep := Sep+|Pnew(J)−P(J)|           criterion}
   │   FOR  J = 1 TO N DO   P(J) := Pnew(J)                  {Update values,
   └── K := K+1                                        increment the counter}
```

```
IF    Sep < Tol  THEN                                                {Output}
│         PRINT "The solution to the linear system is:"
│    ELSE
│         PRINT "Jacobi iteration did not converge:"
ENDIF
```

```
FOR J = 1 TO N DO  PRINT "P(" ; J ; ") = " ; P(J)
```

Algorithm 8.2 [*Gauss–Seidel*] Assume that A is diagonally dominant. Solve the linear system $AX = B$ by iteration. Start with $\mathbf{P}_0 = \mathbf{0}$, and generate a sequence $\{\mathbf{P}_k\}$ that converges to the solution $\mathbf{P}$.

```
      Tol := 10⁻⁶                                              {Tolerance}
      Sep := 1, K := 1, Max := 99                             {Initialize}
      Cond := 1                                       {Condition of matrix}

      INPUT  N, A[1..N, 1..N], B[1..N]                             {Input}

FOR  R = 1 TO N DO
│     Row := 0                                               {Check for
│     FOR  C = 1 TO N DO  Row := Row+|A(R, C)|                 diagonal
└──── IF  Row ≥ 2*|A(R, R)| THEN  Cond := 0                  dominance}

IF    Cond = 0  THEN
│         PRINT "The matrix is not diagonally dominant."
│         TERMINATE ALGORITHM
ENDIF

      FOR  J = 1 TO N DO  P(J) := 0, Pold(J) := 0            {Initialize}
WHILE  K < Max AND Sep > Tol  DO
│     FOR  R = 1 TO N DO                                      {Perform
│     │     Sum := B(R)                                   Gauss–Seidel
│     │     FOR  C = 1 TO N DO                              iteration}
│     │     └──── IF C ≠ R THEN  Sum := Sum−A(R, C)*P(C)
│     └──── P(R) := Sum/A(R, R)
│         Sep := 0                                         {Convergence
│     FOR  J = 1 TO N DO  Sep := Sep+|P(J)−Pold(J)|          criterion}
│     FOR  J = 1 TO N DO  Pold(J) := P(J)              {Update values,
└──── K := K+1                                     increment the counter}

IF    Sep < Tol  THEN                                            {Output}
│         PRINT "The solution to the linear system is:"
│    ELSE
│         PRINT "Gauss–Seidel iteration did not converge:"
ENDIF
```

```
FOR J = 1 TO N DO  PRINT "P(" ; J ; ") = " ; P(J)
```

Exercises for Iterative Methods for Linear Systems

In Exercises 1–8:
  (a) Start with $\mathbf{P}_0 = \mathbf{0}$ and use Jacobi iteration to find $\mathbf{P}_k$ ($k = 1, 2, 3$). Will Jacobi iteration converge to the solution?
  (b) Start with $\mathbf{P}_0 = \mathbf{0}$ and use Gauss–Seidel iteration to find $\mathbf{P}_k$ ($k = 1, 2, 3$). Will Gauss–Seidel iteration converge to the solution?
  (c) Write a computer program that uses iteration, and solve the linear system.

1. $\begin{aligned} 4x - y &= 15 \\ x + 5y &= 9 \end{aligned}$

2. $\begin{aligned} 8x - 3y &= 10 \\ -x + 4y &= 6 \end{aligned}$

3. $\begin{aligned} -x + 3y &= 1 \\ 6x - 2y &= 2 \end{aligned}$

4. $\begin{aligned} 2x + 3y &= 1 \\ 7x - 2y &= 1 \end{aligned}$

5. $\begin{aligned} 5x - y + z &= 10 \\ 2x + 8y - z &= 11 \\ -x + y + 4z &= 3 \end{aligned}$

6. $\begin{aligned} 2x + 8y - z &= 11 \\ 5x - y + z &= 10 \\ -x + y + 4z &= 3 \end{aligned}$

7. $\begin{aligned} x - 5y - z &= -8 \\ 4x + y - z &= 13 \\ 2x - y - 6z &= -2 \end{aligned}$

8. $\begin{aligned} 4x + y - z &= 13 \\ x - 5y - z &= -8 \\ 2x - y - 6z &= -2 \end{aligned}$

9. Consider the following linear system:

$$\begin{aligned} 5x + 3y &= 6 \\ 4x - 2y &= 8. \end{aligned}$$

Can either Jacobi or Gauss–Seidel iteration be used to solve this linear system? Why?

10. Can Jacobi iteration ever be used to find the solution to the following system?

$$\begin{aligned} 2x + y - 5z &= 9 \\ x - 5y - z &= 14 \\ 7x - y - 3z &= 26. \end{aligned}$$

11. Consider the following tridiagonal linear system, and assume that the coefficient matrix is diagonally dominant.

$$\begin{aligned} d_1 x_1 + c_1 x_2 \qquad\qquad\qquad\qquad\qquad &= b_1 \\ a_1 x_1 + d_2 x_2 + c_2 x_3 \qquad\qquad\qquad &= b_2 \\ a_2 x_2 + d_3 x_3 + c_3 x_4 \qquad\qquad &= b_3 \\ \vdots \qquad\qquad\qquad\qquad\qquad &\quad\vdots \\ a_{N-2} x_{N-2} + d_{N-1} x_{N-1} + c_{N-1} x_N &= b_{N-1} \\ a_{N-1} x_{N-1} + d_N x_N &= b_N. \end{aligned}$$

Write an iterative algorithm that will solve this system.

12. Tridiagonal matrices were involved in the construction of a cubic spline. Find the solution to the following systems.

   (a)
$$
\begin{aligned}
4m_1 + m_2 &= 3 \\
m_1 + 4m_2 + m_3 &= 3 \\
m_2 + 4m_3 + m_4 &= 3 \\
m_3 + 4m_4 + m_5 &= 3 \\
&\;\vdots \\
m_{48} + 4m_{49} + m_{50} &= 3 \\
m_{49} + 4m_{50} &= 3
\end{aligned}
$$

   (b)
$$
\begin{aligned}
4m_1 + m_2 &= 1 \\
m_1 + 4m_2 + m_3 &= 2 \\
m_2 + 4m_3 + m_4 &= 1 \\
m_3 + 4m_4 + m_5 &= 2 \\
&\;\vdots \\
m_{48} + 4m_{49} + m_{50} &= 1 \\
m_{49} + 4m_{50} &= 2
\end{aligned}
$$

13. Use Gauss–Seidel iteration to solve the following band system.

$$
\begin{aligned}
12x_1 - 2x_2 + x_3 &= 5 \\
-2x_1 + 12x_2 - 2x_3 + x_4 &= 5 \\
x_1 - 2x_2 + 12x_3 - 2x_4 + x_5 &= 5 \\
x_2 - 2x_3 + 12x_4 - 2x_5 + x_6 &= 5 \\
&\;\vdots \\
x_{46} - 2x_{47} + 12x_{48} - 2x_{49} + x_{50} &= 5 \\
x_{47} - 2x_{48} + 12x_{49} - 2x_{50} &= 5 \\
x_{48} - 2x_{49} + 12x_{50} &= 5.
\end{aligned}
$$

14. Let $\mathbf{X} = (x_1, x_2, \ldots, x_N)$. Prove that the Euclidean length

$$
\|\mathbf{X}\|_2 = \left(\sum_{k=1}^{N} x_k^2 \right)^{1/2}
$$

   satisfies the four properties given in (14)–(17).

15. Let $\mathbf{X} = (x_1, x_2, \ldots, x_N)$. Prove that the function

$$
\|\mathbf{X}\|_\infty = \max_{1 \le k \le N} |x_k|
$$

   satisfies the four properties given in (14)–(17).

16. Write a report on the successive over-relaxation method.

8.2 Iteration for Nonlinear Systems

Iterative techniques will now be discussed that extend the methods of Section 8.1 to the case of nonlinear functions. Consider the two-dimensional functions:

(1)
$$
\begin{aligned}
f_1(x, y) &= x^2 - 2x - y + 0.5, \\
f_2(x, y) &= x^2 + 4y^2 - 4.
\end{aligned}
$$

We seek a method of solution for the system of nonlinear equations

(2)
$$f_1(x, y) = 0 \qquad f_2(x, y) = 0.$$

The equations $f_1(x, y) = 0$ and $f_2(x, y) = 0$ implicitly define curves in the xy-plane. Hence a solution of the system (2) is a point (p, q) where the two curves cross [i.e., both $f_1(p, q) = 0$ and $f_2(p, q) = 0$]. The curves for the system in (1) are well known:

(3)
$$x^2 - 2x - y + 0.5 = 0 \qquad \text{is the graph of a parabola.}$$
$$x^2 + 4y^2 - 4 = 0 \qquad \text{is the graph of an ellipse.}$$

The graphs in Figure 8.1 show that there are two solution points, and that they are in the vicinity of $(-0.2, 1.0)$ and $(1.9, 0.3)$.

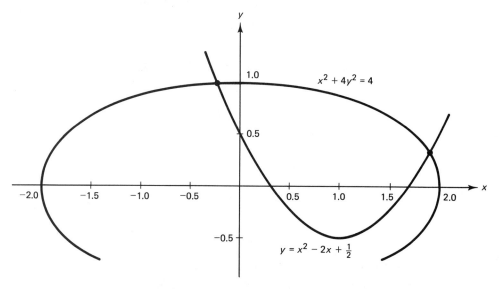

Figure 8.1. Graphs for the nonlinear system $y = x^2 - 2x + 0.5$ and $x^2 + 4y^2 = 4$.

The first technique is fixed-point iteration. A method must be devised for generating a sequence $\{(p_k, q_k)\}$ that converges to the solution (p, q). The first equation in (3) can be used to solve directly for x. However, a multiple of y can be added to each side of the second equation to get $x^2 + 4y^2 - 8y - 4 = -8y$. The choice of adding $-8y$ is crucial and will be explained later. We now have an equivalent system of equations:

(4)
$$x = \frac{x^2 - y + 0.5}{2},$$
$$y = \frac{-x^2 - 4y^2 + 8y + 4}{8}.$$

These two equations are used to write down the recursive formulas. Start with an initial point (p_0, q_0), then compute the sequence $\{(p_{k+1}, q_{k+1})\}$ using:

(5)
$$p_{k+1} = g_1(p_k, q_k) = \frac{p_k^2 - q_k + 0.5}{2},$$

$$q_{k+1} = g_2(p_k, q_k) = \frac{-p_k^2 - 4q_k^2 + 8q_k + 4}{8}.$$

Case (i)   If we use the starting value $(p_0, q_0) = (0, 1)$, then

$$p_1 = \frac{0^2 - 1 + 0.5}{2} = -0.25 \qquad q_1 = \frac{-0^2 - 4 \times 1^2 + 8 \times 1 + 4}{8} = 1.0.$$

Iteration will generate the sequence in case (i) of Table 8.4. In this case the sequence converges to the solution that lies near the starting value $(0, 1)$.

Case (ii)   If we use the starting value $(p_0, q_0) = (2, 0)$, then

$$p_1 = \frac{2^2 - 0 + 0.5}{2} = 2.25 \qquad q_1 = \frac{-2^2 - 4 \times 0^2 + 8 \times 0 + 4}{8} = 0.0.$$

Iteration will generate the sequence in case (ii) of Table 8.4. In this case the sequence diverges away from the solution.

TABLE 8.4   Fixed-point iteration using the formulas in (5)

	Case (i) Start with $(0, 1)$			Case (ii) Start with $(2, 0)$	
k	p_k	q_k	k	p_k	q_k
0	0.00	1.00	0	2.00	0.00
1	−0.25	1.00	1	2.25	0.00
2	−0.21875	0.9921875	2	2.78125	−0.1328125
3	−0.2221680	0.9939880	3	4.184082	−0.6085510
4	−0.2223147	0.9938121	4	9.307547	−2.4820360
5	−0.2221941	0.9938029	5	44.80623	−15.891091
6	−0.2222163	0.9938095	6	1,011.995	−392.60426
7	−0.2222147	0.9938083	7	512,263.2	−205,477.82
8	−0.2222145	0.9938084		This sequence is diverging.	
9	−0.2222146	0.9938084			

Iteration using formulas (5) cannot be used to find the second solution $(1.900677, 0.3112186)$. To find this point a different pair of iteration formulas are needed. Start with equations (3) and add $-2x$ to the first equation and $-11y$ to the second equation and get

$$x^2 - 4x + y + 0.5 = -2x \quad \text{and} \quad x^2 + 4y^2 - 11y - 4 = -11y.$$

These equations can then be used to obtain the iteration formulas

$$p_{k+1} = g_1(p_k, q_k) = \frac{-p_k^2 + 4p_k + q_k - 0.5}{2},$$

(6)

$$q_{k+1} = g_2(p_k, q_k) = \frac{-p_k^2 - 4q_k^2 + 11q_k + 4}{11}.$$

Table 8.5 shows how to use (6) to find the second solution.

TABLE 8.5   Fixed-point iteration using
the formulas in (6)

k	p_k	q_k
0	2.00	0.00
1	1.75	0.00
2	1.71875	0.0852273
3	1.753063	0.1776676
4	1.808345	0.2504410
8	1.903595	0.3160782
12	1.900924	0.3112267
16	1.900652	0.3111994
20	1.900677	0.3112196
24	1.900677	0.3112186

Theory

We want to determine why equations (6) were suitable for finding the solution near $(1.9, 0.3)$ and equations (5) were not. In Section 1.4 the size of the derivative at the fixed point was the necessary idea. When functions of several variables are used, the partial derivatives must be used. The generalization of "the derivative" for systems of functions of several variables is the Jacobian matrix. We will make only a few introductory ideas regarding this topic. More details can be found in any textbook of advanced calculus.

Definition 8.4   [Jacobian matrix] Let $f_1(x, y)$ and $f_2(x, y)$ be two functions of the independent variables x and y; then their Jacobian matrix $J(x, y)$ of partial derivatives is

(7)

$$J(x, y) = \begin{pmatrix} \dfrac{\partial}{\partial x} f_1 & \dfrac{\partial}{\partial y} f_1 \\ \dfrac{\partial}{\partial x} f_2 & \dfrac{\partial}{\partial y} f_2 \end{pmatrix}.$$

Similarly, if $f_1(x, y, z)$, $f_2(x, y, z)$, and $f_3(x, y, z)$ are functions of x, y, and z, their 3 by 3 Jacobian matrix is defined as follows:

(8)
$$J(x, y, z) = \begin{pmatrix} \dfrac{\partial}{\partial x} f_1 & \dfrac{\partial}{\partial y} f_1 & \dfrac{\partial}{\partial z} f_1 \\[2mm] \dfrac{\partial}{\partial x} f_2 & \dfrac{\partial}{\partial y} f_2 & \dfrac{\partial}{\partial z} f_2 \\[2mm] \dfrac{\partial}{\partial x} f_3 & \dfrac{\partial}{\partial y} f_3 & \dfrac{\partial}{\partial z} f_3 \end{pmatrix}.$$

Example 8.5 Find the Jacobian matrix $J(x, y, z)$ of order 3 by 3, at the points $(1, 3, 2)$ and $(2, 0, 3)$, for the three functions

$$f_1(x, y, z) = x^3 - y^2 + y - z^4 + z^2,$$
$$f_2(x, y, z) = xy + yz + xz,$$
$$f_3(x, y, z) = \frac{y}{xz}.$$

Solution. The nine partial derivatives are

$$\frac{\partial}{\partial x} f_1 = 3x^2, \qquad \frac{\partial}{\partial y} f_1 = -2y + 1, \qquad \frac{\partial}{\partial z} f_1 = -4z^3 + 2z,$$

$$\frac{\partial}{\partial x} f_2 = y + z, \qquad \frac{\partial}{\partial y} f_2 = x + z, \qquad \frac{\partial}{\partial z} f_2 = y + x,$$

$$\frac{\partial}{\partial x} f_3 = \frac{-y}{x^2 z}, \qquad \frac{\partial}{\partial y} f_3 = \frac{1}{xz}, \qquad \frac{\partial}{\partial z} f_3 = \frac{-y}{xz^2}.$$

These functions form the nine elements in the Jacobian matrix

$$J(x, y, z) = \begin{pmatrix} 3x^2 & -2y + 1 & -4z^3 + 2z \\[2mm] y + z & x + z & y + x \\[2mm] \dfrac{-y}{x^2 z} & \dfrac{1}{xz} & \dfrac{-y}{xz^2} \end{pmatrix}.$$

The values of $J(1, 3, 2)$ and $J(3, 2, 1)$ are

$$J(1, 3, 2) = \begin{pmatrix} 3 & -5 & -28 \\ 5 & 3 & 4 \\ -\tfrac{3}{2} & \tfrac{1}{2} & -\tfrac{3}{4} \end{pmatrix} \qquad J(3, 2, 1) = \begin{pmatrix} 27 & -3 & -2 \\ 3 & 4 & 5 \\ -\tfrac{2}{9} & \tfrac{1}{3} & -\tfrac{2}{3} \end{pmatrix}.$$

Generalized Differential

For a function of several variables, the differential is used to show how changes of the independent variable affect the change in the dependent variable. Suppose that we have three functions, each of which is a function of three variables,

$$(9) \qquad u = f_1(x, y, z) \qquad v = f_2(x, y, z) \qquad w = f_3(x, y, z).$$

Suppose that the values of the functions in (9) are known at the point (x_0, y_0, z_0) and we wish to predict their value at a nearby point (x, y, z). Let du, dv, and dw denote differential changes in the dependent variables and dx, dy, and dz denote differential changes in the independent variables. These changes obey the relationships

$$du = \frac{\partial}{\partial x} f_1(x_0, y_0, z_0)\, dx + \frac{\partial}{\partial y} f_1(x_0, y_0, z_0)\, dy + \frac{\partial}{\partial z} f_1(x_0, y_0, z_0)\, dz,$$

$$(10) \qquad dv = \frac{\partial}{\partial x} f_2(x_0, y_0, z_0)\, dx + \frac{\partial}{\partial y} f_2(x_0, y_0, z_0)\, dy + \frac{\partial}{\partial z} f_2(x_0, y_0, z_0)\, dz,$$

$$dw = \frac{\partial}{\partial x} f_3(x_0, y_0, z_0)\, dx + \frac{\partial}{\partial y} f_3(x_0, y_0, z_0)\, dy + \frac{\partial}{\partial z} f_3(x_0, y_0, z_0)\, dz.$$

If vector notation is used, (10) can be compactly written by using the Jacobian matrix. The function changes are $d\mathbf{F}$ and the changes in the variables are denoted $d\mathbf{X}$

$$(11) \qquad d\mathbf{F} = \begin{pmatrix} du \\ dv \\ dw \end{pmatrix} = J(x_0, y_0, z_0) \begin{pmatrix} dx \\ dy \\ dz \end{pmatrix} = J(x_0, y_0, z_0)\, d\mathbf{X}.$$

Example 8.6 Use the Jacobian to find the differential changes (du, dv, dw) when the independent variables changes from $(1, 3, 2)$ to $(1.02, 2.97, 2.01)$ for the system of functions

$$u = f_1(x, y, z) = x^3 - y^2 + y - z^4 + z^2,$$

$$v = f_2(x, y, z) = xy + yz + xz,$$

$$w = f_3(x, y, z) = \frac{y}{xz}.$$

Solution. Use equation (11) with $J(1, 3, 2)$ of Example 8.5 and the differential changes $(dx, dy, dz) = (0.02, -0.03, 0.01)$ to obtain

$$\begin{pmatrix} du \\ dv \\ dw \end{pmatrix} = \begin{pmatrix} 3 & -5 & -28 \\ 5 & 3 & 4 \\ -\frac{3}{2} & \frac{1}{2} & -\frac{3}{4} \end{pmatrix} \begin{pmatrix} 0.02 \\ -0.03 \\ 0.01 \end{pmatrix} = \begin{pmatrix} -0.07 \\ 0.05 \\ -0.0525 \end{pmatrix}.$$

Notice that the function values at $(1.02, 2.97, 2.01)$ are close to the linear approximations obtained by adding the differentials $du = -0.07$, $dv = 0.05$, and $dw = -0.0525$ to the corresponding function values $f_1(1, 3, 2) = -17$, $f_2(1, 3, 2) = 11$, and $f_3(1, 3, 2) = 1.5$, that is,

$$f_1(1.02, 2.97, 2.01) = -17.072 \approx -17.07 = f_1(1, 3, 2) + du,$$

$$f_2(1.02, 2.97, 2.01) = 11.0493 \approx 11.05 = f_2(1, 3, 2) + dv,$$

$$f_3(1.02, 2.97, 2.01) = 1.44863916 \approx 1.4475 = f_3(1, 3, 2) + dw.$$

Convergence Near Fixed Points

The extension of the definitions and theorem in Section 1.4 to the case of two and three dimensions are now given. The notation for N-dimensional functions has not been used. The reader can easily find these extensions in many books on numerical analysis.

Definition 8.5 A fixed point for the system of two equations

(12) $$x = g_1(x, y) \qquad y = g_2(x, y)$$

is a point (p, q) such that $p = g_1(p, q)$ and $q = g_2(p, q)$.

Similarly, in three dimensions, a fixed point for the system

(13) $$x = g_1(x, y, z) \qquad y = g_2(x, y, z) \qquad z = g_3(x, y, z)$$

is a point (p, q, r) such that $p = g_1(p, q, r)$, $q = g_2(p, q, r)$, and $r = g_3(p, q, r)$.

Definition 8.6 [Fixed-point iteration] For the functions (12), fixed-point iteration is

(14) $$p_{k+1} = g_1(p_k, q_k) \qquad q_{k+1} = g_2(p_k, q_k) \qquad \text{for } k = 0, 1, \ldots.$$

Similarly, for the functions (13), fixed-point iteration is

(15) $$p_{k+1} = g_1(p_k, q_k, r_k) \qquad q_{k+1} = g_2(p_k, q_k, r_k) \qquad r_{k+1} = g_3(p_k, q_k, r_k).$$

Theorem 8.3 [Fixed-point iteration] Let the functions and their first partial derivatives be continuous on a region that contains the fixed point. Assume that the starting point is chosen sufficiently close to the fixed point and one of the following cases hold.

Case (i) Two Dimensions. (p_0, q_0) is sufficiently close to (p, q).

(16) $$\text{If} \quad \left| \frac{\partial}{\partial x} g_1(p, q) \right| + \left| \frac{\partial}{\partial y} g_1(p, q) \right| < 1$$

$$\text{and} \quad \left| \frac{\partial}{\partial x} g_2(p, q) \right| + \left| \frac{\partial}{\partial y} g_2(p, q) \right| < 1,$$

then the iteration in (14) converges to the fixed point (p, q).

Case (ii) Three Dimensions. (p_0, q_0, r_0) is sufficiently close to (p, q, r).

(17)   If $\left| \dfrac{\partial}{\partial x} g_1(p, q, r) \right| + \left| \dfrac{\partial}{\partial y} g_1(p, q, r) \right| + \left| \dfrac{\partial}{\partial z} g_1(p, q, r) \right| < 1,$

and $\left| \dfrac{\partial}{\partial x} g_2(p, q, r) \right| + \left| \dfrac{\partial}{\partial y} g_2(p, q, r) \right| + \left| \dfrac{\partial}{\partial z} g_2(p, q, r) \right| < 1,$

and $\left| \dfrac{\partial}{\partial x} g_3(p, q, r) \right| + \left| \dfrac{\partial}{\partial y} g_3(p, q, r) \right| + \left| \dfrac{\partial}{\partial z} g_3(p, q, r) \right| < 1,$

then the iteration in (15) converges to the fixed point (p, q, r).

If conditions (16) or (17) are not met, the iteration usually diverges. This will usually be the case if the sum of the magnitude of the partial derivatives is much larger than 1.

Theorem 8.3 can be used to show why the iteration (5) converged to the fixed point near $(-0.2, 1.0)$. The partial derivatives are

$$\frac{\partial}{\partial x} g_1(x, y) = x, \qquad \frac{\partial}{\partial y} g_1(x, y) = \frac{-1}{2},$$

$$\frac{\partial}{\partial x} g_2(x, y) = \frac{-x}{4}, \qquad \frac{\partial}{\partial y} g_2(x, y) = -y + 1,$$

Indeed, for all (x, y) satisfying $-0.5 < x < 0.5$ and $0.5 < y < 1.5$ the partial derivatives satisfy

$$\left| \frac{\partial}{\partial x} g_1(x, y) \right| + \left| \frac{\partial}{\partial y} g_1(x, y) \right| = |x| + |-0.5| < 1,$$

$$\left| \frac{\partial}{\partial x} g_2(x, y) \right| + \left| \frac{\partial}{\partial y} g_2(x, y) \right| = \left| \frac{-x}{4} \right| + |-y + 1| < 0.625 < 1.$$

Therefore, the partial derivative conditions (16) are met and Theorem 8.3 states that fixed-point iteration will converge to $(p, q) \approx (-0.2222146, 0.9938084)$. Notice that near the other fixed point $(1.90068, 0.31122)$ the partial derivatives do not meet conditions (16), hence convergence is not guaranteed, that is,

$$\left| \frac{\partial}{\partial x} g_1(1.90068, 0.31122) \right| + \left| \frac{\partial}{\partial y} g_1(1.90068, 0.31122) \right| = 2.40068 > 1,$$

$$\left| \frac{\partial}{\partial x} g_2(1.90068, 0.31122) \right| + \left| \frac{\partial}{\partial y} g_2(1.90068, 0.31122) \right| = 1.16395 > 1.$$

Seidel Iteration

An improvement of fixed-point iteration can be made. Suppose that p_{k+1} is used in the calculation of q_k (in three dimensions both p_{k+1} and q_{k+1} are used to compute r_k). When these modifications are incorporated in formulas (14) and (15), the method is called *Seidel iteration*:

(18) $p_{k+1} = g_1(p_k, q_k), \quad q_{k+1} = g_2(p_{k+1}, q_k),$

and

(19) $p_{k+1} = g_1(p_k, q_k, r_k), \quad q_{k+1} = g_2(p_{k+1}, q_k, r_k), \quad r_{k+1} = g_3(p_{k+1}, q_{k+1}, r_k).$

Algorithm 8.3 [*Nonlinear Seidel Iteration*] Assume that the Jacobian condition (17) is satisfied for the nonlinear system of equations

$$x = g_1(x, y, z) \qquad y = g_2(x, y, z) \qquad z = g_3(x, y, z).$$

Start with $\mathbf{P}_0 = (p_0, q_0, r_0)$, and generate a sequence $\{\mathbf{P}_k: (p_k, q_k, r_k)\}$ which converges to the solution $\mathbf{P} = (p, q, r)$.

```
        Tol  := 10⁻⁶                                        {Tolerance}
        Sep := 1 , K := 0 , Max := 199                      {Initialize}
        INPUT  P(0), Q(0), R(0)                             {Input}

WHILE  K < Max AND Sep > Tol  DO
        K := K+1
        P(K)  := G1(P(K−1), Q(K−1), R(K−1))                {Perform
        Q(K)  := G2(P(K), Q(K−1), R(K−1))                   Seidel
        R(K)  := G3(P(K), Q(K), R(K−1))                     iteration}
        Sep := |P(K)−P(K−1)| + |Q(K)−Q(K−1)| + |R(K)−R(K−1)|

IF    Sep < Tol   THEN                                      {Output}
            PRINT "After ",K," iterations Seidel's method"
            PRINT "was successful and found the solution:"
        ELSE
            PRINT "Seidel iteration did not converge,"
            PRINT "After ",K," iterations the status is:"
ENDIF

            PRINT "P(",K,") = ", P(K)
            PRINT "Q(",K,") = ", Q(K)
            PRINT "R(",K,") = ", R(K)
```

Exercises for Iteration for Nonlinear Systems

In Exercises 1-6, for each nonlinear system,

$$x = g_1(x, y) \quad \text{and} \quad y = g_2(x, y)$$

(a) Find the Jacobian matrix $J(x, y)$ of partial derivatives

$$
J(x, y) = \begin{pmatrix} \dfrac{\partial}{\partial x} g_1(x, y) & \dfrac{\partial}{\partial y} g_1(x, y) \\[2ex] \dfrac{\partial}{\partial x} g_2(x, y) & \dfrac{\partial}{\partial y} g_2(x, y) \end{pmatrix}.
$$

Also find the matrix $J(p_0, q_0)$.
(b) Compute (p_1, q_1) and (p_2, q_2) using fixed-point iteration and equations (14).
(c) Compute (p_1, q_1) and (p_2, q_2) using Seidel iteration and Equations (18).
(d) Graph the curves $g_1(x, y) = 0$ and $g_2(x, y) = 0$.
(e) Use a computer and find the solution $(p, q) = \lim_{k \to \infty} (p_k, q_k)$.

1. $x = g_1(x, y) = \dfrac{8x - 4x^2 + y^2 + 1}{8}$ \qquad (hyperbola),

$y = g_2(x, y) = \dfrac{2x - x^2 + 4y - y^2 + 3}{4}$ \qquad (circle),

$(p_0, q_0) = (1.1, 2.0)$.

2. $x = g_1(x, y) = \dfrac{2x + x^2 - y}{2}$ \qquad (parabola),

$y = g_2(x, y) = \dfrac{2x - x^2 + 8}{9} + \dfrac{4y - y^2}{4}$ \qquad (ellipse),

$(p_0, q_0) = (-1.2, 1.4)$.

3. $x = g_1(x, y) = \dfrac{2x - x^2 + y}{2}$ \qquad (parabola),

$y = g_2(x, y) = \dfrac{2x - x^2 + 8}{9} + \dfrac{4y - y^2}{4}$ \qquad (ellipse),

$(p_0, q_0) = (1.4, 2.0)$.

4. $x = g_1(x, y) = \dfrac{4x - x^3 + y}{4}$ \qquad (cubic),

$y = g_2(x, y) = -\dfrac{x^2}{9} + \dfrac{4y - y^2}{4} + 1$ \qquad (ellipse),

$(p_0, q_0) = (1.2, 1.8)$.

5. $x = g_1(x, y) = \dfrac{4x - x^2 + y + 3}{4}$ (parabola),

$y = g_2(x, y) = \dfrac{3 - xy + 2y}{2}$ (hyperbola),

$(p_0, q_0) = (2.1, 1.4)$.

6. $x = g_1(x, y) = \dfrac{y - x^3 + 3x^2 + 3x}{7}$ (cubic),

$y = g_2(x, y) = \dfrac{y^2 + 2y - x - 2}{2}$ (parabola),

$(p_0, q_0) = (-0.3, -1.3)$.

7. Show that Jacobi iteration for 2 by 2 linear systems is a special case of fixed-point iteration (14). If A is diagonally dominant, then condition (16) is satisfied.

8. Show that Jacobi iteration for 3 by 3 linear systems is a special case of fixed-point iteration (15). If A is diagonally dominant, then condition (17) is satisfied.

9. We wish to solve the nonlinear system:

$$0 = 7x^3 - 10x - y - 1,$$
$$0 = 8y^3 - 11y + x - 1.$$

Sketch the graphs $y = 7x^3 - 10x - 1$ and $x = -8y^3 + 11y + 1$ and verify that there are nine points where the curves cross. Suppose that we use fixed-point iteration with the formulas

$$x = g_1(x, y) = \frac{7x^3 - y - 1}{10},$$

$$y = g_2(x, y) = \frac{8y^3 + x - 1}{11}.$$

Do some computer experimentation. Discover that no matter what starting value is used, only one of the nine solutions can be found using fixed-point iteration based on these two choices of $g_1(x, y)$ and $g_2(x, y)$. Which solution did you find? In Section 8.3 we will see that Newton's method is able to find all nine solutions.

10. Fixed point-iteration is used to solve the nonlinear system (12). Use the following steps to prove that conditions (16) are sufficient condition to guarantee that $\{(p_k, q_k)\}$ converges to (p, q). Assume that there is a constant K with $0 < K < 1$, so that

$$\left| \frac{\partial}{\partial x} g_1(x, y) \right| + \left| \frac{\partial}{\partial y} g_1(x, y) \right| < K \quad \text{and} \quad \left| \frac{\partial}{\partial x} g_2(x, y) \right| + \left| \frac{\partial}{\partial y} g_2(x, y) \right| < K$$

for all (x, y) in the rectangle $a < x < b, c < y < d$. Also assume that $a < p_0 < b$ and $c < q_0 < d$. Define

$$e_k = p - p_k \qquad E_k = q - q_k, \quad m_k = \max\{|e_k|, |E_k|\}.$$

Use the following form of the mean value theorem applied to functions of two variables:

$$e_{k+1} = \frac{\partial}{\partial x} g_1(a_k^*, q_k) e_k + \frac{\partial}{\partial y} g_1(p, c_k^*) E_k,$$

$$E_{k+1} = \frac{\partial}{\partial x} g_2(b_k^*, q_k) e_k + \frac{\partial}{\partial y} g_2(p, d_k^*) E_k.$$

where a_k^* and b_k^* lie in $[a, b]$ and c_k^* and d_k^* lie in $[c, d]$. Prove the following things:

(a) $|e_1| \leq K m_0$ and $|E_1| \leq K m_0$

(b) $|e_2| \leq K r_1 \leq K^2 r_0$ and $|E_2| \leq K r_1 \leq K^2 r_0$

(c) $|e_k| \leq K r_{k-1} \leq K^k r_0$ and $|E_k| \leq K r_{k-1} \leq K^k r_0$

(d) $\lim_{k \to \infty} p_k = p$ and $\lim_{k \to \infty} p_k = q$

11. Write a report on the "contraction-mapping fixed-point theorem."

12. Rewrite Algorithm 8.3 so that it does not use arrays to store all the calculations.

8.3 Newton's Method for Systems

Transformations in Two Dimensions

Consider the system of nonlinear equations

$$
\begin{aligned}
u &= f_1(x, y), \\
v &= f_2(x, y),
\end{aligned}
$$

(1)

which can be considered a transformation from the xy-plane into the uv-plane. We are interested in the behavior of this transformation near the point (x_0, y_0) whose image is the point (u_0, v_0). If the two functions have continuous partial derivatives, then the differential can be used to write a system of linear approximations that are valid near the point (x_0, y_0):

$$u - u_0 \approx \frac{\partial}{\partial x} f_1(x_0, y_0)(x - x_0) + \frac{\partial}{\partial y} f_1(x_0, y_0)(y - y_0),$$

(2)

$$v - v_0 \approx \frac{\partial}{\partial x} f_2(x_0, y_0)(x - x_0) + \frac{\partial}{\partial y} f_2(x_0, y_0)(y - y_0).$$

The system (2) is a local linear transformation that relates small changes in the independent variables to small changes in the dependent variable. When the Jacobian matrix $J(x_0, y_0)$ is used this relationship is easier to visualize:

(3)

$$
\begin{pmatrix} u - u_0 \\ v - v_0 \end{pmatrix} \approx
\begin{pmatrix} \dfrac{\partial}{\partial x} f_1(x_0, y_0) & \dfrac{\partial}{\partial y} f_1(x_0, y_0) \\[2mm] \dfrac{\partial}{\partial x} f_2(x_0, y_0) & \dfrac{\partial}{\partial y} f_2(x_0, y_0) \end{pmatrix}
\begin{pmatrix} x - x_0 \\ y - y_0 \end{pmatrix}.
$$

If the system in (1) is written as a vector function $\mathbf{V} = \mathbf{F}(\mathbf{X})$, then the Jacobian $J(x, y)$ is the two-dimensional analog of the derivative, because (3) can be written

(4) $$\Delta \mathbf{F} \approx J(x_0, y_0)\, \Delta \mathbf{X}.$$

Newton's Method in Two Dimensions

Consider the system (1) with u and v set equal to zero:

(5) $$\begin{aligned} 0 &= f_1(x, y), \\ 0 &= f_2(x, y). \end{aligned}$$

Suppose that (p, q) is a solution of (5), that is,

(6) $$\begin{aligned} 0 &= f_1(p, q), \\ 0 &= f_2(p, q). \end{aligned}$$

To develop Newton's method for solving (5), we need to consider small changes in the functions near the point (p_0, q_0):

(7) $$\begin{aligned} \Delta u &= u - u_0, & \Delta p &= x - p_0, \\ \Delta v &= v - v_0, & \Delta q &= y - q_0. \end{aligned}$$

Set $(x, y) = (p, q)$ in (1), and use (6) to see that $(u, v) = (0, 0)$. Hence changes in the dependent variables are

(8) $$\begin{aligned} u - u_0 &= f_1(p, q) - f_1(p_0, q_0) = 0 - f_1(p_0, q_0), \\ v - v_0 &= f_2(p, q) - f_2(p_0, q_0) = 0 - f_2(p_0, q_0). \end{aligned}$$

Use the result of (8) in (3) to get the linear approximation

(9) $$\begin{pmatrix} \dfrac{\partial}{\partial x} f_1(p_0, q_0) & \dfrac{\partial}{\partial y} f_1(p_0, q_0) \\[2mm] \dfrac{\partial}{\partial x} f_2(p_0, q_0) & \dfrac{\partial}{\partial y} f_2(p_0, q_0) \end{pmatrix} \begin{pmatrix} \Delta p \\ \Delta q \end{pmatrix} \approx - \begin{pmatrix} f_1(p_0, q_0) \\ f_2(p_0, q_0) \end{pmatrix}.$$

If the Jacobian $J(p_0, q_0)$ in (9) is nonsingular, we can solve for $\Delta \mathbf{P} = (\Delta p, \Delta q) = (p, q) - (p_0, q_0)$ as follows:

(10) $$\Delta \mathbf{P} \approx -J(p_0, q_0)^{-1} \mathbf{F}(p_0, q_0).$$

The next approximation $\mathbf{P}_1$ to the solution $\mathbf{P}$ is

(11) $$\mathbf{P}_1 = \mathbf{P}_0 + \Delta \mathbf{P} = \mathbf{P}_0 - J(p_0, q_0)^{-1} \mathbf{F}(p_0, q_0).$$

Notice that (11) is the generalization of Newton's method for the one-variable case, that is, $p_1 = p_0 - [f'(p_0)]^{-1} f(p_0)$.

Outline of Newton's Method

Suppose that $\mathbf{P}_k$ has been obtained.

Step 1. Evaluate the function

$$\mathbf{F}(\mathbf{P}_k) = \begin{pmatrix} f_1(p_k, q_k) \\ f_2(p_k, q_k) \end{pmatrix}.$$

Step 2. Evaluate the Jacobian

$$J(\mathbf{P}_k) = \begin{pmatrix} \dfrac{\partial}{\partial x} f_1(p_k, q_k) & \dfrac{\partial}{\partial y} f_1(p_k, q_k) \\[2mm] \dfrac{\partial}{\partial x} f_2(p_k, q_k) & \dfrac{\partial}{\partial y} f_2(p_k, q_k) \end{pmatrix}.$$

Step 3. Solve the linear system

$$J(\mathbf{P}_k) \, \Delta\mathbf{P} = -\mathbf{F}(\mathbf{P}_k) \quad \text{for } \Delta\mathbf{P}.$$

Step 4. Compute the next point:

$$\mathbf{P}_{k+1} = \mathbf{P}_k + \Delta\mathbf{P}.$$

Now, repeat the process.

Example 8.7 Consider the nonlinear system:

$$0 = x^2 - 2x - y + 0.5,$$
$$0 = x^2 + 4y^2 - 4.$$

Use Newton's method with the starting value $(p_0, q_0) = (2.00, 0.25)$ and compute (p_1, q_1), (p_2, q_2), and (p_3, q_3).

Solution. The function vector and Jacobian matrix are

$$\mathbf{F}(x, y) = \begin{pmatrix} x^2 - 2x - y + 0.5 \\ x^2 + 4y^2 - 4 \end{pmatrix} \quad J(x, y) = \begin{pmatrix} 2x - 2 & -1 \\ 2x & 8y \end{pmatrix}.$$

At the point $(2.00, 0.25)$ they take on the values

$$\mathbf{F}(2.00, 0.25) = \begin{pmatrix} 0.25 \\ 0.25 \end{pmatrix} \quad J(2.00, 0.25) = \begin{pmatrix} 2.0 & -1.0 \\ 4.0 & 2.0 \end{pmatrix}.$$

The differentials Δp and Δq are solutions of the linear system

$$\begin{pmatrix} 2.0 & -1.0 \\ 4.0 & 2.0 \end{pmatrix} \begin{pmatrix} \Delta p \\ \Delta q \end{pmatrix} = - \begin{pmatrix} 0.25 \\ 0.25 \end{pmatrix}.$$

A straightforward calculation reveals that

$$\Delta\mathbf{P} = \begin{pmatrix} \Delta p \\ \Delta q \end{pmatrix} = \begin{pmatrix} -0.09375 \\ 0.0625 \end{pmatrix}.$$

The next point in the iteration is

$$\mathbf{P}_1 = \mathbf{P}_0 + \Delta\mathbf{P} = \begin{pmatrix} 2.00 \\ 0.25 \end{pmatrix} + \begin{pmatrix} -0.09375 \\ 0.0625 \end{pmatrix} = \begin{pmatrix} 1.90625 \\ 0.3125 \end{pmatrix}.$$

Similarly, the next two points are

$$\mathbf{P}_2 = \begin{pmatrix} 1.900691 \\ 0.311213 \end{pmatrix} \quad \text{and} \quad \mathbf{P}_3 = \begin{pmatrix} 1.900677 \\ 0.311219 \end{pmatrix}.$$

The coordinates of $\mathbf{P}_3$ are accurate to six decimal places. Calculations for finding $\mathbf{P}_2$ and $\mathbf{P}_3$ are summarized in Table 8.6

TABLE 8.6 Function values, Jacobian matrices, and differentials required for each iteration in Newton's solution to Example 8.7

$\mathbf{P}_k$	Solution of the Linear System	$J(\mathbf{P}_k) \Delta\mathbf{P} = -\mathbf{F}(\mathbf{P}_k)$	$\mathbf{P}_k + \Delta\mathbf{P}$
$\begin{pmatrix} 2.00 \\ 0.25 \end{pmatrix}$		$\begin{pmatrix} 2.0 & -1.0 \\ 4.0 & 2.0 \end{pmatrix}\begin{pmatrix} -0.09375 \\ 0.0625 \end{pmatrix} = -\begin{pmatrix} 0.25 \\ 0.25 \end{pmatrix}$	$\begin{pmatrix} 1.90625 \\ 0.3125 \end{pmatrix}$
$\begin{pmatrix} 1.90625 \\ 0.3125 \end{pmatrix}$		$\begin{pmatrix} 1.8125 & -1.0 \\ 3.8125 & 2.5 \end{pmatrix}\begin{pmatrix} -0.005559 \\ -0.001287 \end{pmatrix} = -\begin{pmatrix} 0.008789 \\ 0.024414 \end{pmatrix}$	$\begin{pmatrix} 1.900691 \\ 0.311213 \end{pmatrix}$
$\begin{pmatrix} 1.900691 \\ 0.311213 \end{pmatrix}$	$\begin{pmatrix} 1.801381 & -1.000000 \\ 3.801381 & 2.489700 \end{pmatrix}$	$\begin{pmatrix} -0.000014 \\ 0.000006 \end{pmatrix} = -\begin{pmatrix} 0.000031 \\ 0.000038 \end{pmatrix}$	$\begin{pmatrix} 1.900677 \\ 0.311219 \end{pmatrix}$

Implementation of Newton's method can require the determination of several partial derivatives. It is permissible to use numerical approximations for the values of these partial derivatives, but care must be taken to determine the proper step size. In higher dimensions it is necessary to use the vector form for the functions and the general methods in Chapter 3 for solving the system (10) for $\Delta\mathbf{P}$.

Algorithm 8.4 [*Newton–Raphson Method for Two Dimensions*] Solve the non-linear system $0 = f_1(x, y), 0 = f_2(x, y)$ given an initial approximation (p_0, q_0).

Cond := 0, Max := 99, Delta := 10^{-5}, Epsilon := 10^{-5}, Small := 10^{-5}

INPUT P0, Q0   {Point must be close to the solution}
U0 := F₁(P0, Q0), V0 := F₂(P0, Q0)   {Compute the function values}

```
DO  FOR  K := 1 TO Max   UNTIL   Cond ≠ 0
```

$$D(1, 1) := \frac{\partial}{\partial x}F_1(P0, Q0), \; D(1, 2) := \frac{\partial}{\partial y}F_1(P0, Q0) \qquad \{\text{Compute the partial}$$

$$D(2, 1) := \frac{\partial}{\partial x}F_2(P0, Q0), \; D(2, 2) := \frac{\partial}{\partial y}F_2(P0, Q0) \qquad \text{derivatives}\}$$

```
        Det := D(1, 1)*D(2, 2) − D(1, 2)*D(2, 1)            {Compute the determinant}

        IF Det = 0 THEN                                      {Check division
              DP:0, DQ := 0, Cond := 1                            by zero}
          ELSE
              DP := [U0*D(2, 2) − V0*D(1, 2)]/Det           {Solve the
              DQ := [V0*D(1, 1) − U0*D(2, 1)]/Det           linear system}

        P1 := P0 − DP, Q1 := Q0 − DQ                         {New iterates and
        U1 := F₁(P1, Q1), V1 := F₂(P1, Q1)                  function values}

        RelErr := [|DP|+|DQ|]/[|P1|+|Q1|+Small]             {Relative error}
        FnZero := |U1|+|V1|
        IF  RelErr<Delta  AND  FnZero<Epsilon  THEN         {Check for
              IF Cond ≠ 1 THEN Cond := 2                    convergence}

        P0 := P1, Q0 := Q1, U0 := U1, V0 := V1              {Update values}

PRINT 'The current k-th iterate is ' P1, Q1
PRINT 'The function values are ' U1, V1
IF Cond=0 THEN   PRINT 'The max. number of iterations was exceeded.'
IF Cond=1 THEN   PRINT 'Division by zero was encountered.'
IF Cond=2 THEN   PRINT 'The solution was found with the desired tolerances.'
```

Exercises for Newton's Method for Systems

1. Consider the nonlinear system

$$0 = f_1(x, y) = x^2 - y,$$
$$0 = f_2(x, y) = y^2 - x.$$

(a) Verify that the solutions are $(0, 0)$ and $(1, 1)$.

(b) Start with $(p_0, q_0) = (0.25, -0.25)$ and use Newton's method to find (p_1, q_1) and (p_2, q_2).

(c) Start with $(p_0, q_0) = (0.9, 1.25)$ and use Newton's method to find (p_1, q_1) and (p_2, q_2).

2. Consider the nonlinear system

$$0 = f_1(x, y) = x^2 + y^2 - 2,$$
$$0 = f_2(x, y) = x^2 - y.$$

(a) Verify that the solutions are $(1, 1)$ and $(-1, 1)$.
(b) Start with $(p_0, q_0) = (0.8, 0.75)$ and use Newton's method to find (p_1, q_1) and (p_2, q_2).
(c) Start with $(p_0, q_0) = (-1.25, 0.75)$ and use Newton's method to find (p_1, q_1) and (p_2, q_2).

3. Consider the nonlinear system

$$0 = f_1(x, y) = x^2 + y^2 - 2,$$
$$0 = f_2(x, y) = xy - 1.$$

(a) Verify that the solutions are $(1, 1)$ and $(-1, -1)$.
(b) What difficulties might arise if we try to use Newton's method to find the solutions?

4. Show that Newton's method for two equations can be written in fixed-point iteration form

$$x = g_1(x, y) \qquad y = g_2(x, y).$$

where $g_1(x, y)$ and $g_2(x, y)$ are given by

$$g_1(x, y) = x - \frac{f_1(x, y) \dfrac{\partial}{\partial y} f_2(x, y) - f_2(x, y) \dfrac{\partial}{\partial y} f_1(x, y)}{D(x, y)},$$

$$g_2(x, y) = y - \frac{f_2(x, y) \dfrac{\partial}{\partial x} f_1(x, y) - f_1(x, y) \dfrac{\partial}{\partial x} f_2(x, y)}{D(x, y)},$$

and $D(x, y)$ is the determinant of the Jacobian

$$D(x, y) = |J(x, y)| = \frac{\partial}{\partial x} f_1(x, y) \frac{\partial}{\partial y} f_2(x, y) - \frac{\partial}{\partial x} f_2(x, y) \frac{\partial}{\partial y} f_1(x, y).$$

5. Write a report on the modified Newton method for systems of nonlinear equations.

In Exercises 6–11, use the system of equations $0 = f_1(x, y)$ and $0 = f_2(x, y)$.
(a) Start with the given point (p_0, q_0) and use Newton's method to find (p_1, q_1) and (p_2, q_2).
(b) Use a computer program and find all the solutions.

6. $0 = 2xy - 3, \quad 0 = x^2 - y - 2$
(a) Start with $(p_0, q_0) = (1.5, 0.9)$.

7. $0 = x^2 + 4y^2 - 4, \quad 0 = x^2 - 2x - y + 1$
(a) I. Start with $(p_0, q_0) = (1.5, 0.5)$.
II. Start with $(p_0, q_0) = (-0.25, 1.1)$.

8. $0 = 3x^2 - 2y^2 - 1, \quad 0 = x^2 - 2x + y^2 + 2y - 8$
(a) I. Start with $(p_0, q_0) = (-1.0, 1.0)$.
II. Start with $(p_0, q_0) = (3.0, -3.4)$.

9. $0 = -x + y^2 - 2, \quad 0 = x^3 - 3x^2 + 4x - y$
(a) I. Start with $(p_0, q_0) = (0.5, 1.2)$.
II. Start with $(p_0, q_0) = (-0.25, -1.30)$.

10. $0 = 2x^3 - 12x - y - 1,\quad 0 = 3y^2 - 6y - x - 3$

(a)   I. $(p_0, q_0) = (2.5, 2.5)$.
III. $(p_0, q_0) = (0.0, 0.0)$.
   V. $(p_0, q_0) = (-2.5, 0.0)$.

II. $(p_0, q_0) = (2.5, -1.0)$.
IV. $(p_0, q_0) = (-2.5, 2.5)$.
VI. $(p_0, q_0) = (0.0, 2.5)$.

11. $0 = 3x^2 - 2y^2 - 1,\quad 0 = x^2 - 2x + 2y - 8$

(a)   I. $(p_0, q_0) = (2.5, 3.0)$.
III. $(p_0, q_0) = (5.6, -7.0)$.

II. $(p_0, q_0) = (-1.6, 1.6)$.
IV. $(p_0, q_0) = (-3.0, 3.6)$.

12. Use Newton's method to find all nine solutions to

$$0 = 7x^3 - 10x - y - 1 \qquad 0 = 8y^3 - 11y + x - 1$$

Use the starting points $(0, 0)$, $(1, 0)$, $(0, 1)$, $(-1, 0)$, $(0, -1)$, $(1, 1)$, $(-1, 1)$, $(1, -1)$, and $(-1, -1)$. Compare with Exercise 11 of Section 8.2.

In Exercises 13–18, extend Newton's method to three dimensions and find all solutions of the system.

13. $0 = x^2 - x + y^2 + z^2 - 5,$

$0 = x^2 + y^2 - y + z^2 - 4,$

$0 = x^2 + y^2 + z^2 + z - 6$

I. Start with $(p_0, q_0, r_0) = (-0.8, 0.2, 1.8)$.
II. Start with $(p_0, q_0, r_0) = (1.2, 2.2, -0.2)$.

14. $0 = x^2 - x + 2y^2 + yz - 10,$

$0 = 5x - 6y + z,$

$0 = z - x^2 - y^2$

Start with $(p_0, q_0, r_0) = (1.1, 1.5, 3.5)$.

15. $0 = x^2 + y^2 - z^2 - 1,$

$$0 = \frac{x^2}{4} + \frac{y^2}{9} + \frac{z^2}{4} - 1,$$

$0 = x^2 + (y - 1)^2 - 1$

I. Start with $(p_0, q_0, r_0) = (0.8, 1.6, 1.5)$.
II. Start with $(p_0, q_0, r_0) = (0.8, 1.6, -1.5)$.
III. Start with $(p_0, q_0, r_0) = (-0.8, 1.6, 1.5)$.
IV. Start with $(p_0, q_0, r_0) = (-0.8, 1.6, -1.5)$.

16. $0 = (x + 1)^2 + (y + 1)^2 - z,$

$0 = (x - 1)^2 + y^2 - z,$

$0 = 4x^2 + 2y^2 + z^2 - 16$

17. $0 = 2x^2 + y^2 - 4z - 4,$

$0 = 3x^2 - 2y^2 - 5z,$

$0 = 4x^2 - 8x + 2y^2 + 4z^2$

18. $0 = 9x^2 + 36y^2 + 4z^2 - 36,$

$0 = x^2 - 2y^2 - 20z,$

$0 = 16x - x^3 - 2y^2 - 16z^2$

8.4 Eigenvectors and Eigenvalues

Let A be a square matrix of dimension N by N and $\mathbf{X}$ be a vector of dimension N. The product $\mathbf{Y} = A\mathbf{X}$ can be considered as a linear transformation from N-dimensional space into itself. We want to find all nonzero vectors $\mathbf{X}$ and scalars λ such that

(1) $$A\mathbf{X} = \lambda\mathbf{X}.$$

In general, the scalars λ and vectors $\mathbf{X}$ can involve complex numbers. For simplicity, the examples we present will involve real numbers. The product $A\mathbf{X}$ in (1) transforms $\mathbf{X}$ into a scalar multiple of $\mathbf{X}$.

Definition 8.7 A nonzero vector $\mathbf{V}$ with the property that $A\mathbf{V} = \lambda\mathbf{V}$ is called an *eigenvector* of A and λ is called an *eigenvalue* of A that corresponds to $\mathbf{V}$.

The identity matrix of dimension N that was introduced in Chapter 3 can be used to write equation (1) in the form $A\mathbf{X} = \lambda I\mathbf{X}$. Therefore, if λ and $\mathbf{X}$ are solutions of (1), they are solutions of

(2) $$(A - \lambda I)\mathbf{X} = 0.$$

The significance of equation (2) is that the product of the matrix $(A - \lambda I)$ and the nonzero vector $\mathbf{X}$ is the zero vector! So far we have avoided these situations. The material in Chapter 3 depended on Theorem 3.5, which states that if $\det(A - \lambda I) \neq 0$, then the only solution to (2) is the zero vector. Therefore, equation (2) has a nonzero solution $\mathbf{X}$ if and only if λ is a solution of the characteristic equation

(3) $$\det(A - \lambda I) = 0.$$

One method to solve the eigenproblem is first to determine the eigenvalues by solving equation (3); then each eigenvalue is used in equation (2) and its corresponding eigenvector(s) are found. When the determinant in (3) is expanded, it becomes a polynomial of degree N, which is called the *characteristic polynomial*. Therefore, there are at most N eigenvalues of the matrix A. The case of repeated eigenvalues is an advanced topic and can be found in books on linear algebra (see References 6, 80, 113).

Example 8.8 Find the eigenvalues and their corresponding eigenvectors for the matrix

$$A = \begin{pmatrix} 3 & -1 & 0 \\ -1 & 2 & -1 \\ 0 & -1 & 3 \end{pmatrix}.$$

Solution. The characteristic equation $\det(A - \lambda I) = 0$ is

$$\begin{vmatrix} 3 - \lambda & -1 & 0 \\ -1 & 2 - \lambda & -1 \\ 0 & -1 & 3 - \lambda \end{vmatrix} = -\lambda^3 + 8\lambda^2 - 19\lambda + 12$$

$$= -(\lambda - 1)(\lambda - 3)(\lambda - 4) = 0.$$

Therefore, $\lambda = 1$ or $\lambda = 3$ or $\lambda = 4$.

Case (i)   Using $\lambda = 1$ in (2) yields the linear system

$$2x_1 - x_2 \qquad = 0$$
$$-x_1 + x_2 - \quad x_3 = 0$$
$$-x_2 + 2x_3 = 0.$$

When the second equation is replaced by the sum of the first equation plus two times the second equation plus the third equation, we see that this is equivalent to the system of two equations

$$2x_1 - x_2 \qquad = 0$$
$$x_2 - 2x_3 = 0.$$

If we choose $x_2 = 2a$, where a is an arbitrary constant, the eigenvector corresponding to $\lambda = 1$ is $\mathbf{V} = (a, 2a, a) = a(1, 2, 1)$.

Case (ii)   Using $\lambda = 3$ in (2) yields the linear system

$$-x_2 \qquad = 0$$
$$-x_1 - x_2 - x_3 = 0$$
$$-x_2 \qquad = 0.$$

This is equivalent to the systems of two equations

$$x_1 + x_2 + x_3 = 0$$
$$x_2 \qquad = 0.$$

If we choose $x_1 = b$, where b is an arbitrary nonzero constant, the eigenvector corresponding to $\lambda = 3$ is $\mathbf{V} = (b, 0, -b) = b(1, 0, -1)$.

Case (iii) Using $\lambda = 4$ in (2) yields the linear system

$$-x_1 - \quad x_2 \qquad = 0$$
$$-x_1 - 2x_2 - x_3 = 0$$
$$-x_2 - x_3 = 0.$$

This is equivalent to the system of two equations

$$x_1 + x_2 \qquad = 0$$
$$x_2 + x_3 = 0.$$

If we choose $x_3 = c$, where c is an arbitrary nonzero constant, then the eigenvector corresponding to $\lambda = 4$ is $\mathbf{V} = (c, -c, c) = c(1, -1, 1)$.

The example above serves to show how eigenvalues can be found when the dimension N is small. There are two shortcomings of this method of solution. First, the roots (possibly complex) of the characteristic polynomial must be found. Second, the nonzero solutions of the homogeneous linear system $(A - \lambda I)\mathbf{X} = 0$ must be found. Although it is possible to design a computer algorithm to solve a homogeneous linear system, it will not work well in all cases because of accumulated round-off errors. Instead, we will develop the power method, which is useful for finding the largest eigenvalue and its corresponding eigenvector.

It should be evident that if $\mathbf{V}$ is an eigenvector for A that corresponds to the eigenvalue λ, then the vector $c\mathbf{V}$ is also an eigenvector for A. This is shown by writing

$$(4) \qquad A(c\mathbf{V}) = c(A\mathbf{V}) = c(\lambda\mathbf{V}) = \lambda(c\mathbf{V}).$$

Definition 8.8 If λ_1 is an eigenvalue of A that is larger in absolute value than any other eigenvalue, it is called the *dominant eigenvalue*. The eigenvector $\mathbf{V}_1$ corresponding to λ_1 is called a *dominant eigenvector*.

For Example 8.8 the dominant eigenvalue is $\lambda_1 = 4$ and the dominant eigenvector is $\mathbf{V}_1 = (a, -a, a)$.

Definition 8.9 An eigenvector $\mathbf{V}$ is said to be *normalized* if the coordinate of largest magnitude is equal to unity (i.e., the largest coordinate in $\mathbf{V}$ is the number 1).

It is easy to normalize an eigenvector $(x_1, x_2, \ldots, x_N)$ by forming the new vector

$$(5) \qquad \mathbf{V} = c^{-1}(x_1, x_2, \ldots, x_N),$$

where c is the coordinate of largest magnitude.

The Power Method

Suppose that λ is an eigenvalue of A that is larger than any other eigenvalue, and that there is a unique normalized eigenvector $\mathbf{V}$ that corresponds to λ. This eigenvector $\mathbf{V}$ and its eigenvalue λ can be found by the following iterative procedure. Start with the vector

$$(6) \qquad \mathbf{V}_0 = (1, 1, \ldots, 1).$$

Generate the sequence $\{\mathbf{V}_k\}$ recursively, using

$$(7) \qquad \mathbf{U}_k = A\mathbf{V}_k \quad \text{and} \quad \mathbf{V}_{k+1} = c_{k+1}^{-1}\mathbf{U}_k,$$

where c_{k+1} is the coordinate of $\mathbf{U}_k$ of largest magnitude. The sequences $\{\mathbf{V}_k\}$ and $\{c_k\}$ will converge to the $\mathbf{V}$ and λ, respectively:

$$(8) \qquad \lim_{k \to \infty} \mathbf{V}_k = \mathbf{V} \quad \text{and} \quad \lim_{k \to \infty} c_k = \lambda.$$

Example 8.9 Use the power method to find the dominant eigenvalue and eigenvector for the matrix

$$A = \begin{pmatrix} 0 & 11 & -5 \\ -2 & 17 & -7 \\ -4 & 26 & -10 \end{pmatrix}.$$

Solution. Start with $V_0 = (1, 1, 1)$ and use the formulas in (7) to generate the sequence of vectors $\{V_k\}$ and constants $\{c_k\}$. The first iteration produces

$$\begin{pmatrix} 0 & 11 & -5 \\ -2 & 17 & -7 \\ -4 & 26 & -10 \end{pmatrix} \begin{pmatrix} 1 \\ 1 \\ 1 \end{pmatrix} = \begin{pmatrix} 6 \\ 8 \\ 12 \end{pmatrix} = 12 \begin{pmatrix} \frac{1}{2} \\ \frac{2}{3} \\ 1 \end{pmatrix} = c_1 V_1.$$

The second iteration produces

$$\begin{pmatrix} 0 & 11 & -5 \\ -2 & 17 & -7 \\ -4 & 26 & -10 \end{pmatrix} \begin{pmatrix} \frac{1}{2} \\ \frac{2}{3} \\ 1 \end{pmatrix} = \begin{pmatrix} \frac{7}{3} \\ \frac{10}{3} \\ \frac{16}{3} \end{pmatrix} = \frac{16}{3} \begin{pmatrix} \frac{7}{16} \\ \frac{5}{8} \\ 1 \end{pmatrix} = c_2 V_2.$$

Iteration generates the sequences of vectors and constants

$$V_1 = \begin{pmatrix} \frac{1}{2} \\ \frac{2}{3} \\ 1 \end{pmatrix}, \quad V_2 = \begin{pmatrix} \frac{7}{16} \\ \frac{5}{8} \\ 1 \end{pmatrix}, \quad V_3 = \begin{pmatrix} \frac{5}{12} \\ \frac{11}{18} \\ 1 \end{pmatrix}, \quad \ldots$$

$$c_1 = 12, \qquad c_2 = \frac{16}{3}, \qquad c_3 = \frac{9}{2}, \quad \ldots.$$

The sequence of vectors converges to $V = (0.4, 0.6, 1.0)^T$ and the sequence of constants converges to $\lambda = 4$ (see Table 8.7). It can be proven that the rate of convergence is linear.

Application to a Markov Process

Suppose that the vector $V = (v_1, v_2, \ldots, v_N)$ represents the number of people who use different consumer products (i.e., there are v_k people using the kth consumer product). After a fixed time interval (e.g., one year), people may either change the type product or keep using the same one. Let the number $a_{i,j}$ be the probability that a person who uses product j will switch to product i at the end of the time interval. Then the vector AV is the new distribution of the number of people using the products.

To illustrate this Markov process, let V_0 be the distribution of the number of people in a certain city who own the following three types of vehicles: sports

TABLE 8.7 Power method used in Example 8.9 to find the normalized dominant eigenvector and its corresponding eigenvalue

k	$A\mathbf{V}_k$	$=$	u_k	$=$	$c_{k+1}\mathbf{V}_{k+1}$
2	$A * \begin{pmatrix} 0.437500 \\ 0.625000 \\ 1.00000 \end{pmatrix}$	$=$	$\begin{pmatrix} 1.875000 \\ 2.750000 \\ 4.500000 \end{pmatrix}$	$= 4.500000$	$\begin{pmatrix} 0.416667 \\ 0.611111 \\ 1.0 \end{pmatrix}$
3	$A * \begin{pmatrix} 0.416667 \\ 0.611111 \\ 1.0 \end{pmatrix}$	$=$	$\begin{pmatrix} 1.722222 \\ 2.555556 \\ 4.222222 \end{pmatrix}$	$= 4.222222$	$\begin{pmatrix} 0.407895 \\ 0.605263 \\ 1.0 \end{pmatrix}$
4	$A * \begin{pmatrix} 0.407895 \\ 0.605263 \\ 1.0 \end{pmatrix}$	$=$	$\begin{pmatrix} 1.657895 \\ 2.473684 \\ 4.105263 \end{pmatrix}$	$= 4.105263$	$\begin{pmatrix} 0.403846 \\ 0.602564 \\ 1.0 \end{pmatrix}$
5	$A * \begin{pmatrix} 0.403846 \\ 0.602564 \\ 1.0 \end{pmatrix}$	$=$	$\begin{pmatrix} 1.628205 \\ 2.435897 \\ 4.051282 \end{pmatrix}$	$= 4.051282$	$\begin{pmatrix} 0.401899 \\ 0.601266 \\ 1.0 \end{pmatrix}$
6	$A * \begin{pmatrix} 0.401899 \\ 0.601266 \\ 1.0 \end{pmatrix}$	$=$	$\begin{pmatrix} 1.613924 \\ 2.417722 \\ 4.025316 \end{pmatrix}$	$= 4.025316$	$\begin{pmatrix} 0.400943 \\ 0.600629 \\ 1.0 \end{pmatrix}$
7	$A * \begin{pmatrix} 0.400943 \\ 0.600629 \\ 1.0 \end{pmatrix}$	$=$	$\begin{pmatrix} 1.606918 \\ 2.408805 \\ 4.012579 \end{pmatrix}$	$= 4.012579$	$\begin{pmatrix} 0.400470 \\ 0.600313 \\ 1.0 \end{pmatrix}$

car, sedan, and van. Assume that the probability that a sports car owner will switch to a sedan or a van is 0.3 and 0.3, respectively. The probability that a sedan owner will switch to a sports car or van is 0.3 and 0.2, respectively. The probability that a van owner will switch to a sports car or sedan is 0.1 and 0.3, respectively. Initially, the number of people owning a sports car, sedan, or van is 200, 600, and 400, respectively. If the time interval is one year, then the distribution at the start of the second year will be $A\mathbf{V}_0 = \mathbf{V}_1$:

$$\begin{pmatrix} 0.4 & 0.3 & 0.1 \\ 0.3 & 0.5 & 0.3 \\ 0.3 & 0.2 & 0.6 \end{pmatrix} \begin{pmatrix} 200 \\ 600 \\ 400 \end{pmatrix} = \begin{pmatrix} 300 \\ 480 \\ 420 \end{pmatrix} = \mathbf{V}_1.$$

Assume that the probabilities do not change for the next time interval of one year; then at the start of the third year the distribution will be

$$\begin{pmatrix} 0.4 & 0.3 & 0.1 \\ 0.3 & 0.5 & 0.3 \\ 0.3 & 0.2 & 0.6 \end{pmatrix} \begin{pmatrix} 300 \\ 480 \\ 420 \end{pmatrix} = \begin{pmatrix} 306 \\ 456 \\ 438 \end{pmatrix} = V_2.$$

If the iteration is continued, the sequence of vectors will converge to the eigen-vector $V = (300, 450, 450)$, that is,

$$\begin{pmatrix} 0.4 & 0.3 & 0.1 \\ 0.3 & 0.5 & 0.3 \\ 0.3 & 0.2 & 0.6 \end{pmatrix} \begin{pmatrix} 300 \\ 450 \\ 450 \end{pmatrix} = \begin{pmatrix} 300 \\ 450 \\ 450 \end{pmatrix}.$$

This means that after several years, the distribution of cars in the city will reach a steady-state 300 sports cars, 450 sedans, and 450 vans.

Exercises for Eigenvectors and Eigenvalues

In Exercises 1–4:
   (a) Verify that V_1, V_2, and V_3 are eigenvectors of the matrix A.
   (b) Find the eigenvalues λ_1, λ_2, and λ_3 that correspond to V_1, V_2, and V_3.
   (c) Multiply AV_k and determine which vector in part (a) is the dominant eigen-vector? Find the normalized dominant eigenvectors.
   (d) Use a computer to find the normalized dominant eigenvector.

1. $A = \begin{pmatrix} 7 & 6 & -3 \\ -12 & -20 & 24 \\ -6 & -12 & 16 \end{pmatrix}$

   $V_1 = (2, -1, 0)$ $V_2 = (1, -2, -1)$ $V_3 = (-3, 4, 2)$

2. $A = \begin{pmatrix} -14 & -30 & 42 \\ 24 & 49 & -66 \\ 12 & 24 & -32 \end{pmatrix}$

   $V_1 = (1, -2, -1)$ $V_2 = (-3, 4, 2)$ $V_3 = (2, -1, 0)$

3. $A = \begin{pmatrix} 16 & 30 & -42 \\ -24 & -47 & 66 \\ -12 & -24 & 34 \end{pmatrix}$

   $V_1 = (1, -2, -1)$ $V_2 = (-3, 4, 2)$ $V_3 = (2, -1, 0)$

4. $A = \begin{pmatrix} -12 & -72 & -59 \\ 5 & 29 & 23 \\ -2 & -12 & -9 \end{pmatrix}$

$\mathbf{V}_1 = (-7, 3, -2)$ $\mathbf{V}_2 = (5, -2, 1)$ $\mathbf{V}_3 = (1, -1, 1)$

In Exercises 5–8, the three eigenvalues λ_1, λ_2, and λ_3 for the matrix A are given.
  (a) Follow the three cases of Example 8.8 and find the corresponding eigenvectors $\mathbf{V}_1$, $\mathbf{V}_2$, and $\mathbf{V}_3$.
  (b) Multiply $A\mathbf{V}_k$ and determine which vector in part (a) is the dominant eigenvector? Find the normalized dominant eigenvectors.
  (d) Use a computer to find the normalized dominant eigenvector.

5. $A = \begin{pmatrix} -2 & 1 & 1 \\ -6 & 1 & 3 \\ -12 & -2 & 8 \end{pmatrix}$ **6.** $A = \begin{pmatrix} -3 & -7 & -2 \\ 12 & 20 & 6 \\ -20 & -31 & -9 \end{pmatrix}$

$\lambda_1 = 4$ $\lambda_2 = 2$ $\lambda_3 = 1$ $\lambda_1 = 5$ $\lambda_2 = 2$ $\lambda_3 = 1$

7. $A = \begin{pmatrix} 49 & -24 & -6 \\ 88 & -43 & -12 \\ 28 & -14 & -2 \end{pmatrix}$ **8.** $A = \begin{pmatrix} -49 & 25 & 8 \\ -94 & 48 & 16 \\ -28 & 14 & 5 \end{pmatrix}$

$\lambda_1 = -2$ $\lambda_2 = 5$ $\lambda_3 = 1$ $\lambda_1 = -2$ $\lambda_2 = 5$ $\lambda_3 = 1$

9. Find the eigenvalues and eigenvectors of

(a) $\begin{pmatrix} 1 & 2 \\ 3 & 2 \end{pmatrix}$ (b) $\begin{pmatrix} -1 & -3 \\ 4 & 6 \end{pmatrix}$

10. Find the eigenvalues and eigenvectors of

(a) $\begin{pmatrix} 1 & 3 \\ 3 & 1 \end{pmatrix}$ (b) $\begin{pmatrix} -2 & 3 \\ 3 & -2 \end{pmatrix}$

11. Suppose that $\mathbf{V}_1$ is an eigenvector of A that corresponds to the eigenvalue $\lambda_1 = 4$. Prove that $\lambda = 16$ is an eigenvalue of A^2.

12. Suppose that $\mathbf{V}$ is an eigenvector of A that corresponds to the eigenvalue $\lambda = c$. Prove that $\mathbf{V}$ is an eigenvector of A^{-1} that corresponds to the eigenvalue $\lambda = 1/c$.

13. *Translation of Eigenvalues.* Fix a constant b, and consider the eigenvalue problem for the matrix $B = A - bI$. Prove that if λ_j is an eigenvalue of A, then $\lambda_j - b$ is an eigenvalue of B.

14. (a) Show that a second eigenvalue for A can be found when the power method is applied to the matrix $(A - \lambda_1 I)$, and λ_1 is the dominant eigenvalue of A.
  (b) Apply this technique and use the power method to find a second eigenvalue and eigenvector for one of the matrices in Exercises 1–8.

15. Write a report on the inverse power method for finding the eigenvector corresponding to the smallest eigenvalue.

16. Write a report on the QR algorithm for finding eigenvectors.

17. Suppose that the initial distribution of the number of people owning the three types of vehicles: sports car, sedan, and van is $V_0 = (200, 400, 400)$. Assume that the probability that a sports car owner will switch to a sedan or a van is 0.3 and 0.1, respectively. The probability that sedan owner will switch to a sports car or van is 0.2 and 0.3, respectively. The probability that a van owner will switch to a sports car or sedan is 0.1 and 0.1, respectively.
  (a) Find the distribution at the start of the second year.
  (b) Find the distribution at the start of the third year.
  (c) What is the steady-state distribution?

18. Suppose that the initial distribution of the number of people owning the three types of vehicles: sports car, sedan, and van is $V_0 = (600, 600, 800)$. Assume that the probability a sports car owner will switch to a sedan or a van is 0.4 and 0.2, respectively. The probability that sedan owner will switch to a sports car or van is 0.2 and 0.2, respectively. The probability that a van owner will switch to a sports car or sedan is 0.1 and 0.1, respectively.
  (a) Find the distribution at the start of the second year.
  (b) Find the distribution at the start of the third year.
  (c) What is the steady-state distribution?

19. State College offers three majors: arts, humanities, and science. In each year:

10% of the arts majors switch to humanities and
40% of the arts majors switch to science,

20% of the humanities majors switch to arts and
30% of the humanities majors switch to science,

10% of the science majors switch to arts and
30% of the science majors switch to humanities.

Initially there are 3,000 arts majors, 3,000 humanities majors, and 4,000 science majors. Assume that each graduate recruits a new person to take their place in the same major.
  (a) Find the distribution at the start of the second year.
  (b) Find the distribution at the start of the third year.
  (c) What is the steady-state distribution?

9

Solution of Differential Equations

9.1 Introduction to Differential Equations

Consider the equation

(1)
$$\frac{dy}{dt} = 1 - \exp(-t).$$

It is a differential equation because it involves the derivative dy/dt of the "unknown function" $y = y(t)$. Only the independent variable t appears on the right side of equation (1); hence a solution is an antiderivative of $1 - \exp(-t)$. The rules for integration can be used to find $y(t)$:

(2)
$$y(t) = t + \exp(-t) + C,$$

where C is the constant of integration. All the functions in (2) are solutions of (1) because they satisfy the requirement that $y'(t) = 1 - \exp(-t)$. They form the family of curves in Figure 9.1.

Integration was the technique used to find the explicit formula for the functions in (2), and Figure 9.1 emphasizes that there is one degree of freedom involved in the solution, namely the constant of integration C. By varying the value of C one "moves the solution curve" up or down and a particular curve can be found that will pass through any desired point.

The secrets of the world are seldom observed as explicit formulas. Instead, one usually measures how a change in one variable affects another variable.

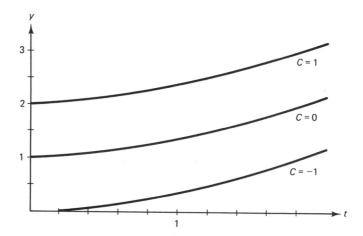

Figure 9.1. The solution curves $y(t) = t + \exp(-t) + C$.

When this is translated into a mathematical model, the result is an equation involving the rate of change of the unknown function and the independent and/or dependent variable.

Consider the temperature $y(t)$ of a cooling object. It might be conjectured that the rate of change of the temperature of the body is related to the temperature difference between its temperature and that of the surrounding medium. Experimental evidence verifies this conjecture. Newton's law of cooling asserts that the rate of change is directly proportional to the difference in these temperatures. If A is the temperature of the surrounding medium and $y(t)$ is the temperature of the body at time t, then

(3)
$$\frac{dy}{dt} = -k(y - A),$$

where k is a positive constant. The negative sign is required because dy/dt will be negative when the temperature of the body is greater than the temperature of the medium.

If the temperature of the object is known at time $t = 0$, we call this an initial condition and include this information in the statement of the problem. Usually we are asked to solve

(4)
$$\frac{dy}{dt} = -k(y - A) \quad \text{with} \quad y(0) = y_0.$$

The technique of separation of variables can be used to find the solution

(5)
$$y = A + (y_0 - A) \exp(-kt).$$

For each choice of y_0, the solution curve will be different, and there is no simple way to move one curve around to get another one. The initial value is a point where the desired solution is "nailed down." Several solution curves are

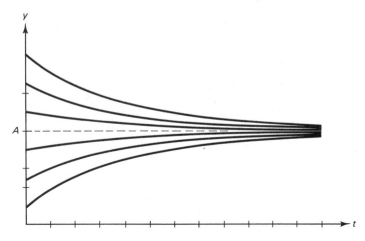

Figure 9.2. The solution curves $y = A + [y_0 - A] \exp(-kt)$ for Newton's law of cooling (and warming).

shown in Figure 9.2, and it can be observed that as t gets large, the temperature of the object approaches room temperature. If $y_0 < A$, then the body is warming instead of cooling.

The Initial Value Problem

Definition 9.1 [Initial value problem] Let $f(t, y)$ be a continuous function of t and y. The *initial value problem* (I.V.P.) is to solve

$$(6) \qquad \frac{dy}{dt} = f(t, y) \quad \text{with} \quad y(t_0) = y_0.$$

A solution to the I.V.P. is a differentiable function $y(t)$ with the property that when t and $y(t)$ are substituted in $f(t, y)$ the result is equal to the derivative $y'(t)$, that is,

$$(7) \qquad y'(t) = f(t, y(t)) \quad \text{and} \quad y(t_0) = y_0.$$

Notice that the solution curve $y = y(t)$ must pass through the initial data point (t_0, y_0).

The Geometric Interpretation

At each point (t, y) in the rectangular region $R: a \leq t \leq b, c \leq y \leq d$, the slope of a solution curve $y = y(t)$ can be found by using the implicit formula $f(t, y(t))$. Hence the values $m_{i,j} = f(t_i, y_j)$ can be computed throughout the rectangle, and each value $m_{i,j}$ represents the slope of the line tangent to a solution curve that passes through the point (t_i, y_j).

A slope field or direction field is a graph that indicates the slopes $\{m_{i,j}\}$ over the region. It can be used to visualize how a solution curves "fits" the slope constraint. To move along a solution curve one must start at the initial point and check the slope field to determine which direction to move. Then take a small step from t_0 to $t_0 + h$ horizontally and move the appropriate vertical distance $hf(t_0, y_0)$ so that the resulting displacement has the required slope. The next point on the solution curve is (t_1, y_1). Repeat the process to continue your journey along the curve. Since a finite number of steps will be used, the method will produce an approximation to the solution.

Example 9.1 The slope field for $y' = (t - y)/2$ over the rectangle $R: 0 \le t \le 5$, $0 \le y \le 3$ is shown in Figure 9.3. The solution curves with the following initial values are also shown:

1. For $y(0) = 1$, the solution is $y(t) = 3 \exp(-t/2) - 2 + t$.
2. For $y(0) = 4$, the solution is $y(t) = 6 \exp(-t/2) - 2 + t$.

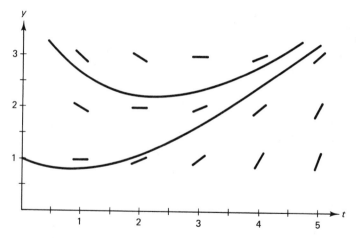

Figure 9.3. The slope field for $y' = f(t, y) = (t - y)/2$.

Definition 9.2 The function $f(t, y)$ is said to satisfy the *Lipschitz condition* in the variable y on the rectangle $R: a \le t \le b, c \le y \le d$, provided that a constant $L > 0$ exists so that

(8)
$$|f(t, y_1) - f(t, y_2)| \le L|y_1 - y_2|$$

whenever $a \le t \le b$ and $c \le y_1 \le y_2 \le d$.

Theorem 9.1 [Existence and uniqueness of a solution] The initial value problem

(9)
$$y' = f(t, y) \quad \text{on } [a, b] \quad \text{with} \quad y(a) = y_0$$

has a unique solution $y(t)$ over the interval $[a, b]$ if $f(t, y)$ is continuous with respect to both variables and satisfies a Lipschitz condition in the variable y on the rectangle $R: a \leq t \leq b, c \leq y \leq d$, where $c \leq y_0 \leq d$.

Proof. See an advanced text on differential equations such as Reference [27].

To apply Theorem 9.1 it will suffice to compute the partial derivative $f_y(t, y)$ and show that its magnitude $|f_y(t, y)|$ is bounded by L on the rectangle R. This is the next result.

Corollary 9.1 The I.V.P. in (9) is well-posed if $f(t, y)$ is differentiable with respect to y and there exists a constant L so that $|f_y(t, y)| \leq L$ throughout the rectangle.

Proof. Hold t fixed and use the Mean Value Theorem to get c_1 with $y_1 < c_1 < y_2$ so that

$$|f(t, y_1) - f(t, y_2)| = |f_y(t, c_1)(y_1 - y_2)|$$
$$= |f_y(t, c_1)| \, |y_1 - y_2| \leq L|y_1 - y_2|.$$

Here $f(t, y)$ satisfies the Lipschitz condition, and the I.V.P. is well-posed.

Let us apply the corollary to the function $f(t, y) = (t - y)/2$ in Example 9.1. The partial derivative is $f_y(t, y) = -\frac{1}{2}$. Hence $|f_y(t, y)| \leq \frac{1}{2}$, and a Lipschitz constant is seen to be $L = \frac{1}{2}$. Therefore, Corollary 9.1 implies that the I.V.P. in Example 9.1 well-posed.

It is important to know the effect of a small change in the initial value $y(t_0) = y_0$. Will a small change in y_0 result in a small change in the solution curve? If the answer to this question is yes, then the I.V.P. is said to be *well-conditioned*.

Definition 9.3 [Conditioning] Consider $y' = f(t, y)$ with $y(t_0) = y_0$ over the interval $[t_0, t_M]$. If $f_y(t, y) \leq 0$ for $t_0 < t < t_M$, then the I.V.P. is said to be well-conditioned over the interval $[t_0, t_M]$. If $f_y(t, y) > 0$ for $t_0 < t < t_M$, then the I.V.P. is ill-conditioned.

If the I.V.P. is well-conditioned, then as t increases, the solution curve that starts with the perturbed value $y_0 + \epsilon$ will approach the curve that started at y_0. The I.V.P. in Example 9.1 is well-conditioned because $f_y(t, y) = -\frac{1}{2} < 0$. Consider two solution curves that arise from different initial conditions:

1. When $y(0) = 1$, the curve is $y = 3 \exp(-t/2) - 2 + t$.
2. When $y(0) = 4$, the curve is $y = 6 \exp(-t/2) - 2 + t$.

The difference between the two curves is $3 \exp(-t/2)$, and this goes to zero as $t \rightarrow \infty$ (see Figure 9.3).

On the other hand, if the I.V.P. is ill-conditioned, then as t increases, the solution curve that starts with the perturbed value $y_0 + \epsilon$ will diverge away from the curve that started at y_0.

Example 9.2 Show that the initial value problem $y' = (y - t)/2$ with $y(t_0) = y_0$ is ill-conditioned over $[0, 5]$.

Solution. Here the function is $f(t, y) = (y - t)/2$ and its partial derivative is $f_y(t, y) = \frac{1}{2}$. Since $f_y(t, y) > 0$, the I.V.P. is ill-conditioned. Consider two solution curves that arise from different initial conditions:

1. When $y(0) = 3$, the curve is $y = \exp(t/2) + 2 + t$.
2. When $y(0) = 4$, the curve is $y = 2\exp(t/2) + 2 + t$.

The difference between the two curves is $\exp(t/2)$ and grows without bound when $t \longrightarrow \infty$. These curves and several other ones are shown in Figure 9.4.

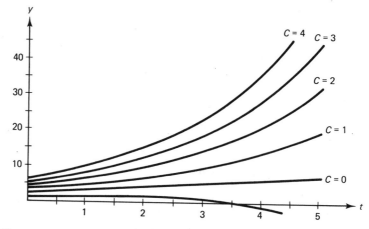

Figure 9.4. Solution curves $y(t) = C \exp(t/2) + 2 + t$ for the ill-conditioned I.V.P. $y' = (y - t)/2$.

Exercises for Introduction to Differential Equations

In Exercises 1–5:
- (a) Show that $y(t)$ is the solution to the differential equation by substituting $y(t)$ and $y'(t)$ into the differential equation $y'(t) = f(t, y(t))$.
- (b) Use Corollary 9.1 to find a Lipschitz constant L for the rectangle R: $0 \leqq t \leqq 3, 0 \leqq y \leqq 5$.
- (c) Is the I.V.P. with $y(0) = 1$ well-posed?
- (d) Is the I.V.P. with $y(0) = 1$ well-conditioned? Why?

1. $y' = t^2 - y, y(t) = C \exp(-t) + t^2 - 2t + 2$
2. $y' = 3y + 3t, y(t) = C \exp(3t) - t - \frac{1}{3}$
3. $y' = -ty, y(t) = C \exp(-t^2/2)$

4. $y' = \exp(-2t) - 2y, y(t) = C \exp(-2t) + t \exp(-2t)$

5. $y' = 2ty^2, y(t) = 1/(C - t^2)$

In Exercises 6-9:
  (a) Draw the slope field $m_{i,j} = f(t_i, y_j)$ over the rectangle $0 \le t \le 4, 0 \le y \le 4$.
  (b) Sketch the indicated solution curves.

6. $y' = f(t, y) = -t/y$. Sketch the solution curves (circles):
$$y(t) = (C - t^2)^{1/2} \quad \text{for} \quad C = 1, 2, 4, 9.$$

7. $y' = f(t, y) = t/y$. Sketch the solution curves (hyperbolas):
$$y(t) = (C + t^2)^{1/2} \quad \text{for} \quad C = -4, -1, 1, 4.$$

8. $y' = f(t, y) = 1/y$. Sketch the solution curves (parabolas):
$$y(t) = (C + 2t)^{1/2} \quad \text{for} \quad C = -4, -2, 0, 2.$$

9. $y' = f(t, y) = y^2$. Sketch the solution curves (hyperbolas):
$$y(t) = \frac{1}{C - t} \quad \text{for} \quad C = 1, 2, 3, 4.$$

10. Here is an example of an initial value problem that has "two solutions": $y' = \frac{3}{2}y^{1/3}$ with $y(0) = 0$.
  (a) Verify that $y(t) = 0$ for $t \ge 0$ is a solution.
  (b) Verify that $y(t) = t^{3/2}$ for $t \ge 0$ is a solution.
  (c) Does this violate Theorem 9.1? Why?

11. Consider the initial value problem
$$y' = (1 - y^2)^{1/2} \qquad y(0) = 0.$$
  (a) Verify that $y(t) = \sin(t)$ is a solution on $[0, \pi/4]$.
  (b) Determine the largest interval over which the solution exists.

12. Show that the definite integral
$$\int_a^b f(t) \, dt$$
can be computed by solving the initial value problem
$$y' = f(t) \quad \text{for} \quad a \le t \le b \quad \text{with} \quad y(a) = 0.$$

In Exercises 13-15, find the solution to the I.V.P.

13. $y' = 3t^2 + \sin(t), \quad y(0) = 2$

14. $y' = 1/(1 + t^2), \quad y(0) = 0$

15. $y' = \exp(-t^2/2), \quad y(0) = 0$. *Hint.* This answer must be expressed as a certain integral.

16. Consider the first-order differential equation
$$y'(t) + p(t)y(t) = q(t).$$
Show that the general solution $y(t)$ can be found by using two special integrals. First define $F(t)$ as follows:
$$F(t) = \exp\left(\int p(t) \, dt\right) \cdot$$

Second, define $y(t)$ as follows:

$$y(t) = \frac{1}{F(t)} \left[\int F(t)q(t)\,dt + C \right].$$

Hint. Differentiate the product $F(t)y(t)$.

17. Consider the decay of a radioactive substance. If $y(t)$ is the amount of substance present at time t, then $y(t)$ decreases and experiments have verified that the rate of change of $y(t)$ is proportional to the amount of undecayed material. Hence the I.V.P. for the decay of a radioactive substance is

$$y' = -ky \quad \text{with} \quad y(0) = y_0.$$

(a) Show that the solution is $y(t) = y_0 \exp(-kt)$.

The half-life of a radioactive substance is the time required for half of an initial amount to decay. The half-life of ^{14}C is 5,730 years.

(b) Find the formula $y(t)$ that gives the amount of ^{14}C present at time t. *Hint.* Find k so that $y(5,730) = 0.5y_0$.

(c) A piece of wood is analyzed and the amount of ^{14}C present is 0.712 of the amount that was present when the tree was alive. How old is the sample of wood?

(d) At a certain instant 10 mg of a radioactive substance is present. After 23 sec only 1 mg is present. What is the half-life of the substance?

9.2 Euler's Method

The reader should be convinced that not all initial value problems can be solved explicitly, and often it is impossible to find a formula for the solution $y(t)$, for example, there is no "closed-form expression" for the solution to $y' = x^3 + y^2$ with $y(0) = 0$. Hence for engineering and scientific purposes it is necessary to have methods for approximating the solution. If a solution with many significant digits is required, then more computing effort and a sophisticated algorithm must be used.

The first approach is called Euler's method and serves to illustrate the concepts involved in the advanced methods. It has limited usage because of the larger error that is accumulated as the process proceeds. However, it is important to study because the error analysis is easier to understand.

Let $[a, b]$ be the interval over which we want to find the solution to the well-posed I.V.P. $y' = f(t, y)$ with $y(a) = y_0$. In actuality, we will *not* find a differentiable function that satisfies the I.V.P. Instead, a set of points $\{(t_k, y_k)\}$ are generated which are used for an approximation [i.e., $y(t_k) \approx y_k$]. How can we proceed to construct a "set of points" that will "satisfy a differential equation approximately"? First we choose the abscissas for the points. For convenience we subdivide the interval $[a, b]$ into M equal subintervals and select the mesh points

(1) $$t_k = a + hk \quad \text{for } k = 0, 1, \ldots, M \quad \text{where} \quad h = \frac{b-a}{M}.$$

The value h is called the *step size*. We now proceed to solve approximately

(2) $$y' = f(t, y) \quad \text{over} \quad [t_0, t_M] \quad \text{with} \quad y(t_0) = y_0.$$

Assume that $y(t)$, $y'(t)$ and $y''(t)$ are continuous and use Taylor's theorem to expand $y(t)$ about $t = t_0$. For each value t, there exists a value c_1 that lies between t_0 and t so that

(3) $$y(t) = y(t_0) + y'(t_0)(t - t_0) + \frac{y''(c_1)(t - t_0)^2}{2}.$$

When $y'(t_0) = f(t_0, y(t_0))$ and $h = t_1 - t_0$ are substituted in equation (3), the result is an expression for $y(t_1)$:

(4) $$y(t_1) = y(t_0) + hf(t_0, y(t_0)) + y''(c_1)\frac{h^2}{2}.$$

If the step size h is chosen small enough, then we may neglect the second-order term (involving h^2) and get

(5) $$y_1 = y_0 + hf(t_0, y_0),$$

which is *Euler's approximation*.

The process is repeated and generates a sequence of points that approximate the solution curve $y = y(t)$. The general step for Euler's method is

(6) $$t_{k+1} = t_k + h, \ y_{k+1} = y_k + hf(t_k, y_k) \quad \text{for } k = 0, 1, \ldots, M - 1.$$

Example 9.3 Use Euler's method to solve approximately the initial value problem

(7) $$y' = Ry \quad \text{over} \quad [0, 1] \quad \text{with} \quad y(0) = y_0 \text{ and } R \text{ constant}.$$

Solution. The step size must be chosen and then the second formula in (6) can be determined for computing the ordinates. This formula is sometimes called a difference equation and in this case it is

(8) $$y_{k+1} = y_k(1 + hR) \quad \text{for } k = 0, 1, \ldots, M - 1.$$

If we trace the solution values recursively, then, we see that

(9)
$$y_1 = y_0(1 + hR)$$
$$y_2 = y_1(1 + hR) = y_0(1 + hR)^2$$
$$\vdots$$
$$y_M = y_{M-1}(1 + hR) = y_0(1 + hR)^M.$$

For most problems there is no explicit formula for determining the solution points and each new point must be computed successively from the previous point. However, for the initial value problem (7) we are fortunate; Euler's method has the explicit solution:

(10) $$t_k = hk \quad y_k = y_0(1 + hR)^k \quad \text{for } k = 0, 1, 2, \ldots, M.$$

Formula (10) can be viewed as the "compound interest" formula, and the Euler approximation gives the future value of a deposit.

Example 9.4 Suppose that $1,000 is deposited and earns 10% interest compounded continuously over 5 years. What is the value at the end of 5 years?

Solution. We choose to use Euler approximations with $h = 1, \frac{1}{12}, \frac{1}{360}$ to approximate $y(5)$ for the I.V.P.:

(11)
$$y' = 0.1y \quad \text{over} \quad [0, 5] \quad \text{with} \quad y(0) = 1,000.$$

Formula (10) with $R = 0.1$ produces Table 9.1.

Table 9.1 Compound interest in Example 9.4

Step Size, h	Number of Iterations, M	Approximation to $y(5)$, y_M
1	5	$1,000\left(1 + \dfrac{0.1}{1}\right)^5 = 1,610.51$
$\frac{1}{12}$	60	$1,000\left(1 + \dfrac{0.1}{12}\right)^{60} = 1,645.31$
$\frac{1}{360}$	1,800	$1,000\left(1 + \dfrac{0.1}{360}\right)^{1800} = 1,648.61$

Think about the different values $y_5, y_{60}, y_{1,800}$ that are used to determine the future value after 5 years. These values are obtained using different step sizes and reflect different amounts of computing effort to obtain an approximation to $y(5)$. The solution to the I.V.P. is $y(5) = 1,000 \exp(0.5) = 1,648.72$. If we did not use the closed-form solution (10), then it would have required 1,800 iterations of Euler's method to obtain $y_{1,800}$, and we still have only about five digits of accuracy in the answer!

If bankers had to approximate the solution to the I.V.P. (7), they would choose Euler's method because of the explicit formula in (10). The more sophisticated methods for approximating solutions do not have an explicit formula for finding y_k, but they will require less computing effort.

The Geometric Description

If you start at the point (t_0, y_0) and compute the value of the slope $m_0 = f(t_0, y_0)$ and move horizontally the amount h and vertically $hf(t_0, y_0)$, then you are moving along the tangent line to $y(t)$ and will end up at the point (t_1, y_1) (see Figure 9.5). Notice that (t_1, y_1) is *not* on the desired solution curve! But this is the approximation that we are generating. Hence we must use (t_1, y_1) as though it were correct and proceed by computing the slope $m_1 = f(t_1, y_1)$ and using it to obtain the next vertical displacement $hf(t_1, y_1)$ to locate (t_2, y_2), and so on.

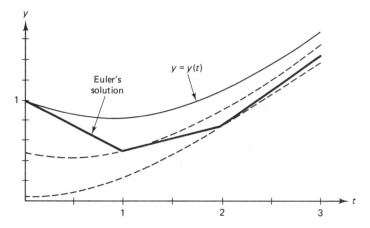

Figure 9.5. Euler's approximations $y_{k+1} = y_k + hf(t_k, y_k)$.

Step Size versus Error

It is evident from Example 9.4 that there can be a significant amount of accumulated error in Euler's method. To understand what is happening, we must start with the concept of *local truncation error* (L.T.E.). When we obtained equation (6) for Euler's method, the neglected term for each step is the local truncation error:

(11) $$y^{(2)}(c_k)\frac{h^2}{2} \qquad \text{(the L.T.E. for Euler's method)}.$$

If the only error at each step is L.T.E., then after M steps the accumulated error would be

(12) $$\sum_{k=1}^{M} y^{(2)}(c_k)\frac{h^2}{2} \approx \frac{b-a}{2}y^{(2)}(c)h \approx O(h^1).$$

There could be more error than indicated by (12) because the L.T.E. assumes that each y_k is exact and $y^{(2)}(c_k)(h^2/2)$ is the only error involved in the kth step. Additional error is introduced because the computed points (t_k, y_k) do not lie on the solution curve (see Figure 9.5).

Definition 9.4 [Global truncation error] If the I.V.P. $y' = f(t, y), y(a) = y_0$ is solved approximately in M steps over the interval $[a, b]$, then the error in the final approximation $y_M \approx y(b)$ is the *global truncation error* (G.T.E.) and is denoted by

(13) $$E_M = y(b) - y_M \qquad \text{(G.T.E. for the I.V.P.)}.$$

The next theorem is important because it states the relationship between G.T.E. and step size. It is used to give us an idea of how much computing effort must be done to obtain an accurate approximation over the interval $[a, b]$.

Theorem 9.2 [G.T.E. for Euler's method] If Euler's method is used to solve the I.V.P. $y' = f(t, y), y(a) = y_0$ over $[a, b]$ and the step size is h, the G.T.E. is of the order $O(h^1)$, that is,

(14) $$E(y(b), h) = y(b) - y_M \approx Ch$$

where C is a constant. The notation $E(y(b), h)$ is used to emphasize the dependence on step size.

Examples 9.5 and 9.6 illustrate the concepts in Theorem 9.2. If approximations are computed using the step sizes h and $h/2$, we should have

(15) $$E(y(b), h) \approx Ch$$

for the larger step size, and

(16) $$E\left(y(b), \frac{h}{2}\right) \approx C\frac{h}{2} \approx \frac{1}{2}Ch \approx \frac{1}{2}E(y(b), h).$$

Hence the idea in Theorem 9.2 is that if the step size in Euler's method is reduced by a factor of $\frac{1}{2}$, we can expect that the overall G.T.E. will be reduced by a factor of $\frac{1}{2}$.

Example 9.5 Use Euler's method to solve the I.V.P.

$$y' = \frac{t - y}{2} \quad \text{on } [0, 3] \text{ with } y(0) = 1.$$

Compare solutions for $h = 1, \frac{1}{2}, \frac{1}{4}$, and $\frac{1}{8}$.

Solution. Figure 9.6 shows graphs of the four Euler solutions and the exact solution curve $y(t) = 3 \exp(-t/2) - 2 + t$. Table 9.2 gives the values for the four solutions at selected abscissas. For the step size $h = 0.25$ the calculations are

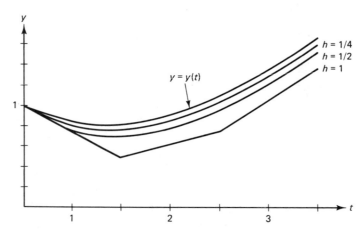

Figure 9.6. Comparison of Euler solutions with different step sizes for $y' = (t - y)/2$ over $[0, 3]$ with $y(0) = 1$.

$$y_1 = 1.0 + 0.25\frac{0.0 - 1.0}{2} = 0.875,$$

$$y_2 = 0.875 + 0.25\frac{0.25 - 0.875}{2} = 0.796875, \quad \text{etc.}$$

This iteration continues until we arrive at the last step:

$$y(3) \approx y_{12} = 1.440573 + 0.25\frac{2.75 - 1.440573}{2} = 1.604252.$$

Table 9.2 Comparison of Euler solutions with different step sizes
for $y' = (t - y)/2$ over $[0, 3]$ with $y(0) = 1$

t_k	$h = 1$	$h = \frac{1}{2}$	$h = \frac{1}{4}$	$h = \frac{1}{8}$	$y(t_k)$ Exact
			y_k		
0	1.0	1.0	1.0	1.0	1.0
0.125				0.9375	0.943239
0.25			0.875	0.886719	0.897491
0.375				0.846924	0.862087
0.50		0.75	0.796875	0.817429	0.836402
0.75			0.759766	0.786802	0.811868
1.00	0.5	0.6875	0.758545	0.790158	0.819592
1.50		0.765625	0.846386	0.882855	0.917100
2.00	0.75	0.949219	1.030827	1.068222	1.103638
2.50		1.211914	1.289227	1.325176	1.359514
3.00	1.375	1.533936	1.604252	1.637429	1.669390

Example 9.6 Compare the G.T.E. when Euler's method is used to solve $y' = (t - y)/2$ over $[0, 3]$ with $y(0) = 1$ using step sizes $1, \frac{1}{2}, \ldots, \frac{1}{64}$.

Solution. Table 9.3 gives the G.T.E. for several step sizes and shows that the error in the approximation to $y(3)$ decreases by about $\frac{1}{2}$ when the step size is reduced by a factor of $\frac{1}{2}$. For the smaller step sizes the conclusion of Theorem 9.2 is easy to see:

$$E(y(3), h) = y(3) - y_M = \mathbf{O}(h^1) \approx Ch \quad \text{where } C = 0.256.$$

Table 9.3 Relation between step size and G.T.E. for Euler solutions to
$y' = (t - y)/2$ over $[0, 3]$ with $y(0) = 1$

Step Size, h	Number of Steps, M	Approximation to $y(3)$, y_M	G.T.E. Error at $t = 3$, $y(3) - y_M$	$\mathbf{O}(h) \approx Ch$ where $C = 0.256$
1	3	1.375	0.294390	0.256
$\frac{1}{2}$	6	1.533936	0.135454	0.128
$\frac{1}{4}$	12	1.604252	0.065138	0.064
$\frac{1}{8}$	24	1.637429	0.031961	0.032
$\frac{1}{16}$	48	1.653557	0.015833	0.016
$\frac{1}{32}$	96	1.661510	0.007880	0.008
$\frac{1}{64}$	192	1.665459	0.003931	0.004

Algorithm 9.1 [*Euler's Method*] Approximate the solution of the initial value problem

$$y' = f(t, y) \quad \text{on } [a, b] \text{ with } y(a) = y_0.$$

INPUT A, B, Y(0)	{Endpoints and initial value}
INPUT M	{Number of steps}
H := [B − A]/M	{Compute the step size}
T(0) := A	{Initialize the variable}

FOR   K = 0   TO   M−1   DO

⎿ Y(K+1) := Y(K) + H*F(T(K),Y(K))   {Euler solution y_{k+1}}

  T(K+1) := A + H*[K+1]   {Compute the mesh point t_{k+1}}

FOR   K = 0   TO   M   DO   {Output}

⎿ PRINT T(K), Y(K)

Exercises for Euler's Method

In Exercises 1–5, use Euler's method to find approximations to the I.V.P.
(a) Compute $y_1, y_2, \ldots, y_M$ for each of the three cases: (i) $h = 0.2, M = 1$, (ii) $h = 0.1, M = 2$, and (iii) $h = 0.05, M = 4$.
(b) Compare the exact solution $y(0.2)$ with the three approximations in part (a).
(c) Does the G.T.E. part (a) behave as expected when h is halved?
(d) Use a computer to carry out the computations on the larger interval $[a, b]$ that is given.

1. (a) Solve $y' = t^2 - y$ over $[0, 0.2]$ with $y(0) = 1$.
   (b) and (c) Compare with $y(t) = -\exp(-t) + t^2 - 2t + 2$.
   (d) Use $[a, b] = [0, 2]$ with $h = 0.2, 0.1, 0.05$.

2. (a) Solve $y' = 3y + 3t$ over $[0, 0.2]$ with $y(0) = 1$.
   (b) and (c) Compare with $y(t) = \frac{4}{3}\exp(3t) - t - \frac{1}{3}$.
   (d) Use $[a, b] = [0, 2]$ with $h = 0.2, 0.1$, and 0.05.

3. (a) Solve $y' = -ty$ over $[0, 0.2]$ with $y(0) = 1$.
   (b) and (c) Compare with $y(t) = \exp(-t^2/2)$.
   (d) Use $[a, b] = [0, 2]$ with $h = 0.2, 0.1$, and 0.05.

4. (a) Solve $y' = \exp(-2t) - 2y$ over $[0, 0.2]$ with $y(0) = \frac{1}{10}$.
   (b) and (c) Compare with $y(t) = \frac{1}{10}\exp(-2t) + t\exp(-2t)$.
   (d) Use $[a, b] = [0, 2]$ with $h = 0.2, 0.1$, and 0.05.

5. (a) Solve $y' = 2ty^2$ over $[0, 0.2]$ with $y(0) = 1$.
   (b) and (c) Compare with $y(t) = 1/(1 - t^2)$.
   (d) Use $[a, b] = [0, 1]$ with $h = 0.1, 0.05$, and 0.025.
   Notice that Euler's method will generate an approximation to $y(1)$ even though the solution curve is not defined at $t = 1$.

6. Consider $y' = 0.12y$ over $[0, 5]$ with $y(0) = 1,000$.
   (a) Apply formula (10) to find Euler's approximation to $y(5)$ using the step sizes $h = 1, \frac{1}{12}$, and $\frac{1}{360}$.
   (b) What is the limit in part (a) when h goes to zero?

7. *Exponential Population Growth.* The population of a certain species grows at a rate that is proportional to the current population, and obeys the I.V.P.

$$y' = 0.02y \quad \text{over } [0, 5] \text{ with } y(0) = 5,000.$$

   (a) Apply formula (10) to find Euler's approximation to $y(5)$ using the step sizes $h = 1, \frac{1}{12}$, and $\frac{1}{360}$.
   (b) What is the limit in part (a) when h goes to zero?

8. *Logistic Population Growth.* The population curve $P(t)$ for the United States is assumed to obey the differential equation for a logistic curve $P' = aP - bP^2$. Let t denote the year past 1900, and let the step be $h = 10$. The values $a = 0.02$ and $b = 0.00004$ produce a model for the population. Find the Euler approximations to $P(t)$ and fill in the table below. Round of each value P_k to the nearest tenth.

Year	t_k	$P(t_k)$ Actual	P_k Euler Approximation
1900	0.0	76.1	76.1
1910	10.0	92.4	89.0
1920	20.0	106.5	____
1930	30.0	123.1	
1940	40.0	132.6	138.2
1950	50.0	152.3	____
1960	60.0	180.7	____
1970	70.0	204.9	202.8
1980	80.0	226.5	____

9. Show that when Euler's method is used to solve the I.V.P.

$$y' = f(t) \quad \text{over } [a, b] \text{ with } y(a) = y_0 = 0,$$

the result is

$$y(b) \approx \sum_{k=0}^{M-1} f(t_k)h,$$

which is a Riemann sum that approximates the definite integral of $f(t)$ taken over the interval $[a, b]$.

10. A skydiver jumps from a plane, and up to the moment he opens the parachute the air resistance is proportional to $v^{3/2}$. Assume that the time interval is $[0, 6]$ and that the differential equation for the downward direction is

$$v' = 32 - 0.032v^{3/2} \quad \text{over } [0, 6] \text{ with } v(0) = 0.$$

Use Euler's method with $h = 0.1$ and find the solution.

11. *Epidemic Model.* The mathematical model for of epidemics is described as follows. Assume that there is a community of L members which contains P infected individuals and Q uninfected individuals. Let $y(t)$ denote the number of infected individuals at time t. For a mild illness such as the common cold, everyone continues to be active, and the epidemic spreads from those who are infected to those uninfected. Since there are PQ possible contacts between these two groups, the rate of change of $y(t)$ is proportional to PQ. Hence the problem can be stated as the I.V.P.

$$y' = ky(L - y) \quad \text{with } y(0) = y_0.$$

Use $L = 25,000$, $k = 0.00003$, and $h = 0.2$ with the initial condition $y(0) = 250$, and compute Euler's approximate solution over $[0, 12]$.

12. Show that Euler's method fails to approximate the solution $y(t) = t^{3/2}$ of the I.V.P.

$$y' = f(t, y) = 1.5y^{1/3} \quad \text{with } y(0) = 0.$$

Justify your answer. What difficulties were encountered?

13. Can Euler's method be used to solve the I.V.P.

$$y' = 1 + y^2 \quad \text{over } [0, 3] \text{ with } y(0) = 0?$$

Hint. The exact solution curve is $y(t) = \tan(t)$.

14. Write a report on the modified Euler method.

9.3 Heun's Method

The next approach is called Heun's method and introduces a new idea for constructing an algorithm to solve the I.V.P.

(1) $$y'(t) = f(t, y(t)) \quad \text{over} \quad [a, b] \quad \text{with} \quad y(t_0) = y_0.$$

To obtain the solution point (t_1, y_1) we can use the fundamental theorem of calculus, and integrate $y'(t)$ over $[t_0, t_1]$ and get

(2) $$\int_{t_0}^{t_1} f(t, y(t)) \, dt = \int_{t_0}^{t_1} y'(t) \, dt = y(t_1) - y(t_0),$$

where the antiderivative of $y'(t)$ is the desired function $y(t)$. When equation (2) is solved for $y(t_1)$, the result is

(3) $$y(t_1) = y(t_0) + \int_{t_0}^{t_1} f(t, y(t)) \, dt.$$

Now a numerical integration method can be used to approximate the definite integral in (3). If the trapezoidal rule is used with step size $h = t_1 - t_0$, then the result is

(4) $$y(t_1) \approx y(t_0) + \frac{h}{2}[f(t_0, y(t_0)) + f(t_1, y(t_1))].$$

Notice that the formula on the right-hand side of (4) involves the yet to be determined value $y(t_1)$. To proceed, we use an estimate for $y(t_1)$. Euler's solution will suffice for this purpose. After it is substituted into (4), the resulting formula for finding (t_1, y_1) is called *Heun's method*:

(5)
$$y_1 = y(t_0) + \frac{h}{2}[f(t_0, y_0) + f(t_1, y_0 + hf(t_0, y_0))].$$

The process is repeated and generates a sequence of points that approximates the solution curve $y = y(t)$. At each step, Euler's method is used as a prediction, and then the trapezoidal rule is used to make a correction to obtain the final value. The general step for Heun's method is

$$p_{k+1} = y_k + hf(t_k, y_k), \qquad t_{k+1} = t_k + h,$$

(6)
$$y_{k+1} = y_k + \frac{h}{2}[f(t_k, y_k) + f(t_{k+1}, p_{k+1})].$$

Notice the role played by differentiation and integration in Heun's method. Draw the line tangent to the solution curve $y = y(t)$ at the point (t_0, y_0), and use it to find the predicted point (t_1, p_1). Now look at the graph $z = f(t, y(t))$ and consider the points (t_0, f_0) and (t_1, f_1), where $f_0 = f(t_0, y_0)$, $f_1 = f(t_1, p_1)$. The area of the trapezoid with vertices (t_0, f_0) and (t_1, f_1) is an approximation to the integral in (3), which is used to obtain the final value in equation (5). The graphs are shown in Figure 9.7.

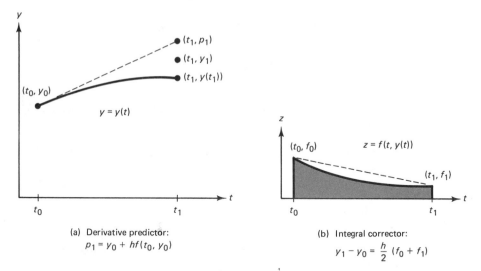

(a) Derivative predictor:
$p_1 = y_0 + hf(t_0, y_0)$

(b) Integral corrector:
$y_1 - y_0 = \frac{h}{2}(f_0 + f_1)$

Figure 9.7. The graphs $y = y(t)$ and $z = f(t, y(t))$ in the derivation of Heun's method.

Step Size versus Error

The error term for the trapezoidal rule used to approximate the integral in (3) is

(7)
$$-y^{(2)}(c_k)\frac{h^3}{12}.$$

If the only error at each step is that given in (7), then after M steps the acumulated error for Heun's method would be

(8)
$$-\sum_{k=1}^{M} y^{(2)}(c_k)\frac{h^3}{12} \approx -\frac{b-a}{12}y^{(2)}(c)h^2 \approx O(h^2).$$

The next theorem is important, because it states the relationship between G.T.E. and step size. It is used to give us an idea of how much computing effort must be done to obtain an accurate approximation using Heun's method.

Theorem 9.3 [G.T.E. for Heun's method] If Heun's method is used to solve the I.V.P. $y' = f(t, y), y(a) = y_0$ over $[a, b]$ and the step size is h, then the G.T.E. is of the order $O(h^2)$, that is,

(9)
$$E(y(b), h) = y(b) - y_M \approx Ch^2,$$

where C is a constant.

Examples 9.7 and 9.8 illustrate Theorem 9.3. If approximations are computed using the step sizes h and $h/2$, then we should have

(10)
$$E(y(b), h) \approx Ch^2$$

for the larger step size, and

(11)
$$E\left(y(b), \frac{h}{2}\right) \approx C\frac{h^2}{4} \approx \frac{1}{4}Ch^2 \approx \frac{1}{4}E(y(b), h).$$

Hence the idea in Theorem 9.3 is that if the step size in Heun's method is reduced by a factor of $\frac{1}{2}$, then we can expect that the overall G.T.E. will be reduced by a factor of $\frac{1}{4}$.

Example 9.7 Use Heun's method to solve the I.V.P.

$$y' = \frac{t-y}{2} \quad \text{on } [0, 3] \text{ with } y(0) = 1.$$

Compare solutions for $h = 1, \frac{1}{2}, \frac{1}{4}$, and $\frac{1}{8}$.

Solution. Figure 9.8 shows graphs of the first two Heun solutions and the exact solution curve $y(t) = 3 \exp(-t/2) - 2 + t$. Table 9.4 gives the values for the four solutions at selected abscissas. For the step size $h = 0.25$ a sample calculation is

$$f(t_0, y_0) = \frac{0-1}{2} = -0.5$$

$$p_1 = 1.0 + 0.25(-0.5) = 0.875,$$

$$f(t_1, p_1) = \frac{0.25 - 0.875}{2} = -0.3125,$$

$$y_1 = 1.0 + 0.125(-0.5 - 0.3125) = 0.8984375.$$

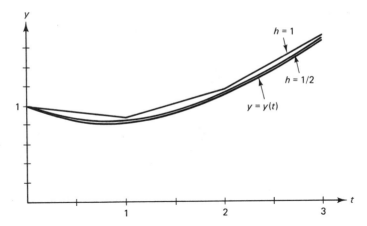

Figure 9.8. Comparison of Heun solutions with different step sizes for $y' = (t - y)/2$ over $[0, 3]$ with $y(0) = 1$.

Table 9.4 Comparison of Heun solutions with different step sizes for $y' = (t - y)/2$ over $[0, 3]$ with $y(0) = 1$

t_k	y_k				$y(t_k)$
	$h = 1$	$h = \frac{1}{2}$	$h = \frac{1}{4}$	$h = \frac{1}{8}$	Exact
0	1.0	1.0	1.0	1.0	1.0
0.125				0.943359	0.943239
0.25			0.898438	0.897717	0.897491
0.375				0.862406	0.862087
0.50		0.84375	0.838074	0.836801	0.836402
0.75			0.814081	0.812395	0.811868
1.00	0.875	0.831055	0.822196	0.820213	0.819592
1.50		0.930511	0.920143	0.917825	0.917100
2.00	1.171875	1.117587	1.106800	1.104392	1.103638
2.50		1.373115	1.362593	1.360248	1.359514
3.00	1.732422	1.682121	1.672269	1.670076	1.669390

This iteration continues until we arrive at the last step:

$$y(3) \approx y_{12} = 1.511508 + 0.125(0.619246 + 0.666840) = 1.672269.$$

Example 9.8 Compare the G.T.E. when Heun's method is used to solve $y' = (t - y)/2$ over $[0, 3]$ with $y(0) = 1$ using step sizes $1, \frac{1}{2}, \ldots, \frac{1}{64}$.

Solution. Table 9.5 gives the G.T.E. and shows that the error in the approximation to $y(3)$ decreases by about $\frac{1}{4}$ when the step size is reduced by a factor of $\frac{1}{2}$:

$$E(y(3), h) = y(3) - y_M = O(h^2) \approx Ch^2 \quad \text{where} \quad C = -0.0432.$$

Table 9.5 Relation between step size and G.T.E. for Heun solutions to $y' = (t - y)/2$ over $[0, 3]$ with $y(0) = 1$

Step Size, h	Number of Steps, M	Approximation to $y(3)$, y_M	G.T.E. Error at $t = 3$, $y(3) - y_M$	$O(h^2) \approx Ch^2$ where $C = -0.0432$
1	3	1.732422	−0.063032	−0.043200
$\frac{1}{2}$	6	1.682121	−0.012731	−0.010800
$\frac{1}{4}$	12	1.672269	−0.002879	−0.002700
$\frac{1}{8}$	24	1.670076	−0.000686	−0.000675
$\frac{1}{16}$	48	1.669558	−0.000168	−0.000169
$\frac{1}{32}$	96	1.669432	−0.000042	−0.000042
$\frac{1}{64}$	192	1.669401	−0.000011	−0.000011

Algorithm 9.2 [*Heun's Method*] Approximate the solution of the initial value problem $y' = f(t, y)$ on $[a, b]$ with $y(a) = y_0$.

```
INPUT A, B, Y(0)                              {Endpoints and initial value}
INPUT M                                       {Number of steps}
H := [B − A]/M                                {Compute the step size}
T(0) := A                                     {Initialize the variable}


FOR  J = 0  TO  M−1  DO
     K₁ := F(T(J),Y(J))                       {Function value at tⱼ}
     P := Y(J) + H∗K₁                         {Euler predictor for yⱼ₊₁}
     T(J+1) := A + H∗[J+1]                    {Compute the mesh point tⱼ₊₁}
     K₂ := F(T(J+1),P)                        {Function value at tⱼ₊₁}
     Y(J+1) := Y(J) + H∗[K₁ + K₂]/2           {Trapezoidal corrector}


FOR  J = 0  TO  M  DO                          {Output}
     PRINT T(J), Y(J)
```

Exercises for Heun's Method

In Exercises 1–5, use Heun's method to find approximations to the I.V.P.
(a) Compute $y_1, y_2, \ldots, y_M$ for each of the three cases: (i) $h = 0.2$, $M = 1$, (ii) $h = 0.1$, $M = 2$, and (iii) $h = 0.05$, $M = 4$.
(b) Compare with the exact solution $y(0.2)$ with the three approximations in part (a).
(c) Does the G.T.E. in part (a) behave as expected when h is halved?
(d) Use a computer to carry out the computations on the larger interval $[a, b]$ that is given.

1. (a) Solve $y' = t^2 - y$ over $[0, 0.2]$ with $y(0) = 1$.
   (b) and (c) Compare with $y(t) = -\exp(-t) + t^2 - 2t + 2$.
   (d) Use $[a, b] = [0, 2]$ with $h = 0.2, 0.1$, and 0.05.

2. (a) Solve $y' = 3y + 3t$ over $[0, 0.2]$ with $y(0) = 1$.
   (b) and (c) Compare with $y(t) = \frac{4}{3}\exp(3t) - t - \frac{1}{3}$.
   (d) Use $[a, b] = [0, 2]$ with $h = 0.2, 0.1$, and 0.05.

3. (a) Solve $y' = -ty$ over $[0, 0.2]$ with $y(0) = 1$.
   (b) and (c) Compare with $y(t) = \exp(-t^2/2)$.
   (d) Use $[a, b] = [0, 2]$ with $h = 0.2, 0.1$, and 0.05.

4. (a) Solve $y' = \exp(-2t) - 2y$ over $[0, 0.2]$ with $y(0) = \frac{1}{10}$.
   (b) and (c) Compare with $y(t) = \frac{1}{10}\exp(-2t) + t\exp(-2t)$.
   (d) Use $[a, b] = [0, 2]$ with $h = 0.2, 0.1$, and 0.05.

5. (a) Solve $y' = 2ty^2$ over $[0, 0.2]$ with $y(0) = 1$.
   (b) and (c) Compare with $y(t) = 1/(1 - t^2)$.
   (d) Use $[a, b] = [0, 1]$ with $h = 0.1, 0.05$, and 0.025.
   Notice that Heun's method will generate an approximation to $y(1)$ even though the solution curve is not defined at $t = 1$.

6. Consider a projectile that is fired straight up and falls straight down. If air resistance is proportional to the velocity, the I.V.P. for the velocity $v(t)$ is

$$v' = -32 - \frac{K}{M}v \quad \text{with} \quad v(0) = v_0,$$

where v_0 is the initial velocity, M the mass, and K the coefficient of air resistance. Suppose that $v_0 = 160$ ft/sec and $K/M = 0.1$. Use Heun's method with $h = 0.5$ to solve

$$v' = -32 - \frac{v}{10} \quad \text{over} \quad [0, 30] \quad \text{with} \quad v(0) = 160.$$

Remark. You can compare your computer solution with the exact solution $v(t) = 480\exp(-t/10) - 320$, which is the velocity component in the vertical direction for Example 2.6. Observe that the limiting velocity is $-320/$ ft/sec.

7. In psychology, the Weber–Fechner law for stimulus–response states that the rate of change dR/dS of the reaction R is inversely proportional to the stimulus. The threshold value is the lowest level of the stimulus that can be consistently detected. The I.V.P. for this model is

$$R' = \frac{k}{S} \quad \text{with} \quad R(s_0) = 0.$$

Suppose that $s_0 = 0.1$ and $R(0.1) = 0$. Use Heun's method with $h = 0.1$ to solve

$$R' = \frac{1}{S} \quad \text{over} \quad [0.1, 5.1] \quad \text{with} \quad R(0.1) = 0.$$

8. Show that when Heun's method is used to solve the I.V.P. $y' = f(t)$ over $[a, b]$ with $y(a) = y_0 = 0$, the result is

$$y(b) \approx \frac{h}{2} \sum_{k=0}^{M-1} [f(t_k) + f(t_{k+1})],$$

which is the trapezoidal rule approximation for the definite integral of $f(t)$ taken over the interval $[a, b]$.

9. The Richardson improvement method discussed in Lemma 7.1 (Section 7.5) can be used in conjunction with Heun's method. If Heun's method is used with step size h, then we have

$$y(b) \approx y_h + Ch^2.$$

If Heun's method is used with $2h$, then we have

$$y(b) \approx y_{2h} + 4Ch^2.$$

The terms involving Ch^2 can be eliminated to obtain an improved approximation for $y(b)$ and the result is

$$y(b) \approx \frac{4y_h - y_{2h}}{3}.$$

This improvement scheme can be used with the values in Example 9.8 to obtain better approximations to $y(3)$. Find the missing entries in the table below.

h	y_h	$(4y_h - y_{2h})/3$
1	1.732422	
1/2	1.682121	1.665354
1/4	1.672269	_____
1/8	1.670076	_____
1/16	1.669558	1.669385
1/32	1.669432	_____
1/64	1.669401	_____

10. Show that Heun's method fails to approximate the solution $y(t) = t^{3/2}$ of the I.V.P.

$$y' = f(t, y) = 1.5y^{1/3} \quad \text{with } y(0) = 0.$$

Justify your answer. What difficulties were encountered?

11. Write a report on the improved Heun's method.

9.4 Taylor Methods

The Taylor method is of general applicability and it gives a means to compare the accuracy of the various numerical methods for solving an I.V.P. It can be constructed to have a high degree of accuracy.

Recall that if $y(t)$ is analytic and has continuous derivatives of all orders, it can be represented in a Taylor series:

(1)
$$y(t + h) = y(t) + y'(t)h + \frac{y^{(2)}(t)h^2}{2} + \frac{y^{(3)}(t)h^3}{6} + \ldots.$$

For numerical purposes, we must fix the number N and use a finite series to approximate $y(t + h)$. Then (1) becomes

(2)
$$y(t + h) \approx y(t) + y'(t)h + \frac{y^{(2)}(t)h^2}{2} + \ldots + \frac{y^{(N)}(t)h^N}{N!}.$$

The error in the approximation (2) is called the local truncation error (L.T.E.) and is given by

(3)
$$\frac{y^{(N+1)}(c)h^{N+1}}{(N + 1)!} \qquad \text{(the L.T.E. for Taylor's method)}$$

where $t < c < t + h$ and $h > 0$.

The solution to $y'(t) = f(t, y)$ over $[t_0, t_M]$ is found by adapting formula (2) to each subinterval $[t_k, t_{k+1}]$. The derivatives $y^{(j)}(t)$ are obtained by differentiating $f(t, y)$. The general step for Taylor's method of order N is

(4)
$$y_{k+1} = y_k + d_1 h + \frac{d_2 h^2}{2} + \frac{d_3 h^3}{6} + \ldots + \frac{d_N h^N}{N!},$$

where $d_j = y^{(j)}(t_k)$ for $j = 1, 2, \ldots, N$, $k = 0, 1, \ldots, M - 1$.

The Taylor method of order N has the desirable feature that the L.T.E. is of order $O(h^{N+1})$, and N can be chosen large so that this error is small. If the order N is fixed, it is possible to determine a priori the step size h so that the G.T.E. will be as small as desired. Although the details of performing this estimate are cumbersome, it is worthwhile to state the theorem that presents this relationship.

Theorem 9.4 [G.T.E. for Taylor's method of order N] If Taylor's method is used to solve $y' = f(t, y)$, $y(a) = y_0$ over $[a, b]$ with step size h, then the G.T.E. satisfies the relation

(5)
$$E(y(b), h) = y(b) - y_M \approx Ch^N = O(h^N),$$

where C is a constant.

Proof. The details involved in this theorem and its proof can be found in Reference 57.

Examples 9.9 and 9.10 illustrate Theorem 9.4 for the case $N = 4$. If approximations are computed using the step sizes h and $h/2$, we should have

(6)
$$E(y(b), h) \approx Ch^4$$

for the larger step size, and

(7)
$$E\left(y(b), \frac{h}{2}\right) \approx C\frac{h^4}{16} \approx \frac{1}{16}Ch^4 \approx \frac{1}{16}E(y(b), h).$$

Hence the idea in Theorem 9.4 is that if the step size in the Taylor method of order 4 is reduced by a factor of $\frac{1}{2}$, the overall G.T.E. will be reduced by about $\frac{1}{16}$.

Example 9.9 Use the Taylor method of order $N = 4$ to solve $y' = (t - y)/2$ on $[0, 3]$ with $y(0) = 1$. Compare solutions for $h = 1, \frac{1}{2}, \frac{1}{4}$, and $\frac{1}{8}$.

Solution. The derivatives of $y(t)$ must first be determined. Recall that the solution $y(t)$ is a function of t, and differentiate the formula $y'(t) = f(t, y(t))$ with respect to t to get $y^{(2)}(t)$. Then continue the process to obtain the higher derivatives:

$$y'(t) = \frac{t - y}{2},$$

$$y^{(2)}(t) = \frac{d}{dt}\frac{t - y}{2} = \frac{1 - y'}{2} = \frac{1 - (t - y)/2}{2} = \frac{2 - t + y}{4},$$

$$y^{(3)}(t) = \frac{d}{dt}\frac{2 - t + y}{4} = \frac{0 - 1 + y'}{4} = \frac{-1 + (t - y)/2}{4} = \frac{-2 + t - y}{8},$$

$$y^{(4)}(t) = \frac{d}{dt}\frac{-2 + t - y}{8} = \frac{-0 + 1 - y'}{8} = \frac{1 - (t - y)/2}{8} = \frac{2 - t + y}{16}.$$

To find y_1, the derivatives given above must be evaluated at the point $(t_0, y_0) = (0, 1)$. Calculation reveals that

$$d_1 = y'(0) = \frac{0.0 - 1.0}{2} = -0.5,$$

$$d_2 = y^{(2)}(0) = \frac{2.0 - 0.0 + 1.0}{4} = 0.75,$$

$$d_3 = y^{(3)}(0) = \frac{-2.0 + 0.0 - 1.0}{8} = -0.375,$$

$$d_4 = y^{(4)}(0) = \frac{2.0 - 0.0 + 1.0}{16} = 0.1875.$$

Next the derivatives $\{d_j\}$ are substituted into (4), with $h = 0.25$, and nested multiplication is used to compute the value y_1:

$$y_1 = 1.0 + 0.25\left\{-0.5 + 0.25\left[\frac{0.75}{2} + 0.25\left(\frac{-0.375}{6} + 0.25\frac{0.1875}{24}\right)\right]\right\}$$

$$= 0.8974915.$$

The computed solution point is $(t_1, y_1) = (0.25, 0.8974915)$.

To find y_2, the derivatives $\{d_j\}$ must now be evaluated at the point $(t_1, y_1) = (0.25, 0.8974915)$. The calculations are starting to require a considerable amount of computational effort and are tedious to do by hand. Calculation reveals that

$$d_1 = y'(0) = \frac{0.25 - 0.8974915}{2} = -0.3237458,$$

$$d_2 = y^{(2)}(0) = \frac{2.0 - 0.25 + 0.8974915}{4} = 0.6618729,$$

$$d_3 = y^{(3)}(0) = \frac{-2.0 + 0.25 - 0.8974915}{8} = -0.3309364,$$

$$d_4 = y^{(4)}(0) = \frac{2.0 - 0.25 + 0.8974915}{16} = 0.1654682.$$

Now these derivatives $\{d_j\}$ are substituted into (4), with $h = 0.25$, and nested multiplication is used to compute the value y_2:

$$y_2 = 0.8974915$$

$$+ 0.25\left\{-0.3237458 + 0.25\left[\frac{0.6618729}{2} + 0.25\left(\frac{-0.3309364}{6} + 0.25\frac{0.1654682}{24}\right)\right]\right\}$$

$$= 0.8364037.$$

The solution point is $(t_1, y_1) = (0.50, 0.8364037)$. Table 9.6 gives solution values at selected abscissas, using various step sizes.

Table 9.6   Comparison of the Taylor solutions of order $N = 4$ for $y' = (t - y)/2$ over $[0, 3]$ with $y(0) = 1$

| t_k | y_k | | | | $y(t_k)$ |
	$h = 1$	$h = \frac{1}{2}$	$h = \frac{1}{4}$	$h = \frac{1}{8}$	Exact
0	1.0	1.0	1.0	1.0	1.0
0.125				0.9432392	0.9432392
0.25			0.8974915	0.8974908	0.8974917
0.375				0.8620874	0.8620874
0.50		0.8364258	0.8364037	0.8364024	0.8364023
0.75			0.8118696	0.8118679	0.8118678
1.00	0.8203125	0.8196285	0.8195940	0.8195921	0.8195920
1.50		0.9171423	0.9171021	0.9170998	0.9170997
2.00	1.1045125	1.1036826	1.1036408	1.1036385	1.1036383
2.50		1.3595575	1.3595168	1.3595145	1.3595144
3.00	1.6701860	1.6694308	1.6693928	1.6693906	1.6693905

Example 9.10   Compare the G.T.E. for the Taylor solutions to $y' = (t - y)/2$ over $[0, 3]$ with $y(0) = 1$ given in Example 9.9.

Solution.   Table 9.7 gives the G.T.E. for these step sizes and shows that the error in the approximation to $y(3)$ decreases by about $\frac{1}{16}$ when the step size is reduced by a factor of $\frac{1}{2}$:

$$E(y(3), h) = y(3) - y_N = O(h^4) \approx Ch^4 \quad \text{where} \quad C = -0.000614.$$

Table 9.7   Relation between step size and G.T.E. for the Taylor solutions to $y' = (t - y)/2$ over $[0, 3]$

Step Size, h	Number of Steps, N	Approximation to $y(3)$, y_N	G.T.E Error at $t = 3$, $y(3) - y_N$	$O(h^2) \approx Ch^4$ where $C = -0.000614$
1	3	1.6701860	−0.0007955	−0.0006140
$\frac{1}{2}$	6	1.6694308	−0.0000403	−0.0000384
$\frac{1}{4}$	12	1.6693928	−0.0000023	−0.0000024
$\frac{1}{8}$	24	1.6693906	−0.0000001	−0.0000001

Algorithm 9.3 [*Taylor Method of Order $N = 4$*] Solve the I.V.P. $y' = f(t, y)$ on $[a, b]$ with $y(a) = y_0$. The derivatives $y^{(k)} = F_k(T, Y)$ must be specified for $k = 1, 2, 3, 4$.

```
INPUT A, B, Y(0)                    {Endpoints and initial value}
INPUT M                             {Number of steps}
H := [B − A]/M                      {Compute the step size}
T(0) := A                           {Initialize the variable}


FOR  K = 0  TO  M−1  DO
    T := T(K) and Y := Y(K)
    D₁ := F₁(T,Y)                   {Slope function f(t,y(t))}
    D₂ := F₂(T,Y)                   {Derivative of f(t,y(t))}
    D₃ := F₃(T,Y)                   {Second derivative of f(t,y(t))}
    D₄ := F₄(T,Y)                   {Third derivative of f(t,y(t))}
    Y(K+1) := Y + H*[D₁ + H*[D₂/2 + H*[D₃/6 + H*D₄/24]]]
    T(K+1) := A + H*[K+1]


FOR  K = 1  TO  M  DO                              {Output}
    PRINT T(K), Y(K)
```

Exercises for Taylor Methods

In Exercises 1–5, use the Taylor method, of order $N = 4$, to find approximations to the I.V.P.

    (a) Compute $y_1, y_2, \ldots, y_M$ for each of the two cases (i) $h = 0.2$, $M = 1$ and (ii) $h = 0.1$, $M = 2$.

    (b) Compare with the exact solution $y(0.2)$ with the two approximations in part (a).

    (c) Does the G.T.E. in part (a) behave as expected when h is halved?

(d) Use a computer to carry out the computations on the larger interval $[a, b]$ that is given.

1. (a) Solve $y' = t^2 - y$ over $[0, 0.2]$ with $y(0) = 1$.
   (b) and (c) Compare with $y(t) = -\exp(-t) + t^2 - 2t + 2$.
   (d) Use $[a, b] = [0, 2]$ with $h = 0.2, 0.1$, and 0.05.

2. (a) Solve $y' = 3y + 3t$ over $[0, 0.2]$ with $y(0) = 1$.
   (b) and (c) Compare with $y(t) = \frac{4}{3} \exp(3t) - t - \frac{1}{3}$.
   (d) Use $[a, b] = [0, 2]$ with $h = 0.2, 0.1$, and 0.05.

3. (a) Solve $y' = -ty$ over $[0, 0.2]$ with $y(0) = 1$.
   (b) and (c) Compare with $y(t) = \exp(-t^2/2)$.
   (d) Use $[a, b] = [0, 2]$ with $h = 0.2, 0.1$, and 0.05.

4. (a) Solve $y' = \exp(-2t) - 2y$ over $[0, 0.2]$ with $y(0) = \frac{1}{10}$.
   (b) and (c) Compare with $y(t) = \frac{1}{10} \exp(-2t) + t \exp(-2t)$.
   (d) Use $[a, b] = [0, 2]$ with $h = 0.2, 0.1$, and 0.05.

5. (a) Solve $y' = 2ty^2$ over $[0, 0.2]$ with $y(0) = 1$.
   (b) and (c) Compare with $y(t) = 1/(1 - t^2)$.
   (d) Use $[a, b] = [0, 1]$ with $h = 0.1, 0.05$, and 0.025.
   Notice that Taylor's method will generate an approximation to $y(1)$ even though the solution curve is not defined at $t = 1$.

6. The Richardson improvement method discussed in Lemma 7.1 (Section 7.5) can be used in conjunction with Taylor's method. If Taylor's method of order $N = 4$ is used with step size h, then $y(b) \approx y_h + Ch^4$. If Taylor's method of order $N = 4$ is used with step size $2h$, then $y(b) \approx y_{2h} + 16Ch^4$. The terms involving Ch^4 can be eliminated to obtain an improved approximation for $y(b)$:

$$y(b) \approx \frac{16y_h - y_{2h}}{15}.$$

This improvement scheme can be used with the values in Example 9.10 to obtain better approximations to $y(3)$. Find the missing entries in the table below.

h	y_h	$(16y_h - y_{2h})/15$
1.0	1.6701860	
0.5	1.6694308	_____
0.25	1.6693928	_____
0.125	1.6693906	_____

7. Show that when Taylor's method of order N is used with step sizes h and $h/2$, then the overall G.T.E. will be reduced by a factor of about 2^{-N} for the smaller step size.

8. Show that Taylor's method fails to approximate the solution $y(t) = t^{3/2}$ of the I.V.P. $y' = f(t, y) = 1.5y^{1/3}$ with $y(0) = 0$. Justify your answer. What difficulties were encountered?

9. (a) Verify that the solution to the I.V.P. $y' = y^2, y(0) = 1$ over the interval $[0, 1)$ is $y(t) = 1/(1 - t)$.
   (b) Verify that the solution to the I.V.P. $y' = 1 + y^2, y(0) = 1$ over the interval $[0, \pi/4)$ is $\tan(t + \pi/4)$.
   (c) Use the result of parts (a) and (b) to argue that the solution to the I.V.P. $y' = t^2 + y^2, y(0) = 1$ has a vertical asymptote between $\pi/4$ and 1. (Its location is near $t = 0.96981$.)

10. Consider the I.V.P. $y' = 1 + y^2, y(0) = 1$.
    (a) Find an expression for $y^{(2)}(t), y^{(3)}(t)$, and $y^{(4)}(t)$.
    (b) Evaluate these derivatives at $t = 0$, and use them find the first five terms in the Maclaurin expansion for $\tan(t)$.

11. The following table of values is the Taylor solution of order $N = 4$ for the I.V.P. $y' = t^2 + y^2, y(0) = 1$ over $[0, 0.8]$. The true solution at $t = 0.8$ is known to be $y(0.8) = 5.8486168$.

Step Size, h	Number of Steps, M	Approximation to $y(0.8)$, y_M	G.T.E. Error at $t = 0.8$, $y(0.8) - y_M$
0.05	16	5.8410780	0.0075388
0.025	32	5.8479637	0.0006531
0.0125	64	5.8485686	0.0000482
0.00625	128	5.8486136	0.0000032
0.003125	256	5.8486166	0.0000002

   (a) Does the G.T.E. behave as expected when h is halved?
   (b) Use Richardson improvement scheme to find improvements for the approximations.

12. The following table of values is the Taylor solution of order $N = 3$ for the I.V.P. $y' = t^2 + y^2, y(0) = 1$ over $[0, 0.8]$.

Step Size, h	Number of Steps, M	Approximation to $y(0.8)$, y_M	G.T.E. Error at $t = 0.8$, $y(0.8) - y_M$
0.05	16	5.8050504	0.0435664
0.025	32	5.8416192	0.0069976
0.0125	64	5.8476250	0.0009918
0.00625	128	5.8484848	0.0001320
0.003125	256	5.8485998	0.0000170

   (a) Does the G.T.E. behave as expected when h is halved?
   (b) Develop the Richardson improvement scheme for an order 3 method and find improvements the approximations.

9.5 Runge–Kutta Methods

The Taylor methods in the preceding section have the desirable feature that the G.T.E. is of order $O(h^N)$, and N can be chosen large so that this error is small. However, the shortcoming of the Taylor methods is the a priori determination of N and the computation of the higher derivatives, which can be very complicated. Each Runge–Kutta method is derived from an appropriate Taylor method in such a way that the G.T.E. is of order $O(h^N)$. A trade-off is made to perform several function evaluations at each step and eliminate the necessity to compute the higher derivatives. These methods can be constructed for any order N. The Runge–Kutta method of order $N = 4$ is most popular. It is a good choice for common purposes because it is quite accurate, stable, and easy to program. Most authorities proclaim that it is not necessary to go to a higher-order method because the increased accuracy is offset by additional computational effort. If more accuracy is required, then either a smaller step size or an adaptive method should be used (see Section 9.7).

The fourth-order Runge–Kutta method (denoted RK4) simulates the accuracy of the Taylor series method of order 4. The proof is algebraically complicated and results in a formula involving a linear combination of function values. The coefficients involved are chosen so that the method has a L.T.E. of order $O(h^5)$, and hence a G.T.E. of order $O(h^4)$. As before, the interval $[a, b]$ is divided into M subintervals of equal width h. Starting with (t_0, y_0), four function evaluations per step are required to generate the discrete approximations (t_k, y_k) as follows:

(1)
$$y_{k+1} = y_k + \frac{h}{6}(f_1 + 2f_2 + 2f_3 + f_4)$$

where

$$f_1 = f(t_k, y_k),$$

$$f_2 = f\left(t_k + \frac{h}{2}, y_k + \frac{h}{2}f_1\right),$$

$$f_3 = f\left(t_k + \frac{h}{2}, y_k + \frac{h}{2}f_2\right),$$

$$f_4 = f(t_k + h, y_k + hf_3).$$

Discussion about the Method

The rigorous development of the equations in (1) is beyond the scope of this book and can be found in advanced texts, but we can get some insights. Consider the graph of the solution curve $y = y(t)$ over the first subinterval $[t_0, t_1]$. The function values in (1) are approximations for slopes to this curve. Here f_1 is the slope at the left, f_2 and f_3 are two estimates for the slope in the middle, and f_4 is the slope at the right [see Figure 9.9(a)]. The next point (t_1, y_1) is obtained by integrating the slope function:

(2)
$$y(t_1) - y(t_0) = \int_{t_0}^{t_1} f(t, y(t)) \, dt.$$

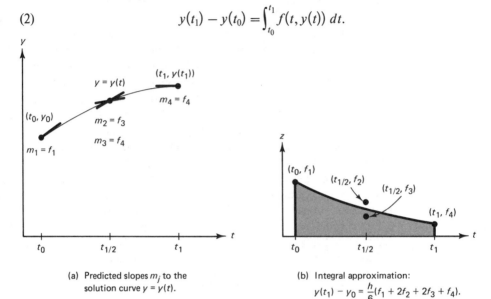

(a) Predicted slopes m_j to the solution curve $y = y(t)$.

(b) Integral approximation:
$$y(t_1) - y_0 = \frac{h}{6}(f_1 + 2f_2 + 2f_3 + f_4).$$

Figure 9.9. The graphs $y = y(t)$ and $z = f(t, y(t))$ in the derivation of the Runge-Kutta method of order $N = 4$.

If Simpson's rule is applied with step size $h/2$, then the approximation to the integral in (2) is

(3)
$$\int_{t_0}^{t_1} f(t, y(t)) \, dt \approx \frac{h}{6} [f(t_0, y(t_0)) + 4f(t_{1/2}, y(t_{1/2})) + f(t_1, y(y_1))],$$

where $t_{1/2}$ is the midpoint of the interval. Three function values are needed; hence we make the obvious choices $f(t_0, y(t_0)) = f_1$ and $f(t_1, y(t_1)) \approx f_4$. For the value in the middle we chose the average of f_2 and f_3, i.e.,

$$f(t_{1/2}, y(t_{1/2})) \approx \frac{f_2 + f_3}{2}.$$

These values are substituted into (3), which is used in equation (2) to get y_1:

(4)
$$y_1 = y_0 + \frac{h}{6} \left[f_1 + \frac{4(f_2 + f_3)}{2} + f_4 \right].$$

When this formula is simplified, it is seen to be the first equation in (1) with $k = 0$. The graph for the integral in (3) is shown in Figure 9.9(b).

Step Size versus Error

The error term for Simpson's rule with step size $h/2$ is

(5)
$$-y^{(4)}(c_1)\frac{h^5}{2,880}.$$

If the only error at each step is that given in (5), then after M steps the accumulated error for the RK4 method would be

(6)
$$-\sum_{k=1}^{M} y^{(4)}(c_k)\frac{h^5}{2,880} \approx -\frac{b-a}{5,760}y^{(4)}(c)h^4 \approx O(h^4).$$

The next theorem states the relationship between G.T.E. and step size. It is used to give us an idea of how much computing effort must be done when using the RK4 method.

Theorem 9.5 [G.T.E. for Runge–Kutta order 4] If the RK4 method is used to solve the I.V.P. $y' = f(t, y), y(a) = y_0$ over $[a, b]$ and the step size is h, the G.T.E. is of the order $O(h^4)$, that is,

(7) $E(y(b), h) = y(b) - y_M \approx Ch^4$ (G.T.E. for the RK4 method),

where C is a constant.

Examples 9.11 and 9.12 illustrate Theorem 9.5. If approximations are computed using the step sizes h and $h/2$, we should have

(8)
$$E(y(b), h) \approx Ch^4$$

for the larger step size, and

(9)
$$E\left(y(b), \frac{h}{2}\right) \approx C\frac{h^4}{16} \approx \frac{1}{16}Ch^4 \approx \frac{1}{16}E(y(b), h).$$

Hence the idea in Theorem 9.5 is that if the step size in the RK4 method is reduced by a factor of $\frac{1}{2}$, we can expect that the overall G.T.E. will be reduced by a factor of $\frac{1}{16}$.

Example 9.11 Use the RK4 method to solve the I.V.P. $y' = (t - y)/2$ on $[0, 3]$ with $y(0) = 1$. Compare solutions for $h = 1, \frac{1}{2}, \frac{1}{4}$, and $\frac{1}{8}$.

Solution. Table 9.8 gives the solution values at selected abscissas. For the step size $h = 0.25$ a sample calculation is

$$f_1 = \frac{0.0 - 1.0}{2} = -0.5,$$

$$f_2 = \frac{0.125 - [1 + 0.25(0.5)(-0.5)]}{2} = -0.40625,$$

$$f_3 = \frac{0.125 - [1 + 0.25(0.5)(-0.40625)]}{2} = -0.4121094,$$

$$f_4 = \frac{0.25 - [1 + 0.25(-0.4121094)]}{2} = -0.3234863,$$

$$y_1 = 1.0 + 0.25\frac{-0.5 + 2(-0.40625) + 2(-0.4121094) - 0.3234863}{6}$$

$$= 0.8974915.$$

Table 9.8 Comparison of the RK4 solutions with different step sizes for $y' = (t - y)/2$ over $[0, 3]$ with $y(0) = 1$

t_k	y_k				$y(t_k)$ Exact
	$h = 1$	$h = \frac{1}{2}$	$h = \frac{1}{4}$	$h = \frac{1}{8}$	
0	1.0	1.0	1.0	1.0	1.0
0.125				0.9432392	0.9432392
0.25			0.8974915	0.8974908	0.8974917
0.375				0.8620874	0.8620874
0.50		0.8364258	0.8364037	0.8364024	0.8364023
0.75			0.8118696	0.8118679	0.8118678
1.00	0.8203125	0.8196285	0.8195940	0.8195921	0.8195920
1.50		0.9171423	0.9171021	0.9170998	0.9170997
2.00	1.1045125	1.1036826	1.1036408	1.1036385	1.1036383
2.50		1.3595575	1.3595168	1.3595145	1.3595144
3.00	1.6701860	1.6694308	1.6693928	1.6693906	1.6693905

Example 9.12 Compare the G.T.E. when the RK4 method is used to solve $y' = (t - y)/2$ over $[0, 3]$ with $y(0) = 1$ using sizes $1, \frac{1}{2}, \frac{1}{4}$, and $\frac{1}{8}$.

Solution. Table 9.9 gives the G.T.E. for the various step sizes and shows that the error in the approximation to $y(3)$ decreases by about $\frac{1}{16}$ when the step size is reduced by a factor of $\frac{1}{2}$:

$$E(y(3), h) = y(3) - y_M = O(h^4) \approx Ch^4 \quad \text{where} \quad C = -0.000614.$$

Table 9.9 Relation between step size and G.T.E. for the RK4 solutions to $y' = (t - y)/2$ over $[0, 3]$ with $y(0) = 1$

Step Size, h	Number of Steps, M	Approximation to $y(3)$, y_M	G.T.E. Error at $t = 3$, $y(3) - y_M$	$O(h^4) \approx Ch^4$ where $C = -0.000614$
1	3	1.6701860	-0.0007955	-0.0006140
$\frac{1}{2}$	6	1.6694308	-0.0000403	-0.0000384
$\frac{1}{4}$	12	1.6693928	-0.0000023	-0.0000024
$\frac{1}{8}$	24	1.6693906	-0.0000001	-0.0000001

A comparison of Examples 9.11 and 9.12 and Examples 9.9 and 9.10 shows what is meant by the statement "The RK4 method simulates the Taylor series method of order 4." For these examples, the two methods generate identical solution sets $\{(t_k, y_k)\}$ over the given interval. The advantage of the RK4 method is obvious; no formulas for the higher derivatives need to be computed nor do they have to be in a program.

It is not easy to determine the accuracy to which a Runge-Kutta solution has been computed. We could estimate the size of $y^{(4)}(c)$ and use formula (6). Another way is to repeat the algorithm using a smaller step size and compare results. A third way is to adaptively determine the step size, which is done in Algorithm 9.5. In Section 9.6 we will see how to change the step size for a multistep method.

Algorithm 9.4 [*Runge-Kutta Method of Order* 4] Solve the I.V.P.

$$y' = f(t, y) \quad \text{on } [a, b] \text{ with } y(a) = y_0.$$

INPUT A, B, Y(0)	{Endpoints and initial value}
INPUT M	{Number of steps}
H := [B − A]/M	{Compute the step size}
T(0) := A	{Initialize the value}

FOR J = 0 TO M−1 DO
    T := T(J) and Y := Y(J)   {Local variables}
    K_1 := H*F(T,Y)   {Function value at t_j}
    K_2 := H*F(T+H/2, Y + .5*K_1)   {Function value at $t_{j+1/2}$}
    K_3 := H*F(T+H/2, Y + .5*K_2)   {Function value at $t_{j+1/2}$}
    K_4 := H*F(T+H, Y + K_3)   {Function value at t_{j+1}}
    Y(J+1) := Y + [K_1+2*K_2+2*K_3+K_4]/6   {Integrate f(t,y)}
    T(J+1) := A + H*[J+1]   {Generate the mesh point}

FOR J = 0 TO M DO   {Output}
    PRINT T(J), Y(J)

Runge–Kutta Methods of Order N = 2

The second-order Runge-Kutta method (denoted RK2) simulates the accuracy of the Taylor series method of order 2. Although this method is not as good to use as the RK4 method, its proof is easier to understand and illustrates the principles involved. To start, we write down the Taylor series formula for $y(t + h)$:

(10) $$y(t + h) = y(t) + hy'(t) + \tfrac{1}{2}h^2 y''(t) + C_T h^3 + \dots,$$

where C_T is a constant involving the third derivative of $y(t)$ and the other terms in the series involve powers of h^j for $j > 3$.

The derivatives $y'(t)$ and $y''(t)$ in equation (10) must be expressed in terms of $f(t,y)$ and its partial derivatives. Recall that

(11)
$$y'(t) = f(t,y).$$

The chain rule for differentiating a function of two variables can be used to differentiate (11) with respect to t, and the result is

$$y''(t) = f_t(t,y) + f_y(t,y)y'(t).$$

Using (11), this can be written

(12)
$$y''(t) = f_t(t,y) + f_y(t,y)f(t,y).$$

The derivatives (11) and (12) are substituted into (10) to give the Taylor expression for $y(t+h)$:

(13)
$$y(t+h) = y(t) + hf(t,y) + \tfrac{1}{2}h^2 f_t(t,y) + \tfrac{1}{2}h^2 f_y(t,y)f(t,y) + C_T h^3 + \dots.$$

Now consider the Runge–Kutta method of order $N = 2$ which uses a linear combination of two function values to express $y(t+h)$:

(14)
$$y(t+h) = y(t) + Ahf_0 + Bhf_1,$$

where

(15)
$$f_0 = f(t,y),$$
$$f_1 = f(t + Ph, y + Qhf_0).$$

Next the Taylor polynomial approximation for a function of two independent variables is used to expand $f(t,y)$ (see Exercises 13 and 14 of Section 1.3). This gives the following representation for f_1:

(16)
$$f_1 = f(t,y) + Phf_t(t,y) + Qhf_y(t,y)f(t,y) + C_P h^2 + \dots,$$

where C_P involves the second-order partial derivatives of $f(t,y)$. Then (16) is used in (14) to get the RK2 expression for $y(t+h)$:

(17)
$$y(t+h) = y(t) + (A+B)hf(t,y) + BPh^2 f_t(t,y)$$
$$+ BQh^2 f_y(t,y)f(t,y) + BC_P h^3 + \dots.$$

A comparison of similar terms in equations (13) and (17) will produce the following conclusions:

$$hf(t,y) = (A+B)hf(t,y) \quad \text{implies that } 1 = A + B,$$
$$\tfrac{1}{2}h^2 f_t(t,y) = BPh^2 f_t(t,y) \quad \text{implies that } \tfrac{1}{2} = BP,$$
$$\tfrac{1}{2}h^2 f_y(t,y)f(t,y) = BQh^2 f_y(t,y)f(t,y) \quad \text{implies that } \tfrac{1}{2} = BQ.$$

Hence if we require that A, B, P, and Q satisfy the relations

(18)
$$A + B = 1 \qquad BP = \tfrac{1}{2} \qquad BQ = \tfrac{1}{2},$$

then the RK2 method in (17) will have the same order of accuracy as the Taylor's method in (13).

Since there are only three equations in four unknowns, the system of equations (18) is underdetermined, and we are permitted to choose one of the coefficients. There are several special choices that have been studied in the literature, we mention two of them.

Case (i) Choose $A = \frac{1}{2}$.

This choice leads to $B = \frac{1}{2}$, $P = 1$, and $Q = 1$. If equation (14) is written with these parameters, the formula is

(19)
$$y(t + h) \approx y(t) + \frac{h}{2}[f(t, y) + f(t + h, y + hf(t, h))].$$

When this scheme is used to generate $\{(t_k, y_k)\}$, the result is Heun's method.

Case (ii) Choose $A = 0$.

This choice leads to $B = 1$, $P = \frac{1}{2}$, and $Q = \frac{1}{2}$. If equation (14) is written with these parameters, the formula is

(20)
$$y(t + h) = y(t) + hf\left(t + \frac{h}{2}, y + \frac{h}{2}f(t, h)\right).$$

When this scheme is used to generate $\{(t_k, y_k)\}$, it is called the *modified Euler-Cauchy method*.

The Runge–Kutta–Fehlberg Method (RKF45)

One way to guarantee accuracy in the solution of an I.V.P. is to solve the problem twice using step sizes h and $h/2$ and compare answers at the mesh points corresponding to the larger step size. But this requires a significant amount of computation for the smaller step size and must be repeated if it is determined that the agreement is not good enough.

The Runge–Kutta–Fehlberg method (denoted RKF45), is one way to try to resolve this problem. It has a procedure to determine if the proper step size h is being used. At each step two different approximations for the solution are made and compared. If the two answers are in close agreement, the approximation is accepted. If the answers do not agree to a specified accuracy, the step size is reduced. If the answers agree to more significant digits than required, the step size is increased.

Each step requires the use of the following six values:

$$k_1 := hf(t_k, y_k),$$

$$k_2 := hf\left(t_k + \frac{1}{4}h, y_k + \frac{1}{4}k_1\right),$$

$$k_3 := hf\left(t_k + \frac{3}{8}h, y_k + \frac{3}{32}k_1 + \frac{9}{32}k_2\right),$$

(21)
$$k_4 := hf\left(t_k + \frac{12}{13}h, y_k + \frac{1,932}{2,197}k_1 - \frac{7,200}{2,197}k_2 + \frac{7,296}{2,197}k_3\right),$$

$$k_5 := hf\left(t_k + h, y_k + \frac{439}{216}k_1 - 8k_2 + \frac{3,680}{513}k_3 - \frac{845}{4,104}k_4\right),$$

$$k_6 := hf\left(t_k + \frac{1}{2}h, y_k - \frac{8}{27}k_1 + 2k_2 - \frac{3,544}{2,565}k_3 + \frac{1,859}{4,104}k_4 - \frac{11}{40}k_5\right).$$

Then an approximation to the solution of the I.V.P. is made using a Runge–Kutta method of order 4:

(22)
$$y_{k+1} = y_k + \frac{25}{216}k_1 + \frac{1,408}{2,565}k_3 + \frac{2,197}{4,104}k_4 - \frac{1}{5}k_5.$$

where the four function values $f_1, f_3, f_4,$ and f_5 are used. Notice that f_2 is not used in formula (22). A better value for the solution is determined using a Runge–Kutta method of order 5:

(23)
$$z_{k+1} = y_k + \frac{16}{135}k_1 + \frac{6,656}{12,825}k_3 + \frac{28,561}{56,430}k_4 - \frac{9}{50}k_5 + \frac{2}{55}k_6.$$

The optimal step sh size can be determined by multiplying the scalar s times the current step size h. The scalar s is

(24)
$$s = \left(\frac{\text{Tol } h}{2|z_{k+1} - y_{k+1}|}\right)^{1/4} \approx 0.84 \left(\frac{\text{Tol } h}{|z_{k+1} - y_{k+1}|}\right)^{1/4}.$$

where Tol is the specified error control tolerance.

The derivation of formula (24) can be found in advanced books on numerical analysis. It is important to learn that a fixed step size is not the best strategy even though it would give a nicer-appearing table of values. If values are needed that are not in the table, polynomial interpolation should be used.

Example 9.13 Compare RKF45 and RK4 solutions to the I.V.P.
$$y' = 1 + y^2 \qquad y(0) = 0 \quad \text{on} \quad [0, 1.4].$$

Solution. An RKF45 program was used with the value Tol $= 2 \times 10^{-5}$ for the error control tolerance. It automatically changed the step size and generated the 10 approximations to the solution in Table 9.10. A RK4 program was used with

Table 9.10 RKF45 solution to $y' = 1 + y^2, y(0) = 0$

k	t_k	RKF45 Approximation y_k	True Solution, $y(t_k) = \tan(t_k)$	Error, $y(t_k) - y_k$
0	0.0	0.0000000	0.0000000	0.0000000
1	0.2	0.2027100	0.2027100	0.0000000
2	0.4	0.4227933	0.4227931	−0.0000002
3	0.6	0.6841376	0.6841368	−0.0000008
4	0.8	1.0296434	1.0296386	−0.0000048
5	1.0	1.5574398	1.5574077	−0.0000321
6	1.1	1.9648085	1.9647597	−0.0000488
7	1.2	2.5722408	2.5721516	−0.0000892
8	1.3	3.6023295	3.6021024	−0.0002271
9	1.35	4.4555714	4.4552218	−0.0003496
10	1.4	5.7985045	5.7978837	−0.0006208

the a priori step size of $h = 0.1$, which required the computer to generate 14 approximations at the equally spaced points in Table 9.11. The approximations at the endpoint are

$$y(1.4) \approx y_{10} = 5.7985045 \quad \text{and} \quad y(1.4) \approx y_{14} = 5.7919748$$

and the errors are

$$E_{10} = -0.0006208 \quad \text{and} \quad E_{14} = 0.0059089$$

for the RKF45 and RK4 methods, respectively. The RKF45 method has the smaller error.

Table 9.11 RK4 solution to $y' = 1 + y^2, y(0) = 0$

k	t_k	RK4 Approximation y_k	True Solution, $y(t_k) = \tan(t_k)$	Error, $y(t_k) - y_k$
0	0.0	0.0000000	0.0000000	0.0000000
1	0.1	0.1003346	0.1003347	0.0000001
2	0.2	0.2027099	0.2027100	0.0000001
3	0.3	0.3093360	0.3093362	0.0000002
4	0.4	0.4227930	0.4227932	0.0000002
5	0.5	0.5463023	0.5463025	0.0000002
6	0.6	0.6841368	0.6841368	0.0000000
7	0.7	0.8422886	0.8422884	−0.0000002
8	0.8	1.0296391	1.0296386	−0.0000005
9	0.9	1.2601588	1.2601582	−0.0000006
10	1.0	1.5574064	1.5574077	0.0000013
11	1.1	1.9647466	1.9647597	0.0000131
12	1.2	2.5720718	2.5721516	0.0000798
13	1.3	3.6015634	3.6021024	0.0005390
14	1.4	5.7919748	5.7978837	0.0059089

Algorithm 9.5 [*Runge-Kutta-Fehlberg Method, RKF45*] Solve the I.V.P.

$$y' = f(t, y) \quad \text{on } [a, b] \text{ with } y(a) = y_0.$$

```
Tol := 2*10⁻⁵                              {Error control tolerance}
INPUT A, B, Y(0)                           {Endpoints and initial value}
INPUT N                                    {Tentative number of steps}
H := [B − A]/N                             {Initial the step size}
Hmin := H/64 and Hmax := h*64    {Minimum and maximum step sizes}
T(0) := A and J := 0 and T := A                          {Initialize}
```

```
WHILE  T < B  DO
   IF  T+H > B  THEN  H := B−T                          {The last step}
   T := T(J) and Y := Y(J)
   K₁ := H*F(T,Y)                                       {Compute
   K₂ := H*F(T + ¼H, Y + ¼K₁)                           the function
   K₃ := H*F(T + ⅜H, Y + 9/32K₁ + 9/32K₂)              values}
   K₄ := H*F(T + 12/13H, Y + 1932/2197K₁ − 7200/2197K₂ + 7296/2197K₃)
   K₅ := H*F(T + H, Y + 439/216K₁ − 8K₂ + 3680/513K₃ − 845/4104K₄)
   K₆ := H*F(T + ½H, Y − 8/27K₁ + 2K₂ − 3544/2565K₃ + 1859/4104K₄ − 11/40K₅)
   Err := |1/360K₁ − 128/4275K₃ − 2197/75240K₄ + 1/50K₅ + 2/55K₆|   {||z_{k+1} − y_{k+1}||}
   IF  Err<Tol OR H<2*Hmin  THEN                        {Accept
      Y(J+1) := Y + 25/216K₁ + 1408/2565K₃ + 2197/4104K₄ − 1/5K₅    the
      T(J+1) := T+H,  J := J+1                          approximation}
   IF  Err=0  THEN
      S := 0                                            {Trap division by 0}
   ELSE
      S := .84*[Tol*H/Err]^{1/4}                        {Step size scalar}
   ENDIF
   IF  S<.75 AND H>2*Hmin  THEN  H := H/2               {Reduce step}
   IF  S>1.5 AND 2*H<Hmax  THEN  H := H*2               {Increase step}
END


FOR  I = 0  TO  J  DO                                   {Output}
   PRINT T(I), Y(I)
```

Exercises for Runge–Kutta Methods

Exercises for Runge-Kutta Methods

In Exercises 1–5, use Runge-Kutta method, of order $N = 4$, to find approximations to the I.V.P.

(a) Compute $y_1, y_2, \ldots, y_M$ for each of the two cases (i) $h = 0.2$, $M = 1$ and (ii) $h = 0.1$, $M = 2$.

(b) Compare with the exact solution $y(0.2)$ with the two approximations in part (a).

(c) Does the G.T.E. in part (a) behave as expected when h is halved?

(d) Use a computer to carry out the computations on the larger interval $[a, b]$ that is given.

1. (a) Solve $y' = t^2 - y$ over $[0, 0.2]$ with $y(0) = 1$.
   (b) and (c) Compare with $y(t) = -\exp(-t) + t^2 - 2t + 2$.
   (d) Use $[a, b] = [0, 2]$ with $h = 0.2, 0.1$, and 0.05.

2. (a) Solve $y' = 3y + 3t$ over $[0, 0.2]$ with $y(0) = 1$.
   (b) and (c) Compare with $y(t) = \frac{4}{3}\exp(3t) - t - \frac{1}{3}$.
   (d) Use $[a, b] = [0, 2]$ with $h = 0.2, 0.1$, and 0.05.

3. (a) Solve $y' = -ty$ over $[0, 0.2]$ with $y(0) = 1$.
   (b) and (c) Compare with $y(t) = \exp(-t^2/2)$.
   (d) Use $[a, b] = [0, 2]$ with $h = 0.2, 0.1$, and 0.05.

4. (a) Solve $y' = \exp(-2t) - 2y$ over $[0, 0.2]$ with $y(0) = \frac{1}{10}$.
   (b) and (c) Compare with $y(t) = \frac{1}{10}\exp(-2t) + t\exp(-2t)$.
   (d) Use $[a, b] = [0, 2]$ with $h = 0.2, 0.1$, and 0.05.

5. (a) Solve $y' = 2ty^2$ over $[0, 0.2]$ with $y(0) = 1$.
   (b) and (c) Compare with $y(t) = 1/(1 - t^2)$.
   (d) Use $[a, b] = [0, 1]$ with $h = 0.1, 0.5$, and 0.025.
   Notice that the Runge–Kutta method will generate an approximation to $y(1)$ even though the solution curve is not defined at $t = 1$.

6. In a chemical reaction, one molecule of A combines with one molecule of B to form one molecule of the chemical product C. It is found that the concentration $y(t)$ of C, at time t, is the solution to the I.V.P.

$$y' = k(a - y)(b - y) \quad \text{with} \quad y(0) = 0,$$

where k is a positive constant and a and b are the initial concentrations of A and B, respectively. Suppose that $k = 0.01$, $a = 70$ millimole/liter and $b = 50$ millimole/liter. Use the Runge–Kutta method of order $N = 4$ with $h = 0.5$ to find the solution over $[0, 20]$.
   Remark. You can compare your computer solution with the exact solution $y(t) = 350[1 - \exp(-0.2t)]/[7 - 5\exp(-0.2t)]$. Observe that the limiting value is 50 as $t \longrightarrow +\infty$.

7. By solving an appropriate initial value problem, make a table of values of the function $f(x)$ given by the following integral:

$$f(x) = (2\pi)^{-1/2} \int_0^x \exp\left(\frac{-t^2}{2}\right) dt \quad \text{for } 0 \leq x \leq 3.$$

Use the Runge–Kutta method of order $N = 4$ with $h = 0.1$ for your computations. Your solution should agree with the values in the table below.
   Remark. This is a good way to generate the table of areas for a standard normal distribution (see Exercise 8).

x	$f(x)$
0.0	0.0
0.5	0.1914625
1.0	0.3413448
1.5	0.4331928
2.0	0.4772499
2.5	0.4937903
3.0	0.4986501

8. Show that when the Runge–Kutta method of order $N = 4$ is used to solve the I.V.P. $y' = f(t)$ over $[a, b]$ with $y(a) = 0$, the result is

$$y(b) \approx \frac{h}{6} \sum_{k=0}^{M-1} [f(t_k) + 4f(t_{k+1/2}) + f(t_{k+1})],$$

where $h = (b - a)/M$ and $t_k = a + kh$, and $t_{k+1/2} = a + (k + \frac{1}{2})h$, which is Simpson's approximation (with step size $h/2$) for the definite integral of $f(t)$ taken over the interval $[a, b]$.

9. The Richardson improvement method discussed in Lemma 7.1 (Section 7.5) can be used in conjunction with the Runge–Kutta method. If the Runge–Kutta method of order $N = 4$ is used with step size h, then we have

$$y(b) \approx y_h + Ch^4.$$

If the Runge–Kutta method of order $N = 4$ is used with $2h$, then we have

$$y(b) \approx y_{2h} + 16Ch^4.$$

The terms involving Ch^4 can be eliminated to obtain an improved approximation for $y(b)$ and the result is

$$y(b) \approx \frac{16y_h - y_{2h}}{15}.$$

This improvement scheme can be used with the values in Example 9.12 to obtain better approximations to $y(3)$. Find the missing entries in the table below.

h	y_h	$\dfrac{16y_h - y_{2h}}{15}$
1	1.6701860	
$\frac{1}{2}$	1.6694308	_____
$\frac{1}{4}$	1.6693928	_____
$\frac{1}{8}$	1.6693906	_____

10. Write a report about the proof of the Runge–Kutta method of order $N = 4$. Be sure to mention the parameters and the system of eight equations involving them.

11. Write a report on the Runge–Kutta–Fehlberg method.

9.6 Predictor–Corrector Methods

The methods of Euler, Heun, Taylor, and Runge–Kutta are called *single-step methods* because they use only the information from one previous point to compute the successive point, that is, only the initial point (t_0, y_0) is used to compute (t_1, y_1) and in general y_k is needed to compute y_{k+1}. After several points have been found it is feasible to use several prior points in the calculation. For illustration, we will develop the Adams–Bashforth four-step method, which requires $y_{k-3}, y_{k-2}, y_{k-1}$, and y_k in the calculation of y_{k+1}. This method is not self-starting; four initial points $(t_0, y_0), (t_1, y_1), (t_2, y_2)$, and (t_3, y_3) must be given in advance in order to generate the points $\{(t_k, y_k)\}$: for $k \geq 4\}$.

A desirable feature of a multistep method is that the local truncation error (L.T.E.) can be determined and a correction term can be included, which improves the accuracy of the answer at each step. Also, it is possible to determine if the step size is small enough to obtain an accurate value for y_{k+1}, yet large enough so that unnecessary and time-consuming calculations are eliminated. Using the combination of a predictor and corrector requires only two function evaluations of $f(t, y)$ per step.

The Adams-Bashforth-Moulton Method

The Adams-Bashforth-Moulton predictor-corrector method is a multistep method based on the fundamental theorem of calculus:

$$(1) \qquad y(t_{k+1}) = y(t_k) + \int_{t_k}^{t_{k+1}} f(t, y(t))\, dt.$$

The predictor uses the Lagrange polynomial approximation for $f(t, y(t))$ based on the points $(t_{k-3}, f_{k-3}), (t_{k-2}, f_{k-2}), (t_{k-1}, f_{k-1})$, and (t_k, f_k). It is integrated over the interval $[t_k, t_{k+1}]$ in (1). This produces the Adams-Bashforth predictor:

$$(2) \qquad p_{k+1} = y_k + \frac{h}{24}(-9f_{k-3} + 37f_{k-2} - 59f_{k-1} + 55f_k).$$

The corrector is developed similarly. The value p_{k+1} can now be used. A second Lagrange polynomial for $f(t, y(t))$ is constructed which is based on the points $(t_{k-2}, f_{k-2}), (t_{k-1}, f_{k-1}), (t_k, f_k)$, and the new point $(t_{k+1}, f(t_{k+1}, p_{k+1}))$. It is integrated over $[t_k, t_{k+1}]$. This produces the Adams-Moulton corrector:

$$(3) \qquad y_{k+1} = y_k + \frac{h}{24}(f_{k-2} - 5f_{k-1} + 19f_k + 9f_{k+1}).$$

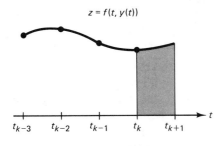

(a) The four nodes for the Adams-Bashforth predictor (extrapolation is used).

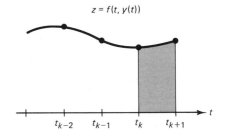

(b) The four nodes for the Adams-Moulton corrector (interpolation is used).

Figure 9.10. Integration over the interval $[t_k, t_{k+1}]$.

Figure 9.10 shows the nodes for the Lagrange polynomials that are used in developing formulas (2) and (3), respectively.

Error Estimation and Correction

The error terms for the numerical integration formulas used to obtain both the predictor and corrector are of the order $O(h^5)$. The L.T.E. for formulas (2) and (3) are

(4) $y(t_{k+1}) - p_{k+1} = \frac{251}{720} y^{(5)}(c_{k+1}) h^5$ (L.T.E. for the predictor),

(5) $y(t_{k+1}) - y_{k+1} = \frac{-19}{720} y^{(5)}(d_{k+1}) h^5$ (L.T.E. for the corrector).

Suppose that h is small and $y^{(5)}(t)$ is nearly constant over the interval; then the terms involving the fifth derivative in (4)-(5) can be eliminated and the result is

(6) $$y(t_{k+1}) - y_{k+1} \approx \frac{-19}{270} (y_{k+1} - p_{k+1}).$$

The importance of the predictor–corrector method should now be evident. Formula (6) gives an approximate error estimate based on the two computed values p_{k+1}, y_{k+1} and does not use $y^{(5)}(t)$.

Practical Considerations

The corrector (3) used the approximation $f_{k+1} \approx f(t_{k+1}, p_{k+1})$ in the calculation of y_{k+1}. Since y_{k+1} is also an estimate for $y(t_{k+1})$, it could be used in the corrector (3) to generate a new approximation for f_{k+1}, which in turn will generate a new value for y_{k+1}. However, when this iteration on the corrector is continued, it will converge to a fixed point of (3) rather than the differential equation. It is more efficient to reduce the step size if more accuracy is needed.

Formula (6) can be used to determine when to change step size. Although elaborate methods are available, we show how to reduce the step size to $h/2$ or increase it to $2h$. Let RelErr $= 5 \times 10^{-6}$ be our relative error criterion, and let Small $= 10^{-5}$.

(7)   IF $\dfrac{19}{270} \dfrac{|y_{k+1} - p_{k+1}|}{|y_{k+1}| + \text{Small}} > \text{RelErr}$ THEN   Set $h = \dfrac{h}{2}$.

(8)   IF $\dfrac{19}{270} \dfrac{|y_{k+1} - p_{k+1}|}{|y_{k+1}| + \text{Small}} < \dfrac{\text{RelErr}}{100}$ THEN   Set $h = 2h$.

When the predicted and corrected values do not agree to five significant digits, then (7) reduces the step size. If they agree to seven or more significant digits, then (8) increases the step size. Fine tuning of these parameters should be made to suit your particular computer.

Reducing the step size requires four new starting values. Interpolation of $f(t, y(t))$ with a fourth-degree polynomial is used to supply the missing values that bisect the intervals $[t_{k-2}, t_{k-1}]$ and $[t_{k-1}, t_k]$. The four mesh points $t_{k-3/2}$, $t_{k-1}, t_{k-1/2}$, and t_k used in the successive calculations are shown in Figure 9.11.

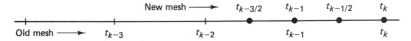

Figure 9.11. Reduction of step size to $h/2$.

The interpolation formulas needed to obtain the new starting values for the step size $h/2$ are

(9)

$$f_{k-1/2} = \frac{-5f_{k-4} + 28f_{k-3} - 70f_{k-2} + 140f_{k-1} + 35f_k}{128},$$

$$f_{k-3/2} = \frac{3f_{k-4} - 16f_{k-3} + 54f_{k-2} + 24f_{k-1} - f_k}{64}.$$

Increasing the step size is an easier task. Seven prior points are needed to double the step size. The four new points are obtained by omitting every second one, as shown in Figure 9.12.

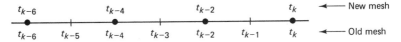

Figure 9.12. Increasing the step size to $2h$.

The Milne–Simpson Method

Another popular predictor–corrector scheme is known as Milne–Simpson's method. Its predictor is based on integration of $f(t, y(t))$ over the interval $[t_{k-3}, t_{k+1}]$:

(10)

$$y(t_{k+1}) = y(t_{k-3}) + \int_{t_{k-3}}^{t_{k+1}} f(t, y(t)) \, dt.$$

The predictor uses the Lagrange polynomial approximation for $f(t, y(t))$ based on the points (t_{k-3}, f_{k-3}), (t_{k-2}, f_{k-2}), (t_{k-1}, f_{k-1}), and (t_k, f_k). It is integrated over the interval $[t_{k-3}, t_{k+1}]$. This produces the Milne predictor:

(11)

$$p_{k+1} = y_{k-3} + \frac{4h}{3}(2f_{k-2} - f_{k-1} + 2f_k).$$

The corrector is developed similarly. The value p_{k+1} can now be used. A second Lagrange polynomial for $f(t, y(t))$ is constructed which is based on the points (t_{k-2}, f_{k-2}), (t_{k-1}, f_{k-1}), and the new point $(t_{k+1}, f(t_{k+1}, p_{k+1}))$. The polynomial is integrated over $[t_{k-1}, t_{k+1}]$ and the result is the familiar Simpson's rule:

(12)

$$y_{k+1} = y_{k-1} + \frac{h}{3}(f_{k-1} + 4f_k + f_{k+1}).$$

Error Estimation and Correction

The error terms for the numerical integration formulas used to obtain both the predictor and corrector are of the order $O(h^5)$. The L.T.E. for the formulas in (11) and (12) are

(13) $$y(t_{k+1}) - p_{k+1} = \tfrac{28}{90} y^{(5)}(c_{k+1}) h^5 \qquad \text{(L.T.E. for the predictor)},$$

(14) $$y(t_{k+1}) - y_{k+1} = \tfrac{-1}{90} y^{(5)}(d_{k+1}) h^5 \qquad \text{(L.T.E. for the corrector)}.$$

Suppose that h is small enough so that $y^{(5)}(t)$ is nearly constant over the interval $[t_{k-3}, t_{k+1}]$. Then the terms involving the fourth derivative can be eliminated in (13) and (14) and the result is

(15) $$y(t_{k+1}) - p_{k+1} \approx \tfrac{28}{29}(y_{k+1} - p_{k+1}).$$

Formula (15) gives an error estimate for the predictor that is based on the two computed values p_{k+1}, y_{k+1} and does not use $y^{(5)}(t)$. It can be used to improve the predicted value. Under the assumption that the difference between the predicted and corrected values at each step changes slowly, we can substitute p_k and y_k for p_{k+1} and y_{k+1} in (15) and get the following modifier:

(16) $$m_{k+1} = p_{k+1} + 28 \frac{y_k - p_k}{29}.$$

This modified value is used in place of p_{k+1} in the correction step, and equation (12) becomes

(17) $$y_{k+1} = y_{k-1} + \frac{h}{3}\left[f_{k-1} + 4f_k + f(t_{k+1}, m_{k+1}) \right].$$

Therefore, the improved (modified) Milne-Simpson method is

$$p_{k+1} = y_{k-3} + \frac{4h}{3}(2f_{k-2} - f_{k-1} + 2f_k) \qquad \text{(predictor)}$$

$$m_{k+1} = p_{k+1} + 28 \frac{y_k - p_k}{29} \qquad \text{(modifier)}$$

(18)

$$f_{k+1} = f(t_{k+1}, m_{k+1})$$

$$y_{k+1} = y_{k-1} + \frac{h}{3}(f_{k-1} + 4f_k + f_{k+1}) \qquad \text{(corrector)}.$$

Hamming's method is another important method. We shall omit its derivation but present it in algorithmic form. As a final precaution we mention that all the predictor–corrector methods have stability problems. Stability is an advanced topic and the serious reader should research this subject.

Algorithm 9.6 [*Adams-Bashforth-Moulton Method*] Solve the I.V.P.

$$y' = f(t, y) \quad \text{on } [a, b] \text{ with } y(a) = y_0.$$

INPUT A, B, Y(0)   {Endpoint and initial value}
INPUT N   {Number of steps, N > 3}
H := [B − A]/N   {Compute the step size}
T(0) := A, F_0 := F(T(0),Y(0))

FOR K = 1 TO 3 DO   {Either input three additional
  T(K) := A + K∗H   starting values or compute them
  Get Y(K)   using the Runge–Kutta method}

  F_1 := F(T(1),Y(1)), F_2 := F(T(2),Y(2)), F_3 := F(T(3),Y(3))
  H2 := H/24   {Saves wasted comutations}

FOR K = 3 TO N−1 DO
  P := Y(K) + H2∗[−9∗F_0+37∗F_1−59∗F_2+55∗F_3]   {Predictor}
  T(K+1) := A + H∗[K+1]   {Next abscissa}
  F_4 := F(T(K+1),P)   {Evaluate f(t,y)}
  Y(K+1) := Y(K) + H2∗[F_1−5∗F_2+19∗F_3+9∗F_4]   {Corrector}
  F_0 := F_1, F_1 := F_2, F_2 := F_3 {Update
  F_3 := F(T(K+1),Y(K+1))   the values}

FOR K = 0 TO N DO   {Output}
  PRINT T(K), Y(K)

Algorithm 9.7 [*Milne-Simpson's Method*] Solve the I.V.P.

$$y' = f(t, y) \quad \text{on } [a, b] \text{ with } y(a) = y_0.$$

INPUT A, B, Y(0)   {Endpoints and initial value}
INPUT N   {Number of steps, n > 3}
H := [B − A]/N   {Compute the step size}
T(0) := A   {Initialize}

FOR ·K = 1 TO 3 DO   {Either input three additional
  T(K) := A + K∗H   starting values or compute them
  Get Y(K)   using the Runge–Kutta method}

  F_1 := F(T(1),Y(1)), F_2 := F(T(2),Y(2)), F_3 := F(T(3),Y(3))
  Pold := 0, Yold := 0   {Initialize}

```
FOR  K = 3  TO  N−1  DO
     Pnew := Y(K−3) + 4*H*[2*F₁−F₂+2*F₃]/3          {Milne predictor}
     Pmod := Pnew + 28*[Yold − Pold]/29             {Modifier}
     T(K+1) := A + H*[K+1]                          {New mesh point}
     F₄ := F(T(K+1),Pmod)                           {Function value}
     Y(K+1) := Y(K−1) + H*[F₂+4*F₃+F₄]/3            {Simpson corrector}
     Pold := Pnew, Yold := Y(K+1)                   {Update
     F₁ := F₂, F₂ := F₃, F₃ := F(T(K+1),Y(K+1))     values}
END
```

```
FOR  K = 0  TO  N  DO                               {Output}
     PRINT T(K), Y(K)
```

Algorithm 9.8 [*Hamming's Method*] Solve the I.V.P.

$$y' = f(t, y) \quad \text{on } [a, b] \text{ with } y(a) = y_0.$$

```
INPUT A, B, Y(0)                                    {Endpoints and initial value}
INPUT N                                             {Number of steps, N > 3}
H := [B − A]/N                                       {Compute the step size}
T(0) := A                                           {Initialize}
```

```
FOR  K = 1  TO  3  DO                               {Either input three additional
     T(K) := A + K*H                                 starting values or compute them
     Get Y(K)                                        using the Runge–Kutta method}
```

```
F₁ := F(T(1),Y(1)), F₂ := F(T(2),Y(2)), F₃ := F(T(3),Y(3))
Pold := 0, Cold := 0                                {Initialize}
```

```
FOR  K = 3  TO  N−1  DO
     Pnew := Y(K−3) + 4*H*[2*F₁−F₂+2*F₃]/3          {Milne predictor}
     Pmod := Pnew + 112*[Cold − Pold]/121           {Modifier}
     T(K+1) := A + H*[K+1]                          {New mesh point}
     F₄ := F(T(K+1),Pmod)                           {Function value}
     Cnew := [9*Y(K)−Y(K−2) + 3*H*[−F₂+2*F₃+F₄]]/8  {Hamming corrector}
     Y(K+1) := Cnew + 9*[Pnew − Cnew]/121           {New value y_{k+1}}
     Pold := Pnew, Cold := Cnew                     {Update
     F₁ := F₂, F₂ := F₃, F₃ := F(T(K+1),Y(K+1))     values}
END
```

```
FOR  K = 0  TO  N  DO                               {Output}
     PRINT T(K), Y(K)
```

Example 9.14 Use the Adams-Bashforth-Moulton, Milne-Simpson, and Hamming methods with $h = \frac{1}{8}$ and compute approximations for the solution of the I.V.P.

$$y' = \frac{t - y}{2} \quad \text{with } y(0) = 1 \text{ over } [0, 3].$$

Solution. A Runge-Kutta method was used to obtain the starting values

$$y_1 = 0.94323919, \quad y_2 = 0.89749071, \quad \text{and} \quad y_3 = 0.86208736.$$

Then a computer implementation of Algorithms 9.6–9.8 produced the values in Table 9.12. The error for each entry in the table is given as a multiple of 10^{-8}. In all entries there are at least six digits of accuracy. In this example, the best answers were produced by Hamming's method.

Table 9.12 Comparison of the Adams-Bashforth-Moulton, Milne-Simpson, and Hamming methods for solving $y' = (t - y)/2, y(0) = 1$

k	Adams-Bashforth-Moulton	Error	Milne-Simpson	Error	Hamming's Method	Error
0.0	1.00000000	0E-8	1.00000000	0E-8	1.00000000	0E-8
0.5	0.83640227	8E-8	0.83640231	4E-8	0.83640234	1E-8
0.625	0.81984673	16E-8	0.81984687	2E-8	0.81984688	1E-8
0.75	0.81186762	22E-8	0.81186778	6E-8	0.81186783	1E-8
0.875	0.81194530	28E-8	0.81194555	3E-8	0.81194558	0E-8
1.0	0.81959166	32E-8	0.81959190	8E-8	0.81959198	0E-8
1.5	0.91709920	46E-8	0.91709957	9E-8	0.91709967	−1E-8
2.0	1.10363781	51E-8	1.10363822	10E-8	1.10363834	−2E-8
2.5	1.35951387	52E-8	1.35951429	10E-8	1.35951441	−2E-8
2.625	1.43243853	52E-8	1.43243899	6E-8	1.43243907	−2E-8
2.75	1.50851827	52E-8	1.50851869	10E-8	1.50851881	−2E-8
2.875	1.58756195	51E-8	1.58756240	6E-8	1.58756248	−2E-8
3.0	1.66938998	50E-8	1.66939038	10E-8	1.66939050	−2E-8

The Right Step

Our selection of methods has a purpose: first, their development is easy enough for a first course; second, more advanced methods have a similar development; third, most undergraduate problems can be solved by one of the methods. However, when a predictor-corrector method is used to solve the I.V.P. $y' = f(t, y), y(t_0) = y_0$ over a large interval, difficulties sometimes occur.

If $f_y(t, y) < 0$ and the step size is too large, a predictor-corrector method might be unstable. As a rule of thumb, stability exists when a small error is propagated as a decreasing error and instability exists when a small error is propagated as an increasing error. When too large a step size is used over a large interval, instability will result and is sometimes manifest by oscillations in the computed solution. They can be attenuated by changing to a smaller step size. Formulas (7)–(9) suggest how to modify the algorithm(s). When step-size control is included, the following error estimate(s) should be used:

(19) $$y(t_k) - y_k \approx 19\frac{p_k - y_k}{270} \qquad \text{(Adams–Bashforth–Moulton)},$$

(20) $$y(t_k) - y_k \approx \frac{p_k - y_k}{29} \qquad \text{(Milne–Simpson)},$$

(21) $$y(t_k) - y_k \approx 9\frac{p_k - y_k}{121} \qquad \text{(Hamming)}.$$

In all methods, the corrector step is a type of fixed-point iteration. It can be proven that the step size h for the methods must satisfy the following conditions:

(22) $$h \ll \frac{2.66667}{|f_y(t,y)|} \qquad \text{(Adams–Bashforth–Moulton)},$$

(23) $$h \ll \frac{3.00000}{|f_y(t,y)|} \qquad \text{(Milne–Simpson)},$$

(24) $$h \ll \frac{2.66667}{|f_y(t,y)|} \qquad \text{(Hamming)}.$$

The notation $\ll$ in (22)–(24) means "much smaller than." The next example shows that more stringent inequalities should be used:

(25) $$h < \frac{0.75}{|f_y(t,y)|} \qquad \text{(Adams–Bashforth–Moulton)},$$

(26) $$h < \frac{0.45}{|f_y(t,y)|} \qquad \text{(Milne–Simpson)},$$

(27) $$h < \frac{0.69}{|f_y(t,y)|} \qquad \text{(Hamming)}.$$

Inequality (27) is found in advanced books on numerical analysis. The other two inequalities seem appropriate for the example.

Example 9.15 Use the Adams–Bashforth–Moulton, Milne–Simpson, and Hamming methods and compute approximations for the solution of

$$y' = 30 - 5y \qquad y(0) = 1 \qquad \text{over the interval } [0, 10].$$

Solution. All three methods are of the order $O(h^4)$. When $N = 120$ steps was used for all three methods, the maximum error for each method occurred at a different place:

$$y(0.41666667) - y_5 \approx -0.00277037 \quad \text{(Adams–Bashforth–Moulton)},$$

$$y(0.33333333) - y_4 \approx -0.00139255 \quad \text{(Milne–Simpson)},$$

$$y(0.33333333) - y_4 \approx -0.00104982 \quad \text{(Hamming)}.$$

At the right endpoint $t = 10$, the error was

$$y(10) - y_{120} \approx 0.00000000 \quad \text{(Adams–Bashforth–Moulton)},$$

$$y(10) - y_{120} \approx 0.00001015 \quad \text{(Milne–Simpson)},$$

$$y(10) - y_{120} \approx 0.00000000 \quad \text{(Hamming)}.$$

Both the Adams–Bashforth–Moulton and Hamming methods gave approximate solutions with eight digits of accuracy at the right endpoint.

It is instructive to see that if the step size is too large, the computed solution oscillates about the true solution. Figure 9.13(a)-(c) illustrate this phenomenon. The small number of steps were determined experimentally so that the oscillations were about the same magnitude. The large number of steps required to attenuate the oscillations were determined with (25)-(27).

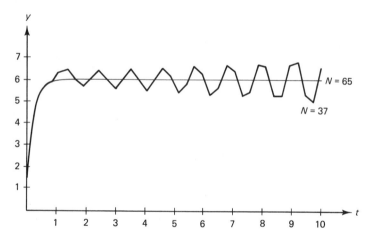

Figure 9.13(a). Adams–Bashforth–Moulton solution to $y' = 30 - 5y$ with $N = 37$ steps produces oscillations. It is stabilized when $N = 65$ because

$$h = \frac{10}{65} = 0.1538 \approx 0.15 = \frac{0.75}{5} = \frac{0.75}{|f_y(t,y)|}.$$

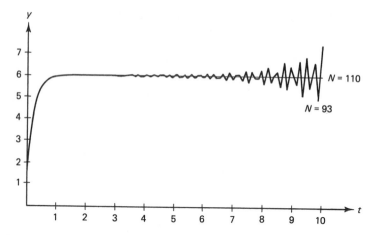

Figure 9.13(b). Milne–Simpson solution to $y' = 30 - 5y$ with $N = 93$ steps produces oscillations. It is stabilized when $N = 110$ because

$$h = \frac{10}{110} = 0.0909 \approx 0.09 = \frac{0.45}{5} = \frac{0.45}{|f_y(t,y)|}.$$

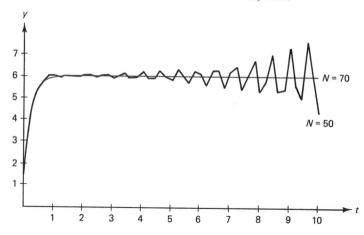

Figure 9.13(c). Hamming's solution to $y' = 30 - 5y$ with $N = 50$ steps produces oscillations. It is stabilized when $N = 70$ because

$$h = \frac{10}{70} = 0.1428 \approx 0.138 = \frac{0.69}{5} = \frac{0.69}{|f_y(t,y)|}.$$

Exercises for Predictor–Corrector Methods

In Exercises 1–10:
   (a) Use any one of the three predictor–corrector methods and the three starting values y_1, y_2, and y_3 and $h = 0.05$ to calculate the next value y_4 for the I.V.P.

(b) Use a computer implementation of any predictor–corrector method and solve the I.V.P. in part (a) over $[a, b]$. Use the three starting values y_1, y_2, and y_3 that are given or generate them with a Runge-Kutta method.

(c) Compare your answer with the true solution $y(t)$.

1. (a) $y' = t^2 - y$ with $y(0) = 1$
   (b) $[a, b] = [0, 5]$

   $y(0.05) = 0.95127058, \quad y(0.10) = 0.90516258, \quad y(0.15) = 0.86179202$

   (c) $y(t) = -\exp(-t) + t^2 - 2t + 2$

2. (a) $y' = y + 3t - t^2$ with $y(0) = 1$
   (b) $[a, b] = [0, 5]$

   $y(0.05) = 1.0550422, \quad y(0.10) = 1.1203418, \quad y(0.15) = 1.1961685$

   (c) $y(t) = 2 \exp(t) + t^2 - t - 1$

3. (a) $y' = -t/y$ with $y(1) = 1$
   (b) $[a, b] = [1, 1.4]$

   $y(1.05) = 0.94736477, \quad y(1.10) = 0.88881944, \quad y(1.15) = 0.82310388$

   (c) $y(t) = (2 - t^2)^{1/2}$

4. (a) $y' = \exp(-t) - y$ with $y(0) = 1$
   (b) $[a, b] = [0, 5]$

   $y(0.05) = 0.99879090, \quad y(0.10) = 0.99532116, \quad y(0.15) = 0.98981417$

   (c) $y(t) = t \exp(-t) + \exp(-t)$

5. (a) $y' = 2ty^2$ with $y(0) = 1$
   (b) $[a, b] = [0, 0.95]$

   $y(0.05) = 1.0025063, \quad y(0.10) = 1.0101010, \quad y(0.15) = 1.0230179$

   (c) $y(t) = 1/(1 - t^2)$

6. (a) $y' = 1 + y^2$ with $y(0) = 1$
   (b) $[a, b] = [0, 0.75]$

   $y(0.05) = 1.1053556, \quad y(0.10) = 1.2230489, \quad y(0.15) = 1.3560879$

   (c) $y(t) = \tan(t + \pi/4)$

7. (a) $y' = 2y - y^2$ with $y(0) = 1$
   (b) $[a, b] = [0, 5]$

   $y(0.05) = 1.0499584, \quad y(0.10) = 1.0996680, \quad y(0.15) = 1.1488850$

   (c) $y(t) = 1 + \tanh(t)$

8. (a) $y' = [1 - y^2]^{1/2}$ with $y(0) = 0$
   (b) $[a, b] = [0, 1.55]$

   $y(0.05) = 0.049979169, \quad y(0.10) = 0.099833417, \quad y(0.15) = 0.14943813$

   (c) $y(t) = \sin(t)$

9. (a) $y' = y^2 \sin(t)$ with $y(0) = 1$
   (b) $[a, b] = [0, 1.55]$

$$y(0.05) = 1.0012513, \quad y(0.10) = 1.0050209, \quad y(0.15) = 1.0113564$$

   (c) $y(t) = \sec(t)$

10. (a) $y' = 1 - y^2$ with $y(0) = 0$
    (b) $[a, b] = [0, 5]$

$$y(0.05) = 0.049958375, \quad y(0.10) = 0.099667995, \quad y(0.15) = 0.14888503$$

    (c) $y(t) = [1 - \exp(-2t)]/[1 + \exp(-2t)]$

11. Write a report on Hamming's method.

12. Modify one of the predictor–corrector algorithms so that it includes step-size control.

13. Use a computer program with step-size control to solve one of Exercises 1–10.

14. Write a report about the stability of predictor–corrector methods.

15. Write a report about stiff differential equations.

9.7 Systems of Differential Equations

This section is an introduction to systems of differential equations. To illustrate the concepts, we consider the initial value problem

(1)
$$\frac{dx}{dt} = f(t, x, y) \qquad \text{with} \qquad \begin{cases} x(t_0) = x_0, \\ \\ y(t_0) = y_0. \end{cases}$$
$$\frac{dy}{dt} = g(t, x, y)$$

A solution to (1) is a pair of differentiable functions $x(t)$ and $y(t)$ with the property that when t, $x(t)$, and $y(t)$ are substituted in $f(t, x, y)$ and $g(t, x, y)$, the result is equal to the derivative $x'(t)$ and $y'(t)$, respectively, that is,

(2)
$$\begin{aligned} x'(t) &= f(t, x(t), y(t)) \\ y'(t) &= g(t, x(t), y(t)) \end{aligned} \qquad \text{with} \qquad \begin{cases} x(t_0) = x_0, \\ y(t_0) = y_0. \end{cases}$$

For example, consider the system of differential equations

(3)
$$\frac{dx}{dt} = x + 2y \qquad \text{with} \qquad \begin{cases} x(0) = 6, \\ \\ y(0) = 4. \end{cases}$$
$$\frac{dy}{dt} = 3x + 2y$$

The solution to the I.V.P. (3) is

(4)
$$\begin{aligned} x(t) &= 4e^{4t} + 2e^{-t}, \\ y(t) &= 6e^{4t} - 2e^{-t}. \end{aligned}$$

This is verified by directly substituting $x(t)$ and $y(t)$ into the right-hand side of (3) and computing the derivatives of (4) and substituting them in the left-hand side of (4) to get

$$16e^{4t} - 2e^{-t} = (4e^{4t} + 2e^{-t}) + 2(6e^{4t} - 2e^{-t}),$$
$$24e^{4t} + 2e^{-t} = 3(4e^{4t} + 2e^{-t}) + 2(6e^{4t} - 2e^{-t}).$$

Numerical Solutions

A numerical solution to (1) over the interval $a \leqq t \leqq b$ is found by considering the differentials

(5) $$dx = f(t, x, y) \, dt \quad \text{and} \quad dy = g(t, x, y) \, dt.$$

Euler's method for solving the system is easy to formulate. The differentials $dt = t_{k+1} - t_k$, $dx = x_{k+1} - x_k$, and $dy = y_{k+1} - y_k$ are substituted into (5) to get

(6)
$$x_{k+1} - x_k \approx f(t_k, x_k, y_k)(t_{k+1} - t_k),$$
$$y_{k+1} - y_k \approx g(t_k, x_k, y_k)(t_{k+1} - t_k).$$

The interval is divided into M subintervals of width $h = (b - a)/M$, and the mesh points are $t_{k+1} = t_k + h$. This is used in (6) to get the recursive formulas for Euler's method:

(7)
$$t_{k+1} = t_k + h,$$
$$x_{k+1} = x_k + hf(t_k, x_k, y_k),$$
$$y_{k+1} = y_k + hg(t_k, x_k, y_k) \quad \text{for } k = 0, 1, \ldots, M - 1.$$

A higher-order method should be used to achieve a reasonable amount of accuracy. For example, the Runge–Kutta formulas of order 4 are

(8)
$$x_{k+1} = x_k + \frac{h}{6}(f_1 + 2f_2 + 2f_3 + f_4)$$

$$y_{k+1} = y_k + \frac{h}{6}(g_1 + 2g_2 + 2g_3 + g_4)$$

where

$$f_1 = f(t_k, x_k, y_k), \qquad\qquad g_1 = g(t_k, x_k, y_k),$$

$$f_2 = f\left(t_k + \frac{h}{2}, x_k + \frac{h}{2}f_1, y_k + \frac{h}{2}g_1\right), \quad g_2 = g\left(t_k + \frac{h}{2}, x_k + \frac{h}{2}f_1, y_k + \frac{h}{2}g_1\right),$$

$$f_3 = f\left(t_k + \frac{h}{2}, x_k + \frac{h}{2}f_2, y_k + \frac{h}{2}g_2\right), \quad g_3 = g\left(t_k + \frac{h}{2}, x_k + \frac{h}{2}f_2, y_k + \frac{h}{2}g_2\right),$$

$$f_4 = f(t_k + h, x_k + hf_3, y_k + hg_3), \qquad g_4 = g(t_k + h, x_k + hf_3, y_k + hg_3).$$

Example 9.16 Use the Runge–Kutta method given in (8) and compute the numerical solution to (3) over the interval $[0.0, 0.2]$ using 10 subintervals and the step size $h = 0.02$.

Solution. For the first point we have $t_1 = 0.02$ and the intermediate calculations required to compute x_1 and y_1 are

$$f_1 = f(0.00, 6.0, 4.0) = 14.0 \qquad g_1 = g(0.00, 6.0, 4.0) = 26.0$$

$$x_0 + \frac{h}{2} f_1 = 6.14 \qquad y_0 + \frac{h}{2} g_1 = 4.26,$$

$$f_2 = f(0.01, 6.14, 4.26) = 14.66 \qquad g_2 = g(0.01, 6.14, 4.26) = 26.94$$

$$x_0 + \frac{h}{2} f_2 = 6.1466 \qquad y_0 + \frac{h}{2} g_2 = 4.2694,$$

$$f_3 = f(0.01, 6.1466, 4.2694) = 14.6854,$$

$$g_3 = g(0.01, 6.1466, 4.2694) = 26.9786,$$

$$x_0 + hf_3 = 6.293708 \qquad y_0 + hg_3 = 4.539572,$$

$$f_4 = f(0.02, 6.293708, 4.539572) = 15.372852,$$

$$g_4 = g(0.02, 6.293708, 4.539572) = 27.960268.$$

These values are used in the final computation:

$$x_1 = 6 + \frac{0.02}{6}(14.0 + 2 \times 14.66 + 2 \times 14.6854 + 15.372852) = 6.29354551,$$

$$y_1 = 4 + \frac{0.02}{6}(26.0 + 2 \times 26.94 + 2 \times 26.9786 + 27.960268) = 4.53932490.$$

The calculations are summarized in Table 9.13.

Table 9.13 Runge–Kutta solution to
$x'(t) = x + 2y, y'(t) = 3x + 2y$
with the initial values $x(0) = 6$
and $y(0) = 4$

k	t_k	x_k	y_k
0	0.00	6.00000000	4.00000000
1	0.02	6.29354551	4.53932490
2	0.04	6.61562213	5.11948599
3	0.06	6.96852528	5.74396525
4	0.08	7.35474319	6.41653305
5	0.10	7.77697287	7.14127221
6	0.12	8.23813750	7.92260406
7	0.14	8.74140523	8.76531667
8	0.16	9.29020955	9.67459538
9	0.18	9.88827138	10.6560560
10	0.20	10.5396230	11.7157807

The numerical solutions contain a certain amount of error at each step. For the example above, the error grows and at the right endpoint $t = 0.2$ it reaches its maximum:

$$x(0.2) - x_{10} = 10.5396252 - 10.5396230 = 0.0000022,$$

$$y(0.2) - y_{10} = 11.7157841 - 11.7157807 = 0.0000034.$$

Higher-Order Differential Equations

Higher-order differential equations involve the higher derivatives $x''(t)$, $x'''(t)$, and so on. They arise in mathematical models for problems in physics and engineering. For example,

$$\frac{m}{g} x''(t) + cx'(t) + kx(t) = f(t),$$

represents a mechanical system in which a spring with spring constant k restores a displaced mass m. Friction is assumed to be proportional to the velocity and the function $f(t)$ is an external force. It is often the case that the position $x(t_0)$ and velocity $x'(t_0)$ are known at a certain time t_0.

By solving for the second derivative, we can write a second-order initial value problem in the form

(9) $$x''(t) = f(t, x(t), x'(t)) \quad \text{with} \quad x(t_0) = x_0 \quad \text{and} \quad x'(t_0) = y_0.$$

The second-order differential equation can be reformulated as a system of two first-order equations if we use the substitution

(10) $$x'(t) = y(t)$$

then $x''(t) = y'(t)$ and the differential equation in (9) becomes a system

(11) $$\frac{dx}{dt} = y \qquad \text{with} \quad \begin{cases} x(t_0) = x_0, \\ \\ y(t_0) = y_0. \end{cases}$$
$$\frac{dy}{dt} = f(t, x, y)$$

A numerical procedure such as the Runge–Kutta method can be used to solve (11) and will generate two sequences, $\{x_k\}$ and $\{y_k\}$. The first sequence is the numerical solution to (9). The next example can be interpreted as damped harmonic motion.

Example 9.17 Consider the second-order initial value problem

$$x''(t) + 4x'(t) + 5x(t) = 0 \quad \text{with} \quad x(0) = 3 \quad \text{and} \quad x'(0) = -5.$$

(a) Write down the equivalent system of two first-order equations.
(b) Use the Runge–Kutta method to solve the reformulated problem over $[0, 5]$ using $M = 50$ subintervals of width $h = 0.1$.

(c) Compare the numerical solution with the true solution:

$$x(t) = 3e^{-2t}\cos(t) + e^{-2t}\sin(t).$$

Solution. The differential equation has the form

$$x''(t) = f(t, x(t), x'(t)) = -4x'(t) - 5x(t).$$

Using the substitutions in (10), we get the reformulated problem:

$$\frac{dx}{dt} = y \qquad \text{with} \quad \begin{cases} x(0) = 3, \\ \\ y(0) = -5. \end{cases}$$
$$\frac{dy}{dt} = -5x - 4y$$

Samples of the numerical computations are given in Table 9.14. The values $\{y_k\}$ are extraneous and are not included. Instead, the true solution values $\{x(t_k)\}$ are included for comparison.

Table 9.14   Runge-Kutta solution to
$x''(t) + 4x'(t) + 5x(t) = 0$ with
the initial conditions $x(0) = 3$
and $x'(0) = -5$

k	t_k	x_k	$x(t_k)$
0	0.0	3.00000000	3.00000000
1	0.1	2.52564583	2.52565822
2	0.2	2.10402783	2.10404686
3	0.3	1.73506269	1.73508427
4	0.4	1.41653369	1.41655509
5	0.5	1.14488509	1.14490455
10	1.0	0.33324302	0.33324661
20	2.0	−0.00620684	−0.00621162
30	3.0	−0.00701079	−0.00701204
40	4.0	−0.00091163	−0.00091170
48	4.8	−0.00004972	−0.00004969
49	4.9	−0.00002348	−0.00002345
50	5.0	−0.00000493	−0.00000490

Exercises for Systems of Differential Equations

In Exercises 1–4:
  (a) Verify that the two functions $x(t)$ and $y(t)$ are solutions to the system of differential equations.
  (b) Use $h = 0.1$ and Euler's method to find x_1, y_1 and x_2, y_2.
  (c) Use $h = 0.1$ and the Runge–Kutta method to find x_1 and y_1.
  (d) Use the Runge–Kutta method to solve the system over the interval $[0, 2]$ using $M = 40$ steps and $h = 0.05$.

1. $\dfrac{dx}{dt} = 2x + 3y$

$\qquad\qquad$ with $\qquad \begin{cases} x(0) = 2, \\[2mm] y(0) = 3. \end{cases}$

$\quad\dfrac{dy}{dt} = 2x + y$

$\quad x(t) = 3e^{4t} - e^{-t}, \qquad y(t) = 2e^{4t} + e^{-t}$

2. $\dfrac{dx}{dt} = 3x - y$

$\qquad\qquad$ with $\qquad \begin{cases} x(0) = 2, \\[2mm] y(0) = 3. \end{cases}$

$\quad\dfrac{dy}{dt} = 4x - y$

$\quad x(t) = 2e^{t} + te^{t}, \qquad y(t) = 3e^{t} + 2te^{t}$

3. $\dfrac{dx}{dt} = x - 4y$

$\qquad\qquad$ with $\qquad \begin{cases} x(0) = 2, \\[2mm] y(0) = 3. \end{cases}$

$\quad\dfrac{dy}{dt} = x + y$

$\quad x(t) = 2e^{t}\cos(2t) - 6e^{t}\sin(2t), \qquad y(t) = e^{t}\sin(2t) + 3e^{t}\cos(2t)$

4. $\dfrac{dx}{dt} = 2x + 9y$

$\qquad\qquad$ with $\qquad \begin{cases} x(0) = -3, \\[2mm] y(0) = 3. \end{cases}$

$\quad\dfrac{dy}{dt} = x + 2y$

$\quad x(t) = 3e^{5t} - 6e^{-t} \qquad y(t) = e^{5t} + 2e^{-t}$

In Exercises 5–8:

$\quad$ (a) Verify that the function $x(t)$ is the solution.

$\quad$ (b) Reformulate the second-order differential equation as a system of two first-order equations.

$\quad$ (c) Use $h = 0.1$ and Euler's method to find x_1 and x_2.

$\quad$ (d) Use $h = 1$ and the Runge–Kutta method to find x_1.

$\quad$ (e) Use the Runge–Kutta method to solve the differential equation over the interval $[0, 2]$ using $M = 40$ steps and $h = 0.05$.

5. $2x''(t) - 5x'(t) - 3x(t) = 45e^{2t}$ with $x(0) = 2$ and $x'(0) = 1$.

$\quad x(t) = 4e^{-t/2} + 7e^{3t} - 9e^{2t}$

6. $x''(t) + 6x'(t) + 9x(t) = 0$ with $x(0) = 4$ and $x'(0) = -4$.

$\quad x(t) = 4e^{-3t} + 8te^{-3t}$

7. $x''(t) + x(t) = 6\cos(t)$ with $x(0) = 2$ and $x'(0) = 3$.

$\quad x(t) = 2\cos(t) + 3\sin(t) + 3t\sin(t)$

8. $x''(t) + 3x'(t) = 12$ with $x(0) = 5$ and $x'(0) = 1$.

$\quad x(t) = 4 + 4t + e^{-3t}$

9. A certain resonant spring system with a periodic forcing function is modeled by
$$x''(t) + 25x(t) = 8 \sin(5t) \quad \text{with} \quad x(0) = 0 \quad \text{and} \quad x'(0) = 0.$$
Use the Runge–Kutta method to solve the differential equation over the interval $[0, 2]$ using $M = 40$ steps and $h = 0.05$.

10. The mathematical model of a certain RLC electrical circuit is
$$Q''(t) + 20Q'(t) + 125Q(t) = 9 \sin(5t) \quad \text{with} \quad Q(0) = 0 \quad \text{and} \quad Q'(0) = 0.$$
Use the Runge–Kutta method to solve the differential equation over the interval $[0, 2]$ using $M = 40$ steps and $h = 0.05$.
    Remark. $I(t) = Q'(t)$ is the current at time t.

11. At time t, a pendulum makes an angle $x(t)$ with the vertical axis. Assuming that there is no friction, the equation of motion is
$$mlx''(t) = -mg \sin(x(t)),$$
where m is the mass and l is the length of the string. Use the Runge–Kutta method to solve the differential equation over the interval $[0, 2]$ using $M = 40$ steps and $h = 0.05$ if $g = 32$ ft/sec^2 and
    (a) $l = 3.2$ ft and $x(0) = 0.3$ and $x'(0) = 0$
    (b) $l = 0.8$ ft and $x(0) = 0.3$ and $x'(0) = 0$

12. *Predator–Prey Model.* An example of a system of nonlinear differential equations is the predator–prey problem. Let $x(t)$ and $y(t)$ denote the population of rabbits and foxes, respectively, at time t. The predator–prey model asserts that $x(t)$ and $y(t)$ satisfy
$$x'(t) = Ax(t) - Bx(t)y(t),$$
$$y'(t) = Cx(t)y(t) - Dy(t).$$
A typical computer simulation might use the coefficients
$$A = 2 \qquad B = 0.02 \qquad C = 0.0002 \qquad D = 0.8.$$
Use the Runge–Kutta method to solve the differential equation over the interval $[0, 5]$ using $M = 50$ steps and $h = 0.1$ if
    (a) $x(0) = 3{,}000$ rabbits and $y(0) = 120$ foxes
    (b) $x(0) = 5{,}000$ rabbits and $y(0) = 100$ foxes

13. Write a report about the shooting method.

14. Write a report about the finite-difference method.

Appendix A

Formulas from Algebra and Calculus

$|ab| = |a| \, |b|$ and $|a + b| \leq |a| + |b|$ and $|a| - |b| \leq |a - b|$

$0^0 = 0$ and $a^0 = 1$ when $a \neq 0$

$a^N a^M = a^{N+M}$ and $\dfrac{1}{a^M} = a^{-M}$ and $\dfrac{a^N}{a^M} = a^{N-M}$

$(ab)^N = a^N b^N$ and $(a^M)^N = a^{MN}$

$y = x^{1/N}$ if and only if $y^N = x$

We use the notation $\exp(x) = e^x$

$\exp(\ln(x)) = x$ for $x > 0$ and $\ln(\exp(x)) = x$ for all x

$\ln(ab) = \ln(a) + \ln(b)$ and $\ln(a^N) = N \ln(a)$

$e^{a+b} = e^a e^b$ and $a^x = e^{\ln(a)x}$

$1 + r + r^2 + \ldots + r^{N-1} = \dfrac{1 - r^N}{1 - r}$ for $r \neq 1$

$1 + 2 + 3 + 4 + \ldots + N = \dfrac{N(N + 1)}{2}$

$1^2 + 2^2 + 3^2 + \ldots + N^2 = \dfrac{N(N + 1)(2N + 1)}{6}$

$1^3 + 2^3 + 3^3 + \ldots + N^3 = \dfrac{N^2(N + 1)^2}{4}$

$0! = 1$ and for any positive integer N, $N! = N(N - 1)(N - 2) \ldots (2)(1)$

$$\binom{N}{J} = \frac{N!}{J!(N-J)!} \qquad \textit{binomial coefficient}$$

Quadratic formula: If $a \neq 0$, then the solutions of $ax^2 + bx + c = 0$ are given by

$$x_1 = \frac{-b + (b^2 - 4ac)^{1/2}}{2a} \qquad x_2 = \frac{-b - (b^2 - 4ac)^{1/2}}{2a}$$

Formulas from Trigonometry

$$\tan(x) = \frac{\sin(x)}{\cos(x)} \quad \text{and} \quad \cot(x) = \frac{\cos(x)}{\sin(x)}$$

$$\sec(x) = \frac{1}{\cos(x)} \quad \text{and} \quad \csc(x) = \frac{1}{\sin(x)}$$

$$\sin^2(x) + \cos^2(x) = 1 \quad \text{and} \quad 1 + \tan^2(x) = \sec^2(x)$$

$$\sin(-x) = -\sin(x) \quad \text{and} \quad \cos(-x) = \cos(x)$$

$$\sin\left(\frac{\pi}{2} - x\right) = \cos(x) \quad \text{and} \quad \cos\left(x - \frac{\pi}{2}\right) = \sin(x)$$

$$\arcsin(x) + \arccos(x) = \frac{\pi}{2}$$

$$\sin(x + y) = \sin(x)\cos(y) + \cos(x)\sin(y)$$

$$\cos(x + y) = \cos(x)\cos(y) - \sin(x)\sin(y)$$

$$\tan(x + y) = \frac{\tan(x) + \tan(y)}{1 - \tan(x)\tan(y)}$$

$$\sin^2(x) = \tfrac{1}{2}[1 - \cos(2x)] \quad \text{and} \quad \cos^2(x) = \tfrac{1}{2}[1 + \cos(2x)]$$

$$\sin(x)\sin(y) = \tfrac{1}{2}[\cos(x - y) - \cos(x + y)]$$

$$\cos(x)\cos(y) = \tfrac{1}{2}[\cos(x - y) + \cos(x + y)]$$

$$\sin(x)\cos(y) = \tfrac{1}{2}[\sin(x + y) + \sin(x - y)]$$

Law of cosines: $C^2 = A^2 + B^2 - 2AB\cos(\theta)$

Hyperbolic Functions

$$\sinh(x) = \frac{e^x - e^{-x}}{2} \quad \text{and} \quad \cosh(x) = \frac{e^x + e^{-x}}{2}$$

$$\sinh(-x) = -\sinh(x) \quad \text{and} \quad \cosh(-x) = \cosh(x)$$

$$\cosh^2(x) - \sinh^2(x) = 1$$

$$\sinh(x + y) = \sinh(x)\cosh(y) + \cosh(x)\sinh(y)$$

$$\cosh(x + y) = \cosh(x)\cosh(y) + \sinh(x)\sinh(y)$$

Formulas from Geometry

Area of a circle: $A = \pi R^2$ where R is the radius

Circumference of a circle: $L = 2\pi R$ (R is the radius)

Area of a triangle: $A = \frac{1}{2}BH$ (B is the base, H is the height)

Area of a trapezoid: $A = \frac{1}{2}H(B_1 + B_2)$

Volume of a sphere: $V = \frac{4}{3}\pi R^3$ where R is the radius

Surface area of a sphere: $S = 4\pi R^2$ (R is the radius)

Formulas from Analytic Geometry

Slope of a line: $m = \dfrac{y_2 - y_1}{x_2 - x_1}$

Equation of a line: $y - y_1 = m(x - x_1)$

Equation of a circle: $(x - x_0)^2 + (y - y_0)^2 = R^2$

Equation of an ellipse: $\dfrac{x^2}{A^2} + \dfrac{y^2}{B^2} = 1$

Equation of a hyperbola: $\dfrac{x^2}{A^2} - \dfrac{y^2}{B^2} = 1$

Equation of a plane: $Ax + By + Cz = D$

Equation of a sphere: $(x - x_0)^2 + (y - y_0)^2 + (z - z_0)^2 = R^2$

Equation of an ellipsoid: $\dfrac{x^2}{A^2} + \dfrac{y^2}{B^2} + \dfrac{z^2}{C^2} = 1$

Equation of an elliptic paraboloid: $\dfrac{x^2}{A^2} + \dfrac{y^2}{B^2} = Cz$

Definitions from Calculus

Limit (i). The notation $\lim_{x \to a} f(x) = L$ means that for any $\epsilon > 0$ there exists a $\delta > 0$ such that

$$0 < |x - a| < \delta \qquad \text{implies that } |f(x) - L| < \epsilon.$$

Limit (ii). The notation $\lim_{h \to 0} f(a + h) = L$ means that for any $\epsilon > 0$ there exists a $\delta > 0$ such that

$$0 < |h| < \delta \qquad \text{implies that } |f(a + h) - L| < \epsilon.$$

Continuity (i). The function $f(x)$ is continuous at $x = a$ provided that

$$\lim_{x \to a} f(x) = f(a).$$

Continuity (ii). The function $f(x)$ is continuous at $x = a$ provided that

$$\lim_{h \to 0} f(a + h) = f(a).$$

Derivative (i). The derivative of $f(x)$ at $x = a$ is denoted by $f'(a)$ and is defined by the limit:

$$f'(a) = \lim_{x \to a} \frac{f(x) - f(a)}{x - a}.$$

Derivative (ii). The derivative of $f(x)$ at $x = a$ is denoted by $f'(a)$ and is defined by the limit:

$$f'(a) = \lim_{h \to 0} \frac{f(a + h) - f(a)}{h}.$$

Properties of Derivatives

Notation. If $y = f(x)$, then we write $y' = \dfrac{dy}{dx} = f'(x)$.

$[f(x) + g(x)]' = f'(x) + g'(x)$	addition rule
$[Cf(x)]' = Cf'(x)$ where C is a constant	scalar multiple
$[f(x)g(x)]' = f(x)g'(x) + f'(x)g(x)$	product rule
$\left[\dfrac{f(x)}{g(x)}\right]' = \dfrac{f'(x)g(x) - f(x)g'(x)}{[g(x)]^2}$	quotient rule
$[f(g(x))]' = f'(g(x))g'(x)$	chain rule

Derivatives of Common Functions

$$\frac{d}{dx}(x^N) = Nx^{N-1}$$

$$\frac{d}{dx}[\exp(x)] = \exp(x) \quad \text{and} \quad \frac{d}{dx}(\ln|x|) = \frac{1}{x}$$

$$\frac{d}{dx}[\sin(x)] = \cos(x) \quad \text{and} \quad \frac{d}{dx}[\cos(x)] = -\sin(x)$$

$$\frac{d}{dx}[\tan(x)] = \sec^2(x) \quad \text{and} \quad \frac{d}{dx}[\sec(x)] = \sec(x)\tan(x)$$

$$\frac{d}{dx}[\sinh(x)] = \cosh(x) \quad \text{and} \quad \frac{d}{dx}[\cosh(x)] = \sinh(x)$$

$$\frac{d}{dx}[\arctan(x)] = (1 + x^2)^{-1} \quad \text{and} \quad \frac{d}{dx}[\arcsin(x)] = (1 - x^2)^{-1/2}$$

$$\frac{d}{dx}(a^x) = \ln(a)a^x \quad \text{and} \quad \frac{d}{dx}\log_a|x| = \frac{1}{\ln(a)x} \quad \text{where } a > 0$$

Properties of Integrals

Definition of the definite integral: Consider a continuous function $f(x)$ on the interval $[a, b]$, and let $a = x_0 < x_1 < \ldots < x_N = b$ be a partition of $[a, b]$. For each $k = 1, 2, \ldots, N$, select an arbitrary point t_k in the subinterval $[x_{k-1}, x_k]$. Let $\Delta x_k = x_k - x_{k-1}$.

Then the sum

(1)
$$\sum_{k=1}^{N} f(t_k)\, \Delta x_k$$

is called a *Riemann sum approximation* for the definite integral of $f(x)$ over $[a, b]$.

The definite integral

$$\int_a^b f(x)\, dx$$

is the limit of the Riemann sums (1) as the number of subintervals in the partition tend to infinity and the widths of the subintervals tend to zero.

First fundamental theorem of calculus:

$$\int_a^b f(x)\, dx = F(b) - F(a) \quad \text{where} \quad F'(x) = f(x)$$

Second fundamental theorem of calculus:

$$\frac{d}{dx}\int_a^x f(t)\, dt = f(x)$$

Properties of integrals:

$$\int_a^b [Cf(x)]\, dx = C\int_a^b f(x)\, dx \quad \text{where } C \text{ is a constant}$$

$$\int_a^b [f(x) + g(x)]\, dx = \int_a^b f(x)\, dx + \int_a^b g(x)\, dx$$

$$\int_b^a f(x)\, dx = -\int_a^b f(x)\, dx$$

$$\int_a^b f(x)\, dx = \int_a^c f(x)\, dx + \int_c^b f(x)\, dx \quad \text{where } a < c < b$$

Integration by parts:

$$\int_a^b u(x)v'(x)\, dx = u(x)v(x)\Big|_{x=a}^{x=b} - \int_a^b v(x)u'(x)\, dx$$

Change of variable:

$$\int_a^b f(g(x))g'(x)\, dx = \int_{g(a)}^{g(b)} f(u)\, du$$

Integrals of Common Functions

$$\int x^N \, dx = \frac{x^{N+1}}{N+1} + C \quad \text{provided that} \quad N \neq -1$$

$$\int x^{-1} \, dx = \ln|x| + C$$

$$\int e^x \, dx = e^x + C$$

$$\int \cos(x) \, dx = \sin(x) + C$$

$$\int \sin(x) \, dx = -\cos(x) + C$$

$$\int \sec^2(x) \, dx = \tan(x) + C$$

$$\int \sec(x) \tan(x) \, dx = \sec(x) + C$$

$$\int \tan(x) \, dx = -\ln|\cos(x)| + C$$

$$\int \sec(x) \, dx = \ln|\sec(x) + \tan(x)| + C$$

$$\int \frac{dx}{a^2 + x^2} = \frac{1}{a} \arctan\left(\frac{x}{a}\right) + C \quad \text{where } a > 0$$

$$\int (a^2 - x^2)^{-1/2} \, dx = \arcsin\left(\frac{x}{a}\right) + C \quad \text{where } a > 0$$

$$\int (x^2 + a^2)^{-1/2} \, dx = \ln|x + (x^2 + a^2)^{1/2}| + C$$

$$\int (x^2 + a^2)^{1/2} \, dx = \frac{x}{2}(x^2 + a^2)^{1/2} + \frac{a^2}{2}\ln|x + (x^2 + a^2)^{1/2}| + C$$

$$\int \frac{dx}{x^2 - a^2} = \frac{1}{2a} \ln\left|\frac{x-a}{x+a}\right| + C$$

$$\int \ln|x| \, dx = x \ln|x| - x + C$$

$$\int x \ln|x| \, dx = \frac{-x^2}{4} + \frac{x^2}{2} \ln|x| + C$$

$$\int a^x \, dx = \frac{a^x}{\ln(a)} + C \quad \text{where } a > 0 \text{ and } a \neq 1$$

$$\int x \, e^x \, dx = x \, e^x - e^x + C$$

$$\int x \sin (x) \, dx = \sin (x) - x \cos (x) + C$$

$$\int x \cos (x) \, dx = \cos (x) + x \sin (x) + C$$

$$\int \sin^2 (x) \, dx = \frac{x}{2} - \frac{\sin (2x)}{4} + C$$

$$\int \cos^2 (x) \, dx = \frac{x}{2} + \frac{\sin (2x)}{4} + C$$

Appendix B

Some Suggested References for Reports

Adaptive Integration (Quadrature) [3, 20, 23, 28, 34, 43, 70, 72, 112]

Band Systems of Equations [20, 23, 29, 94, 112, 139]

Basic Splines (B-Splines) [23, 70, 72, 107, 112]

Calculus and Computers [9, 21, 24, 39, 41, 79, 87, 89, 98, 114, 115, 125, 127, 146]

Choleski's Factorization [3, 20, 28, 29, 36, 66, 71, 109, 112]

Condition Number of a Matrix [3, 10, 20, 28, 29, 43, 47, 69, 70, 72, 84, 94, 105, 108, 109, 112, 139]

Economization of Power Series [1, 3, 20, 29, 36, 47, 55, 62, 84, 108, 109, 132]

Eigenvectors and Eigenvalues [1, 3, 20, 28, 29, 47, 66, 84, 100, 102, 105, 148]

Engineering Usage of Numerical Methods [11, 38, 45, 68, 96, 116, 123, 131, 137]

Error Propagation [3, 4, 28, 29, 34, 36, 57, 58, 60, 102, 105, 109, 149]

Extrapolation [10, 20, 23, 28, 29, 57, 84, 109]

Fast Fourier Transform [20, 28, 36, 47, 58, 70, 105, 107, 109]

Finite-Difference Method [20, 23, 28, 29, 47, 66, 70, 84]

Floating-Point Arithmetic [3, 4, 23, 28, 29, 36, 43, 47, 66, 72, 73, 94, 95, 102, 109, 129, 132, 152]

Forward-Difference Formulas [3, 20, 28, 29, 36, 55, 57, 60, 62, 66, 69, 74, 84, 94, 103, 105, 109, 129, 132]

Gauss–Jordan Method [20, 31, 36, 47, 58, 62, 66, 84]

Gradient Search for a Minimum [1, 2, 10, 19, 23, 28, 29, 31, 34, 35, 43, 48, 54, 61, 64, 65, 70, 101, 108]

Bibliography and References

[1] Acton, Forman S. (1970). *Numerical Methods That Work*, Harper & Row, Publishers, Inc., New York.

[2] Adby, P. R., and M. A. H. Dempster (1974). *Introduction to Optimization Methods*, Halsted Press, New York.

[3] Aitkinson, Kendall E. (1978). *An Introduction to Numerical Analysis*, John Wiley & Sons, Inc., New York.

[4] Aitkinson, Kendall E. (1985). *Elementary Numerical Analysis*, John Wiley & Sons, Inc., New York.

[5] Aitkinson, L. V., and P. J. Harley (1983). *An Introduction to Numerical Methods with Pascal*, Addison-Wesley Publishing Company, Inc., Reading, Mass.

[6] Anton, Howard (1984). *Elementary Linear Algebra*, 4th ed., John Wiley & Sons, Inc., New York.

[7] Bender, Carl M., and Steven A. Orszag (1978). *Advanced Mathematical Methods for Scientists and Engineers*, McGraw-Hill Book Company, New York.

[8] Bender, Edward A. (1978). *An Introduction to Mathematical Modeling*, John Wiley & Sons, Inc., New York.

[9] Bitter, Gary G. (1983). *Microcomputer Applications for Calculus*, Prindle, Weber & Schmidt, Boston.

[10] Blum, E. K. (1972). *Numerical Analysis and Computation: Theory and Practice*, Addison-Wesley Publishing Company, Inc., Reading, Mass.

[11] Borse, G. J. (1985). *FORTRAN 77 and Numerical Methods for Engineers*, Prindle, Weber & Schmidt, Boston.

[12] Boyce, William E., and Richard C. DiPrima (1977). *Elementary Differential Equations and Boundary Value Problems*, 3rd ed., John Wiley & Sons, Inc., New York.

[13] Brainerd, Walter S., and Lawrence H. Landweber (1974). *Theory of Computation*, John Wiley & Sons, Inc., New York.

[14] Brams, Steven J., William F. Lucas, and Philip D. Straffin, eds. (1983). *Political and Related Models*, Springer-Verlag, New York.

[15] Braun, Martin, Courtney S. Coleman, and Donald A. Drew, eds. (1983). *Differential Equation Models*, Springer-Verlag, New York.

[16] Brent, Richard P. (1973). *Algorithms for Minimization without Derivatives*, Prentice-Hall, Inc., Englewood Cliffs, N.J.

[17] Buck, R. Creighton (1978). *Advanced Calculus*, 3rd ed., McGraw-Hill Book Company, New York.

[18] Bunday, Brian D. (1984). *Basic Linear Programming*, Edward Arnold (Publishers) Ltd., London.

[19] Bunday, Brian D. (1984). *Basic Optimisation Methods*, Edward Arnold (Publishers) Ltd., London.

[20] Burden, Richard L., J. Douglas Faires, and Albert C. Reynolds, (1981). *Numerical Analysis*, 2nd ed., Prindle, Weber & Schmidt, Boston.

[21] Burgmeier, James W., and Larry L. Kost (1985). *EPIC Exploration Programs in Calculus*, Prentice-Hall, Inc., Englewood Cliffs, N.J.

[22] Cheney, E. W. (1966). *Introduction to Approximation Theory*, McGraw-Hill Book Company, New York.

[23] Cheney, Ward, and David Kincaid (1985). *Numerical Mathematics and Computing*, 2nd ed., Brooks/Cole Publishing Co., Monterey, Calif.

[24] Christensen, Mark J. (1981). *Computer for Calculus*, Academic Press, Inc., New York.

[25] Churchill, Ruel V., James W. Brown, and Roger F. Verhey (1974). *Complex Variables and Applications*, 3rd ed., McGraw-Hill Book Company, New York.

[26] Chvatal, Vasek (1980). *Linear Programming*, W. H. Freeman and Company, Publishers, New York.

[27] Coddington, Earl A., and Norman Levinson (1955). *Theory of Ordinary Differential Equations*, McGraw-Hill Book Company, New York.

[28] Conte, S. D., and Carl de Boor (1980). *Elementary Numerical Analysis: An Algorithmic Approach*, McGraw-Hill Book Company, New York.

[29] Dahlquist, Germund, and Ake Bjorck (1974). *Numerical Methods*, Prentice-Hall, Inc., Englewood Cliffs, N.J.

[30] Daniels, Richard W. (1978). *An Introduction to Numerical Methods and Optimization Techniques*, North-Holland, New York.

[31] Davis, Philip J. (1963). *Interpolation and Approximation*, Blaisdell Publishing Company, Waltham, Mass.

[32] Davis, Philip J., and Philip Rabinowitz (1984). *Methods of Numerical Integration*, 2nd ed., Academic Press, Inc., New York.

[33] deBoor, C. (1978). *A Practical Guide to Splines*, Springer-Verlag, New York.

[34]   Dew, P. M., and K. R. James (1983). *Introduction to Numerical Computation in Pascal*, Springer-Verlag, New York.

[35]   Dixon, L. C. W. (1972). *Nonlinear Optimisation*, Crane, Russak & Co., Inc., New York.

[36]   Dodes, Irving Allen (1978). *Numerical Analysis for Computer Science*, North-Holland, New York.

[37]   Dorn, William S., and Daniel D. McCracken (1976). *Introductory Finite Mathematics with Computing*, John Wiley & Sons, Inc., New York.

[38]   Dorn, William S., and Daniel D. McCracken (1972). *Numerical Methods with FORTRAN IV Case Studies*, John Wiley & Sons, Inc., New York.

[39]   Edwards, C. H. (1986). *Calculus and the Personal Computer*, Prentice-Hall, Englewood Cliffs, N.J.

[40]   Fike, C. T. (1968). *Computer Evaluation of Mathematical Functions*, Prentice-Hall, Inc., Englewood Cliffs, N.J.

[41]   Finney, Ross L., Dale R. Hoffman, Judah L. Schwartz, and Carroll O. Wilde (1984). *The Calculus Toolkit*, Addison-Wesley Publishing Company, Inc., Reading, Mass.

[42]   Fogiel, M. (1983). *The Numerical Analysis Problem Solver*, Research and Education Association, New York.

[43]   Forsythe, George E., Michael A. Malcolm, and Cleve B. Moler (1977). *Computer Methods for Mathematical Computations*, Prentice-Hall, Inc., Englewood Cliffs, N.J.

[44]   Forsythe, George E., and Cleve B. Moler (1967). *Computer Solution of Linear Algebraic Systems*, Prentice-Hall, Inc., Englewood Cliffs, N.J.

[45]   Fox L., and D. F. Mayers (1968). *Computing Methods for Scientists and Engineers*, Clarendon Press, Oxford.

[46]   Gear, C. William (1971). *Numerical Initial Value Problems in Ordinary Differential Equations*, Prentice-Hall, Inc., Englewood Cliffs, N.J.

[47]   Gerald, Curtis F., and Patrick O. Wheatley (1984). *Applied Numerical Analysis*, 3rd ed., Addison-Wesley Publishing Company, Inc., Reading, Mass.

[48]   Gill, Philip E., Walter Murray, and Margaret H. Wright (1981). *Practical Optimization*, Academic Press, Inc., New York.

[49]   Giordano, Frank R., and Maurice D. Weir (1985). *A First Course in Mathematical Modeling*, Brooks/Cole Publishing Co., Monterey, Calif.

[50]   Goldstine, Herman H. (1977). *A History of Numerical Analysis from the 16th through the 19th Century*, Springer-Verlag, New York.

[51]   Greenspan, Donald (1971). *Introduction to Numerical Analysis and Applications*, Markham Publishing Co., Chicago.

[52]   Grove, Wendell E. (1966). *Brief Numerical Methods*, Prentice-Hall, Inc., Englewood Cliffs, N.J.

[53]   Haberman, Richard (1977). *Mathematical Models: Mechanical Vibrations, Population Dynamics, and Traffic Flow*, Prentice-Hall, Inc., Englewood Cliffs, N.J.

[54]   Hamming, Richard W. (1971). *Introduction to Applied Numerical Analysis*, McGraw-Hill Book Company, New York.

[55] Hamming, Richard W. (1973). *Numerical Methods for Scientists and Engineers*, 2nd ed., McGraw-Hill Book Company, New York.

[56] Henrici, Peter (1974). *Applied and Computational Complex Analysis*, Vol. 1, John Wiley & Sons, Inc., New York.

[57] Henrici, Peter (1964). *Elements of Numerical Analysis*, John Wiley & Sons, Inc., New York.

[58] Henrici, Peter (1982). *Essentials of Numerical Analysis with Pocket Calculator Demonstrations*, John Wiley & Sons, Inc., New York.

[59] Hildebrand, Francis B. (1976). *Advanced Calculus for Applications*, 2nd ed., Prentice-Hall, Inc., Englewood Cliffs, N.J.

[60] Hildebrand, Francis B. (1974). *Introduction to Numerical Analysis*, 2nd ed., McGraw-Hill Book Company, New York.

[61] Hillier, Frederick S., and Gerald J. Lieberman (1974). *Operations Research*, 2nd ed., Holden-Day, Inc., San Francisco, Calif.

[62] Hornbeck, Robert W. (1975). *Numerical Methods*, Quantum Publishers, Inc., New York.

[63] Householder, Alston S. (1970). *The Numerical Treatment of a Single Nonlinear Equation*, McGraw-Hill Book Company, New York.

[64] Householder, Alston S. (1953). *Principles of Numerical Analysis*, McGraw-Hill Book Company, New York.

[65] Hundhausen, Joan R., and Robert A. Walsh (1985). Unconstrained Optimization, *The UMAP Journal*, Vol. 6, No. 4, pp. 57–90.

[66] Isaacson, Eugene, and Herbert Bishop Keller (1966). *Analysis of Numerical Methods*, John Wiley & Sons, Inc., New York.

[67] Jacobs, D., ed. (1977). *The State of the Art in Numerical Analysis*, Academic Press, Inc., New York.

[68] James, M. L., G. M. Smith, and J. C. Wolford, (1985). *Applied Numerical Methods for Digital Computation*, 3rd ed., Harper & Row, Publishers, Inc., New York.

[69] Jensen, Jens A., and John H. Rowland (1975). *Methods of Computation: The Linear Space Approach to Numerical Analysis*, Scott, Foresman and Company, Glenview, Ill.

[70] Johnson, Lee W., and R. Dean Riess (1982). *Numerical Analysis*, 2nd ed., Addison-Wesley Publishing Company, Inc., Reading, Mass.

[71] Johnston, R. L. (1982). *Numerical Methods a Software Approach*, John Wiley & Sons, Inc., New York.

[72] King, J. Thomas (1984). *Introduction to Numerical Computation*, McGraw-Hill Book Company, New York.

[73] Knuth, Donald E. (1981). *The Art of Computer Programming*, Vol. 2, *Seminumerical Algorithms*, 2nd ed., Addison-Wesley Publishing Company, Inc., Reading, Mass.

[74] Kunz, Kaiser S. (1957). *Numerical Analysis*, McGraw-Hill Book Company, New York.

[75] Lambert, J. D. (1973). *Computational Methods in Ordinary Differential Equations*, John Wiley & Sons, Inc., New York.

[76] Lancaster, Peter (1976). *Mathematics: Models of the Real World*, Prentice-Hall, Inc., Englewood Cliffs, N.J.

[77] Lapidus L., and J. H. Seinfeld (1971). *Numerical Solution of Ordinary Differential Equations*, Academic Press, Inc., New York.

[78] Lawson, C. L., and R. J. Hanson (1974). *Solving Least-Squares Problems*, Prentice-Hall, Inc., Englewood Cliffs, N.J.

[79] Lax, Peter, Samuel Burstein, and Anneli Lax (1976). *Calculus with Applications and Computing*, Springer-Verlag, New York.

[80] Leon, Steven J. (1980). *Linear Algebra with Applications*, Macmillan Publishing Co., New York.

[81] Lucas, William F., Fred S. Roberts, and Robert M. Thrall, eds. (1983). *Discrete and System Models*, Spring-Verlag, New York.

[82] Maki, Daniel P., and Maynard Thompson (1973). *Mathematical Models and Applications*, Prentice-Hall, Inc., Englewood Cliffs, N.J.

[83] Marcus-Roberts, Helen, and Maynard Thompson, eds. (1983). *Life Science Models*, Springer-Verlag, New York.

[84] Maron, Melvin J. (1982), *Numerical Analysis: A Practical Approach*, Macmillan Publishing Co., New York.

[85] Mathews, John H. (1982), *Basic Complex Variables for Mathematics and Engineering*, Allyn and Bacon, Inc., Boston.

[86] McCalla, Thomas Richard (1967). *Introduction to Numerical Methods and FORTRAN Programming*, John Wiley & Sons, Inc., New York.

[87] McCarty, George (1975). *Calculator Calculus*, Page-Ficklin Publications, Palo Alto.

[88] McCormick, John M., and Mario G. Salvadori (1964). *Numerical Methods in FORTRAN*, Prentice-Hall, Inc., Englewood Cliffs, N.J.

[89] McNeary, Samuel S. (1973). *Introduction to Computational Methods for Students of Calculus*, Prentice-Hall, Inc., Englewood Cliffs, N.J.

[90] Miel, George J. (1981). Calculator Demonstrations of Numerical Stability, *The UMAP Journal*, Vol. 2, No. 2, pp. 3–7.

[91] Miller, Webb (1984). *The Engineering of Numerical Software*, Prentice-Hall, Inc., Englewood Cliffs, N.J.

[92] Milne, William Edmund (1949). *Numerical Calculus*, Princeton University Press, Princeton, N.J.

[93] Moore, Ramon E. (1966). *Interval Analysis*, Prentice-Hall, Inc., Englewood Cliffs, N.J.

[94] Morris, John Ll. (1983). *Computational Methods in Elementary Numerical Analysis*, John Wiley & Sons, Inc., New York.

[95] Moursund, David G., and Charles S. Duris (1967). *Elementary Theory and Application of Numerical Analysis*, McGraw-Hill Book Company, New York.

[96] Noble, Ben (1967), *Applications of Undergraduate Mathematics in Engineering*, The Macmillan Company, New York.

[97] Noble, Ben, and James W. Daniel (1977). *Applied Linear Algebra*, 2nd ed., Prentice-Hall, Inc., Englewood Cliffs, N.J.

[98] Oldknow, Adrian, and Derek Smith (1983). *Learning Mathematics with Micros*, Halsted Press, New York.

[99] Olinick, Michael (1978). *An Introduction to Mathematical Models in the Social and Life Sciences*, Addison-Wesley Publishing Company, Inc., Reading, Mass.

[100] Ortega, James M. (1972). *Numerical Analysis*, Academic Press, New York.

[101] Ortega, James M., and W. C. Rheinboldt (1970). *Iterative Solution of Nonlinear Equations in Several Variables*, Academic Press, Inc., New York.

[102] Pennington, Ralph H. (1970). *Introductory Computer Methods and Numerical Analysis*, 2nd ed., Macmillan Publishing Co., New York.

[103] Pettofrezzo, Anthony J. (1967). *Introductory Numerical Analysis*, D. C. Heath and Company, Boston.

[104] Phillips, G. M., and P. J. Taylor (1974). *Theory and Applications of Numerical Analysis*, Academic Press, Inc., New York.

[105] Pizer, Stephen M. (1975). *Numerical Computing and Mathematical Analysis*, Science Research Associates, Inc., Chicago, Ill.

[106] Pizer, Stephen M., with Victor L. Wallace (1983). *To Compute Numerically: Concepts and Strategies*, Little, Brown & Co., Boston.

[107] Powell, M. J. D. (1981). *Approximation Theory and Methods*, Cambridge University Press, Cambridge.

[108] Ralston, Anthony (1965). *A First Course in Numerical Analysis*, McGraw-Hill Book Company, New York.

[109] Ralston, Anthony, and Philip Rabinowitz (1978). *A First Course in Numerical Analysis*, 2nd ed., McGraw-Hill Book Company, New York.

[110] Ralston, Anthony, and Herbert S. Wilf (1960). *Mathematical Methods for Digital Computers*, John Wiley & Sons, Inc., New York.

[111] Rheinboldt, Werner C. (1981). Algorithms for Finding Zeros of Functions, *The UMAP Journal*, Vol. 2, No. 1, pp. 43–72.

[112] Rice, John R. (1983). *Numerical Methods, Software and Analysis: IMSL Reference Edition*, McGraw-Hill Book Company, New York.

[113] Roman, Steven (1985). *An Introduction to Linear Algebra with Applications*, Saunders College Publishing, New York.

[114] Rosser, J. Barkley, and Carl de Boor (1979). *Pocket Calculator Supplement for Calculus*, Addison-Wesley Publishing Company, Inc., Reading, Mass.

[115] Sagan, Hans (1984). *Calculus Accompanied on the Apple*, Reston Publishing Company, Reston, Va.

[116] Salvadori, Mario G., and Melvin L. Baron (1961). *Numerical Methods in Engineering*, 2nd ed., Prentice-Hall, Inc., Englewood Cliffs, N.J.

[117] Scalzo, Frank, and Rowland Hughes (1977). *A Computer Approach to Introductory College Mathematics*, Mason/Charter Publishers, Inc., New York.

[118] Scarborough, James B. (1966). *Numerical Mathematical Analysis*, The John Hopkins Press, Baltimore.

[119] Scheid, Francis (1968). *Theory and Problems of Numerical Analysis*, McGraw-Hill Book Company, New York.

[120] Schultz, M. H. (1966). *Spline Analysis*, Prentice-Hall, Inc., Englewood Cliffs, N.J.

[121] Scraton, R. E. (1984). *Basic Numerical Methods: An Introduction to Numerical Mathematics on a Microcomputer*, Edward Arnold (Publishers) Ltd., London.

[122] Shampine, L. F., and M. K. Gordon (1975). *Computer Solution of Ordinary Differential Equations*, W. H. Freeman and Company Publishers, New York.

[123] Shoup, Terry E. (1979). *A Practical Guide to Computer Methods for Engineers*, Prentice-Hall, Inc., Englewood Cliffs, N.J.

[124] Shoup, Terry E. (1983). *Numerical Methods for the Personal Computer*, Prentice-Hall, Inc., Englewood Cliffs, N.J.

[125] Sicks, Jon L. (1985). *Investigating Secondary Mathematics with Computers*, Prentice-Hall, Inc., Englewood Cliffs, N.J.

[126] Simmons, George F. (1972). *Differential Equations: with Applications and Historical Notes*, McGraw-Hill Book Company, New York.

[127] Smith, David A. (1976). *INTERFACE: Calculus and the Computer*, Houghton Mifflin Company, Boston.

[128] Smith, G. D. (1978). *The Numerical Solution of Partial Differential Equations*, 2nd ed., Oxford University Press, Oxford.

[129] Smith, W. Allen (1979). *Elementary Numerical Analysis*, Harper & Row, Publishers, Inc., New York.

[130] Snyder, Martin Avery (1966). *Chebyshev Methods in Numerical Approximation*, Prentice-Hall, Inc., Englewood Cliffs, N.J.

[131] Stanton, Ralph G. (1961). *Numerical Methods for Science and Engineering*, Prentice-Hall, Inc., Englewood Cliffs, N.J.

[132] Stark, Peter A. (1970). *Introduction to Numerical Methods*, The Macmillan Company, Toronto, Ontario.

[133] Strang, G., and G. Fix (1973). *An Analysis of the Finite Element Method*, Prentice-Hall, Inc., Englewood Cliffs, N.J.

[134] Strecker, George E. (1982). Round Numbers: An Introduction to Numerical Expression, *The UMAP Journal*, Vol. 3, No. 4, pp. 425–454.

[135] Stroud, A. H. (1971). *Approximate Calculation of Multiple Integrals*, Prentice-Hall, Inc., Englewood Cliffs, N.J.

[136] Stroud, A. H., and Don Secrest (1966). *Gaussian Quadrature Formulas*, Prentice-Hall, Inc., Englewood Cliffs, N.J.

[137] Thompson, William J. (1984). *Computing in Applied Science*, John Wiley & Sons, Inc., New York.

[138] Todd, John (1979). *Basic Numerical Mathematics*, Vol. 1: *Numerical Analysis*, Academic Press, Inc., New York.

[139] Todd, John (1977). *Basic Numerical Mathematics*, Vol. 2: *Numerical Algebra*, Academic Press, Inc., New York.

[140] Tompkins, Charles B., and Walter L. Wilson (1969). *Elementary Numerical Analysis*, Prentice-Hall, Inc., Englewood Cliffs, N.J.

[141] Traub, J. F. (1964). *Iterative Methods for the Solution of Equations*, Prentice-Hall, Inc., Englewood Cliffs, N.J.

[142] Vandergraft, James S. (1983). *Introduction to Numerical Computations*, Academic Press, Inc., New York.

[143] VanIwaarden, John L. (1985). *Ordinary Differential Equations with Numerical Techniques*, Harcourt Brace Jovanovich, Publishers, San Diego, Calif.

[144] Varga, Richard S. (1962). *Matrix Iterative Analysis*, Prentice-Hall, Inc., Englewood Cliffs, N.J.

[145] Wachspress, Eugene L. (1966). *Iterative Solution of Elliptic Systems*, Prentice-Hall, Inc., Englewood Cliffs, N.J.

[146] Wattenberg, Frank, and Martin Wattenberg (1984). *Interactive Experiments in Calculus*, Prentice-Hall, Inc., Englewood Cliffs, N.J.

[147] Wendroff, Burton (1966). *Theoretical Numerical Analysis*, Academic Press, Inc., New York.

[148] Wilkinson, J. H. (1965). *The Algebraic Eigenvalue Problem*, Clarendon Press, Oxford.

[149] Wilkinson, J. H. (1963). *Rounding Errors in Algebraic Processes*, Prentice-Hall, Inc., Englewood Cliffs, N.J.

[150] Wilkinson, J. H., and C. Reinsch (1971). *Handbook for Automatic Computation*, Vols. 1–2, Springer-Verlag, New York.

[151] Young, David M. (1971). *Iterative Solution of Large Linear Systems*, Academic Press, Inc., New York.

[152] Young, David M., and Robert Todd Gregory (1972). *A Survey of Numerical Mathematics*, Vols. 1–2, Addison-Wesley Publishing Company, Inc., Reading, Mass.

Answers to Selected Exercises

Section 1.2 Binary Numbers: page 13

1. (a) The computer's answer is not 0 because 0.1 is not an exact binary fraction
(b) 0 (exactly)

2. (a) 21   (c) 109

3. (a) 0.84375   (c) 0.6640625

4. (a) 1.4140625

5. (a) $2^{1/2} - 1.4140625 = 0.000151062\ldots$

6. (a) 10111_{two} (c) 101101000_{two}

7. (a) 0.0111_{two} (c) 0.10111_{two}

8. (a)
$$2R = 0.2, \quad d_1 = 0 = \text{int}\,(0.2), \quad F_1 = 0.2 = \text{frac}\,(0.2)$$
$$2F_1 = 0.4, \quad d_2 = 0 = \text{int}\,(0.4), \quad F_2 = 0.4 = \text{frac}\,(0.4)$$
$$2F_2 = 0.8, \quad d_3 = 0 = \text{int}\,(0.8), \quad F_3 = 0.8 = \text{frac}\,(0.8)$$
$$2F_3 = 1.6, \quad d_4 = 1 = \text{int}\,(1.6), \quad F_2 = 0.6 = \text{frac}\,(1.6)$$
$$2F_4 = 1.2, \quad d_5 = 1 = \text{int}\,(1.2), \quad F_5 = 0.2 = \text{frac}\,(1.2)$$

$$\frac{1}{10} = 0.d_1 d_2 d_3 d_4 d_5 \ldots {}_{two} = 0.0001\overline{10011}\ldots {}_{two}$$

9. (a) $\frac{1}{10} - 0.0001100_{two} = 0.006250000\ldots$

11. Use $c = \frac{3}{16}$ and $r = \frac{1}{16}$ to get $S = \dfrac{\frac{3}{16}}{1 - \frac{1}{16}} = \frac{1}{5}$

14. (a)
$$\begin{array}{ll}
\frac{1}{3} \approx 0.1011_{two} \times 2^{-1} = 0.1011_{two} \times 2^{-1} \\
\frac{1}{5} \approx 0.1101_{two} \times 2^{-2} = 0.01101_{two} \times 2^{-1} \\
\frac{8}{15} \qquad\qquad\qquad\qquad\qquad \overline{0.100011_{two} \times 2^{0}}
\end{array}$$

$$\begin{array}{ll}
\frac{8}{15} \approx 0.1001_{two} \times 2^{0} = 0.1001_{two} \times 2^{0} \\
\frac{1}{6} \approx 0.1011_{two} \times 2^{-2} = 0.001011_{two} \times 2^{0} \\
\frac{7}{10} \qquad\qquad\qquad\qquad\qquad \overline{0.101111_{two} \times 2^{0}} \approx \boxed{0.1100_{two}}
\end{array}$$

15. (b) $10 = 101_{three}$ (d) $49 = 1211_{three}$

16. (c) $\frac{1}{2} = 0.11111111\overline{1}\ldots {}_{three}$

17. (b) $10 = 20_{five}$ (d) $49 = 144_{five}$

18. (c) $\frac{1}{2} = 0.2222222\overline{2}\ldots$ five

24. (a) $213_{16} = 531_{ten}$ (c) $1ABE_{16} = 6846_{ten}$

     (e) $0.2_{16} = 0.125_{ten}$ (g) $0.A4B_{16} = 0.643310546875_{ten}$

25. (a) $512_{ten} = 200_{16}$ (c) $51264_{ten} = C840_{16}$

26. (a) $\frac{1}{3} = 0.555555\ldots_{16}$ (c) $\frac{1}{10} = 0.199999\ldots_{16}$

Section 1.3 Taylor Series and Calculation of Functions: page 28

1. (a) $P_5(x) = x - x^3/3! + x^5/5!$

      $P_7(x) = x - x^3/3! + x^5/5! - x^7/7!$

      $P_9(x) = x - x^3/3! + x^5/5! - x^7/7! + x^9/9!$

   (b) $P_5(0.5) = 0.4794270834,\quad E_5(0.5) = -0.0000015448$

      $P_7(0.5) = 0.4794255333,\quad E_7(0.5) = 0.0000000053$

      $P_9(0.5) = 0.4794255387,\quad E_9(0.5) = -0.0000000001$

      $\sin(0.5) = 0.4794255386$

      $P_5(1.0) = 0.8416666667,\quad E_5(1.0) = -0.0001956818$

      $P_7(1.0) = 0.8414682539,\quad E_7(1.0) = 0.0000027309$

      $P_9(1.0) = 0.8414710096,\quad E_9(1.0) = -0.0000000248$

      $\sin(1.0) = 0.8414709848$

   (c) $|E_9(x)| = \left|\sin(c)x^{10}/10!\right|,\ 1 \times 1^{10}/10! = 0.0000002755\ldots$

   (d) $P_5(x) = 2^{-1/2}[1 + (x - \pi/4) - (x - \pi/4)^2/2 - (x - \pi/4)^3/6 + (x - \pi/4)^4/24 + (x - \pi/4)^5/120]$

   (e) $P_5(0.5) = 0.4794260467,\quad E_5(0.5) = -0.0000005081$

      $\sin(0.5) = 0.4794255386$

      $P_5(1.0) = 0.8414710840,\quad E_5(1.0) = -0.0000000992$

      $\sin(1.0) = 0.8414709848$

3. No. Because the derivatives of $f(x)$ are undefined at $x_0 = 0$.

5. $P_3(x) = 1 + 0x - x^2/2 + 0x^3$

6. (a) $f(2) = 2,\ f'(2) = \frac{1}{4},\ f^{(2)}(2) = -\frac{1}{32},\ f^{(3)}(2) = \frac{3}{256}$

      $P_3(x) = 2 + (x - 2)/4 - (x - 2)^2/64 + (x - 2)^3/512$

   (b) $P_3(1) = 1.732421875$; compare with $3^{1/2} = 1.732050808$

   (c) $f^{(4)}(x) = -15(2 + x)^{-7/2}/16$; the maximum of $|f^{(4)}(x)|$ on the interval $1 \le x \le 3$ occurs when $x = 1$ and

      $|f^{(4)}(x)| \le |f^{(4)}(1)| \le 3^{-7/2} \times 15/16 \approx 0.020046.\ldots$ Therefore,

$$|E_3(x)| \le \frac{(0.020046\ldots) \times 1^4}{4!} = 0.00083529.\ldots$$

8. (d) $P_3(0.5) = 0.41666667$ **10.** (d) $P_2(0.5) = 1.21875000$

     $P_6(0.5) = 0.40468750$ $P_4(0.5) = 1.22607422$

     $P_9(0.5) = 0.40553230$ $P_6(0.5) = 1.22660828$

     $\ln(1.5) = 0.40546511$ $(1.5)^{1/2} = 1.22474487$

11. $f(x) = x^3 - 2x^2 + 2,\ f'(x) = 3x^2 - 4x,\ f^{(2)}(x) = 6x - 4,\ f^{(3)}(x) = 6$

     $f(1) = 1,\ f'(1) = 1,\ f^{(2)}(1) = 2,\ f^{(3)}(1) = 6$

     $P_3(x) = 1 + (x - 1) + (x - 1)^2 + (x - 1)^3$

13. (a) $P_3(2.1) = 2.254,\ P_3(1.9) = 1.646$

   (b) $[P_3(2.1) - P_3(1.9)]/0.2 = 3.04$

   (c) $f(2) = P_3(2) = 2,\ f'(2) = P_3'(2) = 3$

18. (b) $P_2(1.05, 1.1) = 1 - (0.05) + (0.1) + (0.05)^2 - (0.05)(0.1) = 1.0475$. Compare with the function value $f(1.05, 1.1) = 1.047619048$.

Section 1.4 Fundamental Iterative Algorithm: page 40

1. (a) $g(2) = -4 + 8 - 2 = 2,\ g(4) = -4 + 16 - 8 = 4$

(b) $p_0 = 1.9,$ $E_0 = 0.1,$ $R_0 = 0.05$
     $p_1 = 1.795,$ $E_1 = 0.205,$ $R_1 = 0.1025$
     $p_2 = 1.5689875,$ $E_2 = 0.4310125,$ $R_2 = 0.21550625$
     $p_3 = 1.04508911,$ $E_3 = 0.95491089,$ $R_3 = 0.477455444$

(e) The sequence in part (b) does not converge to $P = 2$. The sequence in part (c) converges to $P = 4$.

4. (a) $p_0 = 1.9,$ $E_0 = 0.1,$ $R_0 = 0.05$
     $p_1 = 1.939,$ $E_1 = 0.061,$ $R_1 = 0.0305$
     The sequence converges to $P = 2$ because $|g'(2)| = 0.6 < 1$.

(b) $p_0 = -1.9,$ $E_0 = -0.1,$ $R_0 = 0.05$
     $p_1 = -1.861,$ $E_1 = -0.139,$ $R_1 = 0.0695$
     The sequence does not converge because $|g'(-2)| = 1.4 > 1$.

5. $P = 2,$ $g'(2) = 5,$ iteration will not converge to $P = 2$.
     $P = -2,\ g'(-2) = 5,$ iteration will not converge to $P = -2$.

10. (d) $p_1 = 0.99990000,$ $p_2 = 0.99980002$
     $p_3 = 0.99970006,$ $p_4 = 0.99960012$
     $p_5 = 0.99950020,$ $\{p_k\}$ converges very slowly to zero.

13. (b) $p_0 = 4.4$ (c) $p_0 = 4.6$
     $p_1 = 4.39886349$ $p_1 = 4.59891292$
     $p_2 = 4.39770070$ $p_2 = 4.59784909$
     $p_3 = 4.39651069$ $p_3 = 4.59680778$
     $p_4 = 4.39529251$ $p_4 = 4.59578829$
     The sequence does not   The sequence converges
     converge to $P = 4.5$ slowly to $P = 4.5$.

15. (a) $r_1 = -3,\ r_2 = 3$

(c) One real root, $r_1 = 3$; ignore the complex roots.

16. (a) $g'(x) = (1 - 9x^{-2})/2,\ g'(3) = 0$

(c) $g'(x) = 3x^2,\ g'(3) = 27$

17. (a) $p_1 = 3.00161290,$ $p_2 = 3.00000043$

(c) $p_1 = 5.79100000,$ $p_2 = 170.205129$

Section 2.1 Bracketing Methods for Locating a Root: page 51

1. $I_0 = (0.11 + 0.12)/2 = 0.115,$ $A(0.115) = 254{,}403$
     $I_1 = (0.11 + 0.115)/2 = 0.1125,$ $A(0.1125) = 246{,}072$
     $I_2 = (0.1125 + 0.125)/2 = 0.11375,$ $A(0.11375) = 250{,}198$

4. There are many choices for intervals $[a, b]$ on which $f(a)$ and $f(b)$ have opposite sign. The following answers are one such choice.

(a) $f(1) < 0$ and $f(2) > 0$, so there is a root in $[1, 2]$;
     also
     $f(-1) < 0$ and $f(-2) > 0$, so there is a root in $[-2, -1]$.

(c) $f(3) < 0$ and $f(4) > 0$, so there is a root in $[3, 4]$.

5. (a) $[1.0, 1.8], [1.0, 1.4], [1.0, 1.2], [1.1, 1.2], [1.10, 1.15]$

7. $[3.2, 4.0], [3.6, 4.0], [3.6, 3.8], [3.6, 3.7], [3.65, 3.70]$

8. (a) $[3.2, 4.0], [3.2, 3.6], [3.4, 3.6], [3.5, 3.6], [3.55, 3.60]$

(b) $[6.0, 6.8], [6.4, 6.8], [6.4, 6.6], [6.4, 6.5], [6.40, 6.45]$

9. (a) $r_1 = 1.14619322,\ r_2 = -1.84140566,$ (c) $r = 3.69344136$

10. $c_0 = -1.8300782,\ c_1 = -1.8409252,\ c_2 = -1.8413854,\ c_3 = -1.8414048$

12. $c_0 = 3.6979549,\ c_1 = 3.6935108,\ c_2 = 3.6934424,\ c_3 = 3.6934414$

14. (a) $r_1 = 1.14619322,\ r_2 = -1.84140566,$ (c) $r = 3.69344136$

Section 2.2 Initial Approximations and Convergence Criteria: page 58

1. $g(x) = x^2,\ h(x) = \exp(x)$ for $-2 \leqq x \leqq 2$
Approximate root location -0.7. Computed root -0.7034674225.

3. $g(x) = \sin(x)$, $h(x) = 2\cos(2x)$ for $-2 \leq x \leq 2$
   Approximate root locations $-1.0, 0.6$.
   Computed roots $-1.002966954, 0.6348668712$.
5. $g(x) = \ln(x)$, $h(x) = (x-2)^2$ for $0.5 \leq x \leq 4.5$
   Approximate root locations $1.4, 3.0$.
   Computed roots $1.412391172, 3.057103550$.

7. (a) Ymin := Y(0)
       DO FOR K = 1 TO N
       └── IF Y(K) < Ymin THEN Ymin := Y(K)

Section 2.3 Newton-Raphson and Secant Methods: page 74

1. (a) $p_{k+1} = (p_k + 8/p_k)/2$
   $p_0 = 3$
   $p_1 = 2.833333334$
   $p_2 = 2.828431373$
   $p_3 = 2.828427125$

   (c) $p_{k+1} = (p_k + 91/p_k)/2$
   $p_0 = 10$
   $p_1 = 9.55$
   $p_2 = 9.539397905$
   $p_3 = 9.539392015$

3. (a) $p_{k+1} = (2p_k + 7/p_k^2)/3$
   $p_0 = 2$
   $p_1 = 1.916666667$
   $p_2 = 1.912938458$
   $p_3 = 1.912931183$

   (c) $p_{k+1} = (2p_k + 200/p_k^2)/3$
   $p_0 = 6$
   $p_1 = 5.851851853$
   $p_2 = 5.848037967$
   $p_3 = 5.848035477$

7. (a) $g(p_{k-1}) = (p_{k-1}^2 + 1)/(2p_{k-1} - 2)$
   (b) $p_0 = 2.5$
   $p_1 = 2.41666667$
   $p_2 = 2.41421569$
   $p_3 = 2.41421356$

   (c) $p_0 = -0.5$
   $p_1 = -0.41666667$
   $p_2 = -0.41421569$
   $p_3 = -0.41421356$

10. (a) $f'(t) = 320\exp(-t/5) - 160$
    $p_1 = 7.96819027$, $p_2 = 7.96812129$, $p_3 = 7.96812130$
    (b) Range $= r(7.96812130) = 637.4497040$

12. (a) $g(p_{k-1}) = p_{k-1} - 0.25(p_{k-1} - 2) = 0.5 + 0.75p_{k-1}$
    (b) $p_0 = 2.1$
    $p_1 = 2.075$
    $p_2 = 2.05625$
    $p_3 = 2.0421875$
    $p_4 = 2.031640625$

    (c) Convergence is linear.
    The error is reduced
    by a factor of $\frac{3}{4}$
    with each iteration.

15. (c) $g(p_{k-1}) = (4p_{k-1}^3 + 3)/(6p_{k-1}^2 - 1)$
    $p_0 = 1.0$
    $p_1 = 1.4$
    $p_2 = 1.29888476$
    $p_3 = 1.28969699$
    $p_4 = 1.28962390$

    The point on the parabola is
    $(p, p^2) = (1.28962390, 1.66312980)$.

18. No, because $f'(x)$ is not continuous at the root $p = 0$. You could also try computing terms
    with $g(p_{k-1}) = -2p_{k-1}$ and see that the sequence diverges.

20. There are two solutions, $x = 0.839018884$ and $x = 3.401748648$.

22. (a) $g(p_{k-1}) = p_{k-1} - p_{k-1}\exp(-p_{k-1})/[(1 - p_{k-1})\exp(-p_{k-1})]$
    $g(p_{k-1}) = p_{k-1}^2/(p_{k-1} - 1)$
    (b) $p_0 = 0.20$
    $p_1 = -0.05$
    $p_2 = -0.002380952$
    $p_3 = -0.000005655$
    $p_4 = -0.000000000$
    $\lim\limits_{k \to \infty} p_k = 0.0$

    (c) $p_0 = 20.0$
    $p_1 = 21.05263158$
    $p_2 = 22.10250034$
    $p_3 = 23.14988809$
    $p_4 = 24.19503505$
    $\lim\limits_{k \to \infty} p_k = \infty$

    (d) The value of the function in part (c) is $f(p_4) = 7.515 \times 10^{-10}$.

24. $g(p_{k-1}) = [2 + \sin(p_{k-1}) - p_{k-1}\cos(p_{k-1})]/[2 - \cos(p_{k-1})]$
$$p_0 = 1.5$$
$$p_1 = 1.49870157$$
$$p_2 = 1.49870113$$

31. $p_0 = 2.6$
$$p_1 = 2.5$$
$$p_2 = 2.41935484$$
$$p_3 = 2.41436464$$
$$p_4 = 2.41421384$$
$$p_5 = 2.41421356$$

34. (b) $g(x) = x - \tan(x^3)/x^2$
$$p_0 = 1.0$$
$$p_1 = -0.557407725$$
$$p_2 = 0.005640693$$
$$p_3 = 0.000000000$$

Section 2.4 Synthetic Division and Bairstow's Method: page 92

1. (a) $Q_0(x) = x^3 + 4x^2 - x - 4$ (b) $Q_1(x) = x^2 + 3x - 4$
  (c) $z_2 = -4$, $z_3 = 1$
  (d) $p_1 = 1.1 - (-2.0349)/(-20.646) = 1.001438535$
    $p_2 = 1.000000412$

3. (a) $Q_0(x) = x^3 + x^2 - 4x - 4$ (b) $Q_1(x) = x^2 + 3x + 2$
  (c) $z_2 = -2$, $z_3 = -1$
  (d) $p_1 = 2.1 - (-1.1439)/(-10.816) = 1.994240015$
    $p_2 = 1.999986430$

5. (a) $Q_0(x) = x^3 - 4x^2 + x + 6$ (b) $Q_1(x) = x^2 - 5x + 6$
  (c) $z_2 = 2$, $z_3 = 3$
  (d) $p_1 = 2.1 - (-0.0279)/(-0.536) = 2.047947761$
    $p_2 = 2.023551461$, $p_3 = 2.011678787$, $p_4 = 2.005816089$

8. (a) $P(2) = -8$, $P'(2) = -14$, $P^{(2)}(2) = -6$, $P^{(3)}(2) = 24$, $P^{(4)}(2) = 24$
  (b) $P(x) = -8 - 14(x - 2) - 3(x - 2)^2 + 4(x - 2)^3 + (x - 2)^4$

10. (a) $P(2) = 0$, $P'(2) = 0$, $P^{(2)}(2) = -6$, $P^{(3)}(2) = 12$, $P^{(4)}(2) = 24$
  (b) $P(x) = -3(x - 2)^2 + 2(x - 2)^3 + (x - 2)^4$

12. $-\frac{1}{4}$, -1, 1, $\frac{1}{3}$

14. (a) ± 0.861136312, ± 0.339981044
  (c) ± 0.932469514, ± 0.661209386, ± 0.238619186

18. (a) $P(x) = (x^2 - x + 1)(x^2 + 3x + 1) - 4(x - 1) + 5$
  (b) $r_1 = 2.125$, $s_1 = -1.125$
    $r_2 = 1.87336512$, $s_2 = -1.56352059$
  (c) $P(x) = (x^2 - 2x + 2)(x^2 + 4x + 5)$

20. (a) $P(x) = (x^2 + 4x + 4)(x^2 - 2x + 6) - 6(x + 4) + 25$
  (b) $r_1 = -3.9625$, $s_1 = -4.8375$
    $r_2 = -4.00067756$, $s_2 = -5.00239775$
  (c) $P(x) = (x^2 + 4x + 5)(x^2 - 2x + 5)$

22. (a) $P(x) = (x^2 + 2x + 1)(x^3 - 5x^2 + 12x - 22) + 24(x + 2) - 56$
  (b) $r_1 = -1.84$, $s_1 = -1.544$
    $r_2 = -2.01909009$, $s_2 = -2.04491905$
  (c) $P(x) = (x^2 + 2x + 2)(x^2 - 2x + 5)(x - 3)$

Section 2.5 Aitken's Process and Steffensen's and Muller's Methods: page 103

4. $p_n = 1/(4^n + 4^{-n})$

n	p_n	q_n Aitken's
0	0.5	-0.26437542
1	0.23529412	-0.00158492
2	0.06225681	-0.00002390
3	0.01562119	-0.00000037
4	0.00390619	
5	0.00097656	

5. $g(x) = (6 + x)^{1/2}$

n	p_n	q_n Aitken's
0	2.5	3.00024351
1	2.91547595	3.00000667
2	2.98587943	3.00000018
3	2.99764565	3.00000001
4	2.99960758	
5	2.99993460	

7. Solution of $\cos(x) - 1 = 0$.

n	p_n Steffensen's
0	0.5
1	0.24465808
2	0.12171517
3	0.00755300
4	0.00377648
5	0.00188824
6	0.00000003

9. Solution of $\sin(x^3) = 0$. The accuracy of the table values depends on the correct evaluation of $\sin(x)$ and $\cos(x)$!

n	p_n Steffensen's
0	0.5
1	0.33245982
2	0.22158997
3	0.00468419
4	0.00312280
5	0.00208187
6	0.00000000

11. The sum of the infinite series is $S = 99$.

n	S_n	T_n
1	0.99	98.9999988
2	1.9701	99.0000017
3	2.940399	98.9999988
4	3.90099501	98.9999992
5	4.85198506	
6	5.79346521	

13. The sum of the infinite series is $S = 4$.
15. Muller's method for $f(x) = x^3 - x - 2$.

n	p_n	$f(p_n)$
0	1.0	-2.0
1	1.2	-1.472
2	1.4	-0.656
3	1.52495614	0.02131598
4	1.52135609	-0.00014040
5	1.52137971	-0.00000001

Section 3.1 Introduction to Vectors and Matrices: page 114

1. (a) $(1, 4)$ (b) $(5, -12)$ (c) $(9, -12)$ (d) 5   (e) -38 (f) $(-5, 12)$ (g) 13
3. (a) $(5, -20, -10)$ (b) $(3, 4, 12)$ (c) $(12, -24, 3)$ (d) 9   (e) 89
   (f) $(-3, -4, -12)$ (g) 13
9. (a) $\theta = \arccos(-16/21) \approx 2.437045$ radians
11. (a) $\mathbf{V}_1 = (-2, 5, 12)$, $\mathbf{V}_2 = (1, 4, -1)$, $\mathbf{V}_3 = (7, 0, 6)$, $\mathbf{V}_4 = (11, -3, 8)$

$$\mathbf{C}_1 = \begin{pmatrix} -2 \\ 1 \\ 7 \\ 11 \end{pmatrix}, \quad \mathbf{C}_2 = \begin{pmatrix} 5 \\ 4 \\ 0 \\ -3 \end{pmatrix}, \quad \mathbf{C}_3 = \begin{pmatrix} 12 \\ -1 \\ 6 \\ 8 \end{pmatrix}$$

13. $A + B = \begin{pmatrix} -3 & 7 \\ 5 & 8 \end{pmatrix}$, $E + F = \begin{pmatrix} 3 & 2 & 5 \\ 8 & 4 & 6 \\ 0 & 0 & 0 \end{pmatrix}$

19. No, because none of the dimensions agree.

Section 3.2 Properties of Vectors and Matrices: page 128

1. $AB = \begin{pmatrix} -11 & -12 \\ 13 & -24 \end{pmatrix}$, $BA = \begin{pmatrix} -15 & 10 \\ -12 & -20 \end{pmatrix}$

3. (a) $(AB)C = A(BC) = \begin{pmatrix} 2 & -5 \\ -88 & -56 \end{pmatrix}$

5. (a) 33   (c) The determinant does not exist because the matrix is not square.

6. (a) $A^{-1} = \begin{pmatrix} 0.8 & -0.2 \\ -1.4 & 0.6 \end{pmatrix}$, $\quad \mathbf{X} = \begin{pmatrix} 2.8 \\ -5.4 \end{pmatrix}$

7. (a) $x_1 = \dfrac{12+2}{12-7} = 2.8, \quad x_2 = \dfrac{-6-21}{12-7} = -5.4$

9. No. For example,

$$AB = \begin{pmatrix} 1 & 2 \\ 3 & 6 \end{pmatrix}\begin{pmatrix} -2 & 0 \\ 1 & 0 \end{pmatrix} = \begin{pmatrix} 0 & 0 \\ 0 & 0 \end{pmatrix}$$

But $A \not\equiv 0$ and $B \not\equiv 0$.

11. (a) MN (b) $M(N-1)$

15. $XX^T = (6), \quad X^TX = \begin{pmatrix} 1 & -1 & 2 \\ -1 & 1 & -2 \\ 2 & -2 & 4 \end{pmatrix}$

Section 3.3 Upper-Triangular Linear Systems: page 134

1. $x_1 = 2$, $x_2 = -2$, $x_3 = 1$, $x_4 = 3$ and det $(A) = 120$
6. $x_1 = 3$, $x_2 = 2$, $x_3 = 1$, $x_4 = -1$ and det $(A) = -24$
8. The Forward Substitution Algortithm. $x_1 = b_1/a_{1,1}$. Then compute

$$x_k = \frac{b_k - \displaystyle\sum_{j=1}^{k-1} a_{k,j}x_j}{a_{k,k}} \quad \text{for } k = 2, 3, \ldots, N.$$

Section 3.4 Gaussian Elimination and Pivoting: page 143

1. $x_1 = -3$, $x_2 = 2$, $x_3 = 1$
5. $y = 5 - 3x + 2x^2$
10. $x_1 = 1$, $x_2 = 3$, $x_3 = 2$, $x_4 = -2$
15. Algorithm for the solution of a tridiagonal linear system.

```
FOR  R = 2  TO  N  DO                    {Start the upper-triangularization}
        T := A(R−1)/D(R−1)
        D(R) := D(R) − T*C(R−1)
        B(R) := B(R) − T*B(R−1)
END                                      {End upper-triangularization}


        X(N) := B(N)/D(N)                {Start the back substitution}
FOR  R = N−1  DOWNTO  1  DO
        X(R) := [B(R) − C(R)*X(R+1)]/D(R)    {End back substitution}
```

21. (a) Solution for Hilbert matrix A:
    $x_1 = 25$, $x_2 = -300$, $x_3 = 1{,}050$, $x_4 = -1{,}400$, $x_5 = 630$
(b) Solution for the other matrix A:
    $x_1 = 28.02304$, $x_2 = -348.5887$, $x_3 = 1{,}239.781$
    $x_4 = -1{,}666.785$, $x_5 = 753.5564$

Section 3.5 Matrix Inversion: page 150

1. (a) $\begin{pmatrix} 2 & -1 & 0 \\ -1 & -3 & 2 \\ 0 & 2 & -1 \end{pmatrix}$

3. (a) $\begin{pmatrix} -\frac{13}{3} & -3 & -\frac{2}{3} \\ \frac{1}{3} & 0 & -\frac{1}{3} \\ 2 & 1 & 0 \end{pmatrix}$

5. (a)
$$\begin{pmatrix} -0.20 & -0.76 & -2.36 & 1.08 \\ 0.40 & 0.72 & 2.92 & -0.76 \\ -0.20 & -0.16 & -0.76 & 0.28 \\ 0.80 & 1.44 & 4.84 & -1.52 \end{pmatrix}$$

12. The inverse of the Hilbert matrix of dimension 5 by 5 is
$$\begin{pmatrix} 25 & -300 & 1{,}050 & -1{,}400 & 630 \\ -300 & 4{,}800 & -18{,}900 & 26{,}880 & -12{,}600 \\ 1{,}050 & -18{,}900 & 79{,}380 & -117{,}600 & 56{,}700 \\ -1{,}400 & 26{,}880 & -117{,}600 & 179{,}200 & -88{,}200 \\ 630 & -12{,}600 & 56{,}700 & -88{,}200 & 44{,}100 \end{pmatrix}$$

Section 3.6 Triangular Factorization: page 162

1. (a) $Y^T = (-4, 12, 3)$, $X^T = (-3, 2, 1)$
 (b) $Y^T = (20, 39, 9)$, $X^T = (5, 7, 3)$

4. (a)
$$\begin{pmatrix} -5 & 2 & -1 \\ 1 & 0 & 3 \\ 3 & 1 & 6 \end{pmatrix} = \begin{pmatrix} 1 & 0 & 0 \\ -0.2 & 1 & 0 \\ -0.6 & 5.5 & 1 \end{pmatrix} \begin{pmatrix} -5 & 2 & -1 \\ 0 & 0.4 & 2.8 \\ 0 & 0 & -10 \end{pmatrix}$$

6. (a) $Y^T = (8, -6, 12, 2)$, $X^T = (3, -1, 1, 2)$
 (b) $Y^T(28, 6, 12, 1)$, $X^T = (3, 1, 2, 1)$

9. The triangular factorization $A = LU$ is
$$\begin{pmatrix} 1 & 1 & 0 & 4 \\ 2 & -1 & 5 & 0 \\ 5 & 2 & 1 & 2 \\ -3 & 0 & 2 & 6 \end{pmatrix} = \begin{pmatrix} 1 & 0 & 0 & 0 \\ 2 & 1 & 0 & 0 \\ 5 & 1 & 1 & 0 \\ -3 & -1 & -1.75 & 1 \end{pmatrix} \begin{pmatrix} 1 & 1 & 0 & 4 \\ 0 & -3 & 5 & -8 \\ 0 & 0 & -4 & -10 \\ 0 & 0 & 0 & -7.5 \end{pmatrix}$$

19. (a) $I_1 = 3$, $I_2 = 5$, $I_3 = 1$
  (c) $I_1 = 4$, $I_2 = 3$, $I_3 = -1$

Section 4.1 Introduction to Interpolation: page 172

1. $P(x) = -0.02x^3 + 0.1x^2 - 0.2x + 1.66$
 (a) Use $x = 4$ and get
  $b_3 = -0.02$, $b_2 = 0.02$, $b_1 = -0.12$, $b_0 = 1.18$
  Hence $P(4) = 1.18$.
 (b) Use $x = 4$ and get
  $d_2 = -0.06$, $d_1 = -0.04$, $d_0 = -0.36$.
  Hence $P'(4) = -0.36$.
 (c) Use $x = 4$ and get $i_4 = -0.005$, $i_3 = 0.01333333$,
  $i_2 = -0.04666667$, $i_1 = 1.47333333$, $i_0 = 5.89333333$.
  Hence $I(4) = 5.89333333$.
  Similarly, use $x = 1$ and get $I(1) = 1.58833333$.

$$\int_1^4 P(x)\, dx = I(4) - I(1) = 5.89333333 - 1.58833333 = 4.305$$

 (d) Use $x = 5.5$ and get
  $b_3 = -0.02$, $b_2 = -0.01$, $b_1 = -0.255$, $b_0 = 0.2575$.
  Hence $P(5.5) = 0.2575$.

4. Approximations for $f(x) = \exp(x)$

		$P(x)$	$\exp(x)$	$\exp(x) - P(x)$
(a)	$x = \;\;\; 0.3$	1.34985810	1.34985881	0.00000071
	$x = \;\;\; 0.4$	1.49182470	1.49182470	0.00000000
	$x = \;\;\; 0.5$	1.64872179	1.64872127	-0.00000052
(b)	$x = -0.1$	0.90481537	0.90483742	0.00002205
	$x = \;\;\; 1.1$	3.00413986	3.00416602	0.00002616

Section 4.2 Linear and Lagrange Approximation: page 183

1. (a) $P_1(x) = -1(x - 0)/(-1 - 0) + 0 = x + 0 = x$
   (b) $P_2(x) = -1(x - 0)(x - 1)/(-1 - 0)(-1 - 1) + 0 + 1(x + 1)(x - 0)/(1 + 1)(1 - 0)$
   $\qquad = -0.5(x)(x - 1) + 0.5(x)(x + 1) = 0x^2 + x + 0 = x$
   (c) $P_3(x) = -1(x)(x - 1)(x - 2)/(-1)(-2)(-3) + 0$
   $\qquad\quad + 1(x + 1)(x)(x - 2)/(2)(1)(-1) + 8(x + 1)(x)(x - 1)/(3)(2)(1)$
   $\qquad P_3(x) = x^3 + 0x^2 + 0x + 0 = x^3$
   (d) $P_1(x) = 1(x - 2)/(1 - 2) + 8(x - 1)/(2 - 1)$
   $\qquad P_1(x) = -(x - 2) + 8(x - 1) = 7x - 6$
   (e) $P_2(x) = 0 + (x)(x - 2)/(1)(-1) + 8(x)(x - 1)/(2)(1)$
   $\qquad P_2(x) = -(x)(x - 2) + 4(x)(x - 1) = 3x^2 - 2x + 0$
3. (a) $P_2(x) = -2x^2 + 6x + 0$, $P_2(1.5) = 4.5$
   (b) $P_3(x) = 0.5x^3 - 3.5x^2 + 7x + 0$, $P_3(1.5) = 4.3125$
6. (c) $f^{(4)}(c) = 120(c - 1)$ for all c, so that
   $\qquad E_3(x) = 5(x + 1)(x)(x - 3)(x - 4)(c - 1)$
10. $|f^{(2)}(c)| \le |-\sin(1)| = 0.84147098 = M_2$
    (a) $h^2 M_2/8 = h^2 \times 0.84147098/8 < 5 \times 10^{-7}$
    $\qquad h^2 < 4.753580 \times 10^{-6}$ implies that $h < 0.00218027$
12. (a) $z = 3 - 2x + 4y$
14. $\sin(h) - h = h^3(-1/6 + h^2/120 - \ldots) = O(h^3)$

Section 4.3 Newton Polynomials: page 194

1. $P_1(x) = 4 - (x - 1)$
   $P_2(x) = 4 - (x - 1) + 0.4(x - 1)(x - 3)$
   $P_3(x) = 4 - (x - 1) + 0.4(x - 1)(x - 3) + 0.01(x - 1)(x - 3)(x - 4)$
   $P_4(x) = P_3(x) - 0.002(x - 1)(x - 3)(x - 4)(x - 4.5)$
   $P_1(2.5) = 2.5$, $P_2(2.5) = 2.2$, $P_3(2.5) = 2.21125$
   $P_4(2.5) = 2.21575$
5. $f(x) = 3 \times 2^x$
   $P_4(x) = 1.5 + 1.5(x + 1) + 0.75(x + 1)(x) + 0.25(x + 1)(x)(x - 1)$
   $\qquad\quad + 0.0625(x + 1)(x)(x - 1)(x - 2)$
   $P_1(1.5) = 5.25$, $\qquad P_2(1.5) = 8.0625$
   $P_3(1.5) = 8.53125$, $P_4(1.5) = 8.47265625$
7. $f(x) = 3.6/x$
   $P_4(x) = 3.6 - 1.8(x - 1) + 0.6(x - 1)(x - 2) - 0.15(x - 1)(x - 2)(x - 3)$
   $\qquad\quad + 0.03(x - 1)(x - 2)(x - 3)(x - 4)$
   $P_1(2.5) = 0.9$, $\qquad P_2(2.5) = 1.35$
   $P_3(2.5) = 1.40625$, $P_4(2.5) = 1.423125$

11. $f(x) = x^{1/2}$
$P_4(x) = 2.0 + 0.23607(x - 4) - 0.01132(x - 4)(x - 5)$
$\qquad + 0.00091(x - 4)(x - 5)(x - 6) - 0.00008(x - 4)(x - 5)(x - 6)(x - 7)$
$P_1(4.5) = 2.11804, \; P_2(4.5) = 2.12086$
$P_3(4.5) = 2.12121, \; P_4(4.5) = 2.12128$
20. $P_2(x) = -2 + 4(x) - (x)(x - 1) = -x^2 + 5x - 2$
22. $P_3(x) = 5 - (x)(x - 1) + (x)(x - 1)(x - 2) = x^3 - 4x^2 + 3x + 5$

Section 4.4 Chebyshev Polynomials: page 206

9. (a) $\ln(x + 2) \approx 0.69549038 + 0.49905042x - 0.14334605x^2 + 0.04909073x^3$
(b) $|f^{(4)}(x)|/(2^3 \times 4!) \leq |-6|/(2^3 \times 4!) = 0.03125000$
11. (a) $\cos(x) \approx 1 - 0.46952087x^2$
(b) $|f^{(3)}(x)|/(2^2 \times 3!) \leq |\sin(1)|/(2^2 \times 3!) = 0.03506129$
13. The error bound for Taylor's polynomial is

$$\frac{|f^{(8)}(x)|}{8!} \leq \frac{|\sin(1)|}{8!} = 0.00002087$$

The error bound for the minimax approximation is

$$\frac{|f^{(8)}(x)|}{2^7 \times 8!} \leq \frac{|\sin(1)|}{2^7 \times 8!} = 0.00000016$$

17. $f(x) = \sin(x)$, Chebyshev coefficients:
(a) $C(0) = 0, \; C(1) = 0.88010417, \; C(2) = 0, \; C(3) = -0.03962622$
(b) $C(0) = 0, \; C(1) = 0.88010116, \; C(2) = 0, \; C(3) = -0.03912370$
$\qquad C(4) = 0$
(c) $C(0) = 0, \; C(1) = 0.88010117, \; C(2) = 0, \; C(3) = -0.03912672$
$\qquad C(4) = 0, \; C(5) = 0.00050252$
(d) $C(0) = 0, \; C(1) = 0.88010117, \; C(2) = 0, \; C(3) = -0.03912671$
$\qquad C(4) = 0, \; C(5) = 0.00049951, \; C(6) = 0$

Section 4.5 Padé Approximations: page 212

1. $1 = p_0, \; 1 + q_1 = p_1, \; \frac{1}{2} + q_1 = 0, \; q_1 = -\frac{1}{2}, \; p_1 = \frac{1}{2}$
$\exp(x) \approx R_{1,1}(x) = (2 + x)/(2 - x)$
3. $1 = p_0, \; \frac{1}{3} + 2q_1/15 = p_1, \; \frac{2}{15} + q_1/3 = 0, \; q_1 = -\frac{2}{5}, \; p_1 = -\frac{1}{15}$
5. $1 = p_0, \; 1 + q_1 = p_1, \; \frac{1}{2} + q_1 + q_2 = p_2$

$$\left. \begin{array}{c} \dfrac{1}{6} + \dfrac{q_1}{2} + q_2 = 0 \\[2mm] \dfrac{1}{24} + \dfrac{q_1}{6} + \dfrac{q_2}{2} = 0 \end{array} \right\} \quad \text{First solve this system.}$$

$q_1 = -\frac{1}{2}, \; q_2 = \frac{1}{12}, \; p_1 = \frac{1}{2}, \; p_2 = \frac{1}{12}$
7. $1 = p_0, \; \frac{1}{3} + q_1 = p_1, \; \frac{2}{15} + q_1/3 + q_2 = p_2$

$$\left. \begin{array}{c} \dfrac{17}{315} + \dfrac{2q_1}{15} + \dfrac{q_2}{3} = 0 \\[2mm] \dfrac{62}{2,835} + \dfrac{17q_1}{315} + \dfrac{2q_2}{15} = 0 \end{array} \right\} \quad \text{First solve this system.}$$

$q_1 = -\frac{4}{9}, \; q_2 = \frac{1}{63}, \; p_1 = -\frac{1}{9}, \; p_2 = \frac{1}{945}$

11.

x_k	Taylor Polynomial Approximation	Padé Approximation	Exact Value, $f(x) = \exp(x)$
-1.6	0.270400	0.205298	0.201897
-1.2	0.318400	0.302326	0.301194
-0.8	0.451733	0.449541	0.449329
-0.4	0.670400	0.670330	0.670320
0.0	1.000000	1.000000	1.000000
0.4	1.491733	1.491803	1.491825
0.8	2.222400	2.224490	2.225541
1.2	3.294400	3.307692	3.320117
1.6	4.835733	4.870968	4.953032

13.

x_k	Taylor Polynomial Approximation	Padé Approximation	Exact Value, $f(x) = \tan(x)$
0.0	0.000000	0.000000	0.000000
0.2	0.202710	0.202710	0.202710
0.4	0.422793	0.422793	0.422793
0.6	0.684099	0.684137	0.684137
0.8	1.028611	1.029639	1.029639
1.0	1.542504	1.557407	1.557408
1.2	2.413996	2.572147	2.572152
1.4	4.052510	5.797765	5.797884

Section 5.1 Least-Squares Line: page 220

1. (a) $10A + 0B = 7$
$0A + 5B = 13$
$y = 0.70x + 2.60$, $E_2(f) \approx 0.2449$

2. (a) $40A + 0B = 58$
$0A + 5B = 31.2$
$y = 1.45x + 6.24$, $E_2(f) \approx 0.8958$

3. (a) $60A + 10B = 120$
$10A + 5B = 31$
$y = 1.45x + 3.3$, $E_2(f) \approx 0.7348$

4. (a) $\sum_{k=1}^{5} x_k y_k \Big/ \sum_{k=1}^{5} x_k^2 = 86.9/55 = 1.58$
$y = 1.58x$, $E_2(f) \approx 0.1720$

11. (a) $y = 1.6866x^2$, $E_2(f) \approx 1.3$
$y = 0.5902x^3$, $E_2(f) \approx 0.29$. This is the best fit.

13. (a) $k = 39.98/2.2 \approx 18.1727$

Section 5.2 Curve Fitting: page 234

1. (a) $164A + 20C = 186$
$20B = -34$
$20A + 4C = 26$
$y = 0.875x^2 - 1.70x + 2.125 = 7/8\, x^2 - 17/10\, x + 17/8$

3. (a) $y = 0.5102x^2$, $E_1(f) \approx 0.20$ is the best fit.
$y = 0.4304\, 2^x$, $E_1(f) \approx 0.62$

4. (a) $55A + 15B = 25.9297$
$15A + 5B = 6.1515$
$A = 0.7475, B = -1.0123, C = \exp(B) = 0.3634$
$y = 0.3634e^{0.7475x}, E_1(f) \approx 0.87$
(b) $6.1993A + 4.7874B = 8.9366$
$4.7874A + 5B = 6.1515$
$A = 1.8859, B = -0.5755, C = \exp(B) = 0.5625$
$y = 0.5625x^{1.8859}, E_1(f) \approx 0.27$ is the best fit.
6. (a) $15A + 5B = -0.8647$
$5A + 5B = 4.2196$
$A = -0.5084, B = 1.3524, C = \exp(B) = 3.8665$
$y = 3.8665e^{-0.5084x}, E_1(f) \approx 0.10$

9.

	Using Linearization	Minimizing Least Squares
(a)	$\dfrac{1{,}000}{1 + 4.3018 \exp(-1.0802t)}$	$\dfrac{1{,}000}{1 + 4.2131 \exp(-1.0456t)}$
(b)	$\dfrac{5{,}000}{1 + 8.9991 \exp(-0.81138t)}$	$\dfrac{5{,}000}{1 + 8.9987 \exp(-0.81157t)}$

13. (a) $14A + 15B + 8C = 82,$ $A = 2.4, B = 1.2, C = 3.8$
$15A + 19B + 9C = 93$ Yields $z = 2.4x + 1.2y + 3.8$
$8A + 9B + 5C = 49$

Section 5.3 Interpolation by Spline Functions: page 249

1. (b) $b_0 = -4$ (c) $b_1 = -1$ (d) $S'(1) = -1, S''(1) = 6$
3. (i) $m_0 = -3.6, m_1 = 7.2, m_2 = -7.2, m_3 = 3.6$
$S_0(x) = 1.8x^3 - 1.8x^2 - 5x + 7$
$S_1(w) = -2.4w^3 + 3.6w^2 - 3.2w + 2, w = x - 1$
$S_2(w) = 1.8w^3 - 3.6w^2 - 3.2w, w = x - 2$
(ii) $m_0 = 0, m_1 = 6, m_2 = -6, m_3 = 0$
$S_0(x) = x^3 - 6x + 7$
$S_1(w) = -2w^3 + 3w^2 - 3w + 2, w = x - 1$
$S_2(w) = w^3 - 3w^2 - 3w, w = x - 2$
(iii) $m_0 = 9, m_1 = 3, m_2 = -3, m_3 = -9$
$S_0(x) = -x^3 + 4.5x^2 - 8.5x + 7$
$S_1(w) = -w^3 + 1.5w^2 - 2.5w + 2, w = x - 1$
$S_2(w) = -w^3 - 1.5w^2 - 2.5w, w = x - 2$
(iv) $m_0 = 4.5, m_1 = 4.5, m_2 = -4.5, m_3 = -4.5$
$S_0(x) = 2.25x^2 - 7.25x + 7$
$S_1(w) = -1.5w^3 + 2.25w^2 - 2.75w + 2, w = x - 1$
$S_2(w) = -2.25w^2 - 2.75w, w = x - 2$
(v) $m_0 = 3, m_1 = 4.8, m_2 = -4.2, m_3 = -6$
$S_0(x) = 0.3x^3 + 1.5x^2 - 6.8x + 7$
$S_1(w) = -1.5w^3 + 2.4w^2 - 2.9w + 2, w = x - 1$
$S_2(w) = -0.3w^3 - 2.1w^2 - 2.6w, w = x - 2$
8. (i) $m_0 = -4.25, m_1 = 2.5, m_2 = 6.25, m_3 = -9.5, m_4 = 7.75$
$S_0(x) = 1.125x^3 - 2.125x^2 - 2x + 5$
$S_1(w) = 0.625w^3 + 1.25w^2 - 2.875w + 2, w = x - 1$
$S_2(w) = -2.625w^3 + 3.125w^2 + 1.5w + 1, w = x - 2$
$S_3(w) = 2.875w^3 - 4.75w^2 - 0.125w + 3, w = x - 3$

(ii) $m_0 = 0$, $m_1 = 1.5$, $m_2 = 6$, $m_3 = -7.5$, $m_4 = 0$
$S_0(x) = 0.25x^3 - 3.25x + 5$
$S_1(w) = 0.75w^3 + 0.75w^2 - 2.5w + 2$, $w = x - 1$
$S_2(w) = -2.25w^3 + 3w^2 + 1.25w + 1$, $w = x - 2$
$S_3(w) = 1.25w^3 - 3.75w^2 + 0.5w + 3$, $w = x - 3$

(iii) $m_0 = -1$, $m_1 = 2$, $m_2 = 5$, $m_3 = -4$, $m_4 = -13$
$S_0(x) = 0.5x^3 - 0.5x^2 - 3x + 5$
$S_1(w) = 0.5w^3 + w^2 - 2.5w + 2$, $w = x - 1$
$S_2(w) = -1.5w^3 + 2.5w^2 + w + 1$, $w = x - 2$
$S_3(w) = -1.5w^3 - 2w^2 + 1.5w + 3$, $w = x - 3$

(iv) $m_0 = \frac{19}{15}$, $m_1 = \frac{19}{15}$, $m_2 = \frac{17}{3}$, $m_3 = -\frac{89}{15}$, $m_4 = -\frac{89}{15}$
$S_0(x) = 0.6333\overline{3}x^2 - 3.6333\overline{3}x + 5$
$S_1(w) = 0.7333\overline{3}w^3 + 0.6333\overline{3}w^2 - 2.3666\overline{6}w + 2$, $w = x - 1$
$S_2(w) = -1.9333\overline{3}w^3 + 2.8333\overline{3}w^2 + 1.1w + 1$, $w = x - 2$
$S_3(w) = -2.9666\overline{6}w^2 + 0.9666\overline{6}w + 3$, $w = x - 3$

(v) $m_0 = 0.5$, $m_1 = 1.4$, $m_2 = 5.9$, $m_3 = -7$, $m_4 = -1.9$
$S_0(x) = 0.15x^3 + 0.25x^2 - 3.4x + 5$
$S_1(w) = 0.75w^3 + 0.7w^2 - 2.45w + 2$, $w = x - 1$
$S_2(w) = -2.15w^3 + 2.95w^2 + 1.2w + 1$, $w = x - 2$
$S_3(w) = 0.85w^3 - 3.5w^2 + 0.65w + 3$, $w = x - 3$

14. (i) $m_0 = -2.08$, $m_1 = 1.76$, $m_2 = 1.04$, $m_3 = 0.08$
$m_4 = -1.36$, $m_5 = -0.64$, $m_6 = -2.08$
$S_0(x) = 0.64x^3 - 1.04x^2 - 0.6x + 1$
$S_1(w) = -0.12w^3 + 0.88w^2 - 0.76w$, $w = x - 1$
$S_2(w) = -0.16w^3 + 0.52w^2 + 0.64w$, $w = x - 2$
$S_3(w) = -0.24w^3 + 0.04w^2 + 1.2w + 1$, $w = x - 3$
$S_4(w) = 0.12w^3 - 0.68w^2 + 0.56w + 2$, $w = x - 4$
$S_5(w) = -0.24w^3 - 0.32w^2 - 0.44w + 2$, $w = x - 5$

(ii) $m_0 = 0$, $m_1 = 1.2$, $m_2 = 1.2$, $m_3 = 0$
$m_4 = -1.2$, $m_5 = -1.2$, $m_6 = 0$
$S_0(x) = 0.2x^3 - 1.2x + 1$
$S_1(w) = 0.6w^2 - 0.6w$, $w = x - 1$
$S_2(w) = -0.2w^3 + 0.6w^2 + 0.6w$, $w = x - 2$
$S_3(w) = -0.2w^3 + 1.2w + 1$, $w = x - 3$
$S_4(w) = -0.6w^2 + 0.6w + 2$, $w = x - 4$
$S_5(w) = 0.2w^3 - 0.6w^2 - 0.6w + 2$, $w = x - 5$

(iii) $m_0 = 0.75$, $m_1 = 1$, $m_2 = 1.25$, $m_3 = 0$
$m_4 = -1.25$, $m_5 = -1$, $m_6 = -0.75$
$S_0(x) = 0.0416\overline{6}x^3 + 0.375x^2 - 1.416\overline{6}x + 1$
$S_1(w) = 0.0416\overline{6}w^3 + 0.5w^2 - 0.5416\overline{6}w$, $w = x - 1$
$S_2(w) = -0.2083\overline{3}w^3 + 0.625w^2 + 0.5833\overline{3}w$, $w = x - 2$
$S_3(w) = -0.2083\overline{3}w^3 + 1.208\overline{3}w + 1$, $w = x - 3$
$S_4(w) = 0.0416\overline{6}w^3 - 0.625w^2 + 0.5833\overline{3}w + 2$, $w = x - 4$
$S_5(w) = -0.0416\overline{6}w^3 - 0.5w^2 - 0.5416\overline{6}w + 2$, $w = x - 5$

Section 5.4 Fourier Series and Trigonometric Polynomials: page 261

1. $f(x) = \dfrac{4}{\pi}\left[\sin(x) + \dfrac{\sin(3x)}{3} + \dfrac{\sin(5x)}{5} + \dfrac{\sin(7x)}{7} + \cdots\right]$

5. $f(x) = \dfrac{\pi}{4} + \displaystyle\sum_{j=1}^{\infty}\left[\dfrac{(-1)^j - 1}{\pi j^2}\right]\cos(jx) - \sum_{j=1}^{\infty}\left[\dfrac{(-1)^j}{j}\right]\cos(jx)$

7. $f(x) = \dfrac{4}{\pi}\left[\sin(x) - \dfrac{\sin(3x)}{3^2} + \dfrac{\sin(5x)}{5^2} - \dfrac{\sin(7x)}{7^2} + \cdots\right]$

17. $a_0 = 0.82246703$, $a_1 = -1$, $a_2 = 0.25$, $a_3 = -0.11111111$
$a_4 = 0.0625$, $a_5 = -0.04$

19. (a) *Remark.* Use $f(-\pi) = 0$, $f(0) = 0$, $f(\pi) = 0$.

$$P(x) = 1.244017 \sin(x) + 0.333333 \sin(3x) + 0.089316 \sin(5x)$$

(c) *Remark.* Use $f(-\pi) = \pi/2$, $f(\pi) = \pi/2$.

$$P(x) = 0.785398 - 0.651366 \cos(x) + 0.977049 \sin(x)$$
$$- 0.453450 \sin(2x) - 0.087266 \cos(3x) + 0.261799 \sin(3x)$$
$$- 0.151150 \sin(4x) - 0.046766 \cos(5x) + 0.070149 \sin(5x)$$

21. Average temperature in Fairbanks, Alaska (4-week intervals)

$$P(x) = 27.3077 - 37.9050(\cos 2\pi x/13) - 7.260548 \sin(2\pi x/13)$$

Date	x	y	$P(x)$
Jan. 1	0	-14	-10.6
Jan. 29	1	-9	-9.6
Feb. 26	2	2	-0.2
Mar. 26	3	15	15.5
Apr. 23	4	35	34.0
May 21	5	52	50.9
June 18	6	62	62.4
July 16	7	63	65.8
Aug. 13	8	58	60.5
Sept. 10	9	50	47.5
Oct. 8	10	34	30.0
Nov. 5	11	12	11.8
Dec. 3	12	-5	-2.9
Jan. 1	13	-14	-10.6

Section 6.1 Approximating the Derivative: page 274

1. $f(x) = \sin(x)$

h	Approximate $f'(x)$ Formula (3)	Error in the Approximation	Bound for the Truncation Error
0.1	0.695546112	0.001160597	0.001274737
0.01	0.696695100	0.000011609	0.000012747
0.001	0.696706600	0.000000109	0.000000127

3. $f(x) = \sin(x)$

h	Approximate $f'(x)$ Formula (10)	Error in the Approximation	Bound for the Truncation Error
0.1	0.696704390	0.000002320	0.000002322
0.01	0.696706710	-0.000000001	0.000000000

5. $f(x) = x^3$ (a) $f'(2) \approx 12.0025000$ (b) $f'(2) \approx 12.0000000$
(c) for part (a): $O(h^2) = -(0.05)^2 f^{(3)}(c)/6 = -0.0025000$
for part (b): $O(h^4) = -(0.05)^4 f^{(5)}(c)/30 = -0.0000000$

7. $f(x, y) = xy/(x + y)$
(a) $f_x(x, y) = [y/(x + y)]^2$, $f_x(2, 3) = 0.36$

h	Approximation to $f_x(2, 3)$	Error in the Approximation
0.1	0.360144060	−0.000144060
0.01	0.360001400	−0.000001400
0.001	0.360000000	0.000000000

$f_y(x, y) = [x/(x + y)]^2$, $f_y(2, 3) = 0.16$

h	Approximation to $f_y(2, 3)$	Error in the Approximation
0.1	0.160064030	−0.000064030
0.01	0.160000600	−0.000000600
0.001	0.160000000	0.000000000

9. (a) Formula (3) gives $I'(1.2) \approx -13.5840$ and $E(1.2) \approx 11.3024$.
Formula (10) gives $I'(1.2) \approx -13.6824$ and $E(1.2) \approx 11.2975$.
(b) Using differentiation rules from calculus, we obtain
$I'(1.2) \approx -13.6793$ and $E(1.2) \approx 11.2976$.

11. $f(x) = \cos(x)$, $f^{(3)}(x) = \sin(x)$
Use the bound $|f^{(3)}(x)| \leqq \sin(1.3) \approx 0.96356$.

h	App. $f'(x)$ Equation (17)	Error in the Approximation	Equation (19), Total Error Bound $\lvert$ Round-Off$\rvert$ + $\lvert$ Trunc.$\rvert$
0.1	−0.93050	−0.00154	0.00005 + 0.00161 = 0.00166
0.01	−0.93200	−0.00004	0.00050 + 0.00002 = 0.00052
0.001	−0.93000	−0.00204	0.00500 + 0.00000 = 0.00500

15. $f(x) = \cos(x)$, $f^{(5)}(x) = -\sin(x)$
Use the bound $|f^{(5)}(x)| \leq \sin(1.4) \approx 0.98545$.

h	App. $f'(x)$ Equation (22)	Error in the Approximation	Equation (24), Total Error Bound $\lvert$ Round-Off$\rvert$ + $\lvert$ Trunc.$\rvert$
0.1	−0.93206	0.00002	0.00008 + 0.00000 = 0.00008
0.01	−0.93208	0.00004	0.00075 + 0.00000 = 0.00075
0.001	−0.92917	−0.00287	0.00750 + 0.00000 = 0.00750

Section 6.2 Numerical Differentiation Formulas: page 287

1. $f(x) = \ln(x)$ (a) $f''(5) \approx -0.040001600$
(b) $f''(5) \approx -0.040007900$ (c) $f''(5) \approx -0.039999833$
(d) $f''(5) = -0.04000000 = -1/5^2$
The answer in part (b) is more accurate.

3. $f(x) = \ln(x)$ (a) $f''(5) \approx 0.0000$
(b) $f''(5) \approx -0.0400$ (c) $f''(5) \approx 0.0133$
(d) $f''(5) = -0.0400 = -1/5^2$
The answer in part (b) is more accurate.

5. (a) $f(x) = x^2$, $f''(1) \approx 2.0000$
(b) $f(x) = x^4$, $f''(1) \approx 12.0002$

9. (a)

x	$f'(x)$
0.0	0.141345
0.1	0.041515
0.2	-0.058275
0.3	-0.158025

Section 6.3 Minimization of a Function: page 308

4. $f(x) = 3x^2 - 2x + 5$
$f'(x) = 6x - 2$. Local minimum at $\frac{1}{3}$.

6. $f(x) = 4x^3 - 8x^2 - 11x + 5$
$f'(x) = 12x^2 - 16x - 11$
Critical points $\frac{11}{6}$, $-\frac{1}{2}$. Local minimum at $\frac{11}{6}$.

8. $f(x) = (x + 2.5)/(4 - x^2)$
$f'(x) = (x^2 + 5x + 4)/(4 - x^2)^2$
Critical points $-1, -4$. Local minimum at -1.

10. $f(x) = -\sin(x) - \sin(3x)/3$ on $[0, 2]$
$f'(x) = -[\cos(x) + \cos(3x)]$. Local minimum at 0.785398163.

14. $f(x, y) = x^3 + y^3 - 3x - 3y + 5$
$f_x(x, y) = 3(x^2 - 1)$, $f_y(x, y) = 3(y^2 - 1)$
Critical points $(1, 1)$, $(1, -1)$, $(-1, 1)$, $(-1, -1)$.
Local minimum at $(1, 1)$.

16. $f(x, y) = x^2 y + xy^2 - 3xy$
$f_x(x, y) = y(2x + y - 3)$, $f_y(x, y) = x(x + 2y - 3)$
Critical points $(0, 0)$, $(0, 3)$, $(3, 0)$, $(1, 1)$.
Local minimum at $(1, 1)$.

18. $f(x, y) = (x - y)/(2 + x^2 + y^2)$
$f_x(x, y) = (y^2 + 2xy + 2 - x^2)/(2 + x^2 + y^2)^2$
$f_y(x, y) = (y^2 - 2xy - 2 - x^2)/(2 + x^2 + y^2)^2$
Critical points $(1, -1)$, $(-1, 1)$. Local minimum at $(-1, 1)$.

20. $f(x, y) = (x - y)^4 + (x + y - 2)^2$
$f_x(x, y) = 4(x - y)^3 + 2(x + y - 2)$, $f_y(x, y) = -4(x - y)^3 + 2(x + y - 2)$
Local minimum at $(1, 1)$.

32. $f(x, y, z) = 2x^2 + 2y^2 + z^2 - 2xy + yz - 7y - 4z$
$f_x = 0$: $4x - 2y$ $= 0$
$f_y = 0$: $-2x + 4y + z - 7 = 0$
$f_z = 0$: $y + 2z - 4 = 0$.   Local minimum at $(1, 2, 1)$.

34. $f(x, y, z) = x^4 + y^4 + z^4 - 4xyz$
$f_x(x, y, z) = 4(x^3 - yz)$, $f_y(x, y, z) = 4[y^3 - xz]$
$f_z(x, y, z) = 4(z^3 - xy)$.   Local minimum at $(1, 1, 1)$.

36. $f(x, y, z, u) = 2(x^2 + y^2 + z^2 + u^2) + x(y + z - u) + yz - 5x - 6y - 6z - 3u$
$f_x = 0$: $4x + y + z - u - 5 = 0$
$f_y = 0$: $x + 4y + z - 6 = 0$
$f_z = 0$: $x + y + 4z - 6 = 0$
$f_u = 0$: $-x + 4u - 3 = 0$
Local minimum at $(1, 1, 1, 1)$.

Section 7.2 Composite Trapezoidal Rule: page 319

1. Integrate $f(x) = 3x^2$ over $[1, 3]$.
   (a) 30   (b) 27   (c) 26.25   (d) 26
3. Integrate $f(x) = x^{-2}$ over $[1, 3]$.
   (a) $10/9 = 1.111111\overline{1}$ (b) $29/36 = 0.805555\overline{5}$
   (c) $141/200 = 0.705$ (d) $2/3 = 0.666666\overline{6}$
7. $\int_0^{0.8} [1 + \sin^2 (x)]^{1/2} \, dx = 0.86993 \ldots$

   (a) Using $M = 4$, trapezoidal approx. $= 0.87129$
   (b) Using $M = 8$, trapezoidal approx. $= 0.87027$
10. $\frac{1}{2}(10 + 2 \times 6 + 4) + \frac{4}{2}(4 + 2 \times 2 + 1) = 31$
13. (a) Start with $[(\pi/3)h^2 \times 1]/12 \leq 5 \times 10^{-9}$ or $h^2 \leq (18/\pi)10^{-8}$.
    Solve for M using the relation $(\pi^3/162) \times 10^8 \leq M^2$.
    Since M must be an integer, $M = 4375$ and $h \approx 0.000239359$.
16. $T(f, h) + E(f) = \int_{-0.1}^{0.1} \cos (x) \, dx \approx 0.1996668$

h	$T(f, h)$	$0.1996668 - T(f, h)$	$0.01664h^2$
0.2000	0.1990008	0.0006660	0.0006656
0.1000	0.1995004	0.0001664	0.0001664
0.0500	0.1996252	0.0000416	0.0000416
0.0250	0.1996564	0.0000104	0.0000104
0.0125	0.1996642	0.0000026	0.0000026

Section 7.3 Composite Simpson's Rule: page 329

1. Integrate $f(x) = 4x^3$ over $[1, 3]$.
   (a) 80   (b) 80   (c) 80
3. Integrate $f(x) = x^{-2}$ over $[1, 3]$.
   (a) $19/27 = 0.703703704$ (b) $1,813/2,700 = 0.671481482$
   (c) $2/3 = 0.666666667$
7. $(\pi/3)(36 + 4 \times 30.25 + 2 \times 16 + 4 \times 16 + 2 \times 25 + 4 \times 56.25 + 81)$
   $= \pi \times 203 = 637.7433$
8. $2\pi \int_0^{0.8} \sin (x)(1 + [\cos (x)]^2)^{1/2} \, dx = 2.50226 \ldots$

   (a) Using $M = 1$, Simpson's rule approx. $= 2.50658$.
   (b) Using $M = 2$, Simpson's rule approx. $= 2.50252$.
10. $\frac{1}{3}(10 + 4 \times 6 + 4) + \frac{4}{3}(4 + 4 \times 2 + 1) = 30$
13. $f(x) = \cos (x)$
    $[(\pi/3)h^4 \times 1]/180 \leq 5 \times 10^{-9}, h^4 \leq (27/\pi)10^{-7}$
    Solve for M using the relation $M^4 \geq (\pi^5/34992)10^7$.
    Since M must be an integer, $M = 18$ and $h \approx 0.029088821$.
16. $S(f, h) + E(f) = \int_{-0.75}^{0.75} \cos (x) \, dx \approx 1.3632775$

h	$S(f, h)$	$1.3632775 - S(f, h)$	$-0.00770h^4$
0.75000	1.3658444	-0.0025669	-0.0024363
0.37500	1.3634298	-0.0001523	-0.0001523
0.18750	1.3632869	-0.0000094	-0.0000095
0.09375	1.3632781	-0.0000006	-0.0000006

Section 7.4 Sequential Integration Rules: page 336

1. $\int_1^3 6x^5 \, dx = 728$

J	$T(J)$ Trapezoid Rule	$S(J)$ Simpson's Rule	$B(J)$ Boole's Rule
0	1464.0000		
1	924.0000	744.0000	
2	777.7500	729.0000	728.0000
3	740.484375	728.0625	728.0000

3. $\int_1^3 x^{-2} \, dx = 0.666666$

J	$T(J)$ Trapezoid Rule	$S(J)$ Simpson's Rule	$B(J)$ Boole's Rule
0	1.111111		
1	0.805556	0.703704	
2	0.705000	0.671481	0.669333
3	0.676573	0.667097	0.666805
4	0.669166	0.666697	0.666670

Section 7.5 Romberg Integration: page 346

1. $\int_{-2}^2 7x^6 \, dx = 256$

J	$R(J,0)$ Trapezoidal Rule	$R(J,1)$ Simpson's Rule	$R(J,2)$ Boole's Rule	$R(J,3)$ Third Improvement
0	1792.0000			
1	896.00000	597.333334		
2	462.00000	317.333334	298.666667	
3	310.84375	260.458333	256.666667	256.000000

7. (a) $N = 3$ (b) $N = 5$

10. $f(x) = \cos(x)$

$0.001108616 = I - R(2,0) \approx -0.6h^2K/12 = 0.001109250$

$-0.000001667 = I - R(2,1) \approx 0.6h^4K/180 = -0.000001664$

$0.000000014 = I - R(2,2) \approx -1.2h^6K/945 = 0.000000014$

12. (b) For $\int_0^1 x^{1/2} \, dx$ Romberg integration converges slowly because the higher derivatives of the integrand $f(x) = x^{1/2}$ are not bounded near $x = 0$.

Section 7.6 Gauss–Legendre Integration: page 356

1. $\int_0^2 6t^5 \, dt = 64$

(b) $G(f, 2) = 58.6666667$ (c) $G(f, 3) = 64$ (d) $G(f, 4) = 64$

3. $\int_0^1 \sin(t)/t \, dt \approx 0.9460831$

   (b) $G(f, 2) = 0.9460411$ (c) $G(f, 3) = 0.9460831$
   (d) $G(f, 4) = 0.9460831$

6. (a) $N = 4$ (b) $N = 6$

8. (b) $f(x) = \exp(x)$

$$0.0077063 = I - G(f, 2) \approx K/135 \qquad = 0.0076370$$
$$0.0000655 = I - G(f, 3) \approx K/15{,}750 \qquad = 0.0000655$$
$$0.0000003 = I - G(f, 4) \approx K/347{,}2875 = 0.0000003$$

10. If the fourth derivative does not change too much, then

$$\left| \frac{f^{(4)}(c_1)}{135} \right| < \left| \frac{-f^{(4)}(c_2)}{90} \right|.$$

The truncation error term for the Gauss–Legendre rule will be less than the truncation error term for Simpson's rule.

Section 8.1 Iterative Methods for Linear Systems: page 368

1. (a) Jacobi iteration
   $P_1 = (3.75, 1.8)$
   $P_2 = (4.2, 1.05)$
   $P_3 = (4.0125, 0.96)$
   Iteration will converge to
   $P = (4, 1)$.

   (b) Gauss–Seidel iteration
   $P_1 = (3.75, 1.05)$
   $P_2 = (4.0125, 0.9975)$
   $P_3 = (3.999375, 1.000125)$
   Iteration will converge to
   $P = (4, 1)$.

3. (a) Jacobi iteration
   $P_1 = (-1, -1)$
   $P_2 = (-4, -4)$
   $P_3 = (-13, -13)$
   The iteration diverges
   away from the solution
   $P = (0.5, 0.5)$.

   (b) Gauss–Seidel iteration
   $P_1 = (-1, -4)$
   $P_2 = (-13, -40)$
   $P_3 = (-121, -364)$
   The iteration diverges
   away from the solution
   $P = (0.5, 0.5)$.

5. (a) Jacobi iteration
   $P_1 = (2, 1.375, 0.75)$
   $P_2 = (2.125, 0.96875, 0.90625)$ Iteration will converge to
   $P_3 = (2.0125, 0.95703125, 1.0390625)$ $P = (2, 1, 1)$.

   (b) Gauss–Seidel iteration
   $P_1 = (2, 0.875, 1.03125)$
   $P_2 = (1.96875, 1.01171875, 0.989257813)$ Iteration will converge to
   $P_3 = (2.00449219, 0.99753418, 1.0017395)$ $P = (2, 1, 1)$.

9. No. The matrix is not diagonally dominant, and interchanging the rows will not produce a diagonally dominant matrix. However, we can use transformations to get an equivalent linear system such as

$$5x + 3y = 6$$
$$-6x - 8y = -4$$

which can be solved with iteration.

12. (a) $M_1 = M_{50} = 0.633974596$, $M_2 = M_{49} = 0.464101615$
   $M_3 = M_{48} = 0.509618943$, $M_4 = M_{47} = 0.497422612$
   $M_5 = M_{46} = 0.500690609$, $M_6 = M_{45} = 0.499814952$
   $M_7 = M_{44} = 0.500049584$, $M_8 = M_{43} = 0.499986714$
   $M_9 = M_{42} = 0.500003560$, $M_{10} = M_{41} = 0.499999046$
   $M_{11} = M_{40} = 0.500000255$, $M_{12} = M_{39} = 0.499999932$
   $M_{13} = M_{38} = 0.500000009$, $M_{14} = M_{37} = 0.499999995$
   $M_{15} = M_{36} = 0.500000001$

$$M_k = 0.500000000 \quad \text{for } k = 16 \text{ TO } 35$$

Section 8.2 Iteration for Nonlinear Systems: page 378

1. $J(x,y) = \begin{pmatrix} 1-x & y/4 \\ (1-x)/2 & (2-y)/2 \end{pmatrix}$, $J(1.1, 2.0) = \begin{pmatrix} -0.1 & 0.5 \\ -0.05 & 0.0 \end{pmatrix}$

	Fixed-Point Iteration		Seidel Iteration	
k	p_k	q_k	p_k	q_k
0	1.1	2.0	1.1	2.0
1	1.12	1.9975	1.12	1.9964
2	1.1165508	1.9963984	1.1160016	1.9966327
∞	1.1165151	1.9966032	1.1165151	1.9966032

3. $J(x,y) = \begin{pmatrix} 1-x & 1/2 \\ 2(1-x)/9 & (2-y)/2 \end{pmatrix}$, $J(1.4, 2.0) = \begin{pmatrix} -0.4 & 0.5 \\ -0.088889 & 0.0 \end{pmatrix}$

	Fixed-Point Iteration		Seidel Iteration	
k	p_k	q_k	p_k	q_k
0	1.4	2.0	1.4	2.0
1	1.42	1.9822222	1.42	1.9804
2	1.4029111	1.9803210	1.402	1.9819480
∞	1.4076401	1.9814506	1.4076401	1.9814506

9. Fixed-point iteration will find the solution near the origin $(p, q) \approx (-0.090533, -0.099864)$. Eight other solutions are near: $(1, 0), (0, 1), (-1, 0), (0, -1), (1, 1), (1, -1), (-1, 1), (-1, -1)$.

Section 8.3 Newton's Method for Systems: page 384

1. $0 = x^2 - y$, $0 = y^2 - x$

$\mathbf{P}_k$	Solution of the Linear System: $\quad J(\mathbf{P}_k)\, d\mathbf{P} = -\mathbf{F}(\mathbf{P}_k)$			$\mathbf{P}_k + d\mathbf{P}$
$\begin{pmatrix} 0.25 \\ -0.25 \end{pmatrix}$	$\begin{pmatrix} 0.5 & -1.0 \\ -1.0 & -0.5 \end{pmatrix}$	$\begin{pmatrix} -0.275 \\ 0.175 \end{pmatrix}$	$= -\begin{pmatrix} 0.3125 \\ -0.1875 \end{pmatrix}$	$\begin{pmatrix} -0.025 \\ -0.075 \end{pmatrix}$
$\begin{pmatrix} -0.025 \\ -0.075 \end{pmatrix}$	$\begin{pmatrix} -0.05 & -1.00 \\ -1.00 & -0.15 \end{pmatrix}$	$\begin{pmatrix} 0.0194270 \\ 0.0746537 \end{pmatrix}$	$= -\begin{pmatrix} 0.075625 \\ 0.030625 \end{pmatrix}$	$\begin{pmatrix} -0.005573 \\ -0.000346 \end{pmatrix}$

$\mathbf{P}_k$	Solution of the Linear System: $\quad J(\mathbf{P}_k)\, d\mathbf{P} = -\mathbf{F}(\mathbf{P}_k)$			$\mathbf{P}_k + d\mathbf{P}$
$\begin{pmatrix} 0.9 \\ 1.25 \end{pmatrix}$	$\begin{pmatrix} 1.8 & -1.0 \\ -1.0 & 2.5 \end{pmatrix}$	$\begin{pmatrix} 0.125 \\ -0.215 \end{pmatrix}$	$= -\begin{pmatrix} -0.44 \\ 0.6625 \end{pmatrix}$	$\begin{pmatrix} 1.025 \\ 1.035 \end{pmatrix}$
$\begin{pmatrix} 1.025 \\ 1.035 \end{pmatrix}$	$\begin{pmatrix} 2.05 & -1.00 \\ -1.00 & 2.07 \end{pmatrix}$	$\begin{pmatrix} -0.024223 \\ -0.034033 \end{pmatrix}$	$= -\begin{pmatrix} 0.015625 \\ 0.046225 \end{pmatrix}$	$\begin{pmatrix} 1.000777 \\ 1.000967 \end{pmatrix}$

3. (b) The values of the Jacobian determinant at the solution points are $|J(1, 1)| = 0$ and $|J(-1, -1)| = 0$. Newton' method depends on being able to solve a linear system where the matrix is $J(p_n, q_n)$ and (p_n, q_n) is near a solution. For this example, the system equations are ill-conditioned and thus hard to solve with precision. In fact, for some values near a solution we have $J(x_0, y_0) = 0$, for example, $J(1.0001, 1.0001) = 0$.

6. $0 = 2xy - 3$, $\quad 0 = x^2 - y - 2$
$(p_0, q_0) = (1.5, 0.9)$, $\qquad\qquad (p_1, q_1) = (1.7083333, 0.875)$
$(p_2, q_2) = (1.6980622, 0.8833096)$, $\quad (p, q) = (1.6980481, 0.8833672)$

8. $0 = 3x^2 - 2y^2 - 1$, $\quad 0 = x^2 - 2x + y^2 + 2y - 8$
  I. $(p_0, q_0) = (-1.0, 1.0)$, $\qquad\qquad (p_1, q_1) = (-1.2, 1.3)$
$\qquad (p_2, q_2) = (-1.1928571, 1.2785714)$, $\quad (p, q) = (-1.1928731, 1.2784441)$
  II. $(p_0, q_0) = (3.0, -3.4)$, $\qquad\qquad (p_1, q_1) = (2.925, -3.5125)$
$\qquad (p_2, q_2) = (2.9234928, -3.5100168)$, $\quad (p, q) = (2.9234921, -3.5100156)$

10. $0 = 2x^3 - 12x - y - 1$, $\quad 0 = 3y^2 - 6y - x - 3$
  I. $(p, q) = (2.5908586, 2.6922233)$ $\qquad$ II. $(p, q) = (2.4627339, -0.6795569)$
  III. $(p, q) = (-0.0493209, -0.4083890)$ $\quad$ IV. $(p, q) = (-2.3078910, 2.1093705)$
  V. $(p, q) = (-2.4108204, -0.0937976)$ $\quad$ VI. $(p, q) = (-0.2855601, 2.3801497)$

12. $0 = 7x^3 - 10x - y - 1$, $0 = 8y^3 - 11y + x - 1$
  I. $(-0.0905331, -0.0998637)$ $\qquad$ II. $(1.2433858, 0.0221339)$
  III. $(-0.2311448, 1.2250008)$ $\qquad$ IV. $(-1.1531142, -0.2017060)$
  V. $(0.0124852, -1.1248379)$ $\qquad$ VI. $(1.2912933, 1.1591321)$
  VII. $(1.1860751, -1.1809721)$ $\qquad$ VIII. $(-1.0604370, 1.2569430)$
  IX. $(-1.1980104, -1.0558300)$

Section 8.4 Eigenvectors and Eigenvalues: page 392

1. $A = \begin{pmatrix} 7 & 6 & -3 \\ -12 & -20 & 24 \\ -6 & -12 & 16 \end{pmatrix}$, $\quad \begin{matrix} \lambda_1 = 4, & \mathbf{V}_1 = (2, -1, 0) \text{ dominant} \\ \lambda_2 = -2, & \mathbf{V}_2 = (1, -2, -1) \\ \lambda_3 = 1, & \mathbf{V}_3 = (-3, 4, 2) \end{matrix}$

3. $A = \begin{pmatrix} 16 & 30 & -42 \\ -24 & -47 & 66 \\ -12 & -24 & 34 \end{pmatrix}$, $\quad \begin{matrix} \lambda_1 = -2, & \mathbf{V}_1 = (1, -2, -1) \\ \lambda_2 = 4, & \mathbf{V}_2 = (-3, 4, 2) \text{ dominant} \\ \lambda_3 = 1, & \mathbf{V}_3 = (2, -1, 0) \end{matrix}$

5. $A = \begin{pmatrix} -2 & 1 & 1 \\ -6 & 1 & 3 \\ -12 & -2 & 8 \end{pmatrix}$, $\quad \begin{matrix} \lambda_1 = 4, & \mathbf{V}_1 = (1, 2, 4) \text{ dominant} \\ \lambda_2 = 2, & \mathbf{V}_2 = (2, 3, 5) \\ \lambda_3 = 1, & \mathbf{V}_3 = (1, 1, 2) \end{matrix}$

7. $A = \begin{pmatrix} 49 & -24 & -6 \\ 88 & -43 & -12 \\ 28 & -14 & -2 \end{pmatrix}$, $\quad \begin{matrix} \lambda_1 = -2, & \mathbf{V}_1 = (2, 4, 1) \\ \lambda_2 = 5, & \mathbf{V}_2 = (3, 5, 2) \text{ dominant} \\ \lambda_3 = 1, & \mathbf{V}_3 = (1, 2, 0) \end{matrix}$

9. (a) Characteristic polynomial $\lambda^2 - 3\lambda - 4$.

$$\lambda_1 = 4, \quad \mathbf{V}_1 = (2a, 3a), \quad \lambda_2 = -1, \quad \mathbf{V}_2 = (b, -b)$$

17. The transition matrix is

$$\begin{pmatrix} 0.6 & 0.2 & 0.1 \\ 0.3 & 0.5 & 0.1 \\ 0.1 & 0.3 & 0.8 \end{pmatrix}.$$

(a) $\mathbf{V}_1 = (240, 300, 460)$ (b) $\mathbf{V}_2 = (250, 268, 482)$
(c) The steady-state vector is $\mathbf{V} = (250, 250, 500)$.

Section 9.1 Introduction to Differential Equations: page 400

1. (b) $L = 1$ (c) Yes   (d) Yes
3. (b) $L = 3$ (c) Yes   (d) Yes
5. (b) $L = 60$ (c) Yes   (d) Yes
10. (c) No because $f_y(t, y) = \frac{1}{2}y^{-2/3}$ is not continuous when $t = 0$, and

$$\lim_{y \to 0} f_y(t, y) = \infty.$$

13. $y(t) = t^3 - \cos(t) + 3$

15. $y(t) = \int_0^t \exp(-s^2/2)\, ds$

17. (b) $y(t) = y_0 \exp(-0.000120968t)$ (c) 2,808 years   (d) 6.9237 sec

Section 9.2 Euler's Method: page 408

1. $y' = t^2 - y$ with $y(0) = 1$. The Euler solutions are

t_k	y_k			Exact $y(t_k)$
	$h = 0.2$	$h = 0.1$	$h = 0.05$	
0.05			0.95	0.951271
0.1		0.9	0.902625	0.905163
0.15			0.857994	0.861792
0.2	0.8	0.811	0.816219	0.821269
1.0	0.537856	0.586189	0.609438	0.632121
2.0	1.714101	1.790581	1.827913	1.864665

3. $y' = -ty$ with $y(0) = 1$. The Euler solutions are

t_k	y_k			Exact $y(t_k)$
	$h = 0.2$	$h = 0.1$	$h = 0.05$	
0.05			1.0	0.998751
0.1		1.0	0.9975	0.995012
0.15			0.992513	0.988813
0.2	1.0	0.99	0.985069	0.980199
1.0	0.652861	0.628157	0.616984	0.606531
2.0	0.124379	0.130400	0.132980	0.135335

6. $y_5 = 1{,}000(1 + 0.12)^5 = 1{,}762.3417$
$y_{60} = 1{,}000(1 + 0.01)^{60} = 1{,}816.6967$

$$y_{1,800} = 1{,}000\left(1 + \frac{0.12}{360}\right)^{1,800} = 1{,}821.9355$$

8. $P_{k+1} = P_k + (0.02P_k - 0.00004P_k^2)10$ for $k = 1, 2, \ldots, 8$.

			P_k	
Year	t_k	Actual Population At t_k, $P(t_k)$	Euler Rounded at Each Step	Euler with More Digits Carried at Each Step
1900	0.0	76.1	76.1	76.1
1910	10.0	92.4	89.0	89.0035
1920	20.0	106.5	103.6	103.6356
1930	30.0	123.1	120.0	120.0666
1940	40.0	132.6	138.2	138.3135
1950	50.0	152.3	158.2	158.3239
1960	60.0	180.7	179.8	179.9621
1970	70.0	204.9	202.8	203.0000
1980	80.0	226.5	226.9	227.1164

13. No. For any M, Euler's method produces $0 < y_1 < y_2 < \ldots < y_M$. The mathematical solution is $y(t) = \tan(t)$ and $y(3) < 0$.

Section 9.3 Heun's Method: page 414

1. $y' = t^2 - y$ with $y(0) = 1$. The Heun solutions are

	y_k			Exact
t_k	$h = 0.2$	$h = 0.1$	$h = 0.05$	$y(t_k)$
0.05			0.951313	0.951271
0.1		0.9055	0.905245	0.905163
0.15			0.861915	0.861792
0.2	0.824	0.821928	0.821431	0.821269
1.0	0.643244	0.634782	0.632772	0.632121
2.0	1.881720	1.868726	1.865656	1.864665

3. $y' = -ty$ with $y(0) = 1$. The Heun solutions are

	y_k			Exact
t_k	$h = 0.2$	$h = 0.1$	$h = 0.05$	$y(t_k)$
0.05			0.99875	0.998751
0.1		0.995	0.995011	0.995012
0.15			0.988811	0.988813
0.2	0.98	0.980175	0.980196	0.980199
1.0	0.606975	0.606718	0.606586	0.606531
2.0	0.139628	0.136318	0.135571	0.135335

6. Solution of $v' = -32 - 0.1v$ over $[0, 30]$ with $v(0) = 160$.

t_k	v_k
1.00	114.341
2.00	73.025
3.00	35.639
4.00	1.809
5.00	−28.802
10.00	−143.341
20.00	−254.983
30.00	−296.071

9. Richardson improvement for solving $y' = (t - y)/2$ over $[0, 3]$ with $y(0) = 1$. The table entries are approximations to $y(3)$.

h	y_h	$[4y_h - y_{2h}]/3$
1	1.732422	
1/2	1.682121	1.665354
1/4	1.672269	1.668985
1/8	1.670076	1.669345
1/16	1.669558	1.669385
1/32	1.669432	1.669390
1/64	1.669401	1.669391

10. $y' = 1.5y^{1/3}$, $f(t, y) = 1.5y^{1/3}$, $f_y(t, y) = 0.5y^{-2/3}$.
$f_y(0, 0)$ does not exist. The I.V.P. is not well-posed on any rectangle that contains $(0, 0)$.

Section 9.4 Taylor Methods: page 420

1. $y' = t^2 - y, y'' = -t^2 + 2t + y, y''' = t^2 - 2t + 2 - y$
$y^{(4)} = -t^2 + 2t - 2 + y$ with $y(0) = 1$. Taylor's solutions:

t_k	y_k $h = 0.2$	y_k $h = 0.1$	y_k $h = 0.05$	Exact $y(t_k)$
0.05			0.9512706	0.9512706
0.1		0.9051625	0.9051626	0.9051626
0.15			0.8617920	0.8617920
0.2	0.8212667	0.8212691	0.8212692	0.8212692
1.0	0.6321148	0.6321202	0.6321205	0.6321206
2.0	1.8646605	1.8646645	1.8646647	1.8646647

3. $y' = -ty, y'' = t^2 y - y, y''' = -t^3 y + 3ty$
$y^{(4)} = t^4 y - 6t^2 y + 3y$ with $y(0) = 1$. Taylor solutions:

	y_k			Exact
t_k	$h = 0.2$	$h = 0.1$	$h = 0.05$	$y(t_k)$
0.05			0.9987508	0.9987508
0.1		0.9950125	0.9950125	0.9950125
0.15			0.9888131	0.9888130
0.2	0.9802	0.9801988	0.9801987	0.9801987
1.0	0.6065735	0.6065333	0.6065308	0.6065307
2.0	0.1353374	0.1353354	0.1353353	1.1353353

6. Richardson improvement for the Taylor solution to $y' = (t - y)/2$ over $[0, 3]$ with $y(0) = 1$. The table entries are approximations to $y(3)$.

h	y_h	$[16y_h - y_{2h}]/15$
1	1.6701860	
1/2	1.6694308	1.6693805
1/4	1.6693928	1.6693903
1/8	1.6693906	1.6693905

Section 9.5 Runge-Kutta Methods: page 432

1. $y' = t^2 - y$ with $y(0) = 1$. The RK4 solutions are

	y_k			Exact
t_k	$h = 0.2$	$h = 0.1$	$h = 0.05$	$y(t_k)$
0.05			0.9512706	0.9512706
0.1		0.9051627	0.9051626	0.9051626
0.15			0.8617920	0.8617920
0.2	0.8212733	0.8212695	0.8212693	0.8212692
1.0	0.6321380	0.6321216	0.6321206	0.6321206
2.0	1.8646923	1.8646664	1.8646648	1.8646647

3. $y' = -ty$ with $y(0) = 1$. The RK4 solutions are

	y_k			Exact
t_k	$h = 0.2$	$h = 0.1$	$h = 0.05$	$y(t_k)$
0.05			0.9987508	0.9987508
0.1		0.9950125	0.9950125	0.9950125
0.15			0.9888130	0.9888130
0.2	0.9801987	0.9801987	0.9801987	0.9801987
1.0	0.6065314	0.6065307	0.6065307	0.6065307
2.0	0.1353590	0.1353366	0.1353354	0.1353353

6. Solution of $y' = 0.01(70 - y)(50 - y)$ over $[0, 20]$ with $y(0) = 0$.

t_k	v_k
0.50	13.45109
1.00	21.82776
2.00	31.62582
3.00	37.10403
4.00	40.54658
5.00	42.87100
10.00	47.85965
20.00	49.73487

Section 9.6 Predictor–Corrector Methods: page 444

1. $y' = t^2 - y$ with $y(0) = 1$

t_k	y_k		
	Adams–Bashforth–Moulton Solution	Milne–Simpson Solution	Hamming's Solution
0.2	0.8212693	0.8212692	0.8212692
0.25	0.7836992	0.7836992	0.7836992
0.3	0.7491818	0.7491818	0.7491818
0.4	0.6896800	0.6896800	0.6896800
0.5	0.6434694	0.6434693	0.6434693

3. $y' = -t/y$ with $y(1) = 1$

t_k	y_k		
	Adams–Bashforth–Moulton Solution	Milne–Simpson Solution	Hamming's Solution
1.2	0.7483205	0.7483273	0.7483296
1.25	0.6613998	0.6614224	0.6614250
1.3	0.5566583	0.5567251	0.5567253
1.35	0.4208572	0.4210904	0.4210644
1.4	0.1974740	0.1988783	0.1982560

5. $y' = 2ty^2$ with $y(0) = 1$

t_k	y_k		
	Adams–Bashforth–Moulton Solution	Milne–Simpson Solution	Hamming's Solution
0.2	1.0416675	1.0416670	1.0416668
0.25	1.0666688	1.0666673	1.0666671
0.3	1.0989052	1.0989025	1.0989020
0.4	1.1904878	1.1904801	1.1904788
0.5	1.3333631	1.3333439	1.3333404

7. $y' = 2y - y^2$ with $y(0) = 1$

	y_k		
t_k	Adams–Bashforth–Moulton Solution	Milne–Simpson Solution	Hamming's Solution
0.2	1.1972300	1.1973754	1.1973753
0.25	1.2447770	1.2449187	1.2449186
0.3	1.2911748	1.2913127	1.2913126
0.4	1.3798202	1.3799491	1.3799489
0.5	1.4619988	1.4621172	1.4621170

9. $y' = y^2 \sin(t)$ with $y(0) = 1$

	y_k		
t_k	Adams–Bashforth–Moulton Solution	Milne–Simpson Solution	Hamming's Solution
0.2	1.0203389	1.0203389	1.0203388
0.25	1.0320852	1.0320850	1.0320850
0.3	1.0467519	1.0467517	1.0467516
0.4	1.0857051	1.0857046	1.0857045
0.5	1.1394953	1.1394943	1.1394941

Section 9.7 Systems of Differential Equations: page 450

1. $x' = 2x + 3y, y' = 2x + y$ with $x(0) = 2$ and $y(0) = 3$

	Euler's Solution		Runge–Kutta Order 4	
t_k	x_k	y_k	x_k	y_k
0.05	2.65	3.35	2.7129706	3.3940294
0.1	3.4175	3.7825	3.5706165	3.8884734
0.15	4.326625	4.313375	4.6056114	4.5049209
0.2	5.4062938	4.9617062	5.8578317	5.2697724
0.3	8.2228601	6.7070599	9.2193976	7.3809621
0.4	12.2360305	9.2630544	14.1885088	10.5762060
0.5	17.9764723	12.9822098	21.5601371	15.3843092

3. $x' = x - 4y, y' = x + y$ with $x(0) = 2$ and $y(0) = 3$

t_k	Euler's Solution		Runge–Kutta Order 4	
	x_k	y_k	x_k	y_k
0.05	1.5	3.25	1.4623255	3.2430091
0.1	0.925	3.4875	0.8488992	3.4689862
0.15	0.27375	3.708125	0.1598100	3.6731724
0.2	−0.4541875	3.9072188	−0.6038499	3.8505944
0.3	−2.1372317	4.2209468	−2.3449564	4.1044442
0.4	−4.1077233	4.3869750	−4.3422923	4.1882574
0.5	−6.3302172	4.3614593	−6.5424931	4.0597676

5. $2x'' - 5x' - 3x = 45 \exp(2t)$ with $x(0) = 2$ and $x'(0) = 1$
$x' = y, y' = 1.5x + 2.5y + 22.5 \exp(2t)$

t_k	Euler's Solution		Runge–Kutta Order 4	
	x_k	y_k	x_k	y_k
0.05	2.05	2.4	2.0875384	2.5548149
0.1	2.17	4.0970673	2.2612983	4.4593080
0.2	2.6821548	8.6109876	2.9477416	9.6019238
0.3	3.6910454	15.1009428	4.2609483	17.1320033
0.4	5.4118562	24.3286106	6.4858063	28.0250579
0.5	8.1422522	37.3283984	10.0223747	43.6284511

7. $x'' + x = 6 \cos(t)$ with $x(0) = 2$ and $x'(0) = 3$
$x' = y, y' = -x + 6 \cos(t)$

t_k	Euler's Solution		Runge–Kutta Order 4	
	x_k	y_k	x_k	y_k
0.05	2.15	3.2	2.1549349	3.1960425
0.1	2.31	3.3921251	2.3194586	3.3833472
0.2	2.6583626	3.7477773	2.6753427	3.7269090
0.3	3.0411954	4.0572653	3.0632017	4.0213325
0.4	3.4536490	4.3114154	3.4776789	4.2578745
0.5	3.8899723	4.5017244	3.9125799	4.4285471

Index